Law and Regulation of Commercial Mining
of Minerals in Outer Space

SPACE REGULATIONS LIBRARY

VOLUME 7

EDITORIAL BOARD

Managing Editor

PROF. RAM S. JAKHU, *Institute of Air and Space Law, McGill University, Montreal, Canada*

MEMBERS

M. DAVIS, *Adelta Legal, Adelaide, Australia*
S. LE GOUEFF, *Le Goueff Law Office, Luxembourg*
P. NESGOS, *Milbank, Tweed, Hadley & McCloy, New York, U.S.A.*
S. MOSTESHAR, *Chambers of Sa'id Mosteshar, London, U.K. & Mosteshar Mackenzie, California, U.S.A.*
L. I. TENNEN, *Law Offices of Sterns and Tennen, Phoenix, Arizona, U.S.A.*

For further volumes:
http://www.springer.com/series/6573

Ricky J. Lee

Law and Regulation of Commercial Mining of Minerals in Outer Space

Dr. Ricky J. Lee
Sydney, Australia
ricky.lee@activer.com.au

ISBN 978-94-007-2038-1   e-ISBN 978-94-007-2039-8
DOI 10.1007/978-94-007-2039-8
Springer Dordrecht Heidelberg London New York

Library of Congress Control Number: 2011944336

© Springer Science+Business Media B.V. 2012
No part of this work may be reproduced, stored in a retrieval system, or transmitted in any form or by
any means, electronic, mechanical, photocopying, microfilming, recording or otherwise, without written
permission from the Publisher, with the exception of any material supplied specifically for the purpose
of being entered and executed on a computer system, for exclusive use by the purchaser of the work.

Printed on acid-free paper

Springer is part of Springer Science+Business Media (www.springer.com)

In a closed society Malthusianism has the appearance of self-evident truth, and herein lies the danger. It is not enough to argue against Malthusianism in the abstract - such debates are not settled in academic journals. Unless people can see broad vistas of unused resources in front of them, the belief in limited resources tends to follow as a matter of course. And if the idea is accepted that the world's resources are fixed, then each person is ultimately the enemy of every other person, and each race or nation is the enemy of every other race or nation. The extreme result is tyranny, war and even genocide. Only in a universe of unlimited resources can all men be brothers.

Robert Zubrin and Richard Wagner[1]

---

[1] Robert Zubrin and Richard Wagner, THE CASE FOR MARS: THE PLAN TO SETTLE THE RED PLANET AND WHY WE MUST (1997), at 303.

Acknowledgements

The field of space law appealed to me originally as a convenient combination of my casual interest in space and my legal studies. Over the years, however, it grew from a peripheral interest to become a passion and from that passion grew a career path that has been both challenging and rewarding. I still recall that one Thursday afternoon in 1998 at the University of Adelaide when I was at lunchtime seminar on space law and its opportunities presented by Michael E. Davis, now a partner at the law firm of Adelta Legal in Adelaide, Australia. From that time, not content with having introduced me to this new and exciting field of law, Mr. Davis soon became my employer and continues to be an invaluable support, mentor and friend. Mere written words are never enough to express my gratitude to him for all that he has done for me, but I offer them here nonetheless as a very small token of my thanks.

In late 1999, after I was elected a member of the International Institute of Space Law, Mr. Davis introduced me to the late Associate Professor Alexis Goh of the University of Western Sydney as a useful contact in that endeavour. After I moved to Canberra from Adelaide in 2000, she encouraged me to undertake this monograph as a doctoral thesis at the University of Western Sydney and she herself became my extremely supportive supervisor. She had shown tremendous faith and high expectations in my abilities as a doctoral candidate and as a teaching academic, both of which I have tried over the years to meet as much as I can though in truth I know I probably can never meet. She was a generous and understanding boss, an encouraging but critical supervisor, a resourceful and supportive friend without whom this work (and my academic career) would never have become a reality. She is much missed.

My thanks go to Associate Professor Vernon Nase of the City University of Hong Kong, who kindly agreed to become my doctoral supervisor and has peppered me with critical yet invaluable comments and suggestions on this work and allowed for its completion. I am much indebted to him for the kindness and friendship that he has shown to me, both as my doctoral supervisor and generally in the field of space law.

The contribution of Professor David Flint AM of Murdoch University (Australia), Professor Paul B. Larsen of Georgetown University (the United States) and Professor Francis Lyall of the University of Aberdeen (the United Kingdom) in examining this monograph, which was submitted as a doctoral thesis at Murdoch University, must be recognised.

In addition, the kindness and generosity of Prof. Ram Jakhu of the Institute of Air and Space Law at McGill University, Canada, in taking the time to review this monograph. Further, the understanding, encouragement and support (not to mention tolerance) of Norbert Schweizer and Michael Kobras, the partners of my employer, the Sydney commercial law firm of Schweizer Kobras, are also greatly appreciated.

My thanks also go to Mark Sonter, director of Asteroid Enterprises Pty Ltd, who provided me with some of his expert insights into the physical and technical aspects of mining asteroids.

It goes without saying, though worth saying nonetheless, that without the continuing support encouragement and love of my family, I can achieve nothing in my life. I am very grateful, even though I may not appear to show it from time to time.

Of course, all faults and shortcomings in this monograph are exclusively my own and I count on the support of my friends and the vindictiveness of my enemies to show me any errors of my work.

Sydney, Australia   Ricky J. Lee

Contents

1 Introduction and Overview 1
  1.1 The Problem 1
      1.1.1 The Context 1
      1.1.2 Structure of the Monograph 2
  1.2 Historical Background 4
      1.2.1 Technological Evolution 4
      1.2.2 The Post-Cold War World 5
      1.2.3 The Hypothesis 7
  1.3 Economic and Technical Prospects of Mining on Celestial
      Bodies 8
  1.4 Liability and State Responsibility for Compliance
      with International Legal Principles 9
      1.4.1 Phases of a Commercial Space Mining Venture 9
      1.4.2 Applicable Legal Issues 10
      1.4.3 State Responsibility and International Liability 11
  1.5 Exploration and Extraction Rights 13
  1.6 Exploitation Rights: Effects of the Common Heritage
      of Mankind Doctrine 14
  1.7 Meeting the Challenges and Balancing the Competing
      Interests in Creating a Regulatory Framework for Mineral
      Resources in Space 16
      1.7.1 Meeting the Challenges 16
      1.7.2 Balancing the Competing Interests 17
      1.7.3 Structure and Composition 18
      1.7.4 Procedures 18
      1.7.5 Judicial Mechanisms 19
  1.8 Conclusions 20

2 Economic and Technical Prospects of Mining on Celestial Bodies .. 21
  2.1 Introduction 21
  2.2 Economic Feasibility of Space Mining 26
      2.2.1 The Study of Resource Economics 26
      2.2.2 Economic Scarcity of Mineral Resources 32
      2.2.3 Implications of the Hydrogen Economy 45

ix

2.3	Expanding the Economic Resource Base		48
2.4	The Riches of Space		51
	2.4.1	Mining the Moon	51
	2.4.2	Resources from Mars and Other Planets	52
	2.4.3	Geology and Mineralogy of Asteroids and Their Suitability for Mining Activities	55
	2.4.4	Suitability of Near Earth Asteroids	60
	2.4.5	Other Groups of Potential Mining Candidates	66
2.5	Technical Feasibility of Space Mining		69
	2.5.1	Orbital Mechanics	69
	2.5.2	Mission Trajectories	73
	2.5.3	Energy Requirements for the Mining and Processing of Ores	76
2.6	Exploratory Missions to Near Earth Asteroids		77
	2.6.1	Flyby Missions	78
	2.6.2	Rendezvous and Lander Missions	80
	2.6.3	Sample Return Missions	80
2.7	Commercial Feasibility of Space Mining		82
	2.7.1	Advantages of Mining Near Earth Asteroids	82
	2.7.2	Costing an Asteroid Mining Project	83
	2.7.3	Determining Financial Feasibility	85
	2.7.4	Comparing Returns on Investment	86
	2.7.5	Minimisation of Mission Risks	87
	2.7.6	Practical Implications of Risk Profiles	92
2.8	Conclusions		93

3 State Responsibility and Liability for Compliance with International Space Law 95

3.1	Introduction		95
3.2	Sources of Space Law		98
	3.2.1	United Nations Space Treaties	98
	3.2.2	General Assembly Declarations	111
	3.2.3	Jus Cogens: Space Law Principles as Possible Peremptory Norms of International Law	124
	3.2.4	Other Space-Related Treaties	126
	3.2.5	Space Law and the Lex Specialis Principle	127
3.3	State Responsibility and Jurisdiction		128
	3.3.1	State Responsibility	128
	3.3.2	Jurisdiction	134
3.4	Liability		136
	3.4.1	Overview	136
	3.4.2	International Liability	137
	3.4.3	Modern Liability Controversies	142
	3.4.4	Calculation of Damages	148
3.5	Conclusions		150

4 Rights and Duties in the Commercial Exploration and Extraction of Mineral Resources on Celestial Bodies 153

4.1 Introduction 153
4.2 Commercial Use vs. Public Use 154
    4.2.1 Benefit and Interests of All Countries 154
    4.2.2 Lawfulness of Commercial Use Generally 160
4.3 Freedoms of Exploration and Use and the Principle
of Non-appropriation 162
    4.3.1 Freedoms Under Article I of the Outer Space Treaty 162
    4.3.2 Article II of the Outer Space Treaty 166
    4.3.3 Relevant Provisions of the Moon Agreement 179
    4.3.4 Defining a Celestial Body 187
4.4 Environmental Protection of Celestial Bodies 192
    4.4.1 Article IX of the Outer Space Treaty 192
    4.4.2 Article 7 of the Moon Agreement 193
4.5 Legal Implications for a Regulatory Regime for
Exploration of Mineral Resources on Celestial Bodies 194
    4.5.1 Two Types of Exploration 194
    4.5.2 The Exploration Segment Generally 195
    4.5.3 Prospecting Activities 196
    4.5.4 Exploration Activities 196
4.6 Legal Implications for a Regulatory Regime for Extraction
of Mineral Resources on Celestial Bodies 197
    4.6.1 Sovereignty over Mineral Resources 197
    4.6.2 Extraction Methods 198
4.7 Conclusions 202

5 Exploitation Rights: Evolving from the "Province of Mankind" to the "Common Heritage of Mankind" 203

5.1 Introduction 203
5.2 Antarctica and the 1988 Wellington Convention 204
    5.2.1 The Antarctic Treaty Framework 204
    5.2.2 Operation of the Antarctic Treaty System 206
    5.2.3 The Wellington Convention 211
5.3 Outer Space as the Province of All Mankind 216
    5.3.1 Interpreting Article I of the Outer Space Treaty 216
    5.3.2 Correlation with Article II of the Outer Space Treaty ... 217
5.4 The New International Economic Order 219
    5.4.1 Origins of the New International Economic Order 219
    5.4.2 Sixth Special Session of the General Assembly
and the Principles of the New International
Economic Order 223
5.5 Origins of the Common Heritage of Mankind Concept
in the Context of the Deep Seabed 229
    5.5.1 The Old Law of the High Seas 229

	5.5.2	The Old Law of the Deep Seabed	232
	5.5.3	Proposal from Malta that "Detonated the Time Bomb"	234
	5.5.4	Negotiating History on the Deep Seabed	236
5.6	Part XI of the Convention on the Law of the Sea and the Effects of the Common Heritage of Mankind		244
	5.6.1	Overview and the Features of Part XI	244
	5.6.2	Decision-Making in the International Seabed Authority	247
	5.6.3	The Enterprise	248
	5.6.4	Compulsory Transfer of Technology	249
	5.6.5	Limitations on Production	250
	5.6.6	Financial Terms of the Contracts	251
	5.6.7	The Ultimate Compromise	252
5.7	Evolution of Article 11 of the Moon Agreement		256
	5.7.1	Early Controversies	256
	5.7.2	Provisions of the Moon Agreement	260
	5.7.3	Article 6 of the Moon Agreement	264
	5.7.4	Article 11 of the Moon Agreement	264
	5.7.5	Article 18 of the Moon Agreement	266
	5.7.6	Attempts at Resolving the Political Impasse	267
5.8	Conclusions		270

6 Meeting the Challenges and Balancing the Competing Interests in Creating a Legal and Regulatory Framework 273

6.1	Introduction		273
6.2	Need for Balancing Competing Interests		275
	6.2.1	Overview	275
	6.2.2	Industrialised States vs. Developing States	276
	6.2.3	Economic Development vs. Environmental Safeguards	282
	6.2.4	Regulation vs. Free Market	287
	6.2.5	Public Interest vs. Commercial Concerns	290
	6.2.6	Hard Law vs. Soft Law	293
6.3	Practical Implementation of the New Legal Framework		295
	6.3.1	Overview	295
	6.3.2	Implementation Agreement	296
	6.3.3	Organisational Structure	296
	6.3.4	Membership of the Authority	297
6.4	Administrative and Dispute Settlement Mechanisms		303
	6.4.1	Enforcement and Dispute Settlement Mechanisms	303
	6.4.2	Licensing Oversight Panel	304
	6.4.3	Financial Duties Panel	305
	6.4.4	Environmental Protection Panel	307
	6.4.5	Dispute Settlement Panel and the Space Development Appeals Tribunal	307
6.5	Financing the Proposed Authority		311
6.6	Conclusions		313

7 Concluding Observations ... 315

7.1 Quod Erat Demonstrandum ... 315

    7.1.1 Elements of Proving the Hypothesis ... 315

    7.1.2 Economic Scarcity of Resources in the Earth's Crust ... 316

    7.1.3 Physical and Technical Feasibility of Asteroid Mining ... 316

    7.1.4 Economic and Financial Feasibility of Commercial Mining on Near Earth Asteroids ... 317

    7.1.5 Absence of an Appropriate Legal Framework ... 317

7.2 Quod Erat Meliorandum ... 318

    7.2.1 Conceptual Evolution of the Common Heritage of Mankind ... 318

    7.2.2 Past Failures over the Deep Seabed and Antarctica ... 319

    7.2.3 Attempt to Create a Legal Framework for the Exploitation of Mineral Resources on Celestial Bodies ... 319

    7.2.4 Balancing the Interests of Competing Stakeholders ... 320

7.3 Quod Erat Faciendum ... 320

    7.3.1 Creating a New Legal Framework ... 320

    7.3.2 Practical Aspects of the New Legal Framework ... 321

    7.3.3 How the World May Change ... 321

References ... 323

Treaties ... 323

United Nations Documents ... 326

International Cases ... 328

Domestic Legislation and Regulations ... 328

Domestic Cases ... 329

Secondary Sources ... 329

Index ... 369

List of Abbreviations

AAS	American Astronomical Society
ACP	African, Caribbean and Pacific Group of States
AIAA	American Institute of Aeronautics and Astronautics
AN	Ascending node
ARABSAT	Arab Corporation for Space Communications
AU	Astronomical Unit
COPUOS	Committee on Peaceful Uses of Outer Space
COSPAR	Committee on Space Research
DN	Descending node
ECAS	Eight Colour Asteroid Survey
ECSL	European Centre for Space Law
EEC	European Economic Community
ENPV	Expected Net Present Value
ESA	European Space Agency
EUMETSAT	European Organisation for the Exploitation of Meteorological Satellites
EUTELSAT	European Telecommunication Satellite Organisation
GATT	General Agreement on Tariffs and Trade
GEO	Geostationary Earth Orbit
HEEO	Highly Elliptical Earth Orbit
IAU	International Astronomical Union
IBA	International Bar Association
IBRD	International Bank for Reconstruction and Development
ICC	International Chamber of Commerce
IISL	International Institute of Space Law
ILC	International Law Commission
IMF	International Monetary Fund
INMARSAT	International Mobile Satellite Organisation
INTELSAT	International Telecommunication Satellite Organisation
ISA	International Seabed Authority
ISRO	Indian Space Research Organisation
ITLOS	International Tribunal for the Law of the Sea

ITU	International Telecommunication Union
LEO	Low Earth Orbit
NASA	National Aeronautics and Space Administration
NEAR	Near Earth Asteroid Rendezvous
NIEO	New International Economic Order
NPV	Net Present Value
OPEC	Organisation of Petroleum Exporting Countries
RLV	Reusable Launch Vehicle
ROI	Return on Investment
SSI	Space Studies Institute
SSPS	Space Solar Power Satellite
UN	United Nations
UNCLOS III	Third United Nations Conference on the Law of the Sea
UNCTAD	United Nations Conference on Trade and Development
UNESCO	United Nations Education, Scientific and Cultural Organisation
UNIDROIT	International Institute for the Unification of Private Law
USA	United States of America
USACERL	United States Army Construction Engineering Research Laboratory
WIPO	World Intellectual Property Organisation
WTO	World Trade Organisation

List of Figures

Fig. 1.1	Outline of the contents and their correlation with the hypothesis	3
Fig. 1.2	Segments of a commercial space mining venture	11
Fig. 2.1	Change in velocity needed for soft landing on selected objects	25
Fig. 2.2	The stock resource base and its subdivisions	30
Fig. 2.3	Diminishing returns and marginal physical product curves	34
Fig. 2.4	The perfect market response to resource scarcity	37
Fig. 2.5	The onset of economic exhaustion of mineral resources	39
Fig. 2.6	Example: expected life of bauxite reserves	43
Fig. 2.7	Different types of hydrogen fuel cells	46
Fig. 2.8	Government-industry roles in the transition to a hydrogen economy	47
Fig. 2.9	Relative cost of producing 1 ounce of platinum 2005–2030	49
Fig. 2.10	The three groups of Near Earth Asteroids	61
Fig. 2.11	Gravitational perturbations on long period comets by Jupiter	65
Fig. 2.12	Gravitational perturbations on short period comets by Jupiter	66
Fig. 2.13	Orbital geometry of short period comets	68
Fig. 2.14	Orbital properties of an asteroid	69
Fig. 2.15	Trajectory plan for a 2009 mining mission to asteroid 1991JW	76
Fig. 2.16	Image of 113 Amalthea taken by *Voyager 1*	78
Fig. 2.17	Example trajectory for a multiple flyby mission	79
Fig. 3.1	Timeline of international space law instruments	100
Fig. 3.2	State responsibility under Article VI: summary of analysis	132
Fig. 6.1	Competing interests to be balanced in the creation of the new international regulatory framework	275
Fig. 6.2	Summary of proposed compromises in the new legal framework	294
Fig. 6.3	Structure of the proposed international space development authority	297
Fig. 6.4	Balancing some of the opposing interests between the industrialised states and the developing states	305

List of Tables

Table 2.1	Selected technological development enabling mining in space	26
Table 2.2	Resource base and life expectancy estimates for certain mineral resources in the earth's crust	28
Table 2.3	Estimated reserves and depletion dates	42
Table 2.4	Cumulative reserves and production in selected minerals	44
Table 2.5	Bell superclasses of asteroid taxonomy and likely mineralogy	55
Table 2.6	Mineral resources found on asteroids and their uses in space	57
Table 2.7	The Arjunas with low eccentricities	67
Table 2.8	Energy requirements for various missions	73
Table 2.9	Potential benefits for mining of Near Earth Asteroids	83
Table 2.10	Outline of platinum mining project milestones	84
Table 2.11	Summary of mission risks and effects of the passage of time	93
Table 3.1	Principles Declaration	119
Table 3.2	Broadcasting Principles	120
Table 3.3	Remote Sensing Principles	122
Table 3.4	Nuclear Power Sources Principles	123
Table 3.5	Cooperation declaration	124
Table 6.1	Specific competing interests and concerns	276
Table 6.2	Comparative taxation on a model copper mine in selected states	279
Table 6.3	Summary of admission procedures for major organisations	302

Glossary

Absolute visual magnitude (H)	A measure of a celestial body's intrinsic brightness, measured in the standard V photometric band
Amor Asteroid	Asteroid with perihelion 1.017 AU $< q \leq$ 1.3 AU
Aphelion (Q)	The point on an orbit that is most distant from the Sun
Apollo Asteroid	Asteroid with perihelion $q <$ 1.017 AU and semi major axis $a >$ 1.0 AU
Arjuna Asteroid	Asteroid with Earth-like orbits with low inclination, low eccentricity and orbital periods close to one Earth year
Astronomical Unit (AU)	Unit of length equal to the mean distance between the Earth and the Sun, estimated in 2009 at 149,597,870,700 m
Aten Asteroid	Asteroid with semi major axis $a <$ 1.0 AU and aphelion $Q >$ 0.983 AU
Carbonyl	Compound of a metal with carbon monoxide (CO)
Conjunction	Where two objects are at the aphelion or perihelion at the same time
Earth Minimum Orbital Intersection Distance (MOID)	Minimum distance between closest points on the orbit of the Earth and the orbit of an asteroid or comet, usually given in astronomical units (AU). Potentially hazardous objects to the Earth have an Earth MOID of less than 0.05 AU
Eccentricity (e)	Measure of the circularity of the orbit where the more eccentric an orbit, the more oval shaped the orbit is
Ecliptic	The orbital plane on which the Earth orbits the Sun
Escape Velocity	Minimum speed an object without propulsion needs to have to move away infinitely from the gravity of an object
Hohmann Transfer Orbit	Elliptical orbit that is tangential to two coplanar orbits that is most energy efficient transfer trajectory
Hyperbolic Velocity	The velocity ($\Delta\vec{v}$) of an object relative to Earth or another celestial object when it is outside that body's gravity well

Impulse	Change in velocity ($\Delta \vec{v}$) that is given to an object in a short period of time relative to the total duration of the trajectory
Inclination (i)	The angle between the orbital plane of a particular object and the ecliptic
Kerogen	Solid hydrocarbons found in crude oil
Opposition	Where one object is at perihelion and the other is at aphelion
Perihelion (q)	The point on an orbit which is closest to the Sun
Planetesimal	One of a class of bodies that are theorised to have formed the planets after condensing from diffuse matter early in the history of the solar system
Pyrolysis	Generation of chemicals or free metals by heat decomposition
Regolith	The fragmented rocky debris blanketing the surface of the Moon, some asteroids and other small objects in the Solar System
Semi-major Axis (A)	The longest diameter of an elliptical orbit.
Synodic Period	Period of a body relative to the Earth
Transfer Orbit	The trajectory for an object travelling from one body to another
Trojan	An object which is trapped in a stable orbit 60° ahead of or behind the object as it orbits the Sun
Volatiles	Gases that can be released from comet cores by heating, producing gases such as water, carbon dioxide (CO_2), carbon monoxide (CO), methane (CH_4), ammonia (NH_3) and hydrogen cyanide (HCN)

List of Reports, Series and Journal Titles

Source	Full title
A. F. L. Rev.	Air Force Law Review
A.S.I.L.S. Int'l. L. J.	Association of Student International Law Societies International Law Journal
A.T.S.	Australian Treaty Series
Abs. Lunar & Planetary Sci. Conf.	Abstracts of the Lunar and Planetary Science Conference
Acta Astron.	Acta Astronautica (Journal of the International Academy of Astronautics)
Acta Juridica	Acta Juridica (Law Journal of the University of Cape Town)
Ad Astra	Ad Astra (Journal of the National Space Society)
Adel. L. Rev.	Adelaide Law Review
Adv. Space Res.	Advanced Space Research
Air & Sp. L.	Air & Space Law
Akron L. Rev.	Akron Law Review
Am. J. Int'l. L.	American Journal of International Law
Am. J. Int'l. L. Supp.	American Journal of International Law Supplement
Am. Soc. Int'l. L. Proc.	Proceedings of the American Society of International Law
Am. U. J. Int'l. L. & Pol'y.	American University Journal of International Law and Policy
Am. U. L. Rev.	American University Law Review
Ann. Air & Sp. L.	Annals of Air and Space Law
Ann. Assoc. Am. Geog.	Annals of the Association of American Geographers
Ann. Rev. Astron. & Astrop.	Annual Review of Astronomy and Astrophysics
Ann. Rev. Energy	Annual Review of Energy
App. Geog.	Applied Geography
Astron. & Astrophys.	Astronomy and Astrophysics

Astron. Gesell. Abs. Ser.	Astronomische Gesellschaft Abstract Series (Astronomical Society Abstract Series)
Astron. J.	Astronomical Journal
Astrophysics J.	Astrophysics Journal
Aust. Int'l. L. J.	Australian International Law Journal
Aust. J. Astron.	Australian Journal of Astronomy
Aust. Y. B. Int'l. L.	Australian Yearbook of International Law
Az. J. Int'l. & Comp. L.	Arizona Journal of International and Comparative Law
B. C. Envt'l. Aff. L. Rev.	Boston College Environmental Affairs Law Review
B. C. Int'l. & Comp. L. Rev.	Boston College International and Comparative Law Review
B. U. Int'l. L. J.	Boston University International Law Journal
Baylor L. Rev.	Baylor Law Review
Berkeley Tech. L. J.	Berkeley Technology Law Journal
Brit. Y. B. Int'l. L.	British Yearbook of International Law
Brooklyn J. Int'l. L.	Brooklyn Journal of International Law
Brooklyn L. Rev.	Brooklyn Law Review
Buff. L. Rev.	Buffalo Law Review
Bull. Am. Astron. Soc.	Bulletin of the American Astronomical Society
C.F.R.	Code of Federal Regulations
Cable News Network	Cable News Network
Cal. L. Rev.	California Law Review
Cal. W. Int'l. L. J.	California Western International Law Journal
Cam. L. J.	Cambridge Law Journal
Can. Y. B. Int'l. L.	Canadian Yearbook of International Law
Cardozo L. Rev.	Cardozo Law Review
Case W. Res. J. Int'l. L.	Case Western Reserve Journal of International Law
Chi. J. Int'l. L.	Chinese Journal of International Law
Colo. J. Int'l. Envt'l. L. & Pol'y.	Colorado Journal of International Environmental Law and Policy
Colum. J. Envt'l. L.	Columbia Journal of Environmental Law
Colum. J. Transnat'l. L.	Columbia Journal of Transnational Law
Colum. J. World Bus.	Columbia Journal of World Business
Colum. L. Rev.	Columbia Law Review
Com. L. J.	Commercial Law Journal
Conn. J. Int'l. L.	Connecticut Journal of International Law
Cornell Int'l. L. J.	Cornell International Law Journal
Cornell L. Rev.	Cornell Law Review
Cosmic Research	Cosmic Research
Crosslink	Crosslink
De Economist	De Economist
Def. Sci.	Defence Science

List of Reports, Series and Journal Titles

Denver J. Int'l. L. & Pol'y.	Denver Journal of International Law and Policy
Detroit Coll. L. J. Int'l. L. & Prac.	Detroit College of Law Journal of International Law and Practice
Dick. J. Int'l. L.	Dickenson Journal of International Law
Die Friedenswarte	Die Friedenswarte
E.R.	English Reports
Earth Planets & Sp.	Earth Planets and Space
Earth, Moon & Planets	Earth, Moon and Planets
Ec. & Fin. Rev.	Economics and Finance Review
Ecology L. Q.	Ecology Law Quarterly
Econ. J.	Economics Journal
Emory Int'l. L. Rev.	Emory International Law Review
Emory J. Int'l. Disp. Resol.	Emory Journal of International Dispute Resolution
Energy Explor. & Exploit.	Energy Exploration and Exploitation
Energy Policy	Energy Policy
Envt'l. & Res. Ec.	Environmental and Resource Economics
Envt'l. L.	Environmental Law
Eur. J. Int'l. L.	European Journal of International Law
Fl. Coastal L. Rev.	Florida Coastal Law Review
Fl. Int'l. L. J.	Florida International Law Journal
Fordham Envt'l. L. Rep.	Fordham Environmental Law Report
Fordham L. Rev.	Fordham Law Review
Foreign Aff.	Foreign Affairs
Foreign Policy	Foreign Policy
Foreign Service J.	Foreign Service Journal
Ga. J. Int'l. & Comp. L.	Georgia Journal of International and Comparative Law
Geo. Wash. J. Int'l. L. & Ec.	George Washington Journal of International Law and Economics
Geoadria	Geoadria
Geochim. & Cosmochim. Acta	Geochimica et Cosmochimica Acta (Journal of the Geochemical and Meteoritical Societies)
Georgetown Int'l. Envt'l. L. Rev.	Georgetown International Environmental Law Review
Georgetown L. J.	Georgetown Law Journal
Hague Y. B. Int'l. L.	Hague Yearbook of International Law
Harv. Int'l. L. J.	Harvard International Law Journal
Harv. J. Law & Tech.	Harvard Journal of Law and Technology
Harv. J. on Legis.	Harvard Journal on Legislation
Hastings Int'l. & Comp. L. Rev.	Hastings International and Comparative Law Review

Hastings L. J.	Hastings Law Journal
Herts. L. J.	Hertfordshire Law Journal
Hist. & Tech.	History and Technology
Houston J. Int'l. L.	Houston Journal of International Law
Howard L. J.	Howard Law Journal
Human Rights Q.	Human Rights Quarterly
Hyperfine Interactions	Hyperfine Interactions
I.C.J. Rep.	Reports of the International Court of Justice
I.L.M.	International Legal Materials
I.L.R.	International Law Reports
Icarus	Icarus
Ind. J. Global Legal Stud.	Indiana Journal of Global Legal Studies
Ind. Leg. F.	Indiana Legal Forum
Indian J. Int'l. L.	Indian Journal of International Law
Info. Econ. & Pol'y.	Information Economics and Policy
Int'l. & Comp. L. Q.	International and Comparative Law Quarterly
Int'l. Bus. Lawyer	International Business Lawyer
Int'l. Geol. Rev.	International Geology Review
Int'l. J.	International Journal
Int'l. J. Estuarine & Coastal L.	International Journal of Estuarine and Coastal Law
Int'l. J. Marine & Coastal L.	International Journal of Marine and Coastal Law
Int'l. Lawyer	International Lawyer
Int'l. Leg. Persp.	International Legal Perspectives
Int'l. Org.	International Organisations
Int'l. R. & S. Ab.	Internationalrechtliche und staatrechtliche Abhandlungen
Int'l. Trade L. J.	International Trade Law Journal
Iran-U.S.C.T.R.	Iran-U.S. Claims Tribunal Reports
Isr. L. Rev.	Israel Law Review
J. Afr. L.	Journal of African Law
J. Air L. & Com.	Journal of Air Law and Commerce
J. Astron. & Space Sci.	Journal of Astronomy and Space Science
J. Brit. Interplanetary Soc.	Journal of the British Interplanetary Society
J. Contemp. Leg. Issues	Journal of Contemporary Legal Issues
J. Energy Nat. Res. L.	Journal of Energy and Natural Resources Law
J. Envt'l. L. & Lit.	Journal of Environmental Law and Litigation
J. Geophys. Res.	Journal of Geophysical Research
J. Guid. Con. & Dyn.	Journal of Guidance, Control and Dynamics
J. Hist. Int'l. L.	Journal of the History of International Law
J. Int'l. L. & Bus.	Journal of International Law and Business
J. Japan & Int'l. Econ.	Journal of Japanese and International Economics

List of Reports, Series and Journal Titles

J. L. & Env't.	Journal of Law and Environment
J. Land Use & Envt'l. L.	Journal of Land Use and Environmental Law
J. Legis.	Journal of Legislation
J. Marit. L. & Com.	Journal of Maritime Law and Commerce
J. Nat. Res. & Envt'l. L.	Journal of Natural Resources and Environmental Law
J. Pat. Off. Soc'y.	Journal of the Patent Office Society
J. Pol. Econ.	Journal of Political Economy
J. Policy Modelling	Journal of Policy Modelling
J. Pub. Pol'y.	Journal of Public Policy
J. Royal Soc. Arts	Journal of the Royal Society for the Arts
J. Sp. L.	Journal of Space Law
J. Spacecraft & Rockets	Journal of Spacecraft and Rockets
J. World Trade L.	Journal of World Trade Law
Kyklos	Kyklos
L. & Contemp. Probs.	Law and Contemporary Legal Problems
L. & Pol'y. Int'l. Bus.	Law and Policy in International Business
L. Q. Rev.	Law Quarterly Review
L.N.T.S.	League of Nations Treaty Series
La. L. Rev.	Louisiana Law Review
Lawyer Am.	Lawyer of the Americas
Leiden J. Int'l. L.	Leiden Journal of International Law
Loyola L. A. Int'l. & Comp. L. Ann.	Loyola of Los Angeles International and Comparative Law Annals
Loyola L. A. Int'l. & Comp. L. J.	Loyola of Los Angeles International and Comparative Law Journal
Loyola U. Chi. L. J.	Loyola University Law Journal
Lunar Planet. Sci.	Proceedings of the Lunar and Planetary Science Conference
Malaya L. Rev.	Malaya Law Review
Man. J. Int'l. Econ. L.	Manchester Journal of International Economic Law
Max Planck U.N.Y.B.	Max Planck Yearbook of United Nations Law
Melb. J. Int'l. L.	Melbourne Journal of International Law
Meteor. & Planet. Sci.	Meteoritics and Planetary Sciences
Meteoritics	Meteoritics
Mich. J. Int'l. L.	Michigan Journal of International Law
Mill. J. Int'l. Stud.	Millennium Journal of International Studies
Mo. Envt'l. L. & Pol'y. Rev.	Missouri Environmental Law and Policy Review
Monash U. L. Rev.	Monash University Law Review
Mq. J. Int'l. & Comp. Envt'l. L.	Macquarie Journal of International and Comparative Environmental Law
Murdoch U. Elec. J. L.	Murdoch University Electronic Journal of Law
N. C. J. Int'l. L. & Com. Reg.	North Carolina Journal of International Law and Commercial Regulation

N. Ir. Leg. Q.	Northern Ireland Legal Quarterly
N. Y. J. Int'l. & Comp. L.	New York Journal of International and Comparative Law
N. Y. U. Envt'l. L. J.	New York University Environmental Law Journal
N. Y. U. J. Int'l. L. & Pol.	New York University Journal of International Law and Policy
N. Z. J. Envt'l. L.	New Zealand Journal of Environmental Law
Nat. Res. J.	Natural Resources Journal
Nat. Res. Lawyer	Natural Resources Lawyer
Nature	Nature
Naval L. Rev.	Naval Law Review
Ned. Tijd. Int'l. Recht	Nederlands Tijdschrift voor Internationaal Recht (Netherlands Journal of International Law)
New Scientist	New Scientist
Nordisk Tids. Int'l. Ret. Nordic J. Int'l. L.	Nordisk Tidsskrift for International Ret (Nordic Journal of International Law)
Notre Dame. Int'l. & Comp. L. J.	Notre Dame International and Comparative Law Journal
Nw. J. Int'l. L. & Bus.	Northwestern Journal of International Law and Business
O.J.	Official Journal of the European Union
Ocean & Coastal L. J.	Ocean and Coastal Law Journal
Ocean Dev. & Int'l. L. J.	Ocean Development and International Law Journal
Operations Research	Operations Research
Or. L. Rev.	Oregon Law Review
Organisational Dynamics	Organisational Dynamics
Osgoode Hall L. J.	Osgoode Hall Law Journal
Ottawa L. Rev.	Ottawa Law Review
P.C.I.J. Rep.	Permanent Court of International Justice Reports
Past & Present	Past and Present
Pepp. L. Rev.	Pepperdine Law Review
Phys. Rev. Letters	Physics Review Letters
Physics Edu.	Physics Education
Planet. Sp. Sci.	Planetary and Space Science
Polar Rec.	Polar Record
Proc. Coll. L. Outer Sp.	Proceedings of the Colloquium on the Law of Outer Space (from 2008: Proceedings of the International Institute of Space Law)
Pub. L. Forum	Public Law Forum
R.I.A.A.	Reports of International Arbitral Awards
Regent J. Int'l. L.	Regent Journal of International Law
Rev. Aca. Colom. Juris.	Revista de la Academia Colombiana de Jurisprudencia (Colombian Academy of Law Review)

Rev. Cen. Inv. Dif. Aero. Esp.	Revista del Centro de Investigación y Difusión Aeronáutico-Espacial (Journal of the Centre for Aeronautics and Space Research and Outreach)
Rev. Jur. U. P. R.	Revista Juridica de la Universidad de Puerto Rico (University of Puerto Rico Law Review)
Rev. World Econ.	Review of World Economics
Revue Gén. Dr.	Revue Générale de Droit (General Law Review)
Revue Gén. Dr. Int'l. Pub.	Revue Générale de Droit International Public (General Public International Law Review)
Rich. J. Global L. & Bus.	Richmond Journal of Global Law and Business
Rutgers L. J.	Rutgers Law Journal
S. Afr. L. J.	South African Law Journal
S. Cal. L. Rev.	Southern California Law Review
S. Tex. L. Rev.	South Texas Law Review
San Diego L. Rev.	San Diego Law Review
Santa Clara Computer & High Tech. L. J.	Santa Clara Computer and High Technology Law Journal
Scand. J. Econ.	Scandinavian Journal of Economics
Science	Science
Scientific American	Scientific American
Seton Hall L. Rev.	Seton Hall Law Review
Singapore J. Int'l. & Comp. L.	Singapore Journal of International and Comparative Law (from 2005: Singapore Yearbook of International Law)
Solar Energy	Solar Energy
Solar System Dev. J.	Solar System Development Journal
Space News	Space News
Space Power	Space Power
Space Sci. Rev.	Space Science Review
Sri Lanka J. Int'l. L.	Sri Lanka Journal of International Law
St. John's L. Rev.	St. John's Law Review
St. Thom. L. Rev.	St. Thomas Law Review
Stanford J. Int'l. L.	Stanford Journal of International Law
Stanford J. Int'l. Stud.	Stanford Journal of International Studies
Stud. Transnat'l. Leg. Pol'y.	Studies in Transnational Legal Policy
Suffolk Transnat'l. L. J.	Suffolk Transnational Law Journal
Syracuse J. Int'l. L.	Syracuse Journal of International Law
Syracuse J. Int'l. L. & Com.	Syracuse Journal of International Law and Commerce
T.I.A.S.	Treaties and Other International Acts Series
Temple Int'l. & Comp. L. J.	Temple International and Comparative Law Journal

Tex. Int'l. L. F.	Texas International Law Forum
The Astro. J.	The Astronomy Journal
The Economist	The Economist
The Forum	The Forum
The Telegraph	The Telegraph (United Kingdom)
Third World Leg. Stud.	Third World Legal Studies
Tourism Man.	Tourism Management
Tul. L. Rev.	Tulane Law Review
Tulsa J. Comp. & Int'l. L.	Tulsa Journal of Comparative and International Law
Tulsa L. J.	Tulsa Law Journal
Tx. Int'l. L. J.	Texas International Law Journal
U. Dayton L. Rev.	University of Dayton Law Review
U. Det. L. J.	University of Detroit Law Journal
U. Fla. L. Rev.	University of Florida Law Review
U. Ill. L. F.	University of Illinois Law Forum
U. Miami L. Rev.	University of Miami Law Review
U. N. B. L. J.	University of New Brunswick Law Journal
U. Qld. L. J.	University of Queensland Law Journal
U. S. F. L. Rev.	University of San Francisco Law Review
U.C.L.A. L. Rev.	University of California Los Angeles Law Review
U.K.T.S.	United Kingdom Treaty Series
U.M.K.C. L. Rev.	University of Missouri Kansas City Law Review
U.N.T.S.	United Nations Treaty Series
U.S.	United States Law Reports
U.S.A.F.A. J. Leg. Stud.	United States Air Force Academy Journal of Legal Studies
U.S.C.	United States Code
U.S.D.I.L.	Digest of U.S. Practice in International Law
U.S.T.	United States Treaties and Other Documents
Uni. Toronto L. J.	University of Toronto Law Journal
Va. Envt'l. L. J.	Virginia Environmental Law Journal
Va. J. Int'l. L.	Virginia Journal of International Law
Va. L. Rev.	Virginia Law Review
Vand. J. Transnat'l. L.	Vanderbilt Journal of Transnational Law
Wash. & Lee L. Rev.	Washington and Lee Law Review
Wis. Int'l. L. J.	Wisconsin International Law Journal
Wm. & Mary L. Rev.	William and Mary Law Review
World Dev.	World Development
Y. B. World Aff.	Yearbook of World Affairs
Y.B.I.L.C.	Yearbook of the International Law Commission
Yale L. & Pol'y. Rev.	Yale Law and Policy Review
Yale Stud. World Pub. Ord.	Yale Studies in World Public Order

Zeit. Aus. Recht. Völk.	Zeitschrift für ausländisches öffentliches Recht und Völkerrecht (Heidelberg Journal of International Law)
Zeit. Luft. Welt.	Zeitschrift für Luft- und Weltraumrecht (German Journal of Air and Space Law)

Chapter 1
Introduction and Overview

1.1 The Problem

1.1.1 The Context

Since time immemorial, human civilisation has evolved through the history of the world by the continuing development and use of mineral resources from the Earth's crust. The rapid pace of technological development and the exploding human population make the exhaustion of non-renewable mineral resources on Earth only a matter of time.

In recent years, the increasing pressure placed upon governments, international organisations and international lawyers to produce regulatory frameworks for the exploitation of mineral resources in the deep seabed and the Polar Regions are further evidence of this approaching phenomenon. However, such efforts must be balanced with an increasing consciousness of the need to preserve the global environment. Given all of the above, it is only the logical progression of human development that the exploitation of mineral resources from celestial bodies becomes a necessity.

This monograph aims to highlight the absence of an appropriate regime in the existing body of international law for the commercial exploitation of mineral resources from celestial bodies and proposes a new regulatory framework in anticipation of such ventures becoming feasible, desirable and even necessary. In assessing the need for such a regulatory framework, as well as to determine the appropriate legal and administrative aspects of such a framework, it is prudent to consider that:

(1) while presently not feasible, the economic conditions for the exploitation of mineral resources from celestial bodies will eventuate from the continuing depletion of natural resources and the increasing need for environmental conservation on Earth;

(2) technological capabilities for activities in space will continue to evolve to enable the extraction and exploitation of mineral resources from celestial bodies to be a

---

R.J. Lee, *Law and Regulation of Commercial Mining of Minerals in Outer Space*,
Space Regulations Library 7, DOI 10.1007/978-94-007-2039-8_1,
© Springer Science+Business Media B.V. 2012

feasible alternative to extracting mineral reserves from logistically difficult and environmentally detrimental areas on Earth;

(3) the existing body of international space law does not provide a regulatory framework for the commercial exploitation of mineral resources on celestial bodies, in particular, the freedoms of exploration and use, the principle of non-appropriation and the concept of the common heritage of mankind together create much inconsistency and uncertainty in their application to such commercial activities;

(4) various existing models of international regulation and the competing policy interests of the stakeholders must be considered in determining the appropriate international regulatory framework, even though presently no single existing model is or will be appropriate as a precedent; and

(5) accordingly, a new regulatory framework is proposed with legal and policy considerations taken into account along with the financial, practical and administrative aspects of implementing such a framework.

1.1.2 Structure of the Monograph

In pursuit of the creation of a comprehensive, appropriate and viable international regulatory framework, this monograph is divided into the following chapters:

(1) *Chapter 1* is an introduction and overview of the monograph, which outlines its structure and hypothesis;

(2) *Chapter 2* explores the increasing economic and technical feasibility of the commercial exploitation of mineral resources on celestial bodies, especially from Near Earth Asteroids, and considers the financial commitments required for such activities, recognising that the principal obstacle to such commitments being made is the legal uncertainty over such activities;

(3) *Chapter 3* outlines the general principles in the existing body of international space law as applicable to all space activities, particularly on issues of state responsibility and liability;

(4) *Chapter 4* provides an analysis of the legality of commercial exploration and extraction of mineral resources on celestial bodies, in particular the freedoms of exploration and use, and the principle of non-appropriation;

(5) *Chapter 5* discusses the legal and policy aspects of the common heritage of mankind concept as applicable to celestial bodies and the international impasse over the application of this concept to celestial bodies and considers the failure of previous intergovernmental negotiations over similar legal regimes;

(6) *Chapter 6* attempts to balance the various competing interests to create a legal and policy position that may be acceptable to most members of the international community; and

(7) *Chapter 7* contains some concluding observations (Fig. 1.1).

1.1 The Problem

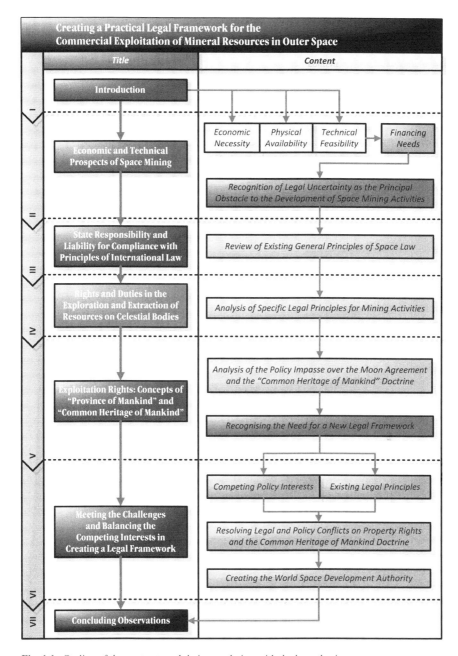

Fig. 1.1 Outline of the contents and their correlation with the hypothesis

1.2 Historical Background

1.2.1 Technological Evolution

From the beginning of recorded history, human civilisation and its gradual evolution have been characterised and driven by the natural resources utilised for the production of tools to increase the productivity of each economic unit of society. For example, the epochs in which humans attained the knowledge and technical skills to produce tools with stone, bronze and iron have been referred to as the Stone Age, Bronze Age and Iron Age respectively. Although such labels have become obsolete in describing civilisations, the technological advancement of different cultures is defined through the use of various resources, such as steel or plastics, to produce tools essential for the prosperity and development of human civilisation.

The computing revolution at the end of the last century has caused the contemporary world to be described as being in the "Information Age".[1] Doing so conceals the fact that this "Information Age" relies on the ability of the human civilisation to utilise the underlying resources. Such resources include the silicon used in the production of computer chips and electronic circuitry, the copper and aluminium used in cabling and wiring, the steel and magnesium used in computer casings and satellites, along with the various resources used to generate the electricity required to fuel this information superhighway, such as coal, natural gas and uranium. Indeed, to refer to contemporary times as the "Information Age" would be akin to calling the Iron Age the "Spear Age" or the "Axe Age". The growing global population and the corresponding growth in the production of various materials have significantly increased the use of and demand for mineral resources found in the Earth's crust.

In addition to being in the "Information Age", the present world has also been described as the "Space Age".[2] Since the launch of *Sputnik-1* by the Soviet Union on 4 October 1957, the rivalry between the antagonists of the Cold War has caused the technological advancements in our space capability to increase exponentially. It is difficult to comprehend that the first artificial satellite was launched just over half a century ago and that it was no more than a steel sphere containing a transmitter that was producing beeps continuously while in orbit.[3] Since then, the Cold War provided the impetus for human civilisation to cross the final frontier into space by launching manned spacecrafts, landing on the Moon, inhabiting orbital space stations and sending unmanned orbital probes and landing crafts to explore the

---

[1] See, for example, David S. Alberts and Daniel S. Papp (eds.), THE INFORMATION AGE: An ANTHOLOGY ON ITS IMPACTS AND CONSEQUENCES (1998).

[2] See, for example, Martin Collins, AFTER SPUTNIK: FIFTY YEARS OF THE SPACE AGE (2007).

[3] John S. Gibson, *Five Days in October: "Tracking" Sputnik I at Redstone Arsenal* (2001), Cold War Museum, at <http://www.coldwar.org/text_files/gibson.pdf>, last accessed on 7 December 2004.

1.2 Historical Background

other planets of the Solar System. By the end of the Cold War, the civilian space businesses have replaced governments as the primary actors in outer space. The commercial applications of space, such as telecommunications, weather forecasting, remote sensing, global positioning and direct television broadcasting, are now being taken for granted by most members of the global community. The recent wars in Afghanistan and Iraq would have been fought very differently if the space technology utilised in these conflicts had not been available to the military in identifying enemy bases, al-Qaeda training camps, tracking troop movements and special operations units, missile targeting and telecommunications. Further, satellites were of key importance in transmitting live television pictures to the anxious world watching outside Afghanistan and Iraq.

1.2.2 The Post-Cold War World

That is not to say that, since the end of the Cold War, government actors no longer play a significant role in the exploration and use of outer space. After all, there is a continuing debate within the United States of America between the need to engage or compete with the space program of China.[4] The present emphasis on private and commercial use of outer space cannot hide the fact that, to date, the governments of the United States of America, the former Soviet Union, Europe and Japan are the only entities that have substantially explored the Moon and other celestial bodies. This governmental exploration has been done traditionally through human missions to the Moon or unmanned probes to the Moon and other celestial bodies in the Solar System.[5] In 2004, the United States of America committed itself to a return of human missions to the Moon, although this commitment by President George W. Bush showed strong commercial motives for the new venture:

> Returning to the Moon is an important step for our space program. Establishing an extended human presence on the Moon could vastly reduce the costs of further space exploration, making possible ever more ambitious missions. Lifting heavy spacecraft and fuel out of the Earth's gravity is expensive. Spacecraft assembled and provisioned on the Moon could escape its far lower gravity using far less energy, and thus, far less cost. Also, the Moon is home to abundant resources. Its soil contains raw materials that might be harvested and processed into rocket fuel or breathable air. We can use our time on the Moon to develop and test new approaches and technologies and systems that will allow us to function in

---

[4] See, for example, Fred Stakelbeck, Jr., *Inconsistent U.S. Policy* (2 June 2006), The Washington Times, at <http://washingtontimes.com/op-ed/20060601-085037-6169r.htm>, last accessed on 11 July 2006; James C. Moltz, *Moonstruck: What's Up with U.S. Space Policy?* (2 February 2004), Centre for Nonproliferation Studies, at <http://cns.miis.edu/pubs/week/040202.htm>, last accessed on 11 July 2006; and Leonard David, *Space Cooperation: The China Factor* (5 January 2003), at <http://www.space.com/news/china_cooperation_030121.html>, last accessed on 11 July 2006.

[5] See David R. Williams, *Chronology of Lunar and Planetary Exploration* (2006), NASA Goddard Space Flight Centre, at <http://nssdc.gsfc.nasa.gov/planetary/chrono.html>, last accessed on 11 July 2006.

other, more challenging environments. The Moon is a logical step toward further progress and achievement.[6]

To the casual observer, three trends are increasingly evident in studying the modern history of human civilisation from the viewpoint of resources. Firstly, the continuing growth of the human population and the corresponding economic development of the Earth, together with technological advances, have increased the consumption of natural resources at an exponential rate.[7] It makes sense that, because the majority of the resources that are utilised are not renewable resources, there must be a finite point in time when the mineral resources of this planet are exhausted. Technological advances in processing techniques, recycling and human innovation in finding and using substitutes have largely alleviated the onset of this physical exhaustion of resources.[8] However, eventually the moment will be reached when extraterrestrial sources of minerals must be found to supplement or even replace the existing deposits known or to be discovered in the Earth's crust.

Secondly, the expansion of the human race into outer space has continued after the end of the Cold War, though principally due to commercial interests playing a more prominent role in financing and operating space ventures. The majority of satellites orbiting the Earth today are now used for commercial applications and there are already several multinational private ventures to build and operate orbital space stations and even interplanetary probes for mineral prospecting purposes.[9] The increased development of orbital and lunar infrastructure, as well as installations or even settlements on celestial bodies, must necessitate the increased use of mineral resources in outer space. Due to the strong gravity of the Earth and the energy required to transport materials from the surface of the Earth to low Earth orbit ("*LEO*") and beyond, a strong commercial and technical case can be made for the extraction and processing of mineral ores either in situ or in LEO. This would provide the necessary materials to reduce the construction costs of orbital space structures.

Thirdly, the global community is becoming increasingly conscious of the impact that increasing industrialisation and mineral extraction activities have on the natural environment of this planet. Global climatic and environmental phenomena, such as acid rain, ozone layer depletion, global warming, desertification and deforestation have intensified the call for mineral resources to be extracted and developed in an ecologically sustainable way. As the mineral deposits found at comparatively

---

[6] The White House, *President Bush Announces New Vision for Space Exploration Program* (14 January 2004), at <http://www.whitehouse.gov/news/releases/2004/01/20040114-3.html>, last accessed on 11 July 2006. The program has been substantially scaled back by the Obama Administration:

[7] Herman E. Daly and Kenneth N. Townsend, VALUING THE EARTH: ECONOMICS, ECOLOGY AND ETHICS (1993), at 267.

[8] Judith Rees, NATURAL RESOURCES: ALLOCATION, ECONOMICS AND POLICY (2nd ed., 1990), at 39.

[9] See, for example, SpaceDev, Inc., *Near Earth Asteroid Prospector* (2004), at <http://www.spacedev.com/newsite/templates/subpage3.php?pid=191&26subNav=11&26subSel=3>, last accessed on 6 December 2004.

1.2 Historical Background

more accessible and ecologically less sensitive areas are being depleted, the global community is likely to be more vocal in resisting the extraction of resources from national parks and other international ecological treasures such as the Great Barrier Reef, the deep seabed and from Antarctica. In recent times, the continuing debate in the United States over exploration and prospecting in Alaska has highlighted this increasing consciousness of humanity to its environmental impact.[10] This resistance would be particularly strengthened when the technology for extracting and processing mineral resources in space become increasingly available, allowing for the retention of such environmentally sensitive areas to be preserved as global sanctuaries for the benefit of present and future generations of the human civilisation.

1.2.3 The Hypothesis

It is clear from the foregoing discussion that the exploitation of natural resources from celestial bodies would occur inevitably as a result of a combination of different factors. There is an increasing number of private firms that have invested heavily into the design, development and construction of asteroid prospecting probes as well as automated robotic mining mechanisms.[11] Research is also gathering pace on the potential infrastructure for deployment on asteroids and other celestial bodies in the Solar System.[12] The recent mineralogical and geological studies conducted on various comets and asteroids have contributed valuable data that may assist in the selection of suitable targets for resource extraction, especially when combined with Earth-based mineralogical investigations.[13]

---

[10] See materials on, for example, Save Alaska, at <http://www.savealaska.com/sa_headlines. html>, last accessed on 6 December 2004 and Thomas M. Power, *The Role of Metal Mining in the Alaskan Economy* (13 February 2002), Southeast Alaska Conservation Council, at <http://www. seacc.org/Publications/MetalMiningReport.doc>, last accessed on 27 January 2007. Further, see generally Earle A. Ripley, ENVIRONMENTAL EFFECTS OF MINING (1996); Jerrold J. Marcus (ed.), MINING ENVIRONMENTAL HANDBOOK: EFFECTS OF MINING ON THE ENVIRONMENT AND AMERICAN ENVIRONMENTAL CONTROLS ON MINING (1997); and George Ledec, *Minimising Environmental Problems from Petroleum Exploration and Development in Tropical Forest Areas*, paper presented at the Proceedings of the First International Symposium on Oil and Gas Exploration and Production Waste Management Practices, 10–13 September 1990 in New Orleans, United States.

[11] See, for example, Island One Society, *LMF Mining Robots* (25 July 1999), at <http://www. islandone.org/MMSG/aasm/AASM5D.html>, last accessed on 7 December 2004.

[12] See, for example, Mark J. Sonter, *Near Earth Objects as Resources for Space Industrialisation* (2001) 1:1 SOLAR SYSTEM DEV. J. 1.

[13] See, for example, Clark R. Chapman, *S-Type Asteroids, Ordinary Chondrites and Space Weathering: The Evidence from Galileo's Fly-bys of Gaspra and Ida* (1996) 31 METEOR. & PLANET. SCI. 699. Consider also the deficiencies in Earth-based mineralogical studies as identified in Robert Jedicke, David Nesvorny, Robert J. Whiteley, Zeljko Ivezic and Mario Juric, *An Age-Colour Relationship for Main Belt S-Complex Asteroids* (2004) 429 NATURE 275.

With the future exploitation of mineral resources in outer space a given certainty, the central hypothesis of this work is that one of the major inhibiting factors of this development is the absence of an appropriate legal framework to govern mining activities in outer space. The existing body of space law already has been subjected to heavy criticism for its failure to provide for the commercial realities of the global space industry.[14] Space law, as a *lex specialis* of international law, was formulated in the 1960s and 1970s and did not anticipate the commercialisation that has now taken place in outer space or the possibility of future commercial exploitation of mineral resources from space. In order for such mining activities to become a practical reality, it is essential for significant reform to take place in international space law.

Accordingly, this work explores the current body of space law and its implications on commercial space mining ventures, taking into account their economic and technical aspects, and then considers the necessary reforms and developments to overcome such legal obstacles to the exploitation of mineral resources in outer space.

1.3 Economic and Technical Prospects of Mining on Celestial Bodies

One of the requisite assumptions in the hypothesis of this work is that commercial space mining is both an economic and a technical possibility, though not necessarily a reality. As a result, the first and most fundamental question to be answered is the feasibility of mining resources on celestial bodies within the Solar System. In this regard, it is interesting to note that the Moon, other planets and their satellites, comets and asteroids in the Solar System are rich in elements, minerals and hydrocarbons that are either abundant or rare on Earth.[15] However, the gravity of large celestial bodies such as the Moon and Mars makes them somewhat expensive and unattractive targets for exploitation, even though the proximity and the size of the Moon does present ideal qualities as an advance base for human expansion within the Solar System.

Within the orbit of Jupiter, there is a large number of asteroids orbiting around the Sun of various shapes, eccentricities and distances. Most asteroids are concentrated in the Main Belt of asteroids between Mars and Jupiter, which are referred to as the

---

[14] See, for example, Bin Cheng, *The Commercial Development of Space: The Need for New Treaties* (1991) 19 J. SP. L. 17; Hanneke L. van Traa-Engelman, COMMERCIAL UTILISATION OF OUTER SPACE: LAW AND PRACTICE (1993); and Patrick Q. Collins, *Implications of Reduced Launch Costs for Commercial Space Law*, in Kunihiko Tatsuzawa (ed.), LEGAL ASPECTS OF SPACE COMMERCIALISATION (1992).

[15] See, for example, Jonathan R. Tate, *Near Earth Objects — A Threat and an Opportunity* (2003) 38 PHYSICS EDU. 218 and Brad R. Blair, *The Role of Near-Earth Asteroids in Long-Term Platinum Supply*, paper presented at the Second Space Resources Roundtable, Colorado School of Mines, 8–10 November 2000, in Boulder, CO, USA, as at <http://www.mines.edu/research/srr/Presentations/blair-platinum.PDF>, last accessed on 8 December 2004.

Main Belt asteroids. The precise origins of the Main Belt remains a mystery, though there is speculation that it may have been formed as a result of a collision of two large planetesimals during the creation of the Solar System or the failure of a large planetesimal to form due to the gravitational tidal forces caused by the proto-Sun, Mars and Jupiter.[16] In addition to the Main Belt asteroids, there are some classes of asteroids that have orbits near, or even crossing, the heliocentric orbit of the Earth. They are collectively referred to as the Near Earth Asteroids. The Apollos, Amors and Atens are groups of asteroids that, along with some other minor classes, belong to the family of Near Earth Asteroids that present themselves as strong candidates for future mining activities as the mineral resources on Earth are being gradually depleted.[17]

The two most important physical considerations in determining the viability of mining a particular asteroid are the time factor and the energy cost required for the mission, assuming that the existing human technology for propulsion, spacecraft design and construction, as well as mining infrastructure, can be adapted for asteroid mining purposes. The ideal asteroid for mining purposes would be one that is close to the Earth and allows for a long mining season without requiring large amounts of energy to reach it or to return to Earth. In this context, the mining of some Near Earth Asteroids would be preferable to other celestial bodies in the Solar System as the utilisation of some energy efficient trajectories, such as Hohmann transfer trajectories, would only allow for very short mining seasons on the asteroid.[18] However, these asteroids would nonetheless make better candidates than the Moon and the Martian satellites or even short period comets in terms of the energy requirements and the associated mass limitations imposed by such requirements on the spacecraft.

1.4 Liability and State Responsibility for Compliance with International Legal Principles

1.4.1 Phases of a Commercial Space Mining Venture

It is conceivable that in the near future, the depletion of terrestrial resources and the continuing technological innovation in spacecraft design, propulsion systems and robotic mining equipment would provide the financial incentives for the technical

---

[16] See George W. Wetherill and Stephen J. Kortenkamp, *Asteroid Belt Formation with an Early Formed Jupiter and Saturn*, paper presented at the 30th Annual Lunar and Planetary Science Conference, 15–29 March 1999, in Houston, TX, USA; Alessandro Morbidelli, William F. Bottke, Christiane Froeschlé and Patrick Michel, *Origin and Evolution of Near-Earth Objects*, in William F. Bottke, Alberto Cellino, Paolo Paolicchi and Richard P. Binzel (eds.), ASTEROIDS III (2002) at 409–422; and Makiko Nagasawa, Shigeru Ida and Hidekazu Tanaka, *Origin of High Orbital Eccentricity and Inclination of Asteroids* (2001) 53 EARTH PLANETS & SP. 1085.

[17] See generally Charles T. Kowal, ASTEROIDS: THEIR NATURE AND UTILISATION (1988).

[18] Mark J. Sonter, *The Technical and Economic Feasibility of Mining the Near-Earth Asteroids*, paper presented at the 49th International Astronautical Congress, 5–9 October 1998, in Melbourne, Australia.

obstacles and associated costs to be overcome. When that reality eventuates, a commercial mining venture in outer space will have to confront the legal issues based on the present body of international and domestic space law. Fundamentally, this is because the enormous financial commitments that would be needed to invest in the technological advances required for such ventures cannot be done in an environment of legal uncertainty.

Therefore, it is prudent to consider the existing *corpus* of international space law and consider what adaptations, if any, are needed in order to provide the legal certainty essential to making space mining a reality. In considering the international legal framework applicable to space mining ventures, it is prudent to note that the typical commercial space mining operation can be divided into seven segments:

(1) the *planning segment* that involves the mineralogical and technical study of the feasibility of mining certain selected celestial bodies;
(2) the *exploration segment* which entails the use of spacecrafts to analyse mineralogical samples and, with the use of remote sensing technology, assess the viability of various deposits;
(3) the *launch segment* which encompasses the operation from conception to the launch of the mining spacecraft or, if the operation involves multiple launches, the last of those launches;
(4) the *transit segment* which encompasses the operation from the launch segment to the arrival of the mining craft to its target celestial body;
(5) the *extraction segment* which begins from the landing of the mining craft on the target celestial body to the full recovery of all mined ores;
(6) the *return segment* which involves the return of mined ores to the Earth, whether with ore processing in situ or otherwise, which may involve the transport of the celestial body being mined or a large portion thereof into the Earth or lunar orbit; and
(7) the *exploitation segment* when the mined ores are sold with a view to profit, in their unprocessed, processed or utilised form (Fig. 1.2).

1.4.2 Applicable Legal Issues

The legal issues arising in the launch, transit and return segments are similar to those of any commercial payload launch and can be divided into issues arising from international law and those arising from domestic law. For example, the absence of a definitive delimitation between airspace and outer space, liability caused to foreign third parties and the use of trajectories that requires flyovers of foreign territories are all legal issues that arise from the operation of international law. On the other hand, legal requirements arising from domestic regulatory frameworks and the application of export controls stem from the operation of domestic law. However, it must be noted that the return segment, if involving the movement of celestial bodies from

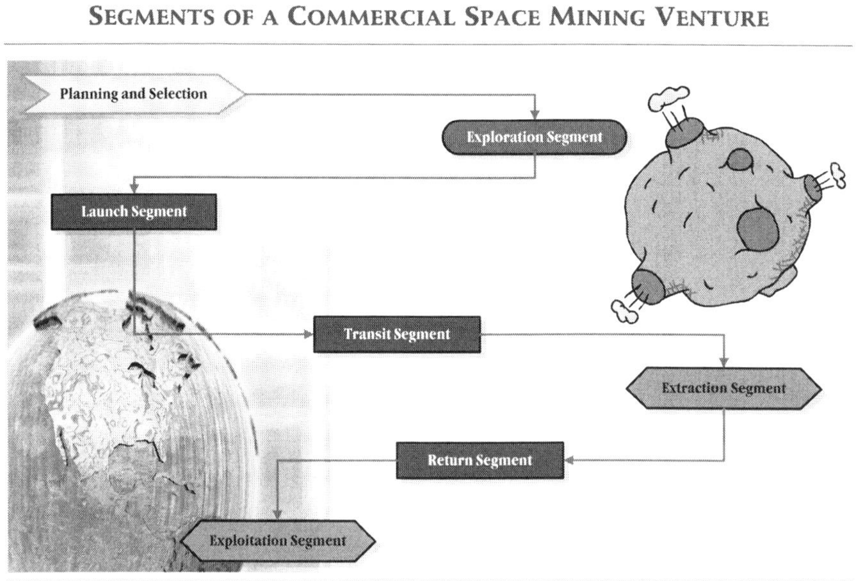

SEGMENTS OF A COMMERCIAL SPACE MINING VENTURE

Fig. 1.2 Segments of a commercial space mining venture

their natural orbit to the Earth or lunar orbit, may invoke the application of some legal principles in relation to the appropriation of celestial bodies.

1.4.3 State Responsibility and International Liability

It has been acknowledged by some commentators that the existing body of space law is very ill-adapted to the commercial realities in the space industry today.[19] When the legal principles contained in the 1967 Treaty on the Principles Governing the Activities of States in the Exploration and Use of Outer Space, including the Moon and other Celestial Bodies (hereinafter the "Outer Space Treaty") were drafted and adopted in the early 1960s, the present commercialisation and development in space was inconceivable to all except the most devoted science fiction writers and film-makers.[20] It was not until the 1979 Agreement Governing the Activities of States

---

[19] See, for example, Nina Tannenwald, *Law Versus Power on the High Frontier: The Case for a Rule-Based Regime in Outer Space* (2004) 29 YALE J. INT'L. L. 363; Andrew T. Park, *Incremental Steps for Achieving Space Security: The Need for a New Way of Thinking to Enhance the Legal Regime for Space* (2006) 28 HOUSTON J. INT'L. L. 871; and Charles M. Dalfen, Andre Bissonnette, Pierre Juneau and Ivan Vlasic, *International Legal Problems of Direct Satellite Broadcasting* (1970) 20 U. TORONTO L. J. 314.

[20] See, for example, Herbert Reis, *Some Reflections on the Liability Convention for Outer Space* (1978) 6 J. SP. L. 161.

on the Moon and other Celestial Bodies (the "Moon Agreement") that the possible mineral exploitation of other celestial bodies was considered. However, even in the Moon Agreement it is quite apparent that its drafters did not anticipate the development of space mining ventures until the distant future, as Article 11 of the Moon Agreement deferred the creation of a specific intergovernmental organisation until such time as deemed necessary by the parties to the Moon Agreement when space mining activities become imminent.

The fundamental principles of space law also pose significant problems for the launch and transit segments of a commercial mining venture in outer space. The United Nations space treaties do not define some important terms, such as "space object", "launching State" and "appropriate State" with sufficient clarity, with important legal ramifications.[21] For example, the 1972 Convention on the International Liability for Damage Caused by Space Objects (the "Liability Convention") imposes liability on "launching States" of space objects for damage caused on the surface of the Earth, in outer space and to aircraft in flight.[22] There is no existing legal demarcation of the boundary between airspace and outer space in international law. Further, the definition of launching States does not reflect the current commercial realities of the launch industry, where more than one State is often involved in the launch process and the parties involved may be multinational corporations with complicated ownership and incorporation structures. These fundamental issues must be resolved before sufficient investor confidence can be found in such inherently risky and unproven ventures.

Further, with the rapid commercial development experienced in space throughout the 1980s, significant legal issues new to the field of space law arose that need to be clarified. For example, the absence of international patent protection for inventions and discoveries that occurred in orbit and for technological developments in space systems, have been a source of significant concern for the commercial space industry.[23] The lack of specialised binding mechanisms for settling disputes between States and between commercial entities from different States in the

---

[21] Treaty on Principles Governing the Activities of States in the Exploration and Use of Outer Space, including the Moon and other Celestial Bodies (the "Outer Space Treaty"), opened for signature on 27 January 1967, 610 U.N.T.S. 205; 18 U.S.T. 2410; T.I.A.S. 6347; 6 I.L.M. 386 (entered into force on 10 October 1967), Articles VI and VII.

[22] Convention on International Liability for Damage Caused by Space Objects (the "Liability Convention"), opened for signature on 29 March 1972, 961 U.N.T.S. 187, 24 U.S.T. 2389, T.I.A.S. 7762; 1975 A.T.S. 5 (entered into force on 1 September 1972), Article I.

[23] See, generally, Anna Maria Balsano, *Industrial Property Rights in Outer Space in the International Governmental Agreement (IGA) on the Space Station and the European Partner* (1992) 35 PROC. COLL. L. OUTER SP. 216; Sa'id Mosteshar, *Intellectual Property Issues in Space Activities*, in Sa'id Mosteshar (ed.), RESEARCH AND INVENTION IN OUTER SPACE: LIABILITY AND INTELLECTUAL PROPERTY RIGHTS (1995), at 189–198; Glenn Harlan Reynolds, *Legislative Comment: The Patents in Space Act* (1990) 3 HARV. J. LAW & TECH. 13; René Oosterlink, *The Intergovernmental Space Station Agreement and Intellectual Property Rights* (1989) 17 J. SP. L. 31; and Bradford L. Smith, *Intellectual Property Rights in Outer Space Activities – Aid or Impediment?*, paper presented at the ISRO-IISL Space Law Conference 2005, 26–29 June 2005 in Bangalore, India.

Liability Convention also has been referred to as a possible source of future problems.[24] In the field of private or commercial space financing, for example, the problems associated with executing and recovering security interests over satellites and other space-based assets have already been highlighted following the recent collapse of some high profile satellite telecommunications ventures, though some measures have been taken outside the arena of intergovermental organisations to provide some international legal rules for such matters.[25]

1.5 Exploration and Extraction Rights

Despite the legal issues already encountered in the launch and transit segments of a commercial space mining venture, it is the exploration and extraction segments of the mining operation that would encounter most of the legal obstacles. In the exploration segment, ore samples will either be robotically analysed in situ or returned to the surface of the Earth for further and more detailed analysis. While this exercise of gathering samples may arguably be no different to the collection of lunar rocks during the Apollo Program missions of the United States, the crucial distinction is that in the case of mineral prospecting, the samples are collected for ultimate private commercial profit rather than public scientific gain. Accordingly, this raises issues on the lawfulness of such activities in outer space.

In the extraction segment, the legal obstacles involved are more complex and difficult to resolve. The freedom of access and the principle of non-appropriation found in Articles I and II of the Outer Space Treaty, in particular, makes mineral extraction activities on celestial bodies difficult, if not impossible, to justify in law. This is because the conduct of mineral extraction activities must have, as a necessary requirement, some degree of exclusionary right in the area of asteroid being mined, a right that would be contrary to those legal principles set out in the Outer Space Treaty. Without the ability to exclude third parties, a commercial miner would have no protection for its financial investment as it would be unable to prevent a third

---

[24] See generally Karl-Heinz Böckstiegel, *Settlement of Disputes Regarding Space Activities* (1993) 21 J. SP. L. 1; Philip D. Bostwick, *Going Private with the Judicial System: Making Creative Use of ADR Procedures to Resolve Commercial Space Disputes* (1995) 23 J. SP. L. 1; Alexis Goh, *Coping with the Lack of a Mechanism for the Settlement of Disputes Arising in Relation to Space Commercialisation* (2001) 5 SINGAPORE J. INT'L. & COMP. L. 180; Particia M. Sterns and Leslie I. Tennen, *Resolution of Disputes in the Corpus Juris Spatialis: Domestic Law Considerations* (1993) 36 PROC. COLL. L. OUTER SP. 172; and Hanneke L. van Traa-Engelman, *Settlement of Space Law Disputes* (1990) 3 LEIDEN J. INT'L. L. 139.

[25] See generally Martin Stanford, *The Cape Town Convention and the Preliminary Draft Space Protocol: An Update*, paper presented at the ISRO-IISL Space Law Conference 2005, 26–29 June 2005 in Bangalore, India; Paul B. Larsen, *Critical Issues in the UNIDROIT Draft Space Protocol* (2002) 45 PROC. COLL. L. OUTER SP. 2; Nicholas Humphrey and Vernon Nase, *The Cape Town Convention 2001: An Australian Perspective* (2006) 31 AIR & SP. L. 5; and Alvaro Fabricio dos Santos, *Developing Countries and the UNIDROIT Protocol on Space Property* (2002) 45 PROC. COLL. L. OUTER SP. 23.

14   1 Introduction and Overview

party from extracting mineral resources from the same site. Further, the act of extraction itself, by its very nature, may contravene the principle of non-appropriation, assuming that the principle extends to prohibit the existence of exclusive property or mining rights. These tensions are symptomatic of the fundamental conflict between the principles of international space law and contemporary commercial applications of space technology.

The idea that States may not own a particular spatial area to the exclusion of other States is not a new one, for as far back as the seventeenth century it was recognised already that the sovereignty of States does not extend to the high seas. However, the idea that a spatial area may be subject to the universal ownership of humankind is certainly a new concept. For example, it was proposed in the early twentieth century that the Antarctic continent should be a sanctuary for all humankind.[26] Although the Antarctic Treaty does not stipulate this explicitly, it is certainly the intention of the Antarctic Treaty System for some form of common management to take place over the exploration and scientific work conducted in Antarctica by deferring the territorial claims of the claimant States.[27]

1.6 Exploitation Rights: Effects of the Common Heritage of Mankind Doctrine

In the 1970s, as the number of developing States increased, the idea of a New International Economic Order ("NIEO") emerged whereby the industrialised States were assigned responsibility to equalise the economic inequality between the

---

[26] Thomas W. Balch, *The Arctic and Antarctic Regions and the Law of Nations* (1910) 4 AM. J. INT'L. L. 265. See also David E. Marko, *A Kinder, Gentler Moon Treaty: A Critical Review of the Current Moon Treaty and a Proposed Alternative* (1992) 8 J. NAT. RES. & ENVT'L. L. 293 at 310–313; and Grier C. Raclin, *From Ice to Ether: The Adoption of a Regime to Govern Resource Exploitation in Outer Space* (1986) 7 J. INT'L. L. & BUS. 727 at 737–738.

[27] Article IX of the Antarctic Treaty provides for consultative meetings in relation to the use of Antarctica for peaceful purposes, the facilitation of scientific research, cooperation and inspection, the exercise of jurisdiction and the preservation and conservation of living resources: Antarctic Treaty, opened for signature on 1 December 1959, 402 U.N.T.S. 71; 12 U.S.T. 794; 19 I.L.M. 860 (entered into force on 23 June 1961). In particular, separate treaties dealing with environmental issues have been formulated, including the Convention for the Conservation of Antarctic Seals, opened for signature on 1 June 1972, 29 U.S.T. 441; 11 I.L.M. 251 (entered into force on 11 March 1978); Convention on the Conservation of Antarctic Marine Living Resources, opened for signature on 20 May 1980, 1329 U.N.T.S. 47; 33 U.S.T. 3476 (entered into force on 7 April 1982); Wellington Convention on the Regulation of Antarctic Mineral Resource Activities, opened for signature on 2 June 1988, 21 I.L.M. 859 (not in force); and Madrid Protocol on Environmental Protection to the Antarctic Treaty, opened for signature on 4 October 1991, 30 I.L.M. 1455 (entered into force on 14 January 1998).

1.6 Exploitation Rights: Effects of the Common Heritage of Mankind Doctrine

"North", or the industrialised States, and the "South" or developing States.[28] As a result, when it was proposed that the deep seabed and celestial bodies be declared as the common heritage of mankind, the industrialised and developing States took very different views as to the substance of this doctrine. The industrialised States saw the concept as providing that all States shall have access to the benefits derived from the resources contained in those spatial areas and nothing more.[29] The developing States, on the other hand, believed that the industrialised States would be required to share their profits derived from the exploitation of these spatial areas with the developing States on an equitable basis, as was adopted in the case of the deep seabed under the 1982 United Nations Convention on the Law of the Sea.[30] This is because, by exploiting resources in the common property of humankind, the industrialised States are depriving the developing States of the mineral resources of which they are proud part owners.

Due to this differing opinion among the international community, both the Moon Agreement and the original Convention on the Law of the Sea received very little support among States and, consequently, the world reached an impasse over the international law relating to both the deep seabed and celestial bodies.[31] The industrialised States refused to be part of any intergovernmental system that set a precedent for international taxation, while the developing States did not want any agreement that did not clarify the extent of their rights and benefits derived from the common heritage of mankind, which from their perspective must be both financial and substantial in character.

---

[28] See, for example, Jagdish N. Bhagwati (ed.), THE NEW INTERNATIONAL ECONOMIC ORDER: THE NORTH-SOUTH DEBATE (1977); and Robert Gilpin and Jean M. Gilpin, GLOBAL POLITICAL ECONOMY: UNDERSTANDING THE INTERNATIONAL ECONOMIC ORDER (2001).

[29] See, for example, Raclin, supra note 26, at 738–739; and Harminderpal Singh Rana, *The "Common Heritage of Mankind" & the Final Frontier: A Revaluation of Values Constituting the International Legal Regime for Outer Space Activities* (1994) 26 RUTGERS L. J. 225 at 231.

[30] United Nations Convention on the Law of the Sea, opened for signature on 10 December 1982, 1833 U.N.T.S. 3; 21 I.L.M. 1261 (entered into force on 16 November 1994). See, for example, Gennady M. Danilenko, *The Concept of the "Common Heritage of Mankind" in International Law* (1988) 13 ANN. AIR & SP. L. 247 at 249; and Christopher Pinto, *The Developing Countries and the Exploitation of the Deep Seabed* (1980) 15 COLUM. J. WORLD BUS. 30.

[31] As at 1 January 2010, the Moon Agreement has 13 ratifications while the Convention on the Law of the Sea, before the adoption of the Agreement Relating to the Implementation of Part XI, had 63 ratifications: United Nations Office of Outer Space Affairs, *Status of International Agreements Relating to Activities in Outer Space as at 1 January 2010*, 1 January 2010, at <http://www.oosa.unvienna.org/pdf/publications/ST_SPACE_11_Rev2_Add3E.pdf>, last accessed on 22 April 2010 and United Nations Division for Ocean Affairs and the Law of the Sea, *Chronological Lists of Ratifications of, Accessions and Successions to the Convention and the Related Agreements as at 1 March 2010*, 1 March 2010, at <http://www.un.org/Depts/los/reference_files/chronological_lists_of_ratifications.htm#The%20United%20Nations%20Convention%20on%20the%20Law%20of%20the%20Sea>, last accessed on 22 April 2010.

1.7 Meeting the Challenges and Balancing the Competing Interests in Creating a Regulatory Framework for Mineral Resources in Space

1.7.1 Meeting the Challenges

Having identified the applicable legal issues to commercial mining activities in outer space and clarifying the scope and content of the application of the common heritage of mankind doctrine to celestial bodies and their mineral resources, the next step is to resolve the present impasse on the Moon Agreement by finding possible common ground between the two polarised positions. In this context, two steps are necessary:

(1) analysing and learning from the past failures in intergovernmental negotiations over similar regimes; and
(2) then identifying and harmonising the relevant entrenched competing interests.

It is pertinent to note that the three previous attempts by the international community to negotiate a legal framework for a minerals regime for areas considered to be global commons, namely the deep seabed, Antarctica and celestial bodies in outer space, have all ended in failure. The 1984 Convention on the Law of the Sea failed to win acceptance by industrialised States as they refused to accept the common heritage of mankind provisions in relation to mining activities on the deep seabed. Similarly, the international community refused to accept the terms of the Moon Agreement, even though the implementation of the common heritage of mankind provisions was deferred until such time as the international community considered the mining of celestial bodies to be imminent.[32] The 1988 Wellington Convention on the Regulation of Antarctic Mineral Resource Activities failed to attain sufficient support among the international community, even though there was no designation of Antarctica, or any part of it, as the common heritage of mankind with similar doctrinal obstacles as those in relation to the deep seabed and celestial bodies.[33]

When the Convention on the Law of the Sea was about to enter into force in the early 1990s, strenuous diplomatic efforts were made to reach a compromise. In 1994, this resulted in the adoption of the Agreement relating to the Implementation of Part XI of the United Nations Convention on the Law of the Sea (the "Implementation Agreement"), which was acceptable to both industrialised and developing States by significantly reducing the obligations of the deep seabed mining States.[34] No similar efforts have been made yet with the implementation of the

---

[32] Moon Agreement, Article 11.

[33] Wellington Convention on the Regulation of Antarctic Mineral Resource Activities, opened for signature on 2 June 1988, 27 I.L.M. 868 (not presently in force).

[34] Agreement Relating to the Implementation of Part XI of the United Nations Convention on the Law of the Sea of 10 December 1982, opened for signature on 28 July 1994, 1836 U.N.T.S. 3; 33 I.L.M. 1309 (entered into force on 28 July 1996).

Moon Agreement. Meanwhile, views continue to differ over the legal ramifications of the present Article 11 in this interim period before it may be implemented.[35]

It is through a detailed analysis of the failures of the minerals regimes relating to the deep seabed, Antarctica and celestial bodies and the eventual acceptance of the Implementation Agreement for the deep seabed that a starting point can be found in the search for a compromise position on such legal and policy issues, especially on the subject of property rights and the common heritage of mankind doctrine.

1.7.2 Balancing the Competing Interests

The next step on the path towards an acceptable legal framework for mining activities in outer space is to consider the competing interests of various stakeholders on the issue and attempt to reconcile and balance their divergent concerns. In particular, it is pertinent and prudent to consider and address the five principal sets of competing interests as follows:

(1) the polarised positions of the industrialised States and the developing States over the application of the common heritage of mankind and the non-appropriation principles on celestial bodies;

(2) the continuing economic need for mineral resources for human development on the one hand and the need to impose sufficient environmental safeguards for the protection of the environment of the Earth and that of outer space and the celestial bodies;

(3) the commercial objectives of private mining ventures to maximise their commercial gain and the need to provide baseline public services to the international community, particularly the least developed States in furtherance of the global public interest;

(4) the desire to maintain a free market approach to the regulation of the space industry while ensuring that a number of commercial issues, such as the avoidance of harmful interference, the "paper tenement" problem, protection of intellectual and industrial property rights and the application of anti-trust or competition principles are adequately regulated in outer space; and

(5) the need for sufficient dispute settlement mechanisms to adequately settle commercial and administrative disputes and enforce the rules of the regulatory framework without creating an overly litigious environment involving commercial space activities.

Through balancing such interests, it will be possible for new legal principles to be adopted in regulating the exploration and exploitation of mineral resources

---

[35] See, for example, Ricky J. Lee, *Property and Mining Rights for Lunar Mining Operations in the Absence of International Consensus on the Moon Agreement* (2003), paper presented at the 54th International Astronautical Congress, 29 September 2003 to 3 October 2003, in Bremen, Germany; and Marko, supra note 26.

from celestial bodies. This is particularly the case in relation to the resolution of the present impasse over the common heritage of mankind doctrine and the provision of temporary property rights for such purposes, notwithstanding the non-appropriation principle, which are the two legal principles that pose the greatest barriers to the creation of a new regulatory framework.

1.7.3 Structure and Composition

The creation of an international legal framework for commercial mining activities in outer space would not be complete without also resolving the practical issues in relation to the implementation and operation of such a new framework. It would become apparent, for example, that the practical implementation of this new legal framework cannot be achieved without the establishment of a new intergovernmental organisation that would administer the framework. Such an organisation, referred to in the interest of convenience hereinafter as the International Space Development Authority (the "Authority"), would need to have within its structure the following types of organs:

(1) a quasi-legislative body;
(2) an administrative secretariat; and
(3) one or more judicial mechanisms for dispute settlement and enforcement.

The elements of the structure, composition, functions and powers of each of these organs will need to be considered in turn to enable the establishment of the Authority and the implementation of revised legal principles in the new framework. This is in addition to considerations that must be taken into account in relation to the financing and budgetary requirements of creating and operating such an authority, potentially even prior to any revenues being generated from the exploitation of resources from celestial bodies.

1.7.4 Procedures

After determining the appropriate structural and operational aspects of the Authority, attention must be focused on the appropriate internal procedures to be implemented. This is particularly pertinent in relation to the following administrative functions:

- determining applications for "Exploration Permits" for prospecting and exploration activities to take place on various celestial bodies;
- determining applications for "Mining Permits" for extraction activities to take place on celestial bodies;
- determining applications for "Occupation Permits" for the provision of temporary property rights for purposes incidental to the exploration and mining activities on celestial bodies; and

- adopting the appropriate conditions and processes for the "equitable sharing" of the "benefits derived" from the exploitation of mineral resources from outer space, including the means by which the quantum of the benefits to be shared is to be determined and how such benefits are to be distributed.

1.7.5 Judicial Mechanisms

After the adoption and implementation of administrative processes for the operation of the Authority, it is prudent to then turn to the need for dispute settlement mechanisms within the Authority. This is because disputes will inevitably arise between applicants or permit-holders and the Authority as well as between permit-holders themselves. There is also the additional need for judicial accountability in the operation and administration of the Authority. Accordingly, it is clear that judicial mechanisms must be created to enable the peaceful, effective and judicious settlement of such disputes.

To that end, it is probably preferable to create separate arbitral panels for the resolution of different types of disputes, allowing for different specialist expertise to be employed in the determination of different types of disputes. It is envisaged that the following bodies would be created within the Authority, each to be vested with judicial and quasi-executive functions relevant to their area of specialist expertise:

- the "Licensing Oversight Panel" to hear and resolve disputes between the Authority and applicants or permit-holders in decisions relating to the grant or refusal of applications, the enforcement of conditions of permits and considerations of contraventions of legal or regulatory provisions of the relevant international law;
- the "Environmental Protection Panel" to provide for surveillance, monitoring and enforcement of environmental protection safeguards in the protection of the Earth environment, the mitigation of space debris, contamination and pollution effects as well as the remediation works required on celestial bodies;
- the "Financial Duties Panel" for the assessment and settlement of disputes over the quantum of financial duties payable to the Authority arising from the requirement for the equitable sharing of benefits derived from the exploitation of mineral resources from celestial bodies;
- the "Dispute Settlement Panel" for the settlement of commercial disputes between applicants or permit-holders in relation to activities subject to the premises of the Authority; and
- the "Space Development Appeals Tribunal" that hears appeals from the four abovementioned Panels.

It is envisaged that the resolution of the practical issues arising from the implementation of a new international legal framework would improve the acceptability of this proposed framework by the international community. Further, this would also reduce the legal uncertainty relating to commercial space mining ventures and incidentally provide an administrative framework for the regulation of other future activities in outer space.

1.8 Conclusions

The worsening scarcity of mineral resources from the Earth's crust means that the future extraction and exploitation of mineral resources on celestial bodies is a mere question of when. With the economic, physical and technical feasibility in outer space having been assessed, the remaining major inhibiting factor for space mining ventures is the legal uncertainty arising from the absence of an internationally accepted legal framework. It is clearly necessary for a new legal framework to be created in order to enable the commercial mining of celestial bodies to take place in the near future.

Through learning from past failed negotiations over other spatial areas and balancing competing interests, it may be possible to find a compromise position on the legal and policy controversies to develop a new legal framework. Once this position can be found, its implementation through a practical international framework will shepherd the commercial exploitation of mineral resources in outer space into the realm of present reality.

Chapter 2
Economic and Technical Prospects of Mining on Celestial Bodies

2.1 Introduction

The current prosperity of the human civilisation has relied heavily on the exploitation of natural resources on the Earth, particularly minerals and fossil fuels, which are not renewable. Since the beginning of the twentieth century, the world has become heavily dependent on oil, coal, natural gas and nuclear fission for its energy needs.[1] This dependency is highlighted by the Middle East oil crisis of the 1970s and the sudden increases in oil prices in the early 1980s after the revolution in Iran and again the early 1990s after the Iraqi invasion of Kuwait and since 2001 when the United States of America began its campaign against international terrorism. It has been suggested that oil prices will continue to increase in the future as demand continues to increase despite depleting reserves.[2] The recent trend towards the development of renewable energy, such as solar energy and hydrogen fuel cells, nevertheless requires the use of various rare or non-renewable mineral resources, such as platinum and related metals, that require extraction from the Earth's crust.

Consequently, logic dictates that the continuing survival and development of humanity will eventually require the exploitation of the same mineral resources elsewhere in the Solar System. In any event, it is clear that the development of space technology in the last century has allowed human activity to be no longer restricted to the confines of the Earth. Demonstrating the increasing economic and technical

---

[1] U.S. President George W. Bush stated in his 2006 State of the Union Address that "America is addicted to oil, which is often imported from unstable parts of the world. The best way to break this addiction is through technology.": The White House, *President Bush Delivers State of the Union Address* (31 January 2006), at <http://www.whitehouse.gov/news/releases/2006/01/20060131-10.html>, last accessed on 27 January 2007. See also David Suzuki, THE SACRED BALANCE: REDISCOVERING OUR PLACE IN NATURE (1997).

[2] Stephen Leeb, THE OIL FACTOR: HOW OIL CONTROLS THE ECONOMY AND YOUR FINANCIAL FUTURE (2004). See also earlier works such as Colin J. Campbell, THE COMING OIL CRISIS (1997) and Roger W. Bentley, R. H. Booth, J. D. Burton, Max L. Coleman, Bruce W. Sellwood and George R. Whitfield, *Perspectives on the Future of Oil* (2000) 18 ENERGY EXPLOR. & EXPLOIT. 147–206.

R.J. Lee, *Law and Regulation of Commercial Mining of Minerals in Outer Space*,
Space Regulations Library 7, DOI 10.1007/978-94-007-2039-8_2,
© Springer Science+Business Media B.V. 2012

feasibility of exploiting mineral resources on celestial bodies is necessary to show that the stage is set for the next step in the human utilisation of outer space and, accordingly, legal regulation of this next human endeavour is essential.

Resource economists have long argued over the future of mineral resource exploitation on the Earth and its implications on the future of the world economy. The optimists are of the opinion that the increasing scarcity of resources would catalyse technological innovation in conservation, recycling and the development of substitutes along with the exploitation of previously uncommercial deposits.[3] The more pessimistic economists have suggested that the resources available on the Earth are limited and this absolute scarcity effectively imposes an expiry date on human civilisation.[4] The continuing technical developments in space technology leave open the possibility of development mineral resources from celestial bodies, providing human civilisation with the potential to overcome resource scarcity and achieve some of the possibilities created by limitless energy and resources.

As the abundance of mineral resources on the Earth continues to decline, the resulting economic conditions will promote the exploitation of natural resources from celestial bodies in outer space.[5] The concept of mining the Moon or Near Earth Asteroids for mineral resources to be used on the Earth is not new and there has been a substantial number of scientific studies undertaken regarding the technical feasibility of such an endeavour.[6] It is generally believed to be only a matter of time before the technical and economic conditions exist for the mining of asteroids or other celestial bodies to take place.[7]

In addition to satisfying demand on Earth, the prospect of constructing large structures in the Earth orbit or on the Moon has also highlighted the desirability of developing mineral resources in space. In the 1970s, the concept of a space solar power satellite ("SSPS") was developed where a giant array of solar power cells would orbit the Earth, presumably in geostationary orbit, and transmit solar power by microwave transmission to collecting antennae on the Earth that would convert it to usable electricity for domestic and industrial consumption.[8]

---

[3] Judith Rees, NATURAL RESOURCES: ALLOCATION, ECONOMICS AND POLICY (2nd ed., 1990), at 39.

[4] Edward Goldsmith, Robert Allen, Michael Allaby, John Davoll and Sam Lawrence, BLUEPRINT FOR SURVIVAL (1972), at 4.

[5] Dennis Wingo, MOONRUSH: IMPROVING LIFE ON EARTH WITH THE MOON'S RESOURCES (2004), at 90.

[6] See Brian T. O'Leary, *Mining the Apollo and Amor Asteroids* (1977) 197 SCIENCE 363; Samuel Herrick, *Exploration and 1994 Exploitation of Geographos* in Tom Gehrels (ed.), ASTEROIDS (1979) at 212–221; David Morrison and John C. Niehoff, *Future Exploration of the Asteroids*, in Tom Gehrels (ed.), ASTEROIDS (1979) at 227–249; and David L. Kuck, *Near-Earth Extraterrestrial Resources* (1979), paper presented at the 4th Princeton/AIAA Conference on Space Manufacturing, May 1979, in Princeton, NJ, USA.

[7] See, for example, Wingo, supra note 5.

[8] Lara Farrar, *How to Harvest Solar Power? Beam it Down From Space!* (2008), CABLE NEWS NETWORK, at <http://edition.cnn.com/2008/TECH/science/05/30/space.solar/index.html>, 1 June 2008, last accessed on 13 November 2009 and Peter E. Glaser, SPACE INDUSTRIALISATION (1982).

2.1 Introduction

While the technology for a solar power satellite is considered to be already at hand, there are significant factors inhibiting the development and deployment of such a satellite. Nevertheless, Japan is currently studying the possible development of a full-scale SPSS in geostationary orbit by 2030 that can generate around 5 GW of electricity.[9] One of the main inhibiting factors for this development is that such a structure would have the mass of several thousand tonnes, making the launch costs somewhat prohibitive.[10] The costs of constructing such large-scale structures in outer space would be significantly reduced if the materials were derived and processed from non-terrestrial sources, eliminating nearly 99% of the Earth-launch costs.[11]

The prospect of space tourism is also contributing to the imminent need for construction and materials-gathering capability in space. Much media attention had focused on the recent exploits by Dennis Tito, Greg Olsen, Anousheh Ansari, Charles Simonyi and Richard Garriott of the United States of America, Mark Shuttleworth of South Africa and Guy Laliberté of Canada, who all paid U.S. $20,000,000.00 or more each to the Russian Space Agency to visit the International Space Station.[12] These pioneering, albeit prohibitively expensive, tourism ventures have fuelled the imagination of many that tourism in space is now only a matter of time, even for the less wealthy and less technically trained members of society. This was especially the case after the successful award of the Ansari X Prize and the launch and promotion of commercial space tourism ventures.[13] The development of various proposed space tourism vehicles has the potential to increase

---

[9] Makoto Nagatomo, *An Approach to Develop Space Solar Power as a New Energy System for Developing Countries* (1996) 56:1 SOLAR ENERGY 111; Tim Hornyak, *Farming Solar Energy in Space* (2008), SCIENTIFIC AMERICAN, <http://www.scientificamerican.com/article. cfm?id=farming-solar-energy-in-space>, July 2008, last accessed on 13 November 2009; and Tom Chivers, *Japan Plans Giant Solar Power Station in Space* (2009), THE TELEGRAPH (United Kingdom), at <http://www.telegraph.co.uk/earth/energy/solarpower/6536752/Japan-plans-solar-power-station-in-space.html>, 10 November 2009, last accessed on 13 November 2009.

[10] Ibid.

[11] Space Research Associates, *Report of Satellite Solar Power Systems* (1986) 6 SPACE POWER 1.

[12] Laura Woodmansee, *Opinion: Space 'Adventurers' Paving the Way for the Rest of Us* (18 September 2006), at <http://www.space.com/adastra/060918_woodsmansee_ansari.html>, last accessed on 27 January 2007 and Clare Moskowitz, *Space Clown Comes Back Down to Earth* (11 October 2009), at <http://www.msnbc.msn.com/id/33262374/ns/technology_and_science-space/>, last accessed on 26 October 2009. Past space tourists that have flown to the International Space Station include: Dennis Tito (2001), Mark Shuttleworth (2002), Gregory Olsen (2005), Anousheh Ansari (2006), Charles Simonyi (2007 and 2009), Richard Garriott (2008) and Guy Laliberté (2009): Space Adventures Ltd, *Our Clients*, at <http://www.spaceadventures.com/index. cfm?fuseaction=orbital.Clients>, last accessed on 1 October 2009.

[13] Cable News Network, *SpaceShipOne Captures X Prize* (4 October 2004), at <http://edition.cnn. com/2004/TECH/space/10/04/spaceshipone.attempt.cnn/>, last accessed on 21 December 2004; Jeff Foust, *Virgin Galactic and the Future of Commercial Spaceflight* (23 May 2005), Ad Astra, at <http://www.space.com/adastra/050523_virgin_nss.html>, last accessed on 27 January 2007; and Peter B. de Selding, *Virgin Galactic Customers Parting with their Cash* (3 April 2006), Space News, at <http://www.space.com/spacenews/businessmonday_060403.html>, last accessed on 27 January 2007.

both the supply and the demand of tourism structures in low-Earth orbits in the not-too-distant future.[14]

The establishment of such large-scale space projects in Earth orbit are inhibited by two significant factors. Firstly, as indicated above, the cost of launching anything from the surface of the Earth to orbital space remains prohibitively high. Present expendable launch systems available cost between U.S. $11,000.00 and U.S. $22,000.00 per kilogram of payload for low Earth orbit ("LEO") systems.[15] For geostationary orbit ("GEO") systems, the cost of launch ranges from U.S. $16,000.00 to U.S. $50,000.00 per kilogram in 2000 values.[16] For this reason, the development of unmanned reusable launch vehicles ("RLV") that are designed to be cost efficient for launching payloads into LEO has tremendous potential for future space applications. It is also worth noting that the space elevator concept, which involves the construction of a carbon fibre cable between the surface of the Earth to beyond the GEO, has the potential for reducing the cost of transfer into orbit to a sum that is negligible when compared to launch costs using expendable launch vehicles.[17]

When considering the costs of designing propulsion systems on transportation vehicles between the Earth and other celestial bodies, it is important to consider also the energy required for a soft landing on the surface as well as the energy required to escape the gravity of the object. Generally, the heavier the mass of the object, the more energy would be required to counter the gravitational forces to slow its descent for a soft landing on and a takeoff from the surface of the object, as indicated in Fig. 2.1. It is apparent from Fig. 2.1 that the energy required for a landing or take off from the Moon would be far more than that required from an average Near Earth Asteroid, though it would nonetheless be significantly smaller

---

[14] Patrick Q. Collins, Yoichi Iwasaki, Hideki Kanayama and Misuzu Ohmuki, *Commercial Implications of Market Research on Space Tourism* (1994), paper presented at the 19th International Symposium on Space Technology and Science, May 1994, in Yokohama, Japan and Partick Q. Collins and Kohki Isozaki, JRS RESEARCH ACTIVITIES FOR SPACE TOURISM (1995), paper presented at the 6th International Conference of Pacific Basin Space Societies, December 1995, in Marina del Rey, CA, USA.

[15] Robert Hickman and Joseph Adams, *Future Launch Systems* (2003) 5:1 CROSSLINK 42, as at Aerospace Corporation, <http://www.aero.org/publications/crosslink/pdfs/V5N1.pdf>, last accessed on 23 April 2007.

[16] Futron Corporation, *Space Transportation Costs: Trends in Price Per Pound to Orbit 1999–2000* (2002), at <http://www.futron.com/pdf/resource_center/white_papers/FutronLaunchCostWP.pdf>, 6 September 2002, last accessed on 24 April 2007.

[17] See Bradley C. Edwards and Eric A. Westling, THE SPACE ELEVATOR: A REVOLUTIONARY EARTH-TO-SPACE TRANSPORTATION SYSTEM (2003) and Leonard David, *The Space Elevator Comes Closer to Reality* (2002), at <http://www.space.com/businesstechnology/technology/space_elevator_020327-1.html>, last accessed on 15 July 2006. This optimism may be contrasted with the discussion of the technical and conceptual difficulties in Vladimir V. Beletskii, M. B. Ivanov and E. I. Otstavnov, *Model Problem of a Space Elevator* (2005) 43 COSMIC RESEARCH 152–156 and David Brody, *Thinking Differently with Space Elevators* (2006) AD ASTRA, Summer 2006, at 34.

2.1 Introduction

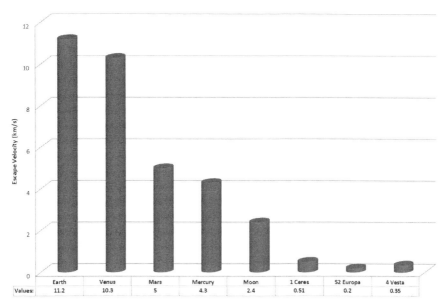

Fig. 2.1 Change in velocity needed for soft landing on selected objects
National Aeronautics and Space Administration, *Solar System Exploration* (2006) at <http://solarsystem.nasa.gov/index.cfm>, last accessed on 15 July 2006, for the escape velocities for Mercury (4.25 km h^{-1}), Venus (10.361 km h^{-1}), Earth (11.18 km h^{-1}), Mars (5.02 km h^{-1}) and the Moon (2.38 km h^{-1}); Calvin J. Hamilton, *Phobos* (2000), at <http://www.solarviews.com/french/phobos.htm>, last accessed on 15 July 2006 for the escape velocity for Phobos (0.0103 km h^{-1}); Calvin J. Hamilton, *Deimos* (2000), at <http://www.solarviews.com/french/deimos.htm>, last accessed on 15 July 2006 for the escape velocity for Deimos (0.0057 km h^{-1}); Bruce McClure, *Escape Velocity on the Moon*, at <http://www.idialstars.com/evme.htm>, last accessed on 15 July 2006 for the escape velocity for Ceres (0.4505 km h^{-1}); and David Whitehouse, *Scientists Get Near the Real Eros* (21 September 2000), British Broadcasting Corporation, at <http://news.bbc.co.uk/1/hi/sci/tech/936149.stm>, last accessed on 15 July 2006 for the escape velocity for Eros (8.33 × 10^{-8} km s^{-1})

than that required to escape the gravity of the Earth or Mars. Only Ceres, the largest known asteroid, has a comparable escape velocity to that of the Moon.[18]

Before considering the technical feasibility of exploiting mineral resources from outer space, although, it is prudent to keep in mind that the present world is driven predominantly by commercial interests, which are in turn driven by the demands of a global market economy. Accordingly, it is clear that sufficient developments in spacecraft design, propulsion systems and robotic mining systems as outlined in Table 2.1 already exist to enable some form of robotic prospecting and mining of asteroids. However, further advancements and definitive developments are unlikely to take place until the economic conditions have ripened and space mining is at

---

[18] See Charles Jaffé, Shane D. Ross, Martin W. Lo, Jerrold Marsden, David Farrelly and T. Uzer, *Statistical Theory of Asteroid Escape Rates* (2002) 89:1 PHYS. REV. LETTERS 11001.

Table 2.1 Selected technological development enabling mining in space

Field	Description of technology	Implications
Robotics	Mining and movement systems, muscle wire and other solutions	Ability to remotely operate and mine without need for human control
Electronics and microsystems	Low cost and improved performance of computer microprocessors and storage technologies	Improved automation at reduced costs and the ability to collect and store large amounts of data
Ballute	Lightweight and inflatable technology for return of materials to the Earth	Ability to deliver large payloads to the surface of the Earth at minimum cost
New materials	Development of carbon nanofibres and advanced composites	Reduced mass of spacecraft with greater materials strength
Power	Improved photovoltaic solar power arrays for spacecraft	Produce more power to operate in an uncertain asteroid environment
Control and automation	Improved software and powerful computer processors	Improved ability to self-repair and react to environmental changes
Simulation	Significant improvements in computer design and modelling software	Ability to test and simulate unproven robotic mining technologies
Propulsion	Development of innovative launch and space propulsion systems	Lower launch costs – being the most significant cost of a mining mission

Source: John E. Tilton (1977, 12–13)

least lucrative, if not a necessary requirement for future of economic survival. It is therefore imperative to investigate the likelihood of the onset of resource exhaustion under present economic and technological conditions.

2.2 Economic Feasibility of Space Mining

2.2.1 The Study of Resource Economics

2.2.1.1 Overview

Resources on Earth are generally divided into stock (non-renewable) and flow (renewable) resources. The essential difference is the timeframe in which they develop by natural processes. Since both stock and flow resources are developed by natural cycles on Earth, they are all technically renewable but at grossly different rates. Stock resources, such as all minerals, fossil fuels and the land itself, are substances that have taken millions of years to form and therefore, from a socio-economic perspective, are fixed in supply. In addition, the ultimate quantity of such resources on Earth is physically limited.

In relation to fossil fuels, because they are consumed by use, they will therefore eventually be exhausted by consumption. With metallic minerals and some non-metallic minerals, on the other hand, the technology exists for most metals to be reused many times over with little loss of quality.[19] However, even the best optimists would concede that the full and total recovery of all used metals is likely to remain a theoretical, rather than a practical, possibility. Price suggested in 1955 that the thermodynamic law of entropy indicates that unavailability is the ultimate tendency of recurring mineral usage as they eventually become too dispersed or impure during each use to be recoverable.[20] In any event, recycling is often an energy-intensive activity that requires heavy reliance on the use of fossil fuels or other sources of energy on Earth.[21]

Many attempts have been made to estimate the ultimate level of available resources using a variety of assessment techniques and assumptions about the future rate of resource consumption, as well as future economic and technological changes.[22] Economists consider the potential availability of a mineral resource by referring to its "resource base", defined as the total quantity of that mineral resource within the geosystem of the Earth.

Among different approaches, the best attempt to calculate the resource base for particular non-fuel minerals is by multiplying their elemental abundance measured in grams per metric ton by the total weight of the Earth's crust to the depth of 1 km or mile.[23] Calculated in this way, the available resources remaining are vast and would be available for millions of years if human consumption levels remained static. However, the availability of resources falls dramatically if there is any increase in human consumption. Even at a growth rate of 10% in consumption levels, all minerals would be exhausted in less than 300 years, as shown in Table 2.2.

The life expectancies of the various minerals listed in Table 2.2 are affected by two factors that alter the reliability of the data. Firstly, life expectancies are calculated on the basis that technological advancements will allow all available elements to be exploited at costs low enough to maintain demand levels or, in other words, that the cost of extraction of the more difficult deposits will not increase to affect demand levels. This is most certainly untrue as only a small fraction of all mineral resources may be extracted at a tolerable financial or environmental cost, resulting in the estimates determined in this way to be overly optimistic. On the other hand, with the exception of uranium, which is used in nuclear fission, elemental minerals

---

[19] The Nickel Institute, *Economics of Recycling* (2004), at <http://www.nickelinstitute.org/index.cfm/ci_id/121.htm>, last accessed on 21 December 2004.

[20] Edward T. Price, *Values and Concepts in Conservation* (1955) 45:1 ANN. ASSOC. AM. GEOG. 65.

[21] David W. Pearce and Ingo Walter, RESOURCE CONSERVATION: THE SOCIAL AND ECONOMIC DIMENSIONS OF RECYCLING (1977).

[22] See, for example, discussion in Wladimir S. Woytinsky and Emma S. Woytinsky, WORLD POPULATION AND PRODUCTION: TRENDS AND OUTLOOK (1953) at 326–333.

[23] Tan Lee and Chi-Lung Yao, *Abundance of Chemical Elements in the Earth's Crust and its Major Tectonic Units* [1970] INT'L. GEOL. REV. 778.

Table 2.2 Resource base and life expectancy estimates for certain mineral resources in the earth's crust

Mineral	Resource base (metric tons)[a]	Life expectancy in years with different consumption growth rates[b]				Average annual growth, 1947–1974 (%)
		0%	2%	5%	10%	
Aluminium	2.0×10^{18}	166×10^9	1107	468	247	9.8
Cadmium	3.6×10^{12}	210×10^6	771	332	177	4.7
Chromium[c]	2.6×10^{15}	1.3×10^9	861	368	196	5.3
Cobalt	600×10^{12}	23.8×10^9	1009	428	227	5.8
Copper	1.5×10^{15}	216×10^6	772	332	177	4.8
Gold	84×10^9	62.8×10^6	709	307	164	2.4
Iron	1.4×10^{18}	2.6×10^9	898	383	203	7.0
Lead	290×10^{12}	83.5×10^6	724	313	164	2.4
Magnesium	672×10^{15}	131.5×10^9	1095	463	244	7.7
Manganese[d]	31.2×10^{15}	3.1×10^9	906	386	205	6.5
Mercury	2.1×10^{12}	223.5×10^6	773	333	178	2.0
Nickel	2.1×10^{12}	3.2×10^6	559	246	133	6.9
Phosphorus	28.8×10^{15}	1.9×10^9	881	376	200	7.3
Potassium	408×10^{15}	22.1×10^9	1005	427	226	9.0
Platinum	1.1×10^{12}	6.7×10^9	944	402	213	9.7
Silver	1.8×10^{12}	194.2×10^6	766	330	176	2.2
Sulphur	9.6×10^{15}	205.3×10^6	769	331	177	6.7
Tin	40.8×10^{12}	172.2×10^6	760	327	175	2.7
Tungsten	26.4×10^{12}	677.2×10^6	829	355	189	3.8
Zinc	2.2×10^{15}	398.6×10^9	1151	486	256	4.7

[a] Calculated by multiplying its elemental abundance measured in grams per metric tons times the total weight of the Earth's crust in metric tons

[b] Calculated based on the average annual production figures for 1972–1974 and these were taken from the U.S. Bureau of Mines, COMMODITY DATA SUMMARIES 1972–1976 (1977) and U.S. Bureau of Mines, MINERALS YEARBOOK (1974)

[c] Production figures assume concentrates are 46% chromium

[d] Production figures assume concentrates are 46% manganese

Source: John E. Tilton (1977, 12–13)

are not destroyed by use and therefore they are recyclable, making the estimates determined to be unduly pessimistic.

2.2.1.2 Proven Reserves

Proven reserves are defined as the deposits of mineral resources that are already discovered and known to be economically extractable under present or similar demand, price and other economic and technological conditions.[24] Despite its label, there is much variance and subjectivity in the data of proven reserves, for what may be considered economically extractable depends on the commercial value,

---

[24] Rees, supra note 3, at 20.

economic conditions, logistical requirements and environmental concerns. When proven reserves are used to estimate the entire life of a physical resource, there is always the implicit assumption in doing so that there would be no new discoveries, no technological advancement and no price changes. Further, as reserves are proved only after considerable expenses are incurred in surveys and borings, investments in exploration are unlikely if sufficient reserves are already held. In fact, there would be an economic incentive on firms to reduce their proven reserves as they may still be taxed in certain countries as company assets even though their actual extraction and exploitation may not be feasible in the short-term to medium-term future.[25]

As a result, the determination of the proven reserves for any given resource generally depends on five important and interrelated factors:

(1) advanced technological knowledge and technical skills involved in exploration and mining methods;
(2) levels of demand;
(3) production, processing and labour costs;
(4) the price of the product;
(5) the availability and price of viable substitutes; and
(6) the availability and price of recycled products.[26]

One other problem with the use of proven resources included in estimating total resource availability levels is that such estimates are skewed heavily towards advanced countries, as exploration activities in most developing countries have not taken place to the same scale as the industrialised States. For example, Jamaica is considered to have sufficient reserves to meet half of its oil requirements.[27] However, it has been dependent totally on large oil companies from industrialised States for whom the emphasis was on establishing large enough deposits for export markets rather than merely satisfying the domestic demand of small Caribbean States (Fig. 2.2).[28]

2.2.1.3 Conditional Reserves

Conditional reserves are deposits that have been discovered but are not economically viable to extract under current demand and price levels or with existing mining technologies and methods.[29] The relationship between economic and uneconomic deposits is a very complex one and is heavily dependent on various social, political and economic factors. For example, in the 1900s, copper ores with less than

---

[25] Michael Tanzer, THE RACE FOR RESOURCES: CONTINUING STRUGGLES OVER MINERALS AND FUELS (1980) at 33.

[26] Rees, supra note 3, at 21.

[27] Richard J. Barnet, THE LEAN YEARS: POLITICS IN THE AGE OF SCARCITY (1980).

[28] Rees, supra note 3, at 22.

[29] Ibid., at 20.

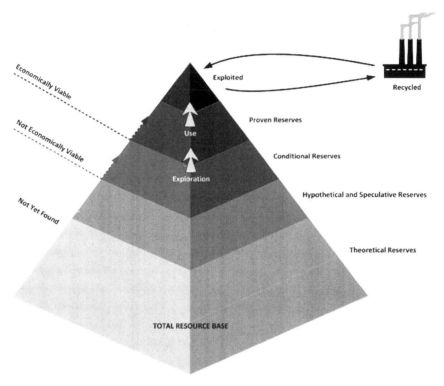

Fig. 2.2 The stock resource base and its subdivisions
Source: Judith Rees (1990, 20)

10% copper content could not be used in most smelters and were therefore useless. However, major demand increases and technological advances have allowed ores with less than 1% copper content to be smelted and thus have dramatically increased the proven reserves for copper.[30] By the 1990s, ores with as little as 0.4% copper content could be exploited with financial feasibility.[31]

Similar to proven resources, the determination of conditional reserves is nevertheless dependent on the industrial strength of the State in which the reserves are located. For example, ore grades in developing States must be higher than those in industrialised States for the reserves to be economically viable, resulting from the additional infrastructure costs required to exploit them in those countries. Further, as with proven reserves, data relating to the abundance of conditional reserves contains much variance and subjectivity. This is partly because of different methodologies used to estimate the size of reserves and the practical difficulties involved in their extraction.

---

[30] Kenneth Warren, MINERAL RESOURCES (1973).

[31] Telephone conversation between the author and an unnamed representative of the Australian Minerals Council, Canberra, Australia, on 23 October 2000.

2.2.1.4 Inferred Resources

Measured reserves are those that have been explored intensively and a margin of error on the size of the deposit is estimated to be around 20%.[32] Both proven reserves and conditional reserves are measured reserves as they are verified after exploration. Less intensively measured deposits with data derived from surveys and geological projections are known as indicated reserves.[33] In such cases where deposits have been located but not explored, they are known as inferred reserves. As a result, the nature of the reserve would carry with it the error margin in its determination, making the calculation of global resource reserves a challenging and dynamic problem.

2.2.1.5 Hypothetical Reserves

Hypothetical reserves are deposits that may be discovered in the future in the areas that have only been partially surveyed and developed. For example, the North Sea is producing significant quantities of oil and natural gas but not all layers of oil-bearing strata have been test drilled and as a result, there may be hypothetical reserves contained therein.[34] The common means of determining hypothetical resources is to extrapolate past rates of growth and production and proven reserves, assuming all past conditional determinants would continue to affect future production of such mineral resources in the same way.[35]

One of the ways to determine hypothetical reserves has been to call together a panel of experts and average their forecasts of reserves.[36] However, this is intrinsically biased because of the expert's perspective of the likely future conditions, causing the estimate to be either overly conservative or optimistic.[37] There is also the suggestion that it would be in the best interests of the oil companies and other resource businesses to be conservative in their estimates, as this would produce an outlook of scarcity and therefore high prices can be maintained, resulting in higher profit margins for the mining companies.[38]

---

[32] Ibid.

[33] Rees, supra note 3, at 20.

[34] Charles D. Masters, David H. Root and Emil D. Attanasi, *World Oil and Gas Resources: Future Production Realities* (1990) 15 ANN. REV. ENERGY 23.

[35] See, for example, M. King Hubbert, *Energy Resources*, in National Academy of Sciences, RESOURCES AND MAN (1969), 157–242.

[36] Exxon Mobil Corporation, *Exploration in Developing Countries* (1978), paper presented at the Energy Committee Seminar, Aspen Institute of Humanistic Studies, 16–20 July 1978, in Boulder, CO, USA.

[37] For example, see Peter R. Odell, *Optimal Development of the North Sea's Oil Fields – A Summary* (1977) 5:4 ENERGY POLICY 282; C. G. Wall, D. C. Wilson and W. Jones, *Optimal Development of the North Sea's Oil Fields – The Criticisms* (1977) 5:4 ENERGY POLICY 284; and Peter R. Odell and Kenneth E. Rosing, *Optimal Development of the North Sea's Oil Fields – The Reply* (1977) 5:4 ENERGY POLICY 295.

[38] See, for example, Tanzer, supra note 25, at 30–40 and Barnet, supra note 27, at 29–33.

2.2.1.6 Speculative Resources

Speculative resources are the deposits that may be found in areas that have not been explored but where favourable geological conditions exist. For example, there are more than 600 sedimentary basins where oil and gas are believed to exist but only around 200 have been explored or developed.[39] Once drilling does take place in explored areas, their potential status may change markedly. For example, the North Sea contained only speculative resources for most of the last century and these were converted to proven, conditional and hypothetical reserves by the end of the 1990s as exploration and exploitation activities were carried out in earnest.[40]

In determining speculative reserves, which is based entirely on the extrapolation from past discovery and development patterns in similar geological conditions, there is the assumption that the future deposits would be equally productive physically and as financially rewarding as past reserves.[41] Of course, there is every possibility that the most profitable reserves are already exploited and as a result, the financial returns on capital may fall over time because of development in the more difficult areas of the Earth for mineral resource exploitation.

2.2.2 Economic Scarcity of Mineral Resources

2.2.2.1 Theoretical Constructs

Because of the problems and inherent uncertainties in estimating hypothetical and speculative reserves with each mineral resource, there are large divergent estimates on the ultimately recoverable reserves of each resource. For example, one significant problem with using claims from the industry is that they would be subject to the company's willingness to disclose resource expectations in each case.[42] However, it is generally not doubted that mineral resources on the Earth will eventually be exhausted and the future of humankind would depend ultimately on our ability to recover mineral resources through recycling or from outside the confines of this planet.[43] A series of popular books published in the 1970s predicted an apocalyptic view on the future of the human civilisation, which coincided with the Arab-Israeli conflict, the 1973 oil crisis and the attempt by the Organisation of Petroleum Exporting Countries ("OPEC") to reduce oil supply in order to hike global oil prices.[44] It was believed that stock resource scarcity would be a major

---

[39] Exxon Mobil Corporation, supra note 36.

[40] Rees, supra note 3, at 24.

[41] Ibid.

[42] Barnet, supra note 27.

[43] See, for example, Stephen P. A. Brown and Daniel Wolk, *Natural Resource Scarcity and Technological Change* [2000:1] EC. & FIN. REV. 2.

[44] See, for example, Jay W. Forrester, WORLD DYNAMICS (1970); Dennis L. Meadows, Donnella L. Meadows, Jørgen Randers and William W. Behrens III, THE LIMITS TO GROWTH (1972); and Goldsmith, Allen, Allaby, Davoll and Lawrence, supra note 4.

barrier to the future development of civilisation and result in the collapse of the developing world or even human society in its entirety.[45]

The concept of the exhaustion of resources being a limit on civilisation growth is not a new one. Although at the time referring only to agricultural production and with particular emphasis on population growth, Malthus wrote as early as 1798 that, as resources are ultimately limited, the exponential growth in population would result in falls in income per capita until poverty and starvation would result as a constricting factor on population.[46] In his view, increasing population would require the cultivation of lower-quality fields that require more capital and labour for same or less output, thus imposing a physical constraint on population growth and subsistence.[47]

Ricardo, in his studies on the concept of resource exhaustion, concluded that as lower grade ores are being exploited, labour costs must increase and productivity must continue to decrease correspondingly.[48] In other words, rather than reaching a threshold of physical constraint as Malthus suggested, Ricardo was of the view that this is a gradual process, stating "every increase of the quantity of labour must augment the value of that commodity on which it is exercised".[49] This was complemented by the analysis made by Mill that technological improvements have the effect of postponing the effects of resource scarcity on growth and that the physical dependence of the economy on mineral resources generates a scarcity effect that is independent of population increases.[50]

Marshall, on the other hand, recognised that the study of mineral resource scarcity cannot be compared perfectly with that of agricultural production, for agricultural produce are "perennial streams" while mines are exhaustible "reservoirs".[51] As a result, the cost involved in the production from mines relates not only to the difficulty involved in extracting the reserves as they are mined near exhaustion, but also to the rate of production itself.[52] Accordingly, even though mining of natural resources would thus not strictly conform to the law of diminishing returns, market

---

[45] Dennis L. Meadows, Donnella L. Meadows and Jørgen Randers, BEYOND THE LIMITS: CONFRONTING GLOBAL COLLAPSE ENVIRONING A SUSTAINABLE FUTURE (1992) at 120.

[46] Thomas Malthus, *An Essay on the Principle of Population* (1798), in Garrett Hardin (ed.), POPULATION, EVOLUTION AND BIRTH CONTROL (1969) at 4–17.

[47] Thomas Malthus, PRINCIPLES OF POLITICAL ECONOMY: CONSIDERED WITH A VIEW TO THEIR PRACTICAL APPLICATION (1820), at 300.

[48] David Ricardo, PRINCIPLES OF POLITICAL ECONOMY AND TAXATION (1817, reprinted 1962).

[49] Ibid., at 7.

[50] John Stuart Mill, PRINCIPLES OF POLITICAL ECONOMY WITH SOME OF THEIR APPLICATION TO SOCIAL PHILOSOPHY (1909). See discussion in, for example, Herman E. Daly, STEADY-STATE ECONOMICS (1977) and Harold J. Barnett and Chandler Morse, SCARCITY AND GROWTH: THE ECONOMICS OF NATURAL RESOURCE AVAILABILITY (1963), at 69–71.

[51] Alfred Marshall, PRINCIPLES OF ECONOMICS: An INTRODUCTORY VOLUME (8th ed., 1949), at 138–139.

[52] Ibid.

prices do increase as a reflection of the reduction in the resources available to be extracted from the mines.[53]

By the end of the twentieth century, even though it is undisputed that lower grade ores are now being exploited, there does not appear to be any rise of labour and capital costs in real terms as the advances in technology have slowed the onset of diminishing returns.[54] In other words, the view that human ability to improve productivity through technological innovation and resource exploration would overcome the deficiencies in ore qualities have so far held true. Assuming Ricardo's law of diminishing return, as modified by Marshall, remains in force, there are two different perspectives on what this means for the future in terms of our continued consumption of natural resources, which are considered below (Fig. 2.3).

The most well known of modelling methodologies on predicting the eventual exhaustion of fossil fuels, particularly oil, is that of Hubbert, who published a paper in 1956 predicting the peak and decline of oil production in the United States.[55] The peak of global oil production when half of the ultimately recoverable reserves of oil will have been produced is expected to be reached sometime between 2010 and

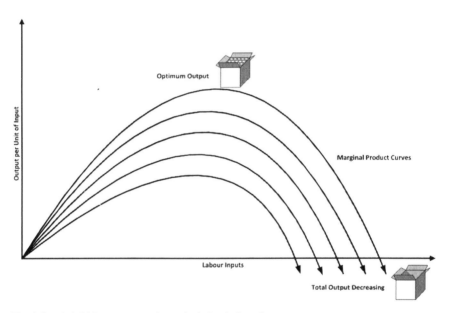

Fig. 2.3 Diminishing returns and marginal physical product curves
Source: Judith Rees (1990, 33)

---

[53] Edward B. Barbier, ECONOMICS, NATURAL RESOURCE SCARCITY AND DEVELOPMENT (1989), at 18.
[54] Harold J. Barnett, G. H. van Muiswinkel, M. Schechter and J. J. Myers, *Global Trends in Non-Fuel Minerals*, in Julian. L. Simon and Herman Kahn (eds.), THE RESOURCEFUL EARTH: A RESPONSE TO GLOBAL 2000 (1984), at 316–338.
[55] M. King Hubbert, *Energy from Fossil Fuels* (1956) 109 SCIENCE 103.

2020, with some analyses suggesting that it may even occur before 2010.[56] Some of the most pessimistic analyses estimate that the United States, for example, has already produced 169 billion barrels of its 195 billion barrels of ultimately recoverable reserves.[57] Meanwhile, even in oil-rich Saudi Arabia, it is estimated that it has already produced 91 billion barrels of its ultimately recoverable reserves of 300 billion barrels by the end of the previous century.[58] In the meantime, the rate of depletion of oil reserves may be increasing, with Saudi Arabian oil production, for example, increasing from 9.5 million barrels per day in 2002 to 12.5 million barrels per day in 2009, with a further potential one million barrels per day in reserve capacity.[59] This increase in production capacity, coupled with declines in worldwide demand growth for oil, has led to the world price of oil fall from its peak at around U.S. $ 150.00 per barrel in July 2008 to around U.S. $60.00 per barrel in July 2009, a price level at which it has been predicted to remain for the near future.[60]

2.2.2.2 The Optimistic Perspective

Perfect Market

In a perfect market economy, the price of any mineral resource that was becoming scarce would inevitably rise. The increased production cost associated with diminishing revenue returns would result in producers being willing to supply less resources for the same price or to supply the same quantity at a higher price. This increase in prices would result in several social, economic and technological responses. Firstly, consumers would turn to cheaper substitutes or introduce conservation measures, such as recycling, to reduce the demand on the scarce mineral resource. Secondly, the scarcity of the resource and the rising prices would provide strong incentives for technological innovation that would improve production, decrease the cost of substitutes and provide for better and more energy efficient recycling methods. Thirdly, the price rise would make it economically viable to exploit less concentrated ores or ores at difficult locations, encouraging increased exploration and production.

As a result, the rise in prices and the corresponding fall in demand do not imply a lowering of living standards for the economic unit. Although there may be a short-term fall in living standards as fewer consumers are able to afford the higher prices of the commodities, suitable substitutes or recycled products would ensue in time. These reactionary developments have the effect of returning the price and demand for the resource to a level similar to its original level. In the long term, there would

---

[56] Jeremy Rifkin, THE HYDROGEN ECONOMY (2002), at 23–24.

[57] Ibid., at 17.

[58] Ibid. See also Organisation of the Petroleum Exporting Countries, WORLD OIL OUTLOOK 2009 (2009), at <http://www.opec.org/library/World%20Oil%20Outlook/pdf/WOO%202009.pdf>, last accessed on 5 November 2009.

[59] Edward L. Morse, *Low and Behold: Making the Most of Cheap Oil* (2009) 88:5 FOREIGN AFF. 36 at 40.

[60] Ibid., at 36–37.

be increases in the marginal cost of production as prophesised by Ricardo, but this would be cushioned by the reactionary factors to allow for an economic and social adjustment to take place.

In the context of minerals, substitution can take place at three levels: the primary extraction level, the secondary production level and the tertiary end-use level. At the primary level, the diminishing reserves for a particular ore may result in the development of suitable extraction methods from alternative ores. For example, the decline in the availability of bauxite, or aluminium oxide (Al_2O_3), has spurred development for processing techniques to extract aluminium from kaolin clays, carbonaceous shales, nemeline and nephelite.[61]

At a secondary level, one metal may directly replace another in the production of consumer products. In the production of high-voltage transmission lines, aluminium has already quite successfully replaced copper. Similarly, stainless steel has also become a good substitute for copper in kitchen pots and pans. The diversity of minerals found on Earth has caused suggestions that there are conceivable substitutes for almost all mineral products, making the scarcity of mineral products a non-issue.[62] However, there remain many strong cases where the costs involved in substitution continue to be too high economically for any effective substitution to take place. For example, the use of manganese in steel production continues to account for 90% of all manganese produced from the crust of the Earth (Fig. 2.4).[63]

At a tertiary level, substitution can also take place where the demand for a particular mineral product is reduced by the utilisation of a different technology. For example, one major use of copper is for undersea transmission cables that are utilised for telecommunications and data transmission.[64] This demand has significantly reduced because of developments in microwave transmission technologies as well as communication satellite constellations.[65] Another example of this is the reduced use of metallic silver in photographic film due to the use of digital cameras.

---

[61] Rees, supra note 3, at 81.

[62] This view was expressed in, for example, Gerald Manners, *Three Issues of Mineral Policy* (1977) 125 J. ROYAL SOC. ARTS 386 at 388.

[63] Michael W. Klass, James C. Burrows and Steven D. Beggs, INTERNATIONAL MINERAL CARTELS AND EMBARGOES: POLICY AND IMPLICATIONS FOR THE U.S. (1980).

[64] See Arthur H. Tuthill, *Guidelines for the Use of Copper Alloys in Seawater* (May 1987), U.S. Copper Development Association, at <http://www.copper.org/applications/marine/seawater/seawater_corrosion.html>, last accessed on 27 January 2007. See, for example, the recent disruption to communications services resulting from damage to the copper undeasea cable as caused by an earthquake in Taiwan: see W. David Gardner, *Telecom Cable Repairs Under Way in Wake of Asian Earthquake* (3 January 2007), Information Week, at <http://www.informationweek.com/management/showArticle.jhtml?articleID=196800845>, last accessed on 27 January 2007 and Sumner Lemon, *Earthquake Disputes Internet Access in Asia* (27 December 2006), at <http://www.computerworld.com/action/article.do?command=viewArticleBasic&articleId=9006819>, last accessed on 27 January 2007.

[65] Raymond F. Mikesell, THE WORLD COPPER INDUSTRY: STRUCTURE AND ECONOMIC ANALYSIS (1979).

2.2 Economic Feasibility of Space Mining

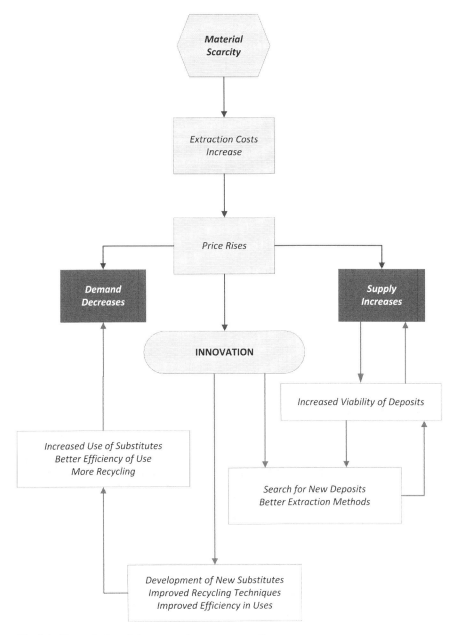

Fig. 2.4 The perfect market response to resource scarcity
Source: Michael W. Klass et al. (1980, 39)

The optimistic view on resource depletion is that the onset of resource exhaustion would be largely, if not totally, offset by the use of substitutes at all three levels of production and consumption.[66] The increase in marginal production costs would cause technological innovations and increased exploration to counteract these market forces and cushion their effects. This perfect market model of resource depletion has been subject to several criticisms, most of which target the basis of the perfect market economy on which the model is based.

Reality Checking the Optimists

The main problem with the view that perfect market forces would rectify any problems of resource scarcity is that the market forces that are relied upon are not perfect. For example, in order to produce the demand, technological and supply responses as indicated in the perfect market model, the market would have to be perfectly competitive, with all participating economic units functioning in a perfectly rational way to maximise profits, not to mention having managers with crystal balls capable of accurately predicting future resource demand and price levels.[67] Because of the impossibility of these factors, the model may be overly optimistic for the purposes of safeguarding our future. On the other hand, the fact that the resource sector tends to be dominated by large firms that engage in monopolistic competition rather than perfect competition may in fact have the opposite effect of producing overly pessimistic projections in contrast. The nature of monopolistic competition in this sector may mean that the resources are currently underexploited because of the desire to restrict production and control artificially high price levels.[68] Consequently, one of the factors that may not have been taken into account is the potential for developing conditional reserves that cannot be exploited profitably except at such high price levels, thus expanding the resource base available.

However, the existence of other market conditions gives rise to the reverse suggestion that the firms are nonetheless overexploiting the resources. After all, it is not possible for any firm to be able to predict perfectly the market conditions that are going to exist in the future. As a result, firms may wish to overproduce in order to sell as much of a mineral resource as possible at ascertained prices before cost or technological changes reduce the value of their proven reserves.[69] Additionally, the prevalence of futures trading in commodities gives further support to the fact that mineral resource prices have remained extremely volatile, with mineral prices rising far more quickly than manufactured goods in world markets.[70] In situations where there is a significant level of uncertainty in the future market, firms would also

---

[66] Rees, supra note 3, at 40–42.

[67] Ibid., at 43.

[68] John A. Kay and James A. Mirrlees, *The Desirability of Natural Resource Depletion*, in David W. Pearce (ed.), THE ECONOMICS OF NATURAL RESOURCE DEPLETION (1975) at 140–176.

[69] Ibid.

[70] Ibid.

increase current output at known price levels and then invest the surplus revenues in anticipation of a time when economic conditions worsen.[71] For resources where there are viable or potential substitutes, firms would have an even stronger incentive to exploit and market as much of the resources as they can effectively produce before the market turns to them.[72]

The paradoxical problem with the perfect market model is that the market cannot impose restrictions to prevent the exhaustion of specific mineral resources or rectify the social and economic problems that would arise when that happens.[73] In fact, the sad reality is that market forces are likely to cause an economic exhaustion of resources before a physical exhaustion of resources would take place. This is because there will eventually be a point when consumers refuse to pay the high prices caused by diminishing resources and increasing costs of extraction, or the producers refuse to sell at government-imposed low prices, even though significant reserves may remain to be exploited.[74] Figure 2.5 illustrates the determination of economic exhaustion of a particular mineral resource.

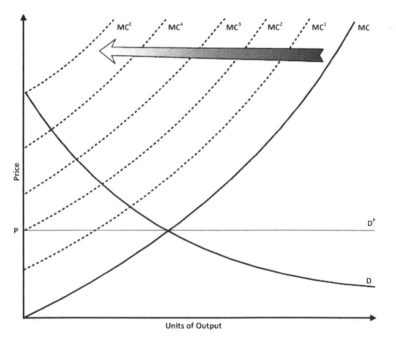

Fig. 2.5 The onset of economic exhaustion of mineral resources
Source: Judith Rees (1990, 48)

---

[71] Ibid.
[72] Ibid.
[73] Rees, supra note 3, at 47.
[74] Ibid.

In Fig. 2.5, as deposits become increasingly difficult to exploit, supply costs will rise and the marginal cost supply curve (MC) will shift upwards. Eventually the prices will reach a point where consumers are not willing to pay at all, as indicated at the point on the demand curve (D) where output is nil.[75] Governmental intervention in order to subsidise indigenous producers or restrict competition from foreign suppliers and the effects of global competition means that the world price for that resource is artificially maintained at a low level (D^p).[76] By the time the D^p curve has adjusted to the rising MC curve, the prices would have lagged behind again, quickening the onset of economic exhaustion of that particular resource as well as introducing the effects of resource scarcity at an earlier time than the market would have generated.[77]

Some commentators have also warned against over-confidence in the ability of technological innovation to overcome resource shortages on Earth. For example, Meadows and his fellow researchers were of the view that:

> The hopes of the technological optimists centre on the ability of technology to remove or extend the limits to growth of population and capital. We have shown that . . . the application of technology to apparent problems of resource depletion or pollution or food shortage has no impact on the essential problem, which is exponential growth in a finite and complex system. Our attempts to use even the most optimistic estimates of the benefits of technology in the model did not prevent the ultimate decline of population and industry and in fact did not in any case postpone the collapse beyond the year 2100.[78]

Even in more recent times, when it is apparent that the projected depletions of the early 1970s have not taken place, the same caution concerning confidence in technology continues to be given, in this case by then Senator Gore:

> We have also fallen victim to a kind of technological hubris, which tempts us to believe that our new powers may be unlimited. We dare to imagine that we will find technological solutions for every technologically induced problem. It is as if civilisation stands in awe of its own technological prowess, entranced by the wondrous and unfamiliar power it never dreamed would be accessible to mortal man. In a modern version of the Greek myth, our hubris tempts us to appropriate for ourselves – not from the gods but from science and technology – awesome powers and to demand from nature godlike privileges to indulge our Olympian appetite for more.[79]

One further inhibitor on the ability of the perfect market being able to solve the resource scarcity problems is the lack of concern by the market forces for the sustainable maintenance of the global environment. There are authors that have suggested that the perfect market model of avoiding resource exhaustion would cause the destruction of the global environment that would render the efforts of the market

---

[75] Ibid.

[76] Ibid.

[77] Ibid.

[78] Meadows, Meadows, Randers and Behrens, supra note 44, at 152.

[79] Albert Gore, Jr., EARTH IN THE BALANCE (1992) at 207.

2.2 Economic Feasibility of Space Mining

somewhat futile.[80] Even if such apocalyptic predictions of global destruction do not eventuate, some scholars have nonetheless suggested that there would be a significant cost to human welfare even though the market forces may successfully protect the material resources of the Earth.[81] As most environmental factors are considered free goods and discounted by the market, environmental damage can be contained in mining processes only through consumer activism or political intervention to highlight the impact of the environment as a negative externality to be factored into the consideration of economic supply and demand.[82] Consequently, as the environmental cost to the extraction of the scarce mineral resources from the crust of the Earth increases, the environmental conscience of the community may pose as another factor that could cause the social exhaustion of mineral resources.

2.2.2.3 The Pessimistic Perspective

The most pessimistic view available on the issue of resource exhaustion is to simply take the current proven reserves and then measure the resource availability by assuming an increase in the consumption level of a particular mineral resource. In this way, it is inferred that a physical or absolute exhaustion of mineral resources would take place regardless of any possible interventions of the market to alleviate its effects.

For example, the proven reserves of bauxite would last less than 100 years if human consumption remained at the same level as it was in the 1970s.[83] However, if the consumption of bauxite increases at the annual rate of around 10%, the proven reserves would be exhausted in the beginning of the present century. These types of estimates almost inevitably indicate that most mineral resources would be exhausted within 30 years.[84] For instance, as detailed in Table 2.3, some researchers in 1971 had projected the then imminent exhaustion of important mineral resources in the Earth's crust.

There is a significant number of resources listed in Table 2.3 that, if correct, would have been depleted already or dangerously close to being depleted, even though this is clearly not the case. The fact is that new discoveries and technological innovation have slowed the onset of exhaustion to the extent that there have

---

[80] See generally Ernest E. Snyder, PLEASE STOP KILLING ME (1971); Paul R. Ehrlich, THE POPULATION BOMB (1970); and Barry Commoner, THE CLOSING CIRCLE: MAN, NATURE AND TECHNOLOGY (1972).

[81] See generally Ezra J. Mishan, THE COSTS OF ECONOMIC GROWTH (1967) and Harold J. Barnett and Chandler Morse, SCARCITY AND GROWTH: THE ECONOMICS OF NATURAL RESOURCE AVAILABILITY (1963).

[82] See generally T. C. Sinclair, *Environmentalism: A la Recherché du Temps Perdu-Bien Perdu?* in H. Samuel D. Cole, Christopher Freeman, Marie Jahoda and Keith L. R. Pavitt (eds.), MODELS OF DOOM: A CRITIQUE OF THE LIMITS TO GROWTH (1973), 175–192.

[83] Ibid., at 33.

[84] Goldsmith, supra note 4, at 4.

Table 2.3 Estimated reserves and depletion dates

		Years to depletion from 1971	
Resource	1971 Reserves	Static growth	Exponential growth
Aluminium	17×10^9 t	100	31
Chromium	7.75×10^8 t	420	95
Coal	5×10^{12} t	2300	111
Cobalt	4.8×10^9 lbs	110	60
Copper	308×10^6 t	36	21
Gold	353×10^6 troy oz	11	9
Iron	1×10^{11} t	240	93
Lead	91×10^6 t	26	21
Manganese	8×10^8 t	97	46
Mercury	3.34×10^6 flasks	13	13
Molybdenum	10.8×10^9 lbs	79	34
Natural Gas	1.14×10^{15} cu ft	38	22
Nickel	147×10^9 lbs	150	53
Petroleum	455×10^9 bbls	31	20
Platinum Group	429×10^6 troy oz	130	47
Silver	5.5×10^9 troy oz	16	13
Tin	4.3×10^6 lg t	17	15
Tungsten	2.9×10^9 lbs	40	28
Zinc	123×10^6 t	23	18

Source: Dennis L. Meadows (1972, 56–59)

been no significant increases in price levels.[85] However, the trend of sharp peaks in the price for crude oil and other mineral resources since 2003 may indicate either the effect of the usual volatility and fluctuations of market prices or the progressive depletion and exhaustion of oil and other mineral resources.[86] In particular, Hubbert predicted that there would be a point in time, referred to as Hubbert's Peak, when the maximum production of oil would be reached and beyond which there will be a continuing decline in global oil production.[87] There are a number of commentators who suggest that the world has now reached, if it has not already done so, the Hubbert's Peak and so the supply of crude oil will continue to decline.[88] In turn, this will raise the price of crude oil and all other mineral resources that require substantial energy for extraction, processing and exploitation.

Regardless of the perspective taken, it is clear that eventually there would be a limit to human innovation or even a time when all possible substitutes will be

---

[85] Brown and Wolk, supra note 43, at 11–13.

[86] World Bank, *Prospects for the Global Economy: Commodity Markets* (2006), at <http://web.worldbank.org/WBSITE/EXTERNAL/EXTDEC/EXTDECPROSPECTS/EXTGBLPROSPECTSAPRIL/0,,contentMDK:20371216~menuPK:2300917~pagePK:2470434~piPK:2470429~theSitePK:659149,00.html>, last accessed on 16 September 2006.

[87] Hubbert, supra note 55.

[88] See, for example, John Vidal, *The End of Oil is Closer than You Think* (21 April 2005), The Guardian, at <http://www.guardian.co.uk/life/feature/story/0,13026,1464050,00.html>, last

2.2 Economic Feasibility of Space Mining

exhausted.[89] In fact, it would be the only logical outcome for this to be the case. This is because the entire study of resource economics is based on the principles of supply and demand being dictated by the scarcity of resources to satisfy all human wants.[90] If there is never a point of physical resource exhaustion then there would be no long-term scarcity of resources, thus rendering the economic concepts of supply and demand as nothing more than an artificial means of price regulation.[91]

Example of a theoretical analysis of the expected life of a mineral resource, in this case bauxite (aluminium ore, mainly $Al(OH)_3$), is illustrated in Fig. 2.6. Using 1974

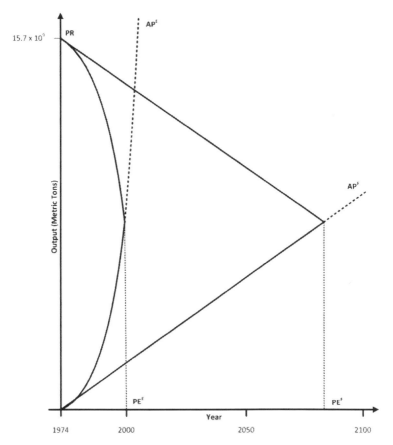

Fig. 2.6 Example: expected life of bauxite reserves
Source: Judith Rees (1990, 35)

---

accessed on 27 January 2007 and Kenneth S. Deffeyes, BEYOND OIL: THE VIEW FROM HUBBERT'S PEAK (2005).
[89] Ibid.
[90] Ibid.
[91] Ibid.

as the starting point, if annual production remained the same at 69.7×10^6 metric tons (AP^F), then the proven reserves (containing at least 52% aluminium) would be exhausted at PE^F. However, if annual production increased at 9.8% per year (AP^E), then the proven reserves of bauxite would be exhausted much earlier at PE^E.[92] Of course, such an analysis does not take into account the increase in proven reserves of bauxite over time trough increased exploration, as discussed above (Table 2.4).

In any event, it is clear from a study of the rise in production rate and the additional discoveries made in the reserves over a fixed period of time that mineral resources nevertheless are being consumed at a faster rate than additional reserves are being found to replace them. Consequently, it only stands to reason that physical exhaustion must follow as a matter of course, although it may be preceded by the onset of economic exhaustion, namely, the point at which the consumers refuse to pay the high prices or when the producers refuse to sell at government-imposed low prices, even though significant reserves may remain to be exploited.

Table 2.4 Cumulative reserves and production in selected minerals

Mineral	1950 reserves	1974 reserves	1950–1974 cumulative production	1950–1974 addition to reserves[a]
Asbestos	3.9×10^7	8.7×10^7	6.2×10^7	1.1×10^8
Bauxite	1.4×10^9	1.6×10^{10}	8.5×10^8	1.5×10^{10}
Chromium	1.0×10^8	1.7×10^9	9.6×10^7	1.7×10^9
Cobalt	7.9×10^5	2.4×10^6	4.4×10^5	2.2×10^6
Copper	1.0×10^8	3.9×10^8	1.1×10^8	4.0×10^8
Gold	3.1×10^4	4.0×10^4	2.9×10^4	3.7×10^4
Iron	1.9×10^{10}	8.8×10^{10}	7.3×10^9	7.6×10^{10}
Lead	4.0×10^7	1.5×10^8	6.3×10^7	1.7×10^8
Manganese	5.0×10^8	1.9×10^9	1.6×10^9	1.6×10^9
Mercury[b]	1.3×10^5	1.8×10^5	1.9×10^5	2.5×10^5
Nickel[c]	1.4×10^7	4.4×10^7	9.4×10^6	3.9×10^7
Phosphates	2.6×10^9	1.3×10^{10}	1.3×10^9	1.2×10^{10}
Platinum	7.8×10^2	1.9×10^4	1.7×10^3	2.0×10^4
Potash	5.0×10^9	8.1×10^{10}	3.0×10^8	7.6×10^{10}
Silver	1.6×10^5	1.9×10^5	2.0×10^5	2.3×10^5
Sulphur	4.0×10^8	2.0×10^9	6.1×10^8	2.2×10^9
Tin	6.0×10^6	1.0×10^7	4.6×10^6	8.6×10^6
Tungsten	2.4×10^6	1.6×10^6	7.6×10^5	-4.3×10^4
Zinc	7.0×10^7	1.2×10^8	9.7×10^7	1.5×10^8

[a] Calculated by adding together the cumulative production to 1974 production to 1974 reserves and subtracting the 1950 reserves
[b] Data on reserves for communist countries (as they then were) incomplete
[c] Data on reserves for communist countries (as they then were) incomplete
Source: Judith Rees (1990, 36)

---

[92] Ibid.

2.2.3 Implications of the Hydrogen Economy

One further factor that must be considered in assessing the mineral resources need of the near future is the impact that increasing popularity in hydrogen fuel cells has on the demand for mineral resources. Modern fuel cells rely on the chemical oxidisation of hydrogen and oxygen to produce electric power and are far more efficient than modern energy generation techniques with far fewer environmental concerns, such as the production of greenhouse gases or radioactive waste.[93] Moreover, the "fuel" used to generate power is hydrogen and oxygen, which are readily available in abundance on Earth and can theoretically be recycled from the water produced from the fuel cells.

Although fuel cells are far more efficient and environmentally friendly than traditional means of power generation, the electrodes of a fuel cell are produced from platinum and other platinum group metals, such as iridium, osmium, rhodium, ruthenium and palladium.[94] There are several types of fuel cells, each type being deployed for specific uses, as described in detail in Fig. 2.7.

The high cost of hydrogen fuel cells lie predominantly in the cost of platinum group metals, which are extremely rare in the Earth's crust and occur in very low concentrations, with a global average of around 4 g per t.[95] Considering a small fuel cell car would require 50 kW of power to be generated and, correspondingly, 57 g of platinum as the catalyst, 14 t of ore would have to be mined in order to produce sufficient platinum for one small fuel cell car.[96] Even without fuel cells, the automobile industry remain the largest users of platinum and palladium in catalytic converters and this is only likely to increase with tightening environmental controls worldwide.[97] To some extent, this is ironic as the production of platinum is a heavily polluting process, involving the release of chlorine, ammonia, hydrochloric acid and various heavy metals such as iron, zinc and nickel.[98]

In any event, some industrialised States have already formulated ambitious plans for the implementation of a "hydrogen economy" in order to increase global energy efficiency and reduce the dependence on oil and other fossil fuels. The United States of America, for example, published a plan for implementation of the hydrogen economy by 2050.[99] This would dramatically increase the global demand for platinum and platinum group metals, to the extent that one million kilograms may be needed

---

[93] See, for example, discussion in Wingo, supra note 5, at 77–82.

[94] Ibid., at 83.

[95] AEA Technologies, *Platinum and Hydrogen for Fuel Cell Vehicles* (2002), U.K. Department for Transportation, at <http://www.dft.gov.uk/stellent/groups/dft_roads/documents/pdf/dft_roads_pdf_024056.pdf>, last accessed on 3 January 2004.

[96] Wingo, supra note 5, at 83.

[97] Johnson Matthey, *Platinum 2004 Interim Report* (2004) at <http://www.platinum.matthey.com/publications/1100682070.html>, last accessed on 21 December 2004.

[98] Johnson Matthey (2004) at <http://www.platinum.matthey.com/production/africa.html>, last accessed on 21 December 2004.

[99] U.S. Department of Energy, HYDROGEN POSTURE PLAN: An INTEGRATED RESEARCH, DEVELOPMENT AND DEMONSTRATION PLAN (2004) at 6.

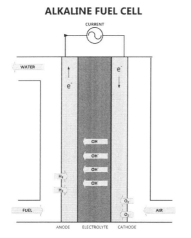

ALKALINE FUEL CELL

Alkaline fuel cells use potassium hydroxide (KOH) or other alkaline electrolytes with platinum or other platinum-group metals used as catalysts at the electrodes.

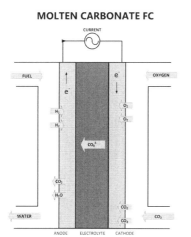

MOLTEN CARBONATE FC

Molten carbonate fuel cells (MCFC) use a molten carbonate salt mixture as the electrolyte. Although non-precious metals can be used as the catalyst, these fuel cells operate at temperatures of 650°C or above.

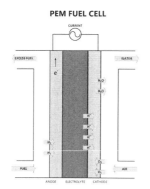

PEM FUEL CELL

Proton exchange membrane (PEM) fuel cells use a special polymer electrolyte membrane to allow for protons to permeate but not electrons, with platinum particles on porous carbon supports to catalyse hydrogen oxidation and oxygen reduction.

Fig. 2.7 Different types of hydrogen fuel cells
Source: U.S. Department of Energy (2004)

2.2 Economic Feasibility of Space Mining   47

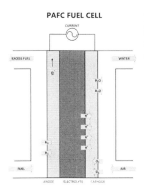

Phosphoric acid fuel cells (PAFC) use liquid phosphoric acid (H_3PO_4) for the electrolyte and carbon paper electrodes coated with a platinum catalyst.

Fig. 2.7 (continued)

each year purely for fuel cell cars alone.[100] Considering the estimated total planetary reserves for platinum and platinum group metals lie somewhere between 43 and 100,000,000 kg and the heavy environmental cost in extraction and refining, the adoption of a "hydrogen economy" may well be simply replacing one rare non-renewable resource for another.[101] Accordingly the only logical solution would be to extract such resources from places other than the crust of the Earth, such as celestial bodies (Fig. 2.8).

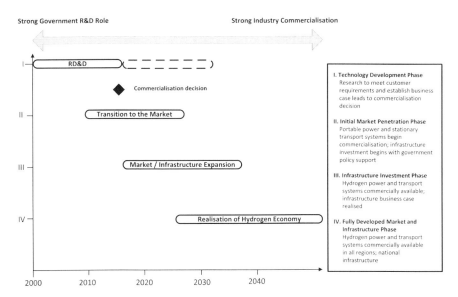

Fig. 2.8 Government-industry roles in the transition to a hydrogen economy
Source: U.S. Department of Energy (2004, 4)

---

[100] AEA Technologies, supra note 95, at 16.
[101] Wingo, supra note 5, at 89.

2.3 Expanding the Economic Resource Base

There is one possibility that the pessimists and doubters of the economic sustainability of resource consumption have not considered. So far, the economic analyses of resource consumption have assumed that available resources are confined to the Earth's crust.[102] If the increasing consumption of resources on the Earth would catalyse technological innovation and additional resource exploration, eventually the technology required to explore and exploit resources from other celestial bodies of the Solar System would be available. As mineral discoveries are made on the Moon, the asteroids, comets and other planets, these additions to the proven or conditional reserves of the minerals on Earth can ultimately support a market-driven resource economy on the Earth virtually into the infinite future. As Lewis suggested somewhat ambitiously:

> Let us recapitulate what we have already found. *Shortage of resources is not a fact; it is an illusion born of ignorance.* Scientifically and technically feasible improvements in launch vehicles will make departure from Earth easy and inexpensive. Once we have a foothold in space, the mass of the asteroid belt will be at our disposal, permitting us to provide for the material needs of a million times as many people as Earth can hold. Solar power can provide all the energy needs of this vast civilisation (10,000,000 billion people) from now until the Sun expires. Using less than one percent of the helium-3 energy resources of Uranus and Neptune for fusion propulsion, we could send a billion interstellar arks, each containing a billion people, to the stars. There are about a billion Sun-like stars in our galaxy. *We have the resources to colonise the entire Milky Way.*[103]

O'Neill, in his testimony before Committee on Science and Technology of the U.S. House of Representatives, said that:

> The fatalism of the limits-to-growth alternative is reasonable only if one ignores all the resources beyond our atmosphere, resources thousands of times greater than we could ever obtain from our beleaguered Earth. As expressed very beautifully in the language of House Concurrent Resolution 451, "This tiny Earth is not humanity's prison, is not a closed and dwindling resource, but is in fact only part of a vast system rich in opportunities, a high frontier which irresistibly beckons and challenges the American genius".[104]

It is pertinent to investigate the economic effects of this potential development in closer detail. As marginal production cost increase, the perfect market model dictates that increased technological innovation and mineral exploration will take place. In addition to developing viable substitutes and improved recycling techniques, the improvements in technology will also lead to evolution of propulsion systems, spacecraft design and robotic mining technology that make mining of the

---

[102] This is apparent in the discussions found in, for example, Rees, supra note 3, *passim* and David W. Pearce and James Rose (eds.), THE ECONOMICS OF NATURAL RESOURCE DEPLETION (1975), *passim*.

[103] John S. Lewis, MINING THE SKY: UNTOLD RICHES FROM THE ASTEROIDS, COMETS AND PLANETS (1997) at 263.

[104] Gerard K. O'Neill, *Testimony on House Concurrent Resolution 451* (1978), U.S. House of Representatives, 25 January 1978.

2.3 Expanding the Economic Resource Base 49

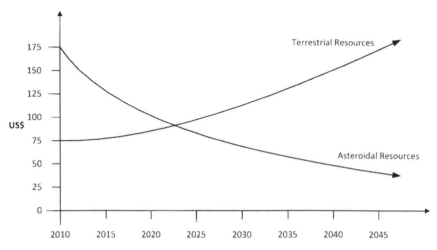

Fig. 2.9 Relative cost of producing 1 ounce of platinum 2005–2030
Source: Charles L. Gerlach (2005, 2)

Moon and asteroids more feasible. In other words, the cost of extracting mineral resources, especially platinum group metals, will continue to increase while the cost of extracting the same mineral resources from celestial bodies will decrease over time, as illustrated in Fig. 2.9.

Combined with increased exploratory and scientific studies into the geology and mineralogy of celestial bodies, the inclusion of asteroidal or lunar resources promises to reduce or even eliminate the onset of economic resource exhaustion on Earth. Consequently, the development of extraterrestrial mineral reserves has particular attraction to a world population that is increasingly conscious of the impact of human activities on the environment. For example, the demand for electricity generated from renewable resources has exceeded what one would expect in a market driven mostly by cost to the consumer.[105] By exploiting resources from other bodies in the Solar System, the pristine environmental balances of the deep seabed and Antarctica can be preserved. Further, the development of space resources would also increase the need for additional infrastructure in the Earth orbit or the surface of the Moon, in turn increasing the demand for mineral resources from space. This need for the development of resource exploitation in outer space appears, according to either the optimistic or the pessimistic school of resource economics, to be only a matter of time.

---

[105] Robert J. S. Beeton, Kristal I. Buckley, Gary J. Jones, Denise Morgan, Russell E. Reichelt and Dennis Trewin, *Australia State of the Environment 2006: Independent Report of the Australian Government Minister for the Environment and Heritage* (6 December 2006), Department of the Environment and Heritage, at <http://www.deh.gov.au/soe/2006/publications/report/index.html>, last accessed on 27 January 2007. See also Mark Z. Jacobsen and Mark A. Delucchi, *A Path to Sustainable Energy by 2030* (2009) 301:5 SCIENTIFIC AMERICAN 58.

This need to exploit resources from celestial bodies is particularly pertinent in contemplating the needs for the implementation of the hydrogen economy, when the following factors are considered together:

(1) the increasing demand for platinum and platinum group metals;
(2) that platinum and platinum group metals are non-renewable and thus finite resources in the Earth's crust;
(3) that platinum and platinum group metals occur rarely and do so in very low concentrations, requiring a large amount of ore to be processed to extract small amounts of the metals;
(4) there is a heavy environmental cost associated with mining large quantities of ore as well as the processes involved in extracting and refining such metals; and
(5) the Moon and most Near Earth Asteroids have platinum group metals in concentrations higher than those contained in the Earth's crust.[106]

In addition to the demand on Earth, the prospect of constructing large space stations, laboratories, factories and hotels in orbit around the Earth and on the Moon creates a substantial demand for mineral resources in outer space. With launch costs being prohibitive, there would logically be a market for various mineral resources in outer space for construction, propellants, manufacturing and material processing.[107]

For the above reasons, the increasing demand and associated production and environmental costs for the terrestrial mining of platinum group metals will in the near future make the mining of the Moon and other celestial bodies, especially Near Earth Asteroids, increasingly appealing. The additional costs involved in transport between the Earth and the celestial bodies concerned are largely offset by the higher concentrations of platinum group metals in the extracted ore as well as rising prices caused by increased production and environmental costs on Earth.[108] This makes mining of celestial bodies in outer space an inevitable and not altogether long-term development. As Levinson suggested:

---

[106] John S. Lewis and Melinda L. Hutson, *Asteroidal Resource Opportunities Suggested by Meteorite Data*, in John S. Lewis, Mildred S. Matthews and Mary L. Guerrieri (eds.), RESOURCES OF NEAR-EARTH SPACE (1993), at 534.

[107] Mark J. Sonter, *The Technical and Economic Feasibility of Mining the Near-Earth Asteroids*, paper presented at the 49th International Astronautical Congress, 5–9 October 1998, in Melbourne, Australia and Shane D. Ross, *Near-Earth Asteroid Mining* (2001), at Laboratory for Spacecraft and Mission Design, California Institute of Technology, <http://www.esm.vt.edu/~sdross/papers/ross-asteroid-mining-2001.pdf>, last accessed on 24 April 2007.

[108] See David R. Wilburn and Donald I. Bleiwas, *Platinum Group Metals – World Supply and Demand* (2004), U.S. Geological Survey, at <http://pubs.usgs.gov/of/2004/1224/2004-1224.pdf>, last accessed on 27 January 2007; David R. Wilburn, *International Mineral Exploration Activities from 1995 through 2004* (25 August 2006), U.S. Geological Survey, at <http://pubs.usgs.gov/ds/2005/139/>, last accessed on 27 January 2007; and Ashok Kumar, *A World Stage: Global Factors are Pushing Platinum Group Metals to Historically High Price Levels* (1 September 2006), Recycling Today, at <http://www.thefreelibrary.com/A+world+stage%3a+global+factors+are+pushing+platinum+group+metals+to...-a0152513170>, last accessed on 27 January 2007.

We of course have our problems, to say the least, in comportment towards ourselves and our environment, but admittance to the cosmos and the spatial infinity and temporal immortality it provides may well be just the remedy for these age-old problems. Access to the boundless resources of the universe may once and for all puncture the pressure of population and politics of scarcity which have generated war, oppression, and plagued our species from the start.[109]

2.4 The Riches of Space

2.4.1 Mining the Moon

The relatively deep gravity well of the Moon vis-à-vis Near Earth Asteroids requires a comparatively higher escape velocity to launch materials from the Moon's surface and, similarly, a high amount of thrust to enable a soft landing on the Moon from lunar orbit. As a result, the propulsion system used for any transportation vehicle between the Moon and the Earth orbit must rely on a chemical or nuclear proponent, imposing a severe constraint on the costs of transport.[110] However, this heavy gravity also means that the design of structures and materials handling processes on the Moon would not have to be dissimilar from that of the Earth and civil engineering would be simpler, making it possible for us to apply the engineering principles as that applied on the Earth.[111] This means that large-scale commercial mining operations can take place on the Moon in order to take advantage of potential economies of scale.

One clear advantage of the mining of lunar resources is its proximity to the Earth. The fact that it is orbiting the Earth rather than the Sun or another planet in the Solar System means that it is accessible at any time. The short distance between the Earth and the Moon means that communications are virtually instantaneous, allowing for real-time remote control of robotic mining operators from the Earth. The substantial water deposits on the Moon that were discovered in the 1990s make the Moon potentially a good location for a permanent lunar settlement as well as providing for in situ production of hydrogen and oxygen that are to be used as fuels for propulsion, provided that a means of replenishing or recycling such water supplies can be found.[112]

As the only celestial body outside the Earth that has been physically visited by humans, there is a large amount of information available about the geology and mineral composition of the Moon. Analysis of lunar soil samples collected from the Apollo missions revealed that the lunar soil contains oxygen, silicon, aluminium,

---

[109] Paul Levinson, *Technology as the Cutting Edge of Cosmic Evolution*, in Paul T. Durbin (ed.), RESEARCH IN PHILOSOPHY AND TECHNOLOGY (1985), at 161–176.

[110] Wingo, supra note 5, at 163.

[111] U.S. Army Corps of Engineers, *Proceedings of a Workshop on Extraterrestrial Mining and Construction* (1990) USACERL SPECIAL REPORT M-92/14.

[112] Wingo, supra note 5, at 204–205.

iron, calcium, magnesium and titanium in various compounds.[113] Even though the Moon has hardly any free metal available for mining purposes, the processes for producing iron and oxygen from ilmenite, as well as aluminium and oxygen from feldspar, have been studied in scientific circles.[114] If solar cells can be manufactured on site, the production of solar cells for the deployment of a SPSS in the Earth orbit is also a possibility for lunar export.[115]

In the short term, it is likely that the most valuable resource to be exploited from the Moon would be the collection of helium-3 (He-3), which is used for various medical and nuclear applications.[116] Helium-3 is extremely rare on Earth as it is lost through dissipation in the upper reaches of the atmosphere, but is collected in abundance from the solar wind on the Moon. In the longer term, the proximity of the Moon and the abundance of mineral resources available would make it a very attractive candidate for exploitation, especially during the development of large-scale infrastructures in the Earth orbit. These infrastructures may form the foundations for future exploration and utilisation of the more distance parts of the Solar System, such as Mars, its satellites, and the Near Earth Asteroids.

2.4.2 Resources from Mars and Other Planets

There is no doubt that the planets close to Earth, namely Mars, Venus and Mercury, may be candidates for exploration and extraction of mineral resources considering the resources that may be available in abundance on these planets. It was recently discovered, for example, that the surface of Mercury is rich in oxides of iron and titanium.[117] Venus has been thought to have similar mineralogy, though there has

---

[113] See Ellen B. Heim, *Exploring the Last Frontiers for Mineral Resources: A Comparison of International Law Regarding the Deep Seabed, Outer Space and Antarctica* (1990) 23 VAND. J. TRANSNAT'L. L. 819 at 831.

[114] William N. Agosto, *Beneficiation and Powder Metallurgical Processing of Lunar Soil Metal*, paper presented in the 5th Princeton/AIAA Conference on Space Manufacturing, May 1981, in Princeton, NJ, USA.

[115] David R. Criswell, *Lunar Solar Power System: Scale and Cost*, paper presented at the 45th International Astronautical Congress, 9–14 October 1995, in Jerusalem, Israel.

[116] U.S. Department of Energy, *Isotope Uses* (2003) <http://www.ne.doe.gov/isotopes/ipuses. asp>, last accessed on 22 December 2004.

[117] Jonathan Amos, *Messenger spies iron on Mercury*, British Broadcasting Corporation, 4 November 2009, at <http://news.bbc.co.uk/2/hi/science/nature/8342000.stm>, last accessed on 5 November 2009 and Andrea Thompson, *NASA Probe See Changing Seasons on Mercury*, MSNBC, 3 November 2009, at <http://www.msnbc.msn.com/id/33609756/ns/technology_and_ science-space/>, last accessed on 5 November 2009. See also Mark S. Robinson and G. Jeffrey Taylor, *Ferrous Oxide in Mercury's Crust and Mantle* (2001) 36 METEOR. & PLANET. SCI. 841 and James W. Head, Clark R. Chapman, Deborah L. Domingue, S. Edward Hawkins III, William E. McClintock, Scott L. Murchie, Louise M. Prockter, Mark S. Robinson, Robert G. Strom and Thomas R. Watters, *The Geology of Mercury: The View Prior to the MESSENGER Mission* (2007) 131 SPACE SCI. REV. 41.

2.4 The Riches of Space

been no direct evidence based on surface mineralogical analysis by landing probes on Venus.[118] Mars, on the other hand, appears to contain abundant reserves of sulphur, potassium, titanium, nickel, zinc, iron and other rare metals.[119] The gas giants, namely Jupiter, Saturn, Uranus and Neptune, contain mostly hydrogen and helium that may be attractive for the extraction of deuterium and helium-3, but they have been found to contain water that may be useful in future expansion of human settlement in the Solar System.[120]

While Mars, Venus and Mercury may contain mineral resources in abundance, the physical extraction of such resources from those planets is very challenging and well beyond present technological capabilities. The absence of an atmosphere and its proximity to the Sun means that Mercury is subject to great variances of surface temperature from $427°C$ to $-180°C$.[121] Venus is definitely not an appealing place, as it has an average surface temperature of $462°C$ that would melt lead, tin and zinc, surface atmospheric pressure of 96 bars with the atmosphere composing primarily of carbon dioxide and clouds of sulphuric acid. The cloud coverage means that there is little or no solar energy on the surface.[122] The outer planets are too cold, too far

---

[118] Göstar Klingelhöfer and Bruce Fegley, Jr., *Iron Mineralogy of Venus's Surface Investigated by Mössbauer Spectroscopy* (2000) 147 ICARUS 1; John A. Woods and Kevin B. Klose, *Mineralogy on Venus and Areas of High Fresnel Reflection Coefficient Detected by Magellan Radar* (1991) 22 ABS. LUNAR & PLANETARY SCI. CONF. 1521; and Andrew M. Davis and Karl K. Turekian, METEORITES, COMETS AND PLANETS (2005), at 497–499.

[119] Göstar Klingelhöfer, Richard V. Morris, Albert S. Yen, Douglas W. Ming, Christian Schröder and Daniel Rodionov, *Mineralogy on Mars at Gusev Crater and Meridiani Planum as seen by Iron Mössbauer Spectroscopy* (2006) 70 Supp 1 GEOCHEM. COSM. ACTA A325; Albert S. Yen, Ralf Gellert, Christian Schröder, Richard V. Morris, James F. Bell III, Amy T. Knudson, Benton C. Clark, Douglas W. Ming, Joy A. Crisp, Raymond E. Arvidson, Diana Blaney, Johannes Brückner, Philip R. Christensen, David J. DesMarais, Paulo A. de Souza, Jr., Thanasis E. Economou, Amitabha Ghosh, Brian C. Hahn, Kenneth E. Herkenhoff, Larry A. Haskin, Joel A. Hurowitz, Bradley L. Joliff, Jeffrey R. Johnson, Göstar Klingelhöfer, Morten Bo Madsen, Scott M. McLennan, Harry Y. McSween, Lutz Richter, Rudi Rieder, Daniel Rodionov, Larry Soderblom, Steven W. Squyres, Nicholas J. Tosca, Alian Wang, Michael Wyatt and Jutta Zipfel, *An Integrated View of the Chemistry and Mineralogy of Martian Soils* (2005) 436 NATURE 49; and Rudi Rieder, Huth J. Wänke, Thanasis E. Economou and Anthony Turkevich, *Determination of the Chemical Composition of Martian Soil and Rocks: The Alpha Proton X Ray Spectrometer* (1997) 102 J. GEOPHYS. RES. 4027.

[120] Thérèse Encrenaz, Pierre Drossart, Helmut Feuchtgruber, Emmanuel Lellouch, Bruno Bézard, Thierry Fouchet and Sushil K. Atreya, *The Atmospheric Composition and Structure of Jupiter and Saturn from ISO Observations: A Preliminary Review* (1999) 47 PLANET. & SPACE SCI. 1225 and Thérèse Encrenaz, *The Chemical Atmospheric Composition of the Giant Planets* (1994) 67 EARTH, MOON & PLANETS 77.

[121] Thomas R. Walters, PLANETS (1995) and Sean R. Solomon, *Return to the Iron Planet*, NEW SCIENTIST, 29 January 2000, at 35.

[122] Geoffrey A. Landis, *Colonisation of Venus*, paper presented at the Conference on Human Space Exploration, Space Technology and Applications International Forum, 2–6 February 2003, in Albuquerque, NM, USA and Kenneth D. Mellott, *Electronics and Sensor Cooling with a Stirling Cycle for Venus Surface Mission* (2004), paper presented at the 2nd International Energy Conversion Engineering Conference, 16–19 August 2004, in Providence, RI, USA.

away, their atmospheres too dense and gravitational fields too strong to make them viable candidates for mining in the near future.[123]

Of all the planets in the Solar System, while Mars presents itself as the most pleasant of the inner planets, significant difficulties due to its volatile weather patterns and heavy gravitational field would nevertheless make Mars an unappealing candidate for mineral exploitation in the near future. However, the satellites of the planets, particular Phobos and Deimos of Mars, present minimal problems and are potential resource targets as well as being a good advance base for the exploration and colonisation of Mars.[124] High-resolution remote sensing photography of the surfaces of Phobos and Deimos show a well-developed regolith layer and are considered likely to be captured asteroids, a view strengthened by the fact that the two bodies are far from being spherical in shape.[125] Scientific studies through absorption spectroscopy have revealed that there is likely to be deep primordial ice in these satellites at depths of around twenty metres at the poles and around 100 ms at the equator, making them prospective ore bodies that are in close proximity to the Earth.[126]

The velocity change or energy required to land and take-off from these satellites is not great, assuming very little energy is needed for the spacecraft to match the heliocentric orbital velocity of Mars.[127] Phobos, as the higher satellite, would have a lower circularisation velocity and as a result would present itself as the better candidate for the first exploration and mining expedition to Mars and its satellites.[128]

However, the proximity of the satellites to Mars would pose significant problems, as extremely delicate navigational controls are required to negotiate the aerocapture of the Mars atmosphere to arrive at the satellites. Since the time for radio communications between the Earth and Mars can amount to several minutes and may be interrupted by occultations by other planets, the Sun and the asteroids,

---

[123] See, for example, Patrick G. J. Irwin, GIANT PLANETS OF OUR SOLAR SYSTEM: AN INTRODUCTION (2006) and Thérèse Encrenaz, Reinald Kallenbach, Tobias C. Owen and Christophe Sotin (eds.), THE OUTER PLANETS AND THEIR MOONS: COMPARATIVE STUDIES OF THE OUTER PLANETS (2005).

[124] See Pascal Lee, Stephen Braham, Greg Mungas, Matt Silver, Peter Thomas and Michael West, *Phobos: A Critical Link between Moon and Mars Exploration*, paper presented at the Space Resources Roundtable VII: Lunar Exploration Analysis Group Conference on Lunar Exploration, 25–28 October 2005, in League City, TX, USA, also at <http://www.lpi.usra.edu/meetings/leag2005/pdf/2049.pdf>, last accessed on 20 April 2007; Pascal Lee, Stephen Braham, Brett J. Gladman, Greg Mungas, Matt Silver, Peter Thomas and Michael West, *Mars Indirect: Phobos as a Critical Step in Human Mars Exploration*, paper presented at the International Space Development Conference, May 2005, in Washington, DC, USA; and Brian O'Leary, *International Manned Missions to Mars and the Resources of Phobos and Deimos* (1992) 26:1 ACTA ASTRONAUTICA 37.

[125] Joseph Veverka and Peter C. Thomas, *Phobos and Deimos: A Preview of What Asteroids Are Like?* in Gehrels, supra note 6 (1979) at 628–651.

[126] Fraser P. Fanale and James R. Salvail, *Evolution of the Water Regime of Phobos* (1990) 88 ICARUS 380.

[127] Veverka and Thomas, supra note 125.

[128] Ibid.

2.4 The Riches of Space

the navigational controls on the spacecrafts must necessarily be autonomous. Such highly advanced space navigational systems on spacecrafts are as yet beyond the means of human technology.[129] As a result, the mining and exploration activities on other planets and their satellites must necessarily wait for further technological advances to take place.

2.4.3 Geology and Mineralogy of Asteroids and Their Suitability for Mining Activities

2.4.3.1 Asteroid Taxonomy

Information about the composition of asteroids has been growing significantly since the 1980s because of a rise in observational and modelling work.[130] Increased use of spectroscopy as well as close flyby observations by Galileo of 243 Ida and by the Near Earth Asteroid Rendezvous ("NEAR") probe of 253 Mathilde and 52 Europa have generated a substantial amount of data for asteroid geologists to determine the hypothetical and observable mineralogy of some asteroids (Table 2.5).

The most widely used system of taxonomy for asteroids is one developed by Tholen using seven colours derived from the Eight Colour Asteroid Survey ("ECAS") at the end of the 1970s and the beginning of the 1980s.[131] In that system,

Table 2.5 Bell superclasses of asteroid taxonomy and likely mineralogy

Superclass	Class	Inferred minerals	Analogous meteorites
Primitive	D	Clays and organics	(None)
	P	Clays and organics	(None)
	C	Clays, carbon and organics	CI and CM chondrites
	K	Olivine, pyroxene, carbon	CV and CO chondrites
Metamorphic	T	?	?
	B, G and F	Clays and opaques	Altered carbonaceous chondrites
	Q	Pyroxene, olivine, grey NiFe	H, L and LL chondrites
Igneous	V	Pyroxene, olivine	Basaltic achondrites
	R	Olivine, pyroxene	Olivine-rich achondrites
	S	Pyroxene, olivine, red NiFe	Pallasites, lodranites and irons
	A	Olivine	Brachinites
	M	NiFe	Irons
	E	Fe-free pyroxene	Aubrites

Source: Thomas H. Burbine and Richard P. Binzel (1993, 255)

---

[129] See generally International Academy of Astronautics, *The International Exploration of Mars* (1996), at <http://www.iaanet.org/p_papers/mars.html>, last accessed on 23 December 2004.

[130] See, for example, discussion in Kowal, supra note 17, at 35–48.

[131] Benjamin Zellner, David J. Tholen and Edward F. Tedesco, *The Eight-Color Asteroid Survey: Results for 589 Minor Planets* (1985) 61 ICARUS 355.

fourteen asteroid classes were created with eleven of the classes distinguished by their ECAS spectra and three classes (E, M and P) distinguished only by visual albedo. Barucci and Tedesco improved on this taxonomic system in the late 1980s, combining the data set from the ECAS with that derived from various infrared spectroscopy studies.[132] In studying the data sets obtained in these different studies, astronomers began to distinguish asteroids based on the degree of metaphoric heating that they are believed to have received during their evolution.[133] During the 1990s, the taxonomic classifications have been further refined to improve the distinctions between the compositional diversity of the classes as our knowledge of the geology of asteroids continues to increase.[134]

2.4.3.2 Overview

Due to their accessibility and relatively low escape velocities, asteroids present themselves to be potentially attractive targets for resources to supplement demand for resources on Earth. Some mineral resources found on these asteroids could be particularly useful for further human development in outer space. Lewis suggested that around half of the reasonably sized Near Earth Asteroids are believed to be carbonaceous and therefore rich in carbon and water.[135] Such asteroids contain important mineral resources for life support and are thus important mining targets for future human development in outer space. The other asteroids are believed to be metallic in nature, containing iron, nickel, troilite, olivine, pyroxene, plagioclase feldspar as well as non-metals such as arsenic, selenium, germanium, phosphorus, carbon and sulphur.[136] These mineral resources are subject to significant demand on Earth and potentially in orbit and on celestial bodies in outer space (Table 2.6).

---

[132] M. Antonella Barucci, Maria Teresa Capria, Angioletta Coradini and Marcello Fulchignoni, *Classification of Asteroids using G-Mode Analysis* (1987) 72 ICARUS 304 and Edward F. Tedesco, Dennis L. Matson, Glenn J. Veeder, Jonathan C. Gradie and Larry A. Lebofsky, *A Three-Parameter Asteroid Taxonomy* (1989) 97 ASTRON. J. 580.

[133] Jeffrey F. Bell, *Mineralogical Evolution of Meteorite Parent Bodies* (1986) 17 LUNAR PLANET. SCI. 985 and Jeffrey F. Bell, Donald R. Davis, William K. Hartmann and Michael J. Gaffey, *Asteroids: The Big Picture*, in Richard P. Binzel, Tom Gehrels and Mildred Shapley Matthews (eds.), ASTEROIDS II (1987), at 921–945.

[134] Thomas H. Burbine and Jeffrey F. Bell, *Asteroid Taxonomy: Problems and Proposed Solutions*, in Andrea Milani, Michel di Martino and Alberto Cellino (eds.), INTERNATIONAL ASTRONOMICAL UNION SYMPOSIUM 160: ASTEROIDS, COMETS AND METEORS (1993), at 49 and Thomas H. Burbine and Jeffrey F. Bell, *How Diverse Is the Asteroid Belt?* (1993) 24 LUNAR PLANET. SCI. 223.

[135] John S. Lewis, *Resources of the Asteroids* (1997) 50 J. BRIT. INTERPLANETARY SOC. 51.

[136] O'Leary, supra note 6 and Charles L. Gerlach, *Profitably Exploiting Near-Earth Object Resources* (2005), paper presented at the 2005 International Space Development Conference, 19-22 May 2005, in Washington, DC, USA, at Gerlach Space Systems L.L.C., <http://gerlachspace.com/resources/NEO%20Resources.pdf>, last accessed on 25 April 2007, at 2.

2.4 The Riches of Space

Table 2.6 Mineral resources found on asteroids and their uses in space

Primary use	Mineral resources
Life support	Water (H_2O), nitrogen (N_2) and oxygen (O_2)
Propellant	Hydrogen (H_2), oxygen (O_2), methane (CH_4), methanol (CH_3OH)
Agriculture	Carbon dioxide (CO_2), ammonia (NH_3), ammonium hydroxide (NH_4OH)
Oxidiser	Hydrogen peroxide (H_2O_2)
Refrigerant	Sulphur dioxide (SO_2)
Metallurgy	Carbon monoxide (CO), hydrogen sulphide (H_2S), nickel carbonyl ($Ni(CO)_4$), iron pentacarbonyl ($Fe(CO)_5$), sulphuric acid (H_2SO_4), sulphite (SO_3)
Construction	Iron (Fe), nickel (Ni)
Semiconductors	Silicon (Si), aluminium (Al), phosphorus (P), gallium (Ga), germanium (Ge), cadmium (Cd), copper (Cu), arsenic (As), selenium (Se), indium (In), antimony (Sb), tellurium (Te)
Precious metals	Gold (Au), platinum (Pt), palladium (Pd), osmium (Os), iridium (Ir), rhodium (Rh), ruthenium (Ru), rhenium (Re), germanium (Ge)

Source: Shane D. Ross (2001)

Particularly of note is the quantity of platinum group metals on asteroids. Platinum group metals include platinum, palladium, rhodium, ruthenium, iridium and osmium. These metals are highly valuable but are rare in the crust of the Earth and in low concentrations. Lewis and Meinel have pointed out that all common classes of meteorites found on Earth have higher concentrations of platinum group metals than even the best mines on Earth. Further, they have suggested that asteroids would similarly have high concentrations of platinum group metals.[137] It is projected that concentrations of 30–60 parts per billion or even 250 to over 1,000 parts per billion of platinum group metals may be found on asteroids, as compared to four to six parts per billion in the best mines on Earth.[138] With the introduction of a hydrogen-based energy economy increasing the global demand for platinum group metals, the extraction and exploitation of mineral resources on asteroids will become increasing feasible over time.

2.4.3.3 The S Class

In studying S class asteroids, high resolution spectra from 0.8 μm to 2.5 μm allow for identification for the presence or absence of absorption features such as the 1.1 μm feature due to feldspar, a 2 μm feature to minerals such as pyroxene and

---

[137] John S. Lewis and C. Meinel, *Asteroid Mining and Space Bunkers* [1983] DEF. SCI. 2000.

[138] Mary F. Horan, Richard J. Walker and John W. Morgan, *High Precision Measurements of Pt and Os in Chondrites*, paper presented at the 30th Annual Lunar and Planetary Science Conference, 15–29 March 1999 in Houston, TX, USA.

spinel ($MgAl_2O_4$) and a 2.2 μm feature attributed to organic compounds.[139] The original S class as designated by Tholen was separated into the K class and two subclasses, being the So and Sp subclasses. K class asteroids are similar to S class in spectra but have flat reflectance in the near infrared.[140] The So subclass of asteroids is rich in olivine with a red continuum slope while Sp subclass asteroids are rich in pyroxene with a less red continuum slope.[141] The remaining S class asteroids have a mixture of pyroxene and olivine in composition and a moderately red continuum slope compared to the So and Sp asteroids.

The diversity of the S class of asteroids has been observed in the past, but spectroscopy has revealed that virtually all S class asteroids contain significant quantities of olivine and pyroxene.[142] Olivine is a group of mineral silicates ranging from fayalite (Fe_2SiO_4) to forsterite (Mg_2SiO_4), providing for a good potential source of iron and magnesium. Pyroxene (($Mg,Fe)Si_2O_6$) and orthopyroxene (($Mg,Fe)_2Si_2O_6$) are bisilicates that similarly provide a good source of metals, along with silicon and oxygen. Considering the varying abundances of olivine, pyroxene and feldspar that may be found,[143] the S class of asteroids can be subdivided into seven compositionally distinct subtypes.[144]

Some S class asteroids are anomalous in composition, such as 387 Aquitania and 980 Anacostia, as indicated in studies of reflective spectra.[145] These two asteroids, for example, have a 2-μm feature stronger than a weak or even non-existent 1-μm feature. The reverse is generally characteristic of assemblages of pyroxene and olivine. It is believed that this is due to abundance in spinel or chromite ($FeCr_2O_4$).[146] The presence of spinel indicates a carbonaceous surface enriched

---

[139] Dale P. Cruikshank, Louis J. Allamandola, William K. Hartmann, David J. Tholen, Robert H, Brown, Clifford N. Matthews and Jeffrey F. Bell, *Solid C≡N Bearing Material on Outer Solar System Bodies* (1991) 94 ICARUS 345.

[140] Ellen S. Howell, Erzsebet Merényi and Larry A. Lebofsky, *Classification of Asteroid Spectra Sing a Neural Network* (1994) 99 J. GEOPHYS. RES. 10.

[141] Thomas H. Burbine and Richard P. Binzel, *Asteroid Spectroscopy and Mineralogy* in Andrea Milani, Michel di Martino and Alberto Cellino (eds.), INTERNATIONAL ASTRONOMICAL UNION SYMPOSIUM 160: ASTEROIDS, COMETS AND METEORS (1993) at 260.

[142] Michael J. Gaffey, Jeffrey F. Bell, Robert H. Brown, Thomas H. Burbine, Jennifer L. Piatek, Kevin L. Reed and Damon A. Chaky, *Mineralogic Variations within the S-Type Asteroid Class* (1993) 106 ICARUS 83.

[143] Feldspar is a large group of minerals that constitute the greatest percentage of minerals in the Earth's crust. The plagioclase feldspars include albite ($NaAlSi_3O_8$), oligoclase (($Na,Ca)[Al(Al,Si)_2SiO_2]$), andesine (($Na,Ca)Al_2Si_3O_8$), labradorite ($(Ca,Na)[(Al,Si)AlSi_2O_8]$), bytownite (($Ca,Na)[(Al,Si)AlSi_2O_8]$) and anorthite ($CaAl_2Si_2O_8$). The alkali feldspars include microcline ($KAlSi_3O_8$), sanidine (($K,Na)AlSi_3O_8$) and orthoclase ($KAlSi_3O_8$).

[144] Gaffey, Bell, Brown, Burbine, Piatek, Reed and Chaky, supra note 143.

[145] Thomas H. Burbine, Michael J. Gaffey and Jeffrey F. Bell, *S-Class Asteroids 387 Aquitania and 980 Anacostia: Possible Fragments of the Breakup of a Spiral-Bearing Parent Body with CO₃/CV₃ Affiliates* (1992) 27 METEORITICS 424.

[146] Ibid., and Edward A. Cloutis and Michael J. Gaffey, *The Constituent Minerals in Calcium-Aluminium Inclusions: Spectral Reflectance Properties and Implications for CO Carbonaceous Chondrites and Asteroids* (1993) 105 ICARUS 568.

2.4 The Riches of Space

in calcium and aluminium while the presence of chromite suggests an achondritic surface enriched in various metallic minerals, including magnesium, aluminium, chromium and iron.[147]

2.4.3.4 The M and E Classes of Asteroids

Astronomers have observed that an absorption feature at 3 mm generally indicates the presence of an abundance of hydrated silicates. Generally, hydrated silicates allow for the recovery of water, silicon and free metals after appropriate means of processing. Of the M class asteroids, 55 Pandora and 92 Undina were found to have 3 mm features that reflect the presence of recoverable hydrated silicates.[148] On the other hand, observations of asteroid 16 Psyche, another M class asteroid, shows the absence of a 3-mm feature. Psyche is generally considered to have a higher metallic iron content than other asteroids because of its high radar albedo.[149]

In addition to the hydrated M class asteroids, a large majority of C class asteroids have also been found to be composed of hydrated silicate surfaces and ice.[150] Recent observations of 44 Nysa, an E class asteroid, showed deposits of pyroxene, low-iron enstatite ($MgSiO_3$) and hydrated silicates.[151] This is in contrast to the previous belief that Nysa, as with other E class asteroids, would be composed of non-absorbing and spectrally neutral materials such as enstatite, forsterite and other low-iron silicates.[152] The only Earth-crossing E class asteroid, 3103 Eger, has been shown to contain mostly enstatite or forsterite. However, these metal rich asteroids are found predominantly in the Hungaria asteroid region between 1.79 and 1.98 AU from the Sun.[153]

---

[147] Hiroshi Takeda, Kazuto Saika, Mayami Otsuki and Takahiro Hiroi, *A New Antarctic Meteorite with Chromite, Orthopyroxene and Metal with Reference to a Formation Model of S Asteroids* (1993) 24 LUNAR PLANET. SCI. 1395.

[148] Andrew S. Rivkin, Daniel T. Britt, Ellen S. Howell and Larry A. Lebofsky, *Hydrated E-Class and M-Class Asteroids* (1995) 117 ICARUS 90.

[149] Steven J. Ostro, Donald B. Campbell and Irwin I. Shapiro, *Mainbelt Asteroids: Dual Polarisation Radar Observations* (1985) 229 SCIENCE 442 and Christopher Magri, Steven J. Ostro, Keith D. Rosema, Michael L. Thomas, David L. Mitchell, Donald B. Campbell, John F. Chandler, Irwin I. Shapiro, Jon D. Giorgini and Donald K. Yeomans, *Mainbelt Asteroids: Results of Arecibo and Goldstone Radar Observations of 37 Objects During 1980–1995* (1999) 140 ICARUS 379.

[150] Thomas D. Jones, Larry A. Lebofsky, John S. Lewis and Mark S. Marley, *The Composition and Origin of the C, P and D Asteroids: Water as a Tracer of Thermal Evolution in the Outer Belt* (1990) 88 ICARUS 172.

[151] Rivkin, Britt, Howell and Lebofsky, supra note 149.

[152] Michael J. Gaffey, Jeffrey F. Bell and Dale P. Cruikshank, *Reflectance Spectroscopy and Asteroid Surface Mineralogy* in Binzel, Gehrels and Matthews, supra note 133, at 98–127.

[153] Michael J. Gaffey, Kevin L. Reed and Michael S. Kelley, *Relationship of E-Type Asteroid 3103 (1982BB) to the Enstatite Achondrite Meteorites and the Hungaria Asteroids* (1992) 100 ICARUS 95.

2.4.3.5 The V Class

The asteroid 4 Vesta, being a V Class asteroid, possesses similar spectra to basaltic achondritic meteorites, such as pyroxene and olivine.[154] Other asteroids in similar orbits have been found to have similar spectra to Vesta, along with three Near Earth Asteroids.[155] The asteroid 3628 Božnêmcová has similar spectra to ordinary chondritis, showing presence of pyroxene, olivine and compounds of nickel and iron.[156] Such v class asteroids are good sources for metals and silicon that, along with other asteroid classes discussed above, show an abundance of metallic compounds throughout the majority of the asteroid population that make them suitable candidates for mineral exploitation.

2.4.4 Suitability of Near Earth Asteroids

Among the large population of asteroids in the Solar System, some groups of asteroids, namely the Apollo, Amor and Aten groups, are categorised as the Near Earth Asteroids. These asteroids are generally very small members of the Solar System with diameters of around 1 km or more. They vary from being metallic or silicate to some form of carbonaceous chondrite that may be hydrated.[157]

The Apollos are asteroids that actually cross the heliocentric orbit of the Earth and therefore represent a collision threat to the Earth. They are considered to be the origins of some recent meteor showers.[158] The Amors do not cross the Earth's orbit but approach it to within 0.3 AU, posing a potential threat to the Earth if gravitational perturbations cause their orbits to migrate across the orbit of the Earth.[159] Atens are asteroids of which the large parts of their orbits are inside the orbit of the Earth, with the semi major axes of their orbits shorter than that of the Earth. As they cross the orbit of the Earth, they also represent a collision threat to the Earth.

The proximity of Near Earth Asteroids to the orbit of the Earth is also their greatest asset (Fig. 2.10). Most of the Near Earth Asteroids are very easy to reach

---

[154] Michael A. Feierberg, Howard P. Larsen, Uwe Fink and Howard A. Smith, *Spectrascopic Evidence for Two Achondrite Parent Bodies: Asteroids, 349 Dembowska and 4 Vesta* (1980) 45 GEOCHIM. & COSMOCHIM. ACTA 971.

[155] Richard P. Binzel and Shui Xu, *Chips Off of Asteroid 4 Vesta: Evidence for the Parent Body of Basaltic Achondrite Meteorites* (1993) 260 SCIENCE 186.

[156] Richard P. Binzel, Shui Xu, Schelte J. Bus, Michael F. Skrutskie, Michael R. Meyer, Patricia M. Knezek and Edwin S. Barker, *Discovery of a Main Belt Asteroid Resembling Ordinary Chondrite Meteorites* (1993) 262 SCIENCE 1541.

[157] Sonter, supra note 107.

[158] David Morrison, *An International Program to Protect the Earth from Impact Catastrophe: Initial Steps* (1993) 30 ACTA ASTRONAUTICA 11 and Duncan I. Olsson-Steel, *Theoretical Meteor Radiants of Recently Discovered Asteroids, Comets and Twin Showers of Known Meteor Streams* [1988] AUST. J. ASTRON. 93.

[159] An astronomical unit (AU) is a measure of distance in space and is defined as length of the semi major axis of the heliocentric orbit of the Earth. One AU is defined as 149,597,870,691 km.

2.4 The Riches of Space 61

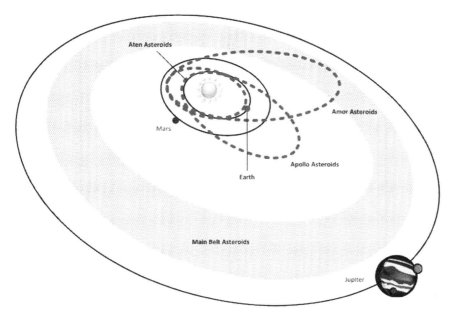

Fig. 2.10 The three groups of Near Earth Asteroids
Source: Charles L. Gerlach (2005, 5)

and return from when compared with missions to the Moon or Mars because of the low thrust required for departing from low Earth orbit and soft landings on the asteroids, as well as departing from the asteroids for the return trajectory back to Earth.[160] The spectral studies discussed above have shown that at least some asteroids have regoliths and have higher concentrations of free metals than the lunar surface, mostly in the form of olivine and pyroxene.[161] At least 20% of all Near Earth Asteroids are considered rich in volatiles and almost all asteroids are considered rich in metallic compounds.[162]

The geological and mineralogical data of asteroids are collected by a number of methods, including visible and infrared spectral studies by astronomers, comparisons with meteorite samples and radar studies from radio telescopes. Useful data

---

[160] Reinhold Bertrand and J. Foliard, *Low-Thrust Optimal Trajectories for Rendezvous with Near Earth Asteroids*, paper presented at the 18th International Symposium on Space Flight Dynamics, 11–15 October 2004, in Munich, Germany, at 2–3.

[161] Andrew S. Rivkin, Ellen S. Howell and Schelte J. Bus, *Diversity of Types of Hydrated Minerals on C-Class Asteroids*, paper presented at the 35th Lunar and Planetary Science Conference, 15–19 March 2004, in League City, TX, USA; Edward F. Tedesco, *Evidence of Hydrated Minerals in E and M Type Asteroids* (1995), paper presented at the 26th Lunar and Planetary Sciences Conference, March 1995, in Houston, TX, USA; and Jeffrey S. Kargel, *Metalliferous Asteroids as Potential Sources of Precious Metals* (1994) 99 J. GEOPHYS. RES. 21129.

[162] Lewis and Hutson, supra note 106, at 523–542.

have also been collected from the visual analysis by the flybys of 951 Gaspra and 243 Ida by the Galileo probe on its way to Jupiter and, more recently, from the flyby of 2867 Steins by the Rosetta probe on its journey to Comet Churyumov-Gerasimenko.[163] The recent observations of 253 Mathilde and 433 Eros by the Near Earth Asteroid Rendezvous probe launched by NASA have also enriched our knowledge about the physical composition of Near Earth Asteroids.[164]

Gaspra, Ida and Mathilde all appear to have a well-developed regolith despite their extremely weak gravity. Previous scientific speculation expected that the weak gravity would cause any original regolith to be lost over time, exposing the bare metal of the asteroid core.[165] Ida was even found to have a satellite, Dactyl, which also retains a regolith despite being less than 1 km in radius.[166] Recent radar reflection studies have shown that two Apollo asteroids, 3554 Amun and 1986DA, are composed of solid metal cores.[167]

In order to review the geological composition of asteroids beyond the study of reflective spectra, it is useful to study the classes and composition of meteorites. Meteorites are generally classified as Stones, Stony-Irons and Irons.[168] Stones comprise two classes: the chondrites and achondrites. Chondrites appear to be samples of the primitive Solar System accretion bodies with very little metamorphism or some minor aqueous alteration or metasomatism and are believed to be sourced from Main Belt Asteroids.[169] It is thought that they have not changed in composition since condensation from the pre-solar nebula that formed the present Solar

---

[163] Kowal, supra note 17, at 35–48.

[164] See generally National Aeronautics and Space Administration, *Near Earth Asteroid Rendezvous* (2001), at <http://nssdc.gsfc.nasa.gov/planetary/near.html>, last accessed on 23 December 2004; A. L. Whipple and P. J. Shelus, *The Orbital Dynamics of the NEAR Mission Asteroids* (1995) 27 BULL. AM. ASTRON. SOC. 1203; and Space Daily, *Rosetta Swings by Asteroid Steins 2867 on Route to Comet Churyumov*, Space Daily, 6 September 2008, at <http://www.spacedaily.com/reports/Rosetta_Swings_By_Asteroid_Steins_2867_On_Route_To_Comet_Churyumov_999.html>, last accessed on 7 September 2008.

[165] Robert J. Sullivan, Ronald Greeley, Robert T. Pappalardo, Erik Asphaug, Jeffrey M. Moore, David Morrison, Michael J. S. Belton, Michael Carr, Clark R. Chapman, Paul E. Geissler, Richard Greenberg, James Granahan, James W. Head III, Randolph L. Kirk, Alfred S. McEwen, Pascal Lee, Peter C. Thomas and Joseph Veverka, *Geology of 243 Ida* (1996) 120 ICARUS 119; and Ronald Greeley, Geoffrey C. Collins, Nicole A. Spaun, Robert J. Sullivan, Jeffrey M. Moore, David A. Senske, Randall B. Tufts, Torrence V. Johnson, Michael J. S. Belton and Kenneth L. Tanaka, *Geologic Mapping of Europa* (2000) 104 J. GEOPHYS. RES. 22559.

[166] Sonter, supra note 107.

[167] Steven J. Ostro, Donald B. Campbell, John F. Chandler, Alice A. Hine, R. Scott Hudson, Keith D. Rosema and Irwin I. Shapiro, *Asteroid 1986DA: Radar Evidence of a Metallic Composition* (1991) 252 SCIENCE 1399.

[168] See, for example, Philip A. Bland, F. J. Berry, A. J. Timothy Jull, T. B. Smith, Adrian W. R. Bevan, John M. Cadogan, Arabella S. Sexton, Ian A. Franchi and Colin T. Pillinger, *^{57}Fe Mössbauer Spectroscopy Studies of Meteorites: Implications for Weathering Rates, Meteorite Flux and Early Solar System Processes* (2003) 142 HYPERFINE INTERACTIONS 481.

[169] Derek W. G. Sears, *The Case for Asteroid Sample Return*, paper presented at the 61st Annual Meteoritical Society Meeting, 27–31 July 1998, in Dublin, Ireland, at 1.

System. Chondrites are further divided into enstatites, being ordinary chondrites that contain free metals, and carbonaceous chondrites that contain no free metals although extraction of volatiles by reduction of magnetite and other iron oxides is possible.[170] In contrast, the achondrites are igneous and basaltic rocks that were formed after being melted and as a result they tend to have dense surfaces with very little or no water. The Stony-Irons and Irons, on the other hand, are believed to be fragments of metallic cores of planetesimals that shattered from a cataclysmic impact of some form during the early formation of the Solar System.[171]

Of all classes of meteorites, ordinary chondrites are by far the most abundant and yet the only spectral matches, as discussed above, are apparently 1864 Apollo and 3628 Božnêmcová.[172] It has been suggested that the meteorites found on Earth may not correspond with the existing population of asteroids due to changes over time. If this is the case, there may be a higher abundance of volatiles in the asteroids than may be deduced from the proportion of chondrites in the meteorite population.[173]

As for carbonaceous chondrites, some of the materials contained in them are considered to be insoluble macromolecular materials, composed of clusters of condensed aromatic, heteroaromatic and hydroaromatic ring systems in clusters that have been cross-linked by short methylene chains, ethers, sulphides and biphenyls.[174] This makes them similar to kerogens from oil shales or low-volatile bituminous coals, resulting in a possibly useful source of hydrocarbons when the existing supplies on Earth become exhausted.[175] These kerogens ($C_{100}H_{71}N_3O_{12}S_2$ or $C_{100}H_{48}NO_{12}S_2$,) have been pyrolysed in various scientific studies and have been found to produce carbon monoxide, carbon dioxide, water, methane, ethane and propane.[176] As a result, asteroids that have similar composition to carbonaceous

---

[170] Lewis and Hutson, supra note 106.

[171] Sears, supra note 170.

[172] Jeffrey F. Bell, *Q-Class Asteroids and Ordinary Chondrites*, paper presented at the 26th Lunar and Planetary Sciences Conference, March 1995, in Houston, TX, USA.

[173] Michael J. Gaffey and Thomas B. McCord, *Mineralogical and Petrological Characterisations of Asteroid Surface Materials*, in Gehrels (ed.), ASTEROIDS (1979), at 688–723.

[174] Ryoichi Hayatsu, Satoshi Matsuoka, Robert G. Scott, Martin H. Studier and Edward Anders, *Origin of Organic Matter in the Early Solar System: The Organic Polymer in Carbonaceous Chondrites* (1977) 41 GEOCHIM. & COSMOCHIM. ACTA 1325; Max P. Bernstein, Jason P. Dworkin, Scott A. Sandford and Louis J. Allamandola, *Ultraviolet Irradiation of Naphthalene in H₂O Ice: Implications for Meteorites and Biogenesis* (2001) 36 METEOR. & PLANET. SCI. 351; and Kyoung Sook Kim and Jong Mann Yang, *Carbon Isotope Analyses of Individual Hydrocarbon Molecules in Bituminous Coal, Oil Shale and Murchison Meteorite* (1998) 15 J. ASTRON. & SPACE SCI. 163.

[175] John R. Cronin, Sandra Pizzarello and Dale P. Cruikshank, *Organic Matter in Carbonaceous Chondrites, Planetary Satellites, Asteroids and Comets*, in John F. Kerridge and Mildred Shapley Matthews (eds.), METEORITES IN THE EARLY SOLAR SYSTEM (1988).

[176] John F. Kerridge, *Carbon, Hydrogen and Nitrogen in Carbonaceous Chondrites: Abundances and Isotopic Composition in Bulk Samples* (1985) 49 GEOCHIM. & COSMOCHIM. ACTA 1707; François Robert and Samuel Epstein, *The Concentration and Isotopic Composition of Hydrogen, Carbon and Nitrogen in Carbonaceous Meteorites* (1982) 46 GEOCHIM. & COSMOCHIM. ACTA 81; François Robert and Samuel Epstein, *Carbon, Hydrogen and Nitrogen Isotopic Composition*

chondrites are very desirable candidates for the extraction of water and various hydrocarbons.

There are increasing indicia that many asteroids are either extinct or dormant comets, which is an observation often made because of the shape of their orbits or due to telescopic and spectroscopic evidence. Steel, for example, suggested that a sizable proportion of asteroids may be linked with the Taurid Complex.[177] The discovery of water in the form of solid ice makes the presence of hydrated volatiles on these types of asteroids a very likely possibility.[178]

Cometary bodies tend to be of high porosity because of the lack of compacting forces and the skeletal structure of the interstellar dust particles.[179] This porosity means that they have very low strength and can therefore be easily broken apart by impact or gravitational tidal forces. For example, the recent fragmentation of Comet Shoemaker-Levy 9 by the gravitational tides of Jupiter suggests that it was a zero-strength celestial body with very little compacting density.[180] These asteroids are considered to have various volatiles trapped at depth, including water ice, dry ice, ammonia and hydrogen cyanide, along with various silicates and hydrocarbons hidden under a layer of remnant non-volatile hydrocarbons and silicate detritus.[181] Although this may prove to be a rich source of hydrocarbons, mining such an object using conventional and terrestrial can be somewhat difficult.

The total number of discovered Near Earth Asteroids is now well above 500 and is increasing at about 50 asteroids per year.[182] The statistically projected number of Near Earth Asteroids with a diameter larger than 100 km is estimated to be several thousands.[183] Theoretically, this number of Near Earth Asteroids is being depleted by collisions with the inner planets and Jupiter, or by being pulled out of

---

of the Renazzo and Orgueil Organic Components (1980) 15 METEORITICS 355; Ram L. Levy, Michael A. Grayson and Clarence J. Wolf, *The Organic Analysis of the Murchison Meteorite* (1973) 37 GEOCHIM. & COSMOCHIM. ACTA 475; and Martin H. Studier, Ryoichi Hayatsu and Edward Anders, *Origin of Organic Matter in the Early Solar System: Further Studies of Meteoritic Hydrocarbons and a Discussion of Their Origin* (1972) 36 GEOCHIM. & COSMOCHIM. ACTA 189.

[177] Duncan Steel, ROGUE ASTEROIDS AND DOOMSDAY COMETS: THE SEARCH FOR THE MILLION MEGATON MENACE THAT THREATENS LIFE ON EARTH (1995).

[178] Ross, supra note 107.

[179] Daniel T. Britt, David A. Kring and James F. Bell, *The Density / Porosity of Asteroids*, paper presented at the 26th Lunar and Planetary Sciences Conference, March 1995, in Houston, TX, USA.

[180] Erik Asphaug and Willy Benz, *The Tidal Evolution of Strengthless Planetesimals* (1995), paper presented at the 26th Lunar and Planetary Sciences Conference, March 1995, in Houston, TX, USA.

[181] Dina Prialnik and Yuri Mekler, *The Formation of an Ice Crust below the Dust Mantle of a Cometary Nucleus* (1991) 366 ASTROPHYSICS J. 318.

[182] Smithsonian Astrophysical Observatory, <http://cfa-www.harvard.edu/cfa/ps/mpc.html>, last accessed in September 2001.

[183] John S. Lewis, *Logistical Implications of Water Extraction from Near Earth Asteroids* (1993), paper presented at the AIAA / SSI Space Manufacturing Conference, May 1993, in Princeton, NJ, USA.

2.4 The Riches of Space

the Solar System or towards the Sun because of gravitational perturbations caused by close encounters with Jupiter.[184] However, there must be some natural means of replenishing the Near Earth Asteroid population or otherwise there would be very few of them remaining after the creation of the Solar System.

This replenishment may result from the orbital decline or capture of comets and collisions between large asteroids in the Main Belt, which catapult fragments into the near Earth orbits. The comets, in turn, are replenished from objects in the Kuiper Belt. Examples of the effects of gravity of a large planet on the mechanics of comets are shown in Figs. 2.11 and 2.12.

In the first example, a comet that had a long period orbit around the Sun loses sufficient orbital energy by passing closely in front of a large planet such as Jupiter to become captured as a short period comet.[185] In the second example, as shown

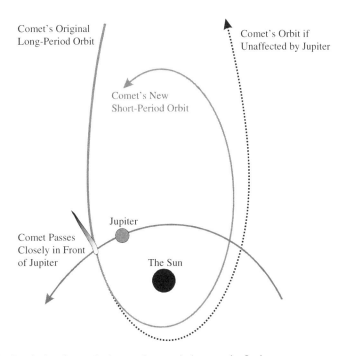

Fig. 2.11 Gravitational perturbations on long period comets by Jupiter

---

[184] Paolo Farinella, Christiane Froeschlé, Claude Froeschlé, Robert Gonczi, Gerhard Hahn, Alessandro Morbidelli and Giovanni B. Valsecchi, *Asteroids Falling into the Sun* (1994) 371 NATURE 314.

[185] See Micha Kamienski, *Researches on the Motion of Comet P/Wolf I: Perturbations Due to Venus, Earth, Mars, Jupiter, Saturn and Uranus During the Period 1950 October – 1959 March* (1957) 7 ACTA ASTRON. 5; Henri Debehogne and Ronaldo Rogério de Freitas Mourão, *Positions of Comet P/Ashbrook-Jackson in 1978 – A Note on Perturbations by Jupiter* (1979) 29 ACTA

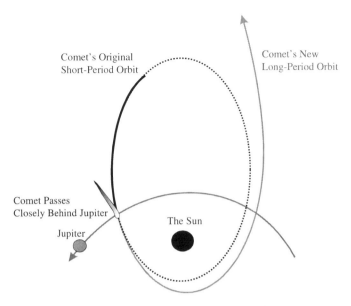

Fig. 2.12 Gravitational perturbations on short period comets by Jupiter

in Fig. 2.12, a short period orbit gains sufficient orbital energy by passing behind a large planet such as Jupiter to become a long period comet. Similar effects would be experienced by asteroids that pass close by a large planet, such as Jupiter, Saturn or even the outer planets, Uranus and Neptune.

2.4.5 Other Groups of Potential Mining Candidates

2.4.5.1 The Arjunas

In addition to the common groups of Near Earth Asteroids, being the Apollos, Amors and Atens, there is also a class of asteroids called the Arjunas that have nearly circular orbits, similar to that travelled by the Earth.[186]

The Arjunas have orbits of around 1 AU in radius with very low orbital eccentricity.[187] Their proximity to the Earth and their small size makes travel to and from the Arjunas a very energy-efficient exercise, provided that the orbital inclination of the

---

ASTRON. 301; Claude Froeschlé and Hans Rickman, *A Monte Carlo Investigation of Jovian Perturbations on Short-Period Comet Orbits* (1981) 46 ICARUS 400; and Klaus Scherer and Wolfram Neutsch, *Dynamical Evolution of Asteroidal Orbits* (1993) 9 ASTRON. GESELL. ABS. SER. 98.

[186] Brett J. Gladman, Patrick Michel and Christiane Froeschlé, *The Near-Earth Object Population* (2000) 146 ICARUS 176 at 176.

[187] Ibid.

2.4 The Riches of Space

Table 2.7 The Arjunas with low eccentricities

Name	E	i	q	Q
4581 Asclepius	0.357	4.9	0.66	1.39
1992BF	0.271	7.25	0.66	1.15
1991JY	0.295	49.00	0.67	1.23
1989UQ	0.265	1.3	0.67	1.16
1989UR	0.356	10.34	0.70	1.47
3554 Amun	0.281	23.4	0.70	1.25
2062 Aten	0.182	18.9	0.79	1.14
1982HR Orpheus	0.322	2.68	0.82	1.60
1991JW	0.118	8.7	0.915	1.161
1994UG	0.246	4.5	0.925	1.527
1991VG	0.049	1.5	0.975	1.077
1992JD	0.032	13.5	1.002	1.067
1993DA	0.094	12.4	0.85	1.02

Source: Minor Planet Centre, Harvard University (2001)

asteroid is minimal. This factor making them highly suitable candidates for asteroid capture and transfer to Earth orbit for exploitation (Table 2.7).

2.4.5.2 Coorbital Asteroids

It has long been a theoretical possibility that there are Near Earth Asteroids that, upon passing near the Earth, are now trapped in strange orbits near the Earth as they orbit the Sun.[188] In 1997, it was discovered that 3753 Cruithne, an Aten asteroid previously discovered in 1986, is indeed one such asteroid.[189] 3753 Cruithne does not orbit the Earth but instead is dragged by the Earth along its highly inclined orbit around the Sun. Since then, there have been other asteroids suspected of being in coorbital motion with the Earth, such as 3362 Khufu.[190] Although numbering only a handful, the fact that these asteroids tend to be close to the orbit of the Earth means that they are suitable candidates for early prospecting and exploitation.

2.4.5.3 Short Period Comets

As discussed above, asteroids that have cometary origins are attractive candidates for the exploitation of hydrocarbons and other volatiles. Similarly, short period comets may present themselves as appealing options for exploitation. Short period comets tend to be composed of water ice, dry ice, ammonia (NH_3),

---

[188] Fathi Namouni, Apostolos A. Christou and Carl D. Murray, *Coorbital Dynamics at Large Eccentricity and Inclination* (1999) 83 PHYS. REV. LTRS. 2506; and Apostolos A. Christou, *Coorbital Objects in the Main Asteroid Belt* (2000) 356 ASTRON. & ASTROPHYS. L71.

[189] Paul Wiegert, Kimmo Innanen and Seppo Mikkola, *Cruithne* (1997) 387 NATURE 685.

[190] Philip Ball, *Many Moons*, NATURE SCIENCE UPDATE, 1 October 1999, at <http://www.nature.com/nsu/991007/991007-2.html#1>, last accessed on 5 October 2001.

hydrogen cyanide (HCN), hydrocarbons and silicates.[191] In addition, cometary nuclei have been found through absorption spectroscopy to contain formaldehyde (H_2CO), methanol (CH_3OH), methane (CH_4) and various hydrocarbon aromatics.[192] Although generally not detected directly by spectroscopy, various nitrogen and sulphur species, such as ammonia and hydrogen sulphide (H_2S), have also been detected in cometary materials from time to time.[193]

It is possible to conduct mining activities on these comets in situ, but one main factor inhibiting such exploitation is the extreme eccentricities of their orbits. Cometary orbits tend to be very eccentric in character, as they have large orbital inclinations.[194] Since travelling to such comets can involve very long transit durations, fast perihelion missions for exploiting resources on these comets may be financially more advantageous, even though it would have a higher energy requirement than an aphelion mission.[195] This is particularly so considering the orbits of such asteroids may lie well beyond the orbits of Jupiter or even Saturn (Fig. 2.13).[196]

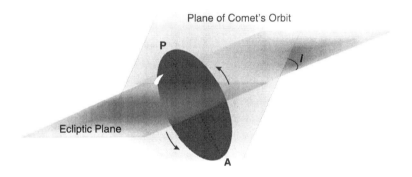

Fig. 2.13 Orbital geometry of short period comets
In this diagram, point P is the perihelion of the orbit, point A is the aphelion and angle *i* is the inclination of the orbit
Source: Drawn by the author

---

[191] Prialnik and Mekler, supra note 138.

[192] Jacques Crovisier, *Molecular Abundances in Comets*, in Milani, Di Mertino and Cellino (eds.), supra note 134, at 313–326 and Damien Hutsemékers, Jean Manfroid, Emmanuël Jehin, Claude Arpigny, Anita L. Cochran, Rita Schulz, Joachim A. Stüwe and Jean-Marc Zucconi, *Isotopic Abundance of Nitrogen and Carbon in Distant Comets* (2005) 432 ASTRON. & ASTROPHYS. 5.

[193] Ibid., at 319–320.

[194] Brian G. Marsden and Gareth V. Williams (eds.), CATALOGUE OF COMETARY ORBITS (1999).

[195] David L. Kuck and Stephen L. Gillett, *Extraterrestrial Resources: Implications from Terrestrial Experience* (1991) in University of Arizona, RESOURCES OF NEAR-EARTH SPACE: ABSTRACTS, at 11 and David L. Kuck, *Exploitation of Space Oases*, in Barbara Faughnan (ed.), SPACE MANUFACTURING PATHWAYS TO THE HIGH FRONTIER: PROCEEDINGS OF THE TWELFTH S.S.I.-PRINCETON CONFERENCE (1995), 136–156.

[196] Ibid.

2.5 Technical Feasibility of Space Mining

2.5.1 Orbital Mechanics

2.5.1.1 Orbital Geometry

Even with the limited knowledge that the scientific community currently possesses on the geology and mineralogy of asteroids, it is clear that the asteroids would nevertheless be attractive targets for future exploration and exploitation vis-à-vis Mars, its moons Phobos and Deimos and other planets. It is pertinent, therefore, to consider the technical challenges facing any serious attempt to achieve the goal of asteroid mining in the future.

The position and orientation of an orbit with reference to the rest of the Solar System is given by its semi major axis (a), the inclination (i) of the orbital plane to the plane of the orbit of the Earth (which is referred to as the ecliptic), the eccentricity of the orbit (e), the longitude of the ascending node (Ω) and the argument of the perihelion (ω).[197] These orbital properties are illustrated in Fig. 2.14.

The longitude of the ascending node of the orbit of an object is the angular distance, measured anticlockwise from the north, at the radius vector giving the Earth's position at vernal equinox to the position at which the object passes south of the ecliptic to north of the ecliptic. The argument of perihelion is the angular distance

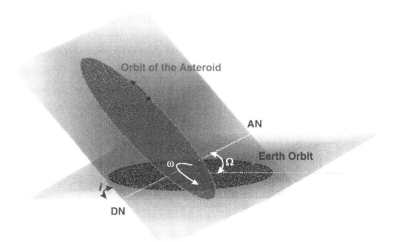

Fig. 2.14 Orbital properties of an asteroid
In this diagram, "AN" denotes the ascending node and "DN" denotes the descending node
Source: Drawn by the author

---

[197] Shane L. Larson, *Determination of Meteor Showers on Other Planets Using Comet Ephemerides* (2001) 121 THE ASTRO. J. 1722 at 1723.

70   2 Economic and Technical Prospects of Mining on Celestial Bodies

around the orbit of the object from its ascending node to its perihelion, measured in the direction of rotation.[198]

In order to travel from the Earth to an asteroid, or vice versa, two velocity changes ($\Delta \vec{v}$) are required. The first is needed to depart from one orbit to intersect another orbit on a transfer trajectory and a second $\Delta \vec{v}$ is required in order to slow down and rendezvous with the target orbit.[199] As a result, it is possible to measure and compare the accessibility of different asteroids by considering the total $\Delta \vec{v}$ required to travel from the Earth to an asteroid as well as the duration of time required in transit between the two.[200]

2.5.1.2 Calculating Energy Requirements

There are several ways of estimating the general velocity requirement for any particular target asteroid, without considering particular launch windows that may give favourable or unfavourable gravitational orbital conditions for the transit trajectory. Shoemaker and Helin presented a set of formulae for calculating the velocity change required to enter into a transfer orbit to a Near Earth Asteroid.[201] The formula for calculating the energy requirements is summarily given below as:[202]

$$\Delta \vec{v} = (30\ F + 0.5)$$

Where:

(1) F is called the "Figure of Merit", an intermediary used by Shoemaker and Helin and defined as:

$$F = U_L + U_R$$

where:

(a) U_L is the impulse required to inject the spacecraft into the transfer orbit from low Earth orbit; and
(b) U_R is the impulse required to rendezvous with the asteroid from the transfer orbit.[203]

Generally, low $\Delta \vec{v}$ trajectories are achieved by a rendezvous missions at, or near, aphelion or perihelion of the target object's orbit. As a result, the most energy

---

[198] Ibid.

[199] Mark J. Sonter, *Near Earth Objects as Resources for Space Industrialisation* (2001) 1:1 SOLAR SYSTEM DEV. J. 1 at 2–3.

[200] Ibid.

[201] Eugene M. Shoemaker and Eleanor F. Helin, *Earth-Approaching Asteroids: Populations, Origin and Compositional Types*, in David Morrison and William C. Wells (eds.), ASTEROIDS: AN EXPLORATION ASSESSMENT (1978), 163–175.

[202] Ibid.

[203] Ibid.

2.5 Technical Feasibility of Space Mining

efficient $\Delta \vec{v}$ missions to the Near Earth Asteroids are commonly achieved by rendezvous near the aphelion.

(2) With some minor qualifications, U_L is defined as:

$$U_L = \sqrt{C_3 + s^2} - U_0$$

where:

(a) s is the Earth's escape velocity, which is 11.2 km s^{-1}; and
(b) U_o is low Earth orbital velocity, which is 8.0 km s^{-1}.[204]

(3) The variable C_3 is the hyperbolic departure velocity squared, is defined as:

$$C_3 = 3 - \frac{2}{Q+1} - 2 \cdot \sqrt{\frac{2Q}{Q+1}} \cdot \cos \frac{i}{2}$$

where:

(a) Q is the aphelion of the asteroid in AU; and
(b) i is the inclination of its orbital plane to the ecliptic.[205]

(4) The required impulse at rendezvous U_R is given as:

$$U_R = \sqrt{U_C^2 - 2U_r \cdot U_c \cdot \cos \frac{i}{2} + U_r^2}$$

which, in short, is the impulse required to match the orbital speed of the target asteroid and then slow down to enter into an orbit around it.[206]

Shoemaker and Helin noted that the rendezvous impulse needed is generally very low for Apollo and Amor asteroids, usually around 1 km s^{-1}.[207] Generally, the outbound $\Delta \vec{v}$ and the inbound $\Delta \vec{v}$ are similar, assuming the launch windows resulting from the gravitational perturbations are ignored. About 10% of all known Near Earth Asteroids have a total $\Delta \vec{v}$ of less than, or equal to, 6 km s^{-1} with a reasonably narrow $\Delta \vec{v}$ launch opportunity occurring once every 2–3 years.[208]

---

[204] Ibid.

[205] Ibid.

[206] Ibid.

[207] Ibid.

[208] Christopher O. Lau and Neal D. Hulkower, *Accessibility of Near-Earth Asteroids* (1987) 10 J. GUID. CON. & DYN. 225 and Christopher O. Lau and Neal D. Hulkower, *On the Accessibility of Near Earth Asteroids*, paper presented at the AAS-AIAA Astrodynamics Conference, 12–15 August 1985, in Vail, CO, USA.

2.5.1.3 Hohmann Transfer Orbits

The most energy efficient transfer trajectories are elliptical transfers between coplanar orbits with similar eccentricity and an alignment of the semi major axis.[209] Of course, this is rarely the case as the two orbits would either not be coplanar or not be aligned along their axes.[210] As a result, the minimum energy transfer between two objects would generally take place at the ascending node or descending node at perihelion or aphelion, when it is possible to take advantage of the tangential momentum carried by the spacecraft at those points.[211] Generally speaking, the $\Delta\vec{v}$ required at aphelion would be less than that required at perihelion since less energy is needed to overcome the gravity of the Sun.

With a Hohmann transfer orbit, $\Delta\vec{v}$ is applied to a spacecraft when at perihelion to boost it to the higher orbit, where it will arrive near aphelion when a second $\Delta\vec{v}$ is applied in order for it to rendezvous with the asteroid.[212] Once the mining operations are complete, the inbound journey begins by applying $\Delta\vec{v}$ to the spacecraft so that it can enter into a transfer orbit, and thus reach the lower orbit of the Earth near the perihelion.[213] Such an orbit has been empirically determined to be the most energy efficient transfer trajectory between the Earth and other celestial bodies at a similar distance from the Sun.[214]

2.5.1.4 Timing Considerations

From the discussion above, there are several factors that influence the choice of trajectories to a Near Earth Asteroid. They are:

- the eccentricity of the orbit;
- whether the perihelion of the orbit of the asteroid is inside or outside the orbit of the Earth;
- the length of time required in transit;
- whether a Hohmann transfer trajectory is possible or a continuously thrusting trajectory would be required;
- the length of the possible mining season on the asteroid and whether repeated missions are possible; and
- whether the mining window is based on a rendezvous at aphelion or perihelion of the object.

---

[209] James Whiting, *Orbital Transfer Trajectory Optimisation* (2004), Massachusetts Institute of Technology, <http://ssl.mit.edu/publications/theses/SM-2004-WhitingJames.html>, last accessed on 23 December 2004.

[210] Ibid.

[211] Ibid.

[212] Ibid.

[213] Ibid.

[214] William Tyrrell Thomson, INTRODUCTION TO SPACE DYNAMICS (1986), at 70–71.

2.5 Technical Feasibility of Space Mining

One of the important considerations in designing trajectories is the synodic period of the asteroid. The synodic period of an object relative to the Earth is the time that elapses between similar configurations in the orbital positions of the objects, such as conjunctions or oppositions.[215] The synodic period for an Arjuna asteroid with a period of twenty months, for example, would be 60 months or 5 years. In other words, if both the Earth and an asteroid with a synodic period of 5 years were at aphelion in September 2001, then the next time this conjunction will occur again would be in September 2006. In designing the mission trajectory to an asteroid, it is important to consider its synodic period because the Earth would have to be at a particular point on its orbit when the spacecraft returns from its rendezvous with the asteroid. This extension of the project time required may possibly have a negative impact on the desirability of the mining operation. On the other hand, the longer mining season may allow for a greater return of mineral resources.

2.5.2 Mission Trajectories

2.5.2.1 Energy Cost of Mining Missions to Celestial Bodies

Generally, accessibility of celestial bodies for mining operations is defined based on energy requirements for the trajectories employed by the mission, expressed in terms of the required velocity to move a mass from one orbit to another. As detailed in Table 2.8, Near Earth Asteroids have relatively low energy requirements to reach them and return mineral resources from them to low Earth orbit, compared to even the surface of the Moon, for example.

Table 2.8 Energy requirements for various missions

Transfer	Energy requirements (km s^{-1})
Surface of the Earth to Low Earth Orbit	8.5
Surface of the Earth to escape velocity	11.2
Surface of the Earth to geostationary orbit	11.8
Low Earth Orbit to escape velocity	3.2
Low Earth Orbit to Mars transfer orbit	3.7
Low Earth Orbit to geostationary orbit	3.5
Low Earth Orbit to highly elliptical Earth orbit	2.5
Low Earth Orbit to landing on the Moon	6.3
Low Earth Orbit to typical Near Earth Asteroid	4.0
Surface of the Moon to Low Earth Orbit (with aerobraking)	2.4
Typical Near Earth Asteroid to Earth transfer orbit	1.0
Phobos/Deimos to Low Earth Orbit	8.0

Source: Charles L. Gerlach (2005, 6)

---

[215] Conjunctions are where the two objects are at the aphelion or perihelion at the same time, while oppositions is where one object is at perihelion and the other is at aphelion, or vice versa.

Gerlach suggested that the energy required to reach a typical Near Earth Asteroid may be less than placing a communications satellite in geostationary orbit.[216] Further, the energy required to place materials from such asteroids on an intercept trajectory with the Earth orbit may be far less than what is required to lift the mass into orbit from the surface of the Earth.[217] Lewis, Matthews and Guerrieri had estimated that 10% of all Near Earth Asteroids are more accessible in terms of the energy requirements than the Moon.[218] Sonter placed that figure as being 6%, applying the approach taken by Shoemaker and Helin.[219] Accordingly, the best targets for initial resource development, considering their relatively low energy requirements, are the Apollo, Amor and Aten asteroids with low eccentricity and low inclination orbits, along with any Trojan asteroids in the orbit of the Earth around the Sun.[220]

2.5.2.2 Apollo Asteroids

For asteroids with highly eccentric but low inclination orbits, mining operations would demand Hohmann transfer trajectories for both outbound and inbound flights because of the high $\Delta \vec{v}$ required. The mining season would therefore be restricted to a short season near aphelion. This type of trajectory is appropriate for the Apollo asteroids or Amor asteroids with comparatively high eccentricities. In terms of energy efficiency, a relatively large hyperbolic $\Delta \vec{v}$ has to be destroyed during the return trajectory, which is negligibly alleviated by a lunar flyby that would remove no more than 1.5 km s^{-1} from its velocity.[221]

This is complicated further by the need to ensure that the mission aligns with the synodic period of the asteroid, or the Earth will not be there when the spacecraft returns from its mining operation. This means that the length of the mission must be calculated in integer years, so that the Earth would be in the same orbital position. Otherwise, the trajectory would have to be adjusted, resulting in some additional use of energy.

2.5.2.3 Short Period Comets

For mining short period comets, it may be more appropriate to undertake the mining activities at perihelion rather than aphelion. This is principally for two reasons: firstly, since solar energy may be too weak at aphelion, the use of solar power may be limited or rendered impossible, and secondly, the transition time between the Earth and aphelion may impose too heavy a financial burden for commercial launchers

---

[216] Ibid., at 7.

[217] Ibid.

[218] Lewis, Matthews and Guerrieri, supra note 106.

[219] Sonter, supra note 200 and Eleanor F. Helin and Eugene M. Shoemaker, *Earth Approaching Asteroids as Targets for Exploration*, in Morrison and Wells, supra note 202.

[220] Ibid.

[221] Sonter, supra note 107.

and operators. The major disadvantage of having the mining season at perihelion is the requirement for a very high energy usage on the return transfer, which must be considered together with the very short mining season available at perihelion.

On the other hand, mining dormant comets with short periods can enable almost total capture of all volatiles, and the equipment and energy costs required for the melting of the cometary core will be significantly less when compared with the mining and processing of regolith on other asteroid types.[222] It may also be possible to move the comet gradually to the Earth orbit as the mass of the cometary nucleus reduces, facilitating the recovery of mineral resources from the comet.

2.5.2.4 Aten Asteroids

Aten asteroids have orbits with low eccentricities. Consequently, the mining mission best suited to them would involve a Hohmann transfer trajectory from low Earth orbit to rendezvous with the target asteroid at its aphelion and, after a long mining season, to depart the asteroid near perihelion. It is not possible to undertake a mission similar to that used for short period comets or Apollo asteroids due to the mining season being too short and the Earth would be out of phase when the spacecraft returned to the Earth.

The determination of whether the rendezvous should be at aphelion or perihelion would depend ultimately on the orbital mechanics of the asteroid and the proposed mining operations. However, in most circumstances, a departure from the asteroid at perihelion would be more energy efficient since the spacecraft would have significantly more mass during the inbound journey vis-à-vis the outbound journey. This would make it desirable to take advantage of the energy savings generated from a departure at perihelion.

On the other hand, the trajectory tends to require demanding propulsion systems and large amounts of fuel in this case (Fig. 2.15). The ideal scenario would be to offset these two characteristics in the selection of potential asteroids as mining targets. If Hohmann transfer trajectories are not used because of their constraints on the length of the mining season, then only asteroids with low orbital eccentricity should be selected to conserve the chemical fuel as well as the transit time required.

2.5.2.5 Arjuna and Amor Asteroids with Low Eccentricities

The Arjuna and some Amor asteroids have very circular orbits and are therefore favourable to spiral, rather than Hohmann, return trajectories. However, the long synodic period of these asteroids means that those with a semi major axis of around 1 AU, as is the case for the majority of these asteroids, would not be ideal candidates for mining. This is because they would move very little along their orbits relative to the Earth and therefore would not provide any ballistic opportunities for reduced

---

[222] David L. Kuck, *The Exploitation of Space Oases*, paper presented at the Princeton-AIAA Conference on Space Manufacturing, May 1995, in Princeton, NJ, USA.

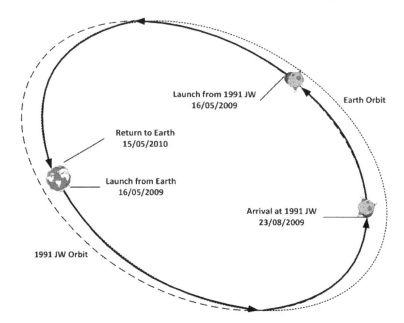

Fig. 2.15 Trajectory plan for a 2009 mining mission to asteroid 1991JW
Source: Mark J. Sonter (1998)

travel distance and time between the Earth and the asteroid.[223] Consequently, the exploitation of these asteroids should be undertaken only if the resources found to be available on the asteroid justify the expense and time involved in the operation.

2.5.3 Energy Requirements for the Mining and Processing of Ores

There are two main energy requirements in any mining venture anywhere: the first being the energy needed to separate the ore from the surrounding regolith and the second being the energy to convert the ore chemically into the desired minerals. The physical ore processing, or beneficiation, generally involves crushing and grinding the material into a fine powder that is then separated by magnetic or aqueous processes, thus improving the concentration of the ore by about 50%.

The chemical processing, on the other hand, uses heat, reactants and catalysts to change the state of the metal from a compound state to a free state. The energy

---

[223] John C. Niehoff, *Asteroid Mission Alternatives*, in Morrison and Wells, supra note 202, at 225–244; John C. Niehoff, *Round-Trip Mission Requirements for Asteroids 1976AA and 1973EC* (1977) 31 ICARUS 430; and Alfred C. Mascy and John C. Niehoff, *Sample Return Missions to the Asteroid Eros*, in Tom Gehrels (ed.), PHYSICAL STUDIES OF MINOR PLANETS: PROCEEDINGS OF INTERNATIONAL ASTRONOMICAL UNION COLLOQUIUM 12, March 1971 in Tucson, AZ, USA, at 513.

required includes the enthalpy (the minimum energy required for the chemical reaction to take place) and additional heat to increase the rate of the reaction. In most cases, this may involve melting the ore, thus significantly increasing the energy required. By undertaking most of the chemical processing in space, the mass of the returned materials is reduced, resulting in lower launch costs from the asteroid.

The total energy requirement for the mining and processing of mined ores from an asteroid can be represented as:

$$F = \frac{E_0}{g \cdot \eta_1} + \frac{\Delta H}{\eta_2}$$

where:

(1) E_0 is the energy needed for ore beneficiation;
(2) g is the "grade" or concentration of the metal in the ore;
(3) η_1 is the efficiency of the beneficiation process;
(4) ΔH is the enthalpy of the reaction during the smelting process; and
(5) η_2 is the efficiency of the smelting process.[224]

As Johnson pointed out, this in effect equates to requiring 500,000,000 joules of energy to produce 1 kg of free metal. As a comparison, 1 kg of methane contains nearly 57,000,000 J of energy.[225] Since the beneficiation process consumes the bulk of the energy, most of the energy source would have to be launched from the Earth and transported to the asteroid along with the mining plant and infrastructure, thus significantly increasing the costs of such a venture. On the other hand, the higher quality ore available on asteroids as compared to those found in the Earth's crust today means that the transportation costs may well be offset by the cost benefits derived from processing higher grade ore from the asteroids. Over time, the energy costs of mining in space will decrease while the corresponding costs of mining terrestrial resources will increase, making the mining of asteroids a comparatively more energy efficient exercise in meeting the increasing demand of mineral resources.

2.6 Exploratory Missions to Near Earth Asteroids

As with mining operations on the Earth, the mining of asteroids and other members of the Solar System requires prior exploration and prospecting. In the case of asteroids, this can only be done with unmanned spacecrafts to determine the mineralogy of potential mining candidates in the Solar System.

---

[224] Fred M. Johnson, *A Comparison of Energy Requirements between Terrestrial Metal Extraction and Recovery of Asteroid Metal Resources*, at <http://www.erie.net/~fjohnson/AsteroidPaper.html>, last accessed on 3 October 2001.

[225] Ibid.

2.6.1 Flyby Missions

In designing exploratory missions to asteroids, the types of missions are similar and analogous to missions concerning planetary exploration: flybys, orbiters, landers and sample return. In scientific writings, the term "rendezvous" is used often to describe missions where the spacecraft meets and then travels in close vicinity to a comet or a small asteroid where the gravity is generally too weak for the spacecraft to enter into the orbit at a safe distance from the surface. In practice, however, for most large asteroids the spacecraft would nevertheless orbit the asteroid.

Flyby missions to asteroids generally involve velocities of around 5 km s^{-1}, similar to the velocity of the asteroids themselves.[226] This type of mission is particularly useful for large asteroids, as the imaging and exploratory capabilities of flyby missions are relatively limited and would be inappropriate for small objects. For example, the *Voyager* exploration of the Galilean satellites suggested that a resolution of around 10 km is the minimum for useful surface geological data to be collected on a flyby mission.[227] For a camera with a focal length of 1,500 mm, this resolution would be reached at a range of 500,000 km. In a typical flyby mission of a large asteroid, this range would last about 2 days. However, as the relatively low-resolution images of 113 Amalthea taken by *Voyager 1* shows in Fig. 2.16, even at a resolution of 8 km, the images taken of a moderately large asteroid would have very limited mineralogical applications. As such, the practical benefits derived from such flyby missions in mineralogical studies are negligible.

As for the physical exploration of asteroids, flyby missions are very limited in their capabilities due to the speed at which the spacecraft would be travelling as well as the distance it would have to maintain from the asteroid. For a large asteroid, mass can be determined from the gravitational perturbation sustained by the spacecraft, which would be too negligible to detect on a small asteroid. Flyby missions are

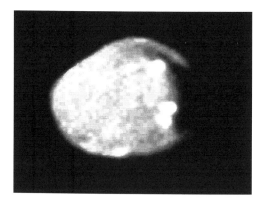

Fig. 2.16 Image of 113 Amalthea taken by *Voyager 1* *Source*: National Aeronautics and Space Administration (2003)

---

[226] Morrison and Niehoff, supra note 6, at 235.

[227] The Galilean satellites are Io, Europa, Ganymede and Callisto, the four largest satellites of Jupiter that were discovered by Galileo.

generally unable to carry out detailed chemical analyses of asteroids. For example, gamma ray spectroscopy can only be effectively carried out when the spacecraft is within one diameter of the asteroid. Depending on the size of the asteroid, this time can range from around 10 s for Eros to around 3 min for Vesta.[228] This is undoubtedly the major deficiency in flyby missions to the asteroids for exploration or exploitation purposes.

The drawback of the limited mineralogical analysis capability of flyby missions is offset greatly by the ability of such missions to visit multiple asteroids.[229] Once the flyby spacecraft enters into a heliocentric orbit near the Main Belt, it is possible to fly past a multiple number of asteroids with the potential of flying by about four asteroids per orbit (Fig. 2.17).[230] Further, the capabilities of a flyby mission

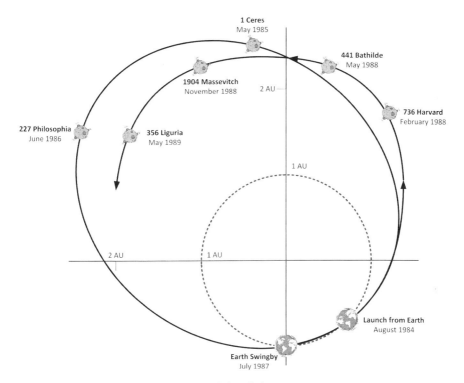

Fig. 2.17 Example trajectory for a multiple flyby mission
Source: John C. Niehoff (1977). Drawn by the author.

---

[228] Ibid., at 236.

[229] See, for example, Masaaki Matsumoto and Junichiro Kawaguchi, *Optimum Trajectory for Low-Thrust Multiple Trojan Asteroid Flybys* (2004), paper presented at the 18th International Symposium on Space Flight Dynamics, 11–15 October 2004 in Munich, Germany.

[230] David R. Brooks and William F. Hampshire II, *Multiple Asteroid Flyby Missions*, in Tom Gehrels (ed.), PHYSICAL STUDIES OF MINOR PLANETS (1981) at 527–537; and Niehoff, supra note 224.

would be increased significantly if the flyby velocity can be reduced. However, a rendezvous or orbiter mission remains a far more attractive option for detailed exploratory activities on such asteroids to take place.

2.6.2 Rendezvous and Lander Missions

In a rendezvous mission such as the Near Earth Asteroid Rendezvous, the spacecraft slows to match the speed of the target asteroid and then starts station keeping or enters into orbit around it.[231] Upon rendezvous, the length of time for exploratory activities on that asteroid is limited only by the life of the spacecraft itself. From an orbit of 1000 km around Vesta, for example, it is possible to map the entire asteroid at a resolution of a few metres, similar to the best pictures taken of Phobos and Deimos by the *Viking* spacecrafts.[232] At close distances, chemical analyses through spectroscopy can also be undertaken on the asteroid.

Another aspect of exploration missions to be considered is the use of landers. There are two general types of landers: hard landers that do not slow when landing on the asteroid, and soft landers that can carry out experiments after having landed softly on the surface. Hard landers are small objects that eventually crash into the asteroid and can provide valuable chemical and geological data in close proximity to the asteroid before impact.[233] Soft landers, on the other hand, are sophisticated spacecrafts that increase significantly in larger mass and higher complexity, but the resulting increases in costs and difficulties due to these factors are offset greatly by the detailed and complex analysis and experiments that can take place on the surface.[234] Such experiments include high-resolution imaging, microscopy and seismic soundings, and can prove invaluable for prospecting purposes.

2.6.3 Sample Return Missions

The capabilities of laboratories on the Earth for chemical and mineralogical analysis are far more powerful than anything a spacecraft robotic laboratory could muster, being limited by size, mass and available power. On the other hand, the cost of deploying a sample return mission is much higher than a flyby, rendezvous or lander

---

[231] Station keeping is where a spacecraft maintains position with a small asteroid which does not have a strong enough gravity to allow the spacecraft to orbit around it.

[232] Veverka and Thomas, supra note 125, at 628–651.

[233] See, for example, European Space Agency, *Not-So-Soft Landings on Other Worlds* (2001), at <http://www.esa.int/export/esaCP/ESAZXCZ84UC_FeatureWeek_0.html>, last accessed on 23 December 2004.

[234] Paul F. Wercinski, *Mars Sample Return – A Direct and Minimum-Risk Design* (1996) 33 J. SPACECRAFT & ROCKETS 381 and Blake Adams, *Mission Design Study for the 2003/2005 Mars Sample Return Mission* (2002), Colorado Centre for Astrodynamics Research, University of Colorado at Boulder, at <http://ccar.colorado.edu/asen5050/projects/projects_2002/adams/>, 12 December 2002, last accessed on 23 December 2004.

2.6 Exploratory Missions to Near Earth Asteroids 81

mission, thus limiting the potential scope of exploration of the asteroid. The lack of detailed prospecting information for most Near Earth Asteroids renders the selection of potential asteroid targets for such an expensive mission somewhat unfeasible.

One proposed mission involves an orbiter spacecraft weighing 600 kg to rendezvous with a suitable asteroid.[235] After orbital capture, the lander weighing about 300–500 kg would soft-land on the asteroid surface and collect a sample of 100 g in a 30 kg canister that is then returned to the orbiter, which subsequently returns to the Earth orbit for possible recovery by the space shuttle or other like vehicle.[236]

On 9 May 2003, the *Hayabusa* (MUSES-C) sample return mission from Japan was launched, with the target asteroid being 25143 Itokawa (1998SF36).[237] After being launched from a Japanese M-V 5 launch vehicle in Kagoshima, a low-thrust solar electric propulsion system was used to enter into a transfer orbit to Itokawa, arriving in June 2005.[238]

When *Hayabusa* actually landed on Itokawa on 23 November 2005, it had failed to collect a sample as planned prior to the landing, but it was still able to determine the mass of the asteroid.[239] The Japan Aerospace Exploration Agency was then able to improvise the possible means to collect a sample before *Hayabusa* left the surface of Itokawa and return the spacecraft to the Earth in 2010.[240] In the meantime, much data has been collected from the mission and much more has been analysed and learnt about the properties of Itokawa.[241]

---

[235] Timothy D. Jones, Duane B. Eppler, Donald R. Davis, Alan L. Friedlander, James V. McAdams and Sergei K. Krikalev, *Human Exploration of Near Earth Asteroids*, in Tom Gehrels, Mildred Shapley Matthews and Alan Schumann (eds.), HAZARDS DUE TO COMETS AND ASTEROIDS (1994) at 683.

[236] Ibid.

[237] Daniel J. Scheeres, Stephen B. Broschart, Steven J. Ostro and Lance A. M. Benner, *The Dynamic Environment about Asteroid 25143 Itokawa: Target of the Hayabusa Mission*, paper presented at the AIAA/AAS Astrodynamics Specialist Conference and Exhibit, 16–19 August 2004, in Providence, RI, USA; Institute of Space and Astronautical Science, Hayabusa (2004) at Japan Aerospace Exploration Agency, <http://www.isas.jaxa.jp/e/enterp/missions/hayabusa/index. shtml>, at 231–232 and Akira Fujiwara, Tadashi Mukai, Junichiro Kawaguchi and Tono Uesugi, *Sample Return Mission to NEA: MUSES-C* (2000) 25 ADV. SPACE RES. 231 at 231–232.

[238] Akira Fujiwara, Junichiro Kawaguchi and Sho Sasaki, *Hayabusa Mission to Asteroid Itokawa: In-Situ Observation and Sample Return*, paper presented at the Workshop on Dust in Planetary Systems, 26–30 September 2005, in Hawaii, USA, at <http://www.lpi.usra.edu/meetings/dust2005/ pdf/4024.pdf>, last accessed on 28 January 2007.

[239] Makato Yoshikawa, *Mass of Asteroid (25143) Itokawa Determined by Hayabusa Spacecraft* (2006), paper presented at the 36th COSPAR Scientific Assembly, 16–23 July 2004 in Beijing, China and Will Knight and Maggie McKee, *Hayabusa Touched Asteroid Itokawa After All* (23 November 2005), NEW SCIENTIST, at <http://www.newscientist.com/article.ns?id=dn8362>, last accessed on 27 January 2007.

[240] Damian Carrington, *Spacecraft Snatches First Samples from Asteroid* (26 November 2005), NEW SCIENTIST, at <http://space.newscientist.com/article.ns?id=dn8380>, last accessed on 28 January 2007.

[241] Akira Fujiwara, Junichiro Kawaguchi, Donald K. Yeomans, Masanao Abe, Tadashi Mukai, Tomomi Okada, Jun Saito, Hajime Yano, Makato Yoshikawa, Daniel J. Scheeres, Olivier S. Barnouin-Jha, Andrew F. Cheng, Hirohide Demura, Robert W. Gaskell, Naru Hirata, Hiroaki

82   2 Economic and Technical Prospects of Mining on Celestial Bodies

The combined effects and results from these different types of exploratory missions will provide a greater understanding of the geological and mineralogical composition of the Near Earth Asteroids with a view to future exploitation. At the present pace of technological development and resource consumption, it is only a matter of time before asteroids of suitable mineralogical content are found for possible human exploitation. When that happens, the only remaining hurdles for the human exploitation of mineral resources in space may be the legal and regulatory obstacles imposed by the existing body of international space law.

2.7 Commercial Feasibility of Space Mining

2.7.1 Advantages of Mining Near Earth Asteroids

Gerlach and Sonter outlined the potential benefits from the mining of Near Earth Asteroids, as adumbrated in Table 2.9. It is worth noting that Gerlach and Sonter also indicated that, in addition to these advantages, mining of asteroids would also benefit from the absence of existing landowners, the need to acquire mining rights, no constraints resulting from environmental conservation and no concerns over disposal of waste products. These factors, along with the physical and mineralogical characteristics of Near Earth Asteroids in general, make them ideal candidates for mineral exploitation. However, as discussed in greater detail later in this monograph, the existing international legal principles of international space law do not accord with these presumed benefits of asteroid mining and, accordingly, are not included in Table 2.9.[242]

Ikeda, T. Kominato, Hideaki Miyamoto, Akiko M. Nakamura, Ryosuke Nakamura, Sho Sasaki and Kentaro Uesugi, *The Rubble-Pile Asteroid Itokawa as Observed by Hayabusa* (2006) 312 SCIENCE 1330; Shinsuke Abe, Tadashi Mukai, Naru Hirata, Olivier S. Barnouin-Jha, Andrew F. Cheng, Hirohide Demura, Robert W. Gaskell, Tatsuaki Hashimoto, Kensuke Hiraoka, Takayuki Honda, Takashi Kubota, Masatoshi Matsuoka, Takahide Mizuno, Ryosuke Nakamura, Daniel J. Scheeres and Makato Yoshikawa, *Mass and Local Topography Measurements of Itokawa by Hayabusa* (2006) 312 SCIENCE 1344; and Akira Fujiwara, Junichiro Kawaguchi, Kentaro Uesugi, Donald K. Yeomans, Jun Saito, Masanao Abe, Tadashi Mukai, Mariko Kato, Tomomi Okada, Makato Yoshikawa, Hajime Yano, Hirohide Demura, Daniel J. Scheeres, Robert W. Gaskell, Olivier S. Barnouin-Jha, Andrew F. Cheng, Hideaki Miyamoto, Naru Hirata, Ryosuke Nakamura, Sho Sasaki and Akiko M. Nakamura, *Global Properties of 25143 Itokawa Observed by Hayabusa*, paper presented at the 37th Annual Lunar and Planetary Science Conference, 13–17 March 2006, in League City, TX, USA.

[242] See Carl Q. Christol, *Protection of Space from Environmental Harms* (1979) 4 ANN. AIR & SP. L. 433; Patricia M. Sterns and Leslie I. Tennen, *Privateering and Profiteering on the Moon and Other Celestial Bodies: Debunking the Myth of Property Rights in Space* (2003) 31 ADV. SPACE RES. 2433; and Ricky J. Lee and Deborah Roach, *The Importance of Private Property Rights for Selected Commercial Applications in Lunar and Martian Settlements*, paper presented at the 58th International Astronautical Congress, 2–6 October 2006, in Valencia, Spain.

2.7 Commercial Feasibility of Space Mining 83

Table 2.9 Potential benefits for mining of Near Earth Asteroids

Benefit	Description
Prospecting	High number of potential targets for mineral resource exploitation
Processing	Comparatively easier metallurgical extraction with higher grade ores
Time	Developing of mine sites can take relatively short time
Capital expenditure	Fewer large capital expenditure and the equipment may be leased
Scalability	Equipment can be mass produced and discarded after use
Flexibility	Feasibility hurdles are lower as mining operation can move from asteroid to asteroid to obtain higher grade ore
Reusability	Equipment can be relocated from asteroid to asteroid
Environment	Removes environmentally destructive mining activities from the Earth environment and ecosystem

Source: Charles L. Gerlach (2005) and Mark J. Sonter (1998).

2.7.2 Costing an Asteroid Mining Project

It is a commonly known fact throughout the space industry and beyond that, commercial space ventures carry a very high level of risk with long lead times and a heavy capital investment cost.[243] Asteroid mining ventures are no different: the mineral ores extracted from space can generate a substantial profit provided that the costs relating to extraction and marketing are less than the revenue generated from sales. Even if the asteroid mining venture is governmental instead of commercial, the government would generally require the operation to maximise its cost-benefit ratio and, consequently, the financial returns.

It is not difficult to anticipate the estimated costs involved in an asteroid mining venture, regardless of whether the venture is by its nature, private or governmental. Kargel had suggested that a typical asteroid mining venture could require a capitalisation of at least U.S. $100 billion, presumably at 1996 values when his analysis was conducted.[244]

Generally, the categories of costs in an asteroid mining venture are:

(1) *research and development costs*, being the costs involved in inventing, designing, constructing and testing new equipment, devices or propulsion systems as well as methodologies that are to be adopted for ore extraction, processing and transportation;

---

[243] Ricky J. Lee, *Costing and Financing a Commercial Asteroid Mining Venture*, paper presented at the 54th International Astronautical Congress, 1–3 October 2003, Bremen, Germany, at 3 and Fabian Eilingsfeld and Daniel Schaetzler, *The Cost of Capital for Space Tourism Ventures*, paper presented at the 51st International Astronautical Congress, 2–6 October 2000, Rio de Janeiro, Brazil, at 6.

[244] Jeffrey S. Kargel, *Market Value of Asteroidal Precious Metals in an Age of Diminishing Terrestrial Resources*, in Stewart W. Johnson (ed.), ENGINEERING, CONSTRUCTION AND OPERATIONS IN SPACE V: PROCEEDINGS OF THE FIFTH INTERNATIONAL CONFERENCE ON SPACE (1996).

(2) *exploration and prospecting costs*, being the steps necessary to undertake a feasibility study into a particular target object to ascertain with reasonable certainty the quality, type and location of ores that are available to be extracted;

(3) *construction and infrastructure development costs*, being the costs involved in drilling, blasting and hauling as well as the transportation costs for plant and equipment to the mine;

(4) *operational and engineering costs*, include costs such as salaries, consumables, fuel, maintenance, taxes, safety systems, surveying, modelling and the acquisition of professional services such as legal and accounting;

(5) *environmental costs*, being those involved in alleviating or minimising the environmental impact on the asteroid; and

(6) *time* as the opportunity cost of the capital invested in the asteroid mining venture would have earned in interest over the estimated duration of the venture if it was instead invested in a deposit account bearing the market rate of commercial lending interest.

In this example, there are seven major stages in the project throughout the estimated 12 years of the commercial mining venture on a Near Earth Asteroid in outer space, as illustrated in Table 2.10.

As Gertsch and Gertsch have suggested, even when compared with the cost of the Henderson Mine in Empire, Colorado, or the Channel Tunnel, this sample asteroid mining venture would have higher risks and a longer payback period than most large terrestrial projects.[245] For example, even a small failure during flight can terminate the entire operation, resulting in no return for the large investment made.

Table 2.10 Outline of platinum mining project milestones

Milestone	Year	Activities
Research and development	1–5	Development and testing of mining and processing equipment to be used on the asteroid
Exploration and prospecting	1–4	Determine mining needs, ore quality, type and location
Construction of infrastructure	2–5	This phase cannot start until exploration and prospecting is complete
Outbound	6–7	Including launch and transit
Mining and Processing	8	Processing to begin at the same time as the mining and extraction process
Inbound	9–11	Continue processing during flight
Recovery and sales	11–12	Completed as soon as possible for maximum return

Source: Jeffrey S. Kargel (1996)

---

[245] Ibid. The Henderson Mine began operation in 1976 after U.S. $500 million was invested in its development: Colorado School of Mines (2004), at <http://cause.mines.edu/media/UNO_Henderson.pdf>, last accessed on 24 July 2007. The Channel Tunnel cost over £10 billion and was completed in 1995: Ian Holliday, Gerard Marcou and Roger Vickerman, THE CHANNEL TUNNEL: PUBLIC POLICY, REGIONAL DEVELOPMENT AND EUROPEAN INTEGRATION (1991).

2.7.3 Determining Financial Feasibility

It has been noted that the parameters for comparing different asteroid mining concepts and mission alternatives are not well developed.[246] Sonter and Ross have suggested the use of net present value ("NPV") is the most appropriate "figure of merit" as the means of determining the financial feasibility of asteroid mining missions.[247] For an asteroid mining mission, the calculation of net present value would need to take into account:

(1) the cost to launch and conduct the mission;
(2) the mass returned and market value for the returned mass; and
(3) the time taken to complete the mission.[248]

Sonter suggested that the formula for determining the net present value of a given asteroid mining mission might be given expression as:

$$NPV = C_{orbit} \cdot M_{equip} \cdot f \cdot t \cdot r \cdot e^{-\Delta v/v_e} \cdot (1 + i)^{-a^{3/2}}$$
$$- \left[C_{op} \cdot \left(M_{equip} + M_{power} + M_{inst} \right) + B \cdot n \right]$$

Where:

(1) NPV is the net present value of the returns on investment;
(2) C_{orbit} is the launch cost from the surface of the Earth to the Earth orbit, expressed as dollar cost per kilogram;
(3) M_{equip} is the total mass of the mining and ore processing equipment, expressed in kilograms;
(4) f is the specific mass throughput ratio for the miner, expressed as kilograms of ores mined per kilogram of equipment per day;
(5) t is the mining period in days;
(6) r is the percentage recovery of valuable mineral from the ore;
(7) Δv is the change in velocity needed for the return trajectory, expressed as kilometres per second (km s^{-1});
(8) v_e is the propulsion exhaust velocity in kilometres per second (km s^{-1});
(9) i is the applicable interest rate on investment as available in the market, expressed as a percentage per annum;
(10) a is the semi-major axis of the transfer orbit, expressed in astronomical units;

---

[246] Knut I. Oxnevad, *An Investment Analysis Model for Space Mining Ventures* (1991), paper presented at the 42nd International Astronautical Congress, 5–11 October 1991, in Montréal, Canada and John S. Lewis, Kumar N. Ramohalli and Terry Triffet, *Extraterrestrial Resource Utilisation for Economy in Space Missions*, paper presented at the 41st International Astronautical Congress, 6–12 October 1990, in Dresden, Germany.

[247] Sonter, supra note 107 and Ross, supra note 107.

[248] Ibid.

86   2 Economic and Technical Prospects of Mining on Celestial Bodies

(11) M_{power} is the mass of the power supply, expressed in kilograms;
(12) M_{inst} is the mass of instrumentation and control systems;
(13) C_{op} is the specific cost of the mining operation;
(14) B is the annual operational budget of the project; and
(15) n is the number of years from launch of the mission to the delivery of mineral resources in low Earth orbit.[249]

It is apparent from the above equation that the per kilogram cost of launch from the surface of the Earth to low Earth orbit is not the most important determinant in a commercial asteroid mining operation, though usually this is the most crucial factor in assessing the financial feasibility of most space applications. It is the cost of the extraterrestrial mining operation itself, the energy cost of returning the mined ores to low Earth orbit, and the duration of years from the commencement of the project to the return of products that must be minimised. In this context, Ross opined that missions that take longer than around 3 years would need to have minimal costs and a good rate of return in order for the net present value of the returns to be positive.[250]

The expected net present value ("ENPV") provides a weighted measure of the NPV takes into account the probability of change in the variables factored into the calculation of NPV as well as the probability of failures. The formula for the ENPV of a given commercial asteroid mining operation would be expressed as:

$$ENPV = \sum_{j=1}^{s} p_j \cdot NPV_j$$

The determination of the probability of failure associated with each of the general and specific risks applicable to the mission is a complex and uncertain task. This is because the use of new and unproven space technology makes it difficult to ascertain the probability of failure associated with each risk. Further, the degree of risk varies at each phase of the mission, particularly where options that vary the relative risks are available at various phases of the mission to adapt to varying situations and to minimise any unfolding risk factors.[251]

2.7.4 Comparing Returns on Investment

For an experimental and high-risk proposition such as asteroid mining, proposals would need to create a very high rate of return on investment (the "ROI"), perhaps in excess of 30% per year, to compete successfully for the financial investment

---

[249] Ibid.

[250] Ross, supra note 107, at 19.

[251] See Andrew A. Gray, Payman Arabshahi, Elisabeth Lamassoure, Clayton Okino and Jason Andringa, *A Real Options Framework for Space Mission Design*, paper presented at the I.E.E.E. Aerospace Conference, 5–12 March 2005, in Big Sky, MT, USA.

2.7 Commercial Feasibility of Space Mining   87

needed.[252] In order to achieve set levels of ROI, or the income as a percentage of the investment after taking into account the projected costs and revenues as well as the project time, certain amounts of platinum metals, for example, must be recovered from the asteroid. Given that currently platinum is valued at around U.S. $1,340.00 per ounce as on 5 September 2008, a 10% return on investment can be achieved by recovering 5,575,000 ounces of platinum from 1,155,000 t of asteroid ore.[253] In order to achieve a 50% return on investment, 54,750,000 ounces of platinum would have to be recovered from an estimated 11,350,000 t of asteroid ore.[254] Clearly, the higher the rate of ROI required, the more economically, technologically and practically unfeasible the venture becomes.

The payback period and the high risk involved make such ventures somewhat uncompetitive vis-à-vis terrestrial projects. However, launching costs for the mining plant and equipment is a significant, if not the most significant, cost barrier that can only reduce over time with increased competition and improved technology. Further, the research and development costs involved in mining the second and subsequent asteroids would decrease, as firms would be able to utilise the existing technology available from previous ventures. Consequently, even though initial ventures may have to rely on governmental participation or implementation through combining with a scientific or water extraction operation, over time asteroid mining ventures will become increasingly viable through the continual reduction of fixed costs.

When considered together with increasing demand and prices for platinum and other platinum-group metals on Earth, the mining of platinum group metals may soon become an attractive commercial investment for space entrepreneurs. This is provided that the risks, being the most important factor in the calculation of the ENPV in the minds of financiers, can be substantially minimised.

2.7.5 Minimisation of Mission Risks

2.7.5.1 Overview

The paramount issue for financiers and venture capital firms in considering such a venture is the level of risk associated with the venture. The duration of time and the amount of initial investment are factors secondary to the issue of risk, otherwise most financiers would be willing to invest a substantial amount of funds provided that the returns exceed that generated from common "terrestrial" investments. With

---

[252] Ibid., at 17.

[253] Richard Gertsch and Leslie Gertsch, *Economic Analysis Tools for Mineral Projects in Space*, Colorado School of Mines, Golden, Colorado, <http://www.mines.edu/research/srr/rgertsch.pdf>, last accessed on 31 October 2001. The figures referred to above have been adjusted to reflect the price of platinum as on 5 September 2008. The authors pointed out that the cost and the time duration are very low estimates and based very much on an optimistic scenario. Current platinum prices can be found at <http://platinumprice.org>, last accessed on 8 September 2008.

[254] Ibid.

significant risks involved, however, the degree of risk must be taken into account when considering the sufficiency of the returns generated from the venture.

There are at least four broad categories of risk that may be generally and specifically associated with mining ventures in outer space as follows:

(1) *technical risks*, being those arising with the application of technology to the venture, especially unproven technology;
(2) *physical risks*, being those associated with scientific and physical uncertainty over the physical, geological and mineralogical nature and characteristics of the celestial body, such as an asteroid;
(3) *economic risks*, being those associated with the economic and market conditions of the mineral products to be recovered by the venture; and
(4) *political and legal risks*, being those associated with the political, legal and regulatory obstacles that would need to be overcome in order for the venture to successfully obtain financial investment and operate with minimal governmental interference.

2.7.5.2 Technical Risks

Traditionally, the technical risks are the greatest and most fundamental to any space mission, as unproven technological applications in space pose significant threat of failure, regardless of whether such failure takes place at the time of launch, orbital operations, landing or return. While risks associated with proven technologies, such as launches, can be minimised, the use of new technology on the surface of the Moon and asteroids will be untested and, unfortunately, computer simulations and scaled models would only partly minimise such technical risks.

The various technical risks specifically associated with a mining venture on celestial bodies in outer space include:

(1) failure during launch from the surface of the Earth;
(2) failure during transfer of robotic mining equipment from low Earth orbit into a transfer orbit to the celestial body;
(3) crashing and other accidents during landing and operations on the celestial body;
(4) system failure while the spacecrafts are in orbit around the Earth or in orbit around the celestial body;
(5) failure in launching containers from the surface of the celestial body to the orbiting spacecraft for return to low Earth orbit;
(6) system failure in the extraction and processing mechanisms on the surface of the celestial body; and
(7) electric power failures while in transit through outer space and on the surface of the celestial body.[255]

---

[255] Gerlach, supra note 137.

2.7 Commercial Feasibility of Space Mining

While these risks may be minimised through the application and use of proven technology and with rigorous testing and simulations, there remains significant threat of failure in the operation of these technical systems during a mining mission on a celestial body.

2.7.5.3 Physical Risks

Despite significant exploration and prospecting activities conducted at most potential mining sites on Earth, there are nevertheless significant risks associated with the uncertain knowledge of mineralogical and geological features of terrestrial mining operations.[256] The lack of precise knowledge in relation to ore concentrations, ore qualities and geological conditions at mining sites are risk factors that have to be considered when planning and financing mining activities on Earth. These risks increase substantially for mining activities that take place in remote areas, such as the Polar Regions and the deep seabed.

For a mining venture in outer space, the degree of risks associated with such uncertainties in relation to the physical characteristics of the celestial body increases. These risks may be summarised as:

(1) uncertainty over mineral compositions and concentrations, because without the collection and analysis of actual samples from the celestial body itself, any such evidence remains theoretical, speculative or circumstantial;
(2) lack of suitable and/or available operating sites for mining equipment;
(3) mechanical characteristics of the celestial body, such as size, shape, geology spin rate, spin state, orientation and angular momentum, that may affect the design and operation of the mining equipment and the amount of solar power available;[257] and
(4) errors and/or miscalculations in the consideration of physical characteristics and orbital mechanics of the celestial body and/or the trajectories to be used.

These physical risks may be minimised through detailed planning as well as exploratory missions to the targeted celestial bodies to provide mapping imagery and data from flyby and orbital missions and to obtain geological and mineralogical samples from these celestial bodies by lander missions. However, such missions are costly and, while laboratory simulations may be conducted, it would

---

[256] See, for example, Ian Lerche, GEOLOGICAL RISK AND UNCERTAINTY IN OIL EXPLORATION: UNCERTAINTY, RISK AND STRATEGY (1997) and Paul D. Newendorp and John R. Schuyler, DECISION ANALYSIS FOR PETROLEUM EXPLORATION (2nd ed., 2000).

[257] Derek W. G. Sears and Daniel J. Scheeres, *Asteroid Constraints on Multiple Near-Earth Asteroid Sample Return* (2001) 36 METEOR. & PLANET. SCI. 186 and Derek W. G. Sears, Daniel T. Britt and Andrew F. Cheng, *Asteroid Sample Return: 433 Eros as an Example of Sample Site Selection* (2001) 36 METEOR. & PLANET. SCI. 30.

not significantly reduce the relevant risks.[258] Although some economic models presently exist for evaluating mining ventures with uncertainties over output yields, these risks are compounded by the difficulties of undertaking substantive exploration and prospecting activities on celestial bodies, such as Near Earth Asteroids.[259]

2.7.5.4 Economic Risks

The exploration, prospecting and development of a mine on the surface of the Earth usually takes a few years between a mining venture obtaining a mining lease or similar entitlement to the site and the production of mineral resources. For example, the Batu Hijau copper and gold deposit on the island of Sumbawa in Indonesia was discovered in 1990 but it was not until 2000 when commercial operation of the mine began.[260] This is the case even when precious minerals such as diamonds are involved, which were first discovered at the remote site of the EKATI Diamond Mine near Wekweti in the Northwest Territories of Canada in 1991. It was not until September 1998 when the first diamond was actually recovered from the mine.[261]

In a mining venture in outer space, the duration of time between exploration missions and the commitment of investment funds on the one hand, and the delivery of processed ores in the low Earth orbit or the surface of the Earth on the other, is even longer than that of terrestrial projects. For instance, Gertsch and Gertsch undertook the analysis of a commercial operation to mine an asteroid for platinum group metals at an estimated cost of around U.S. $5 billion and a duration of 12 years to completion, on a typical Near Earth Asteroid with approximately 150 parts per million in platinum group metals.[262]

In such a context, it is important to factor into account the risk that commodity prices and demands fall significantly from the time of exploration to actual production and delivery of the minerals. This may be the result of various factors including the discovery of substantial deposits on Earth, innovation and improvements in recycling and ore processing technologies, development of alternative substituting products or a decline in the demand for some of the secondary products. However, it is noteworthy that such economic factors may significantly reduce the returns

---

[258] Derek W. G. Sears, Paul H. Benoit, Steven W. S. McKeever, Dipankar P. K. Banerjee, Timothy A. Kral, Wesley E. Stites, L. A. Roe, Pamela E. Jansma and Glen S. Mattioli, *Investigation of Biological, Chemical and Physical Processes on and in Planetary Surfaces by Laboratory Simulation* (2002) 50 PLANET. SP. SCI. 821.

[259] Bardia Kamrad and Ricardo Ernst, *An Economic Model for Evaluating Mining and Manufacturing Ventures with Output Yield Uncertainty* (2001) 49 OPERATIONS RESEARCH 690.

[260] Newmont Mining Corporation, *Batu Hijau, Indonesia,* at <http://www.newmont.com/en/operations/indonesia/batuhijau/index.asp>, last accessed on 28 April 2007.

[261] B.H.P. Billiton Ltd., *BHP Billiton Diamonds: History,* at <http://ekati.bhpbilliton.com/repository/aboutMine/history.asp>, last accessed on 28 April 2007.

[262] Gertsch and Gertsch, supra note 254.

2.7 Commercial Feasibility of Space Mining

generated from the investment but they would not eliminate the returns generated from the investment, unlike most technical and physical factors.

2.7.5.5 Political Risks

Substantial mineral deposits found on the Earth are located in increasingly remote corners of the Earth and, accordingly, the legal and political risks associated with the location of the mining operation are becoming increasingly important. In a recent study in the United States relating to natural gas exploration and production, Hartley and Medlock identified the following relevant social and political risk factors in the establishment of such production operations on the surface of the Earth:

(1) government stability, being a measure of the State's ability to carry out its declared program and its ability to stay in office;
(2) investment receptiveness, being a measure of the State's attitude towards substantial external investments;
(3) internal conflict and tensions, being the existence and activities of armed opposition groups as well as the level of domestic ethnic, cultural and religious tensions and conflicts;
(4) corruption;
(5) rule of law; and
(6) bureaucratic and governmental regulation.[263]

In addition to the risk factors as adumbrated above, one would also consider the state and nature of the international and diplomatic relations of that State. In any event, it is apparent from the risk factors listed above that they are generally inapplicable to a mining venture in outer space, though they may need to be taken into account, at least peripherally, in terms of the corporate and financial structure of the mining venture. However, there are legal risks that must be considered as possible obstacles to mining ventures in outer space.

2.7.5.6 Legal Risks

In order to fully assess the feasibility of mining operations on celestial bodies, it is prudent to undertake a detailed study of the legal issues that apply to the exploration, extraction and exploitation segments of such a mining operation. These legal risk factors include:

---

[263] Peter Hartley and Kenneth B. Medlock, *Political and Economic Influences on the Future World Market for Natural Gas* (2005), at James A. Baker III Institute for Public Policy, Geopolitics of Gas Working Paper Series, at <http://www.rice.edu/energy/publications/docs/GAS_PoliticalEconomicInfluences.pdf>, last accessed on 28 April 2007.

(1) the provision of exclusive property rights for the mining operation;
(2) the right to extract and exploit mineral resources from celestial bodies;
(3) successful procurement of all necessary governmental and international approvals, licences and permits;
(4) potential contractual liabilities; and
(5) potential liability to third parties.

It is through a comprehensive analysis of these risk factors and any legal reforms made to address these risk factors, an attempt at which is undertaken later in this monograph, that a complete picture of risk for mining operations on celestial bodies in outer space can be painted.

2.7.6 Practical Implications of Risk Profiles

It is clear from the above analysis that, given the gravity of the abovementioned risks, it is highly unlikely that any prudent investor in the present day would be willing to finance a commercial mining venture in outer space. What must be considered in assessing these risks, however, is the effects that the passage of time has on most of these risks, as summarised in Table 2.11.

In particular, technical improvements and advancements in launch technologies, propulsion systems, power generation and robotics over time will alleviate the technical risks, in addition to providing significant cost savings. Continuing scientific exploration and studies of celestial bodies, along with improving observation techniques, will reduce the effects of the physical risks. The worsening scarcity of mineral resources on Earth amid rising demand from developing economies, along with increasing concern of the environmental impact of mining activities, will reduce the economic risks that are inherent in terrestrial mining projects in any event. Even political risks can be controlled with case-by-case studies of domestic and regional conditions collected empirically over time.

On the other hand, the passage of time will have negligible effects on the legal risks associated with commercial space mining ventures. Issues such as title and other property rights to mining operations and the extracted ores from celestial bodies are not discussion topics of high priority among intergovernmental institutions, despite being the subject of much debate in academic circles. Similarly, potential liabilities between contractors and to third parties and the need for governmental approvals will remain necessary, regardless of time.

Only through revisions and changes to the body of international and domestic law dealing with outer space and celestial bodies will some of the legal risks be reduced, minimised or eliminated. Therefore, it is important to consider in detail the legal issues associated with commercial space mining ventures and to address the possible need for a new legal framework for the commercial exploitation of mineral resources in outer space.

2.8 Conclusions

Table 2.11 Summary of mission risks and effects of the passage of time

Risks	Effects of the passage of time
Technical risks	Technical improvements and
Launch failure	advancements in launch,
Failure during transit from earth orbit to the celestial body	propulsion, mining and
Accidents when landing on the celestial body	power generation
System failures during orbital operation	technologies
System failures during mining operation	
Accidents when returning mineral ores to the Earth	
Power failures	
Physical risks	General scientific studies of
Uncertainty over mineral compositions and concentrations	those celestial bodies and
Lack of suitable or available sites for mining operations	other similar celestial bodies
Uncertainty and/or miscalculations in the physical and mechanical characteristics of the celestial body	
Uncertainty and/or miscalculations in orbital mechanics of the celestial body	
Uncertainty and/or miscalculations in trajectories	
Economic risks	Worsening scarcity of
Duration of time between exploration and exploitation	resources on Earth and
Fluctuations in demand	increasing demand for
Fluctuations in world prices	mineral resources from
Innovation and improvements in recycling technologies and substituting products	emerging economies
Decline in demand for secondary products	
Political risks	More empirical data available
Government instability	for the selection of the
Receptiveness to substantial foreign investment	appropriate state on a case
Domestic conflicts or tensions	by case basis
Corruption	
Rule of law	
Bureaucratic and governmental regulation	
Changes to trading rules for the relevant mineral resources	
Legal risks	Negligible
Title and other property rights to the mining operations	
Title and other property rights to the extracted ores and derived benefits	
Need for governmental approvals and permits	
Potential contractual liabilities	
Potential liability to third parties	

2.8 Conclusions

It is clear from the present study that the economic and technical conditions for space mining will eventuate in the near future as the next phase in the continual evolution of human civilisation. Further, as the production costs of mineral resources continue to increase due to of depleting reserves, there will eventually be a point where economic or physical exhaustion of resources on Earth takes place or when the continual exploitation of such resources on Earth becomes environmentally undesirable. When that occurs, there will be sufficient technological development to overcome the difficulties involved in extracting resources from the Near Earth Asteroids.

While technological advancements in spacecraft design and exploration in the Solar System continue to increase exponentially, the development of the law has not kept pace. As a result, legal issues in the advancement of commercial asteroid resources have replaced technical and economic issues as the most intractable hurdles to be overcome in the development of space mining. In the following chapters, these issues will be explored to provide a more detailed legal picture of the exploitation of natural resources from celestial bodies in outer space.

Chapter 3
State Responsibility and Liability
for Compliance with International Space Law

3.1 Introduction

Space law is a relatively new branch of public international law, with its origins traced back to no more than a theoretical concept in the 1930s.[1] Since the beginning of the Cold War and the "Space Race" between the United States and the Soviet Union, however, the rapid commercialisation of the Earth orbit has led many scholars and commentators to realise that the existing legal framework for international space law is incapable of dealing with the commercial development of outer space. The framers of the multilateral treaties in the 1960s had not envisaged that private and commercial satellites would orbit the world before the end of the century and deliver many of the services that human civilisation today now takes for granted. These activities, such as remote sensing, weather forecasting, direct television and radio broadcasting and telecommunications, have torn apart the thin fabric of the existing space law framework.

In considering the legal framework applicable to future space activities, such as the launch and transit segments of a commercial asteroid mining venture, it is important to resolve the existing legal issues arising from space commercialisation. Even though the current pace of commercialisation has not been hindered by the ambiguities and uncertainties of space law, it must be kept in mind that the investments required for a space mining venture are far larger than even the largest deployment of telecommunications satellite constellations to date.[2]

---

[1] The first published monograph on space law was published in 1932. See Vladimír Mandl, DAS WELTRAUMRECHT: EIN PROBLEM DER RAUMFAHRT (1932). See also Vladimír Kopal, *Vladimír Mandl – Founder of Space Law* (1968) 11 PROC. COLL. L. OUTER SP. 357; and Gerhard Reintanz, *Vladimír Mandl – The Father of Space Law* (1968) 11 PROC. COLL. L. OUTER SP. 362.

[2] For example, Iridium LLC expended over U.S. $6.5 billion in contracts for satellite design, launch, operations and maintenance: Sydney Finkelstein and Shade H. Sanford, *Learning from Corporate Mistakes: The Rise and Fall of Iridium* (2000) 29 ORGANISATIONAL DYNAMICS 138 and Martin Collins, *One World ... One Telephone: Iridium, One Look at the Making of a Global Age* (2005) 21 HIST. & TECH. 301.

R.J. Lee, *Law and Regulation of Commercial Mining of Minerals in Outer Space*,
Space Regulations Library 7, DOI 10.1007/978-94-007-2039-8_3,
© Springer Science+Business Media B.V. 2012

Indeed, a private commercial space mining venture may be comparable in size and cost to another human landing on the Moon, but without the financial resources and legal protection that was available to the United States Government at the time Neil Armstrong landed on the Moon in 1969. It has been estimated in 1996 that a typical asteroid mining venture would require a capitalisation of at least U.S. $100 billion, or U.S. $120 billion in 2005 values.[3] By comparison, the cost of deployment of the Iridium mobile communications satellite business was estimated at around U.S. $7 billion in 1995 or U.S. $8.6 billion in 2005 values.[4] The entire Apollo Program cost U.S. $25 billion between 1961 and 1972, or projected at U.S. $121.6 billion in 2005 values.[5] Private investment on such a scale can be feasible only with a substantial degree of certainty in the rights to explore, extract and exploit the mineral resources on celestial bodies. In this context, it is important to analyse the predominant legal issues from the existing principles of space law that affect the commercialisation of the space sector, before considering the legal issues that are peculiar to mining activities in outer space.

Firstly, important fundamental terms of the treaties were left vague and ambiguous during the Cold War because of the inability of the two antagonists, the Soviet Union and the United States, to agree on many issues at a detailed level. Not surprisingly, however, private and commercial interests often require relative legal certainty over their rights and liabilities before being able to obtain large-scale investments that commercial activity requires. As a result, these ambiguities created a climate of instability and uncertainty that has been detrimental to the solid development of a private and commercial space sector.

Secondly, the framers of the treaties did not envisage the multinational and commercial nature of the space industry that would evolve when formulating the rules relating to international responsibility and jurisdiction in space. It was believed at the time that the most likely proponent of space activities would be the governments of States or their agencies, a belief that held true for no more than about 20 years after the first General Assembly resolution containing definitive principles of international space law.[6] At the turn of the century, the commercial nature of the space sector has meant that the predominant players in the space industry today are large multinational firms or intergovernmental conglomerates. For example, the

---

[3] Jeffrey S. Kargel, *Market Value of Asteroidal Precious Metals in an Age of Diminishing Terrestrial Resources*, in Stewart W. Johnson (ed.), ENGINEERING, CONSTRUCTION AND OPERATIONS IN SPACE V: PROCEEDINGS OF THE FIFTH INTERNATIONAL CONFERENCE ON SPACE (1996). The value conversion was done based on the annual gross domestic product of the United States as determined by the USA Bureau of Economic Analysis, as this would be the more appropriate indicator than using the official annual consumer price index as calculated by the U.S. Bureau of Labor Statistics, as the latter would consider only consumer goods and, accordingly, is somewhat inappropriate for determining the cost of space missions. For reference, the latter would calculate U.S. $100 billion in 1996 to be U.S. $120 billion in 2005 values.

[4] See Collins, supra note 2.

[5] National Aeronautics and Space Administration, at <http://www1.jsc.nasa.gov/er/seh/apollo_program.pdf>, last accessed on 25 July 2007.

[6] General Assembly Resolution 1962 (XVIII).

3.1 Introduction   97

Sea Launch project involved a joint venture of aerospace and maritime firms from several developed countries and the International Space Station is an intergovernmental effort involving 16 States, for which some contracted various commercial entities to undertake the construction of vital components.[7] These contractors, in turn, are often themselves multinationals with complex ownership, financing and incorporation structures. The ability of the international legal framework to prescribe liability and jurisdiction for the activities of such entities in space is becoming increasingly troublesome, compelling many commercial entities and their governments to prescribe, clarify and limit their liabilities towards each other by private contract.[8] On the other hand, the liability of the launch proponents towards third parties has not been definitively determined by the relevant treaties or the international judicial and arbitral institutions.

Thirdly, there is a developing trend that the line dividing civilian and military activities in the aerospace industry is becoming blurred. Recent commercial activities in space have brought into question the legality of many activities already being conducted in space, such as the military use of civilian communications systems and remote sensing data from satellites. Although the United Nations has taken steps towards the codification of the principles relating to the specific applications of space, the legal rules relating to the military applications in space are no more refined than they were in the early 1960s, except for the continuing work of legal commentators to interpret such rules. The lack of clarity on this issue makes it difficult for any military involvement in any commercial space mining venture, either in the form of investment, personnel, intellectual property or even as a customer utilising the resources mined from outer space.

In order for commercial space mining ventures to take place, it is necessary for these legal issues to be considered. In particular, the creation of any new regulatory

---

[7] The commercial partners of the Sea Launch project are Boeing Commercial Space (USA) (40%), RSC-Energia (Russia) (25%), Akec Kvaerner (Norway) (20%) and SDO Yuzhnoye / PO Yuzhmash (Ukraine) (15%); the governmental partners of the International Space Station are Canada, Japan, Russia, the United States of America and the European Space Agency, of which the Member States are Austria, Belgium, Denmark, Finland, France, Germany, Greece, Ireland, Italy, Luxembourg, the Netherlands, Norway, Portugal, Spain, Sweden, Switzerland and the United Kingdom: European Space Agency, *All About E.S.A.* (2008), at <http://www.esa.int/SPECIALS/About_ESA/SEMW16ARR1F_0.html>, 12 September 2008, last accessed on 1 November 2008.

[8] The 1998 International Space Station Intergovernmental Agreement is an example of a treaty governing an intergovernmental activity; and the 2001 bilateral agreement between Australia and the Russian Federation is an example of a bilateral agreement providing for cooperation between the two States for the regulation of private space launch activities: the Agreement Among the Government of Canada, Governments of the Member States of the European Space Agency, the Government of Japan, the Government of the Russian Federation and the Government of the United States of America Concerning Cooperation on the Civil International Space Station, opened for signature on 29 January 1998, Temp. State Dep't No. 01-52, CTIA No. 10073.000 (entered into force on 27 March 2001) and the Agreement between the Government of Australia and the Government of the Russian Federation on Cooperation in the Field of Exploration and Use of Outer Space for Peaceful Purposes, opened for signature on 23 May 2001, [2004] A.T.S. 17 (entered into force on 12 July 2004).

framework must take into account the entire body of legal principles in space law in order to create a regime that embodies the spirit, if not the content, of all of the relevant legal principles.

3.2 Sources of Space Law

3.2.1 United Nations Space Treaties

In parallel with the advancements made into the final frontier of outer space, the creation of the present corpus of international space law is due substantially to the efforts made in the adoption of multilateral treaties. In particular, much of these efforts were made by the States within the international multilateral framework of the United Nations.

These treaties include:

(1) the Treaty Banning Nuclear Weapon Tests in the Atmosphere, in Outer Space and Under Water (the "Nuclear Test Ban Treaty");[9]
(2) the Treaty on Principles Governing the Activities of States in the Exploration and Use of Outer Space, including the Moon and other Celestial Bodies (the "Outer Space Treaty");[10]
(3) the Agreement on the Rescue of Astronauts, the Return of Astronauts and the Return of Objects Launched into Outer Space (the "Rescue Agreement");[11]
(4) the Convention on International Liability for Damage Caused by Space Objects (the "Liability Convention");[12]
(5) the Convention on Registration of Objects Launched into Outer Space (the "Registration Convention");[13] and

---

[9] Treaty Banning Nuclear Weapon Tests in the Atmosphere, in Outer Space and Under Water (the "Nuclear Test Ban Treaty"), opened for signature on 5 August 1963, 480 U.N.T.S. 43; 14 U.S.T. 1313 (entered into force on 10 October 1963).

[10] Treaty on Principles Governing the Activities of States in the Exploration and Use of Outer Space, including the Moon and other Celestial Bodies (the "Outer Space Treaty"), opened for signature on 27 January 1967, 610 U.N.T.S. 205; 18 U.S.T. 2410; T.I.A.S. 6347; 6 I.L.M. 386 (entered into force on 10 October 1967).

[11] Agreement on the Rescue of Astronauts, the Return of Astronauts and the Return of Objects Launched into Outer Space (the "Rescue Agreement"), opened for signature on 22 April 1968, 672 U.N.T.S. 119; T.I.A.S. 6599; 19 U.S.T. 7570; 1986 A.T.S. 8 (entered into force on 3 December 1968).

[12] Convention on International Liability for Damage Caused by Space Objects (the "Liability Convention"), opened for signature on 29 March 1972, 961 U.N.T.S. 187, 24 U.S.T. 2389, T.I.A.S. 7762; 1975 A.T.S. 5 (entered into force on 1 September 1972).

[13] Convention on Registration of Objects Launched into Outer Space (the "Registration Convention"), opened for signature on 14 January 1975, 1023 U.N.T.S. 15; T.I.A.S. 8480; 28 U.S.T. 695, (entered into force on 15 September 1976).

3.2 Sources of Space Law

(6) the Agreement Governing the Activities of States on the Moon and Other Celestial Bodies (the "Moon Agreement") (Fig. 3.1).[14]

3.2.1.1 Nuclear Test Ban Treaty

Although generally not referred to as one of the United Nations space treaties, the 1963 Nuclear Test Ban Treaty can be regarded as the first multilateral treaty to impose specific obligations on States in relation to outer space. Specifically, the Nuclear Test Ban Treaty requires States to prohibit and prevent the testing of nuclear weapons in space by their agencies and nationals.[15] This requirement is not repeated in other United Nations space treaties and thus only the parties to the Nuclear Test Ban Treaty are prohibited from testing such weapons in space.[16]

3.2.1.2 Outer Space Treaty

It was agreed from the early days of the workings of the Committee on the Peaceful Uses of Outer Space ("COPUOS") that the adoption of a treaty containing basic and general principles of space law was preferable at that time rather than a comprehensive legal code on space activities, similar to that eventually created under the 1982 Convention on the Law of the Sea.[17] The main reason for this was the need to adapt to constantly evolving space technologies and new space applications. As USA Secretary of State, Dean Rusk, stated at the time, the Outer Space Treaty is an "outstanding example of how law and political arrangements can keep pace with science and technology".[18] This is not a feature exclusive to the development of space law – a similar progressive approach was taken with international human

---

[14] Agreement Governing the Activities of States on the Moon and Other Celestial Bodies (the "Moon Agreement"), opened for signature on 18 December 1979, 1363 U.N.T.S. 3; 18 I.L.M. 1434, (entered into force on 11 July 1984).

[15] Nuclear Test Ban Treaty, Article I.

[16] Article IV(1) of the Outer Space Treaty prohibits only the deployment of weapons of mass destruction in space. Article IV(2) of the Outer Space Treaty provides for a complete demilitarisation of celestial bodies and thus effectively prevents the testing of nuclear weapons on celestial bodies.

[17] United Nations Convention on the Law of the Sea, opened for signature on 10 December 1982, 1833 U.N.T.S. 3; 21 I.L.M. 1261 (entered into force on 16 November 1994). To some extent, this debate continues today with the issue being one of four discussion topics of the 2004 Colloquium on the Law of Outer Space in Vancouver, Canada. See Natalia R. Malysheva, *General Convention on Space Law: Some Arguments for Elaboration* (2004) 47 PROC. COLL. L. OUTER SP. 254; Mimi Lytje, *Obstacles on the Way to a General Convention* (2004) 47 PROC. COLL. L. OUTER SP. 267; and Lotta Viikari, *Problems Related to Time in the Development of International Space Law* (2004) 47 PROC. COLL. L. Outer SP. 259.

[18] Dean Rusk, "Letter of Submittal from Secretary Rusk to President Johnson", 27 January 1967, in *Hearings on Treaty on Outer Space Before the Senate Committee on Foreign Relations* (1967), 90th Cong., 1st Sess., at 112.

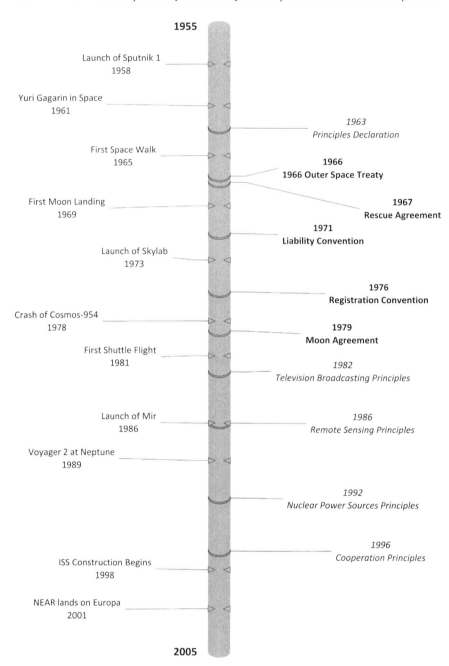

Fig. 3.1 Timeline of international space law instruments

3.2 Sources of Space Law
101

rights instruments.[19] It is in this context that the United Nations adopted the Outer Space Treaty in 1967.

There are some features of the United Nations space treaties that were considered groundbreaking at the time.[20] The Outer Space Treaty, for example, was one of the first multilateral treaties that was open to signature by all States and not only those that were Member States of the United Nations.[21] The space treaties also give explicit access to international organisations, such as the European Space Agency, to "accept" and fall within the scope of the provisions contained in the treaties.[22] The inclusion of review clauses in the later treaties, namely the Registration Convention and the Moon Agreement, make it possible for the international legal community to revise the legal requirements in light of technological and political requirements.[23]

The first document to be submitted by Member States to the General Assembly of the United Nations for consideration was a draft instrument containing the basic principles of space law that was originally proposed by the Soviet Union in 1962.[24] This was followed by draft declarations submitted to the General Assembly by the United Kingdom[25] and the United States.[26] When referred to the Legal Sub-Committee of COPUOS for consideration, the United Arab Republic (now Syria and Egypt) also submitted a draft code of international cooperation in space. However, no consensus was reached at the time on any of the instruments submitted, especially as the issues addressed by the proposed texts were different in both nature and content.[27]

After lengthy discussions, COPUOS was finally able to reach agreement on a text for the Declaration on Legal Principles Governing the Activities of States in the Exploration and Use of Outer Space (the "Principles Declaration"), which was

---

[19] Since the Universal Declaration of Human Rights of 1950, there has been a significant number of international legal instruments dealing with various issues of human rights, such as colonialism, racial discrimination, children, education, religious tolerance, women and slavery: see United Nations Office of the High Commissioner for Human Rights, *International Human Rights Instruments* (2007), at <http://www.unhchr.ch/html/intlinst.htm>, last accessed on 28 January 2007.

[20] Nandasiri Jasentuliyana and Roy S. Lee, MANUAL ON SPACE LAW (1979), vol. III, at xxiv–xxvi.

[21] Outer Space Treaty, Article XIV. Similar provisions can be found in the Rescue Agreement, Article 7; Liability Convention, Article XXIV, Registration Convention, Article VIII and Moon Agreement, Article 19. To some extent, this assertion is difficult to sustain, with Switzerland's accession to the Statute of the International Court of Justice without being a full member of the United Nations.

[22] Outer Space Treaty, Article XIII. See also Rescue Agreement, Article 6; Liability Convention, Article VII; Registration Convention, Article VII; and Moon Agreement, Article 16.

[23] Registration Convention, Article X and Moon Agreement, Article 18.

[24] U.N.Doc. A/AC.105/L.2.

[25] U.N.Doc. A/C.1/879.

[26] U.N.Doc. A/C.1/881.

[27] U.N.Doc. A/AC.105/6.

adopted by the General Assembly on 13 December 1963.[28] On 16 June 1965, the Soviet Union submitted to COPUOS a draft treaty embodying the basic principles contained in the declaration.[29] This was followed by a proposal from the United States containing a treaty for a similar purpose.[30] After consultations over the following years, the Outer Space Treaty was adopted by the United Nations General Assembly on 19 December 1966.[31]

The fundamental legal principles contained in the Outer Space Treaty include:

(1) the freedom of exploration and use of outer space and celestial bodies by all States on a non-discriminatory basis;[32]
(2) the prohibition of national appropriation of outer space and celestial bodies by claim of sovereignty, use, occupation or by any other means;[33]
(3) the application of international law, especially the Charter of the United Nations, to space activities;[34]
(4) the complete demilitarisation of celestial bodies and the prohibition on the deployment of weapons of mass destruction in outer space;[35]
(5) the requirement of States to render assistance to astronauts in distress and repatriate foreign astronauts and space objects found in their territories;[36]
(6) international responsibility of States for "national" space activities and their liability for injury, loss and damage caused to other States;[37]
(7) the jurisdiction and control over the space object by a State through placement of a space object;[38]
(8) the requirement that space activities must be conducted with due regard to the interests of other States and potential harmful interference in the activities of other States is to be avoided;[39] and
(9) the requirement that States are to avoid harmful contamination of the Earth and any adverse changes to the environment of the Earth by the introduction of any extraterrestrial matter.[40]

---

[28] General Assembly Resolution 1962 (XVIII).

[29] U.N.Doc. A/AC.105/C.2/L.13.

[30] U.N.Doc. A/AC.105/C.2/L.12.

[31] General Assembly Resolution 2222 (XXI).

[32] Outer Space Treaty, Article I.

[33] Ibid., Article II.

[34] Ibid., Article III.

[35] Ibid., Article IV.

[36] Ibid., Article V.

[37] Ibid., Articles VI and VII.

[38] Ibid., Article VIII.

[39] Ibid., Article IX.

[40] Ibid., Article IX.

3.2 Sources of Space Law

As at 1 January 2010, 100 States ratified the Outer Space Treaty and a further 26 States have signed it.[41] It has been noted that all States involved in space activities are party to the Outer Space Treaty and that at least some of its provisions are likely to have crystallised into customary international law. However, this is somewhat controversial due to the comparatively little state practice and *opinio juris* on space activities vis-à-vis other subject matters of international law.[42] The implications and effects of the provisions of the Outer Space Treaty as applied to the relevant aspects of commercial space mining ventures in outer space.

3.2.1.3 Rescue Agreement

By 1963, there was agreement among the members of the Legal Sub-Committee of COPUOS that legal principles concerning the rescue and return of astronauts and space objects should be contained in an international treaty, separate to the treaty being drafted at the time that later became the Outer Space Treaty. In 1964, the Legal Sub-Committee created a working group to consider the two draft instruments submitted by the Soviet Union[43] and the United States.[44] Several other States also submitted amendments to the two existing drafts and a new proposal was later jointly submitted by Australia and Canada, based on those discussions and amendments.[45] When two further drafts with some differences were submitted by Italy[46] and Argentina,[47] the Secretariat of the United Nations circulated a consolidated working draft.[48] After debates and several revisions, the text was submitted to the COPUOS and, following its approval, the General Assembly eventually adopted the Rescue Agreement on 19 December 1967.[49]

The Outer Space Treaty already requires all States to regard astronauts as "envoys of mankind in outer space" and render all possible assistance in the event of "accident, distress or emergency landing on the territory of another State Party or on the high seas".[50] This includes the obligation on astronauts in space to render "all possible assistance" to astronauts of other States, which is similar to the requirements of

---

[41] United Nations Office of Outer Space Affairs, *Status of International Agreements Relating to Activities in Outer Space as at 1 January 2010*, 1 January 2010, at <http://www.unoosa.org/pdf/publications/ST_SPACE_11_Rev2_Add3E.pdf>, last accessed on 22 April 2010.

[42] Bin Cheng, *The 1967 Outer Space Treaty: Thirtieth Anniversary* (1998) 23 AIR & SP. L. 156.

[43] U.N.Doc. A/AC.105/C.2/L.2.

[44] U.N.Doc. A/AC.105/C.2/L.9.

[45] U.N.Doc. A/AC.105/C.2/L.2.

[46] U.N.Doc. A/AC.105/C.2/L.21.

[47] U.N.Doc. A/AC.105/C.2/L.23.

[48] U.N.Doc. A/AC.105/C.2/L.28.

[49] General Assembly Resolution 2345 (XXII).

[50] Outer Space Treaty, Article V. It is curious to note that such an obligation is not repeated in the provisions of the Rescue Agreement.

104   3 State Responsibility and Liability for Compliance with International Space Law

international maritime law to render assistance to other vessels in distress at sea.[51] The Rescue Agreement expands on these requirements and provides that:

(1) States are required to notify the launching authority and the United Nations or, if the launching authority cannot be identified, announce publicly any information or discovery of any accident, distress or emergency landing suffered by astronauts onboard a spacecraft;[52]
(2) if astronauts have made an emergency landing within the territory of a State, that State is required to take all possible steps to rescue them, provide all necessary assistance and to return them safely and promptly to the launching authority;[53] and
(3) if astronauts onboard a spacecraft have alighted in the high seas, all States in a position to do so must extend assistance in search and rescue operations and, if they are discovered within or moved into their territory, to return them to the launching authority.[54]

The term "launching authority" refers to the State that was responsible for the launch. In the case of an international organisation, "launching authority" refers to that organisation provided that it has accepted the rights and obligations of the Rescue Agreement and a majority of its members are party to the Outer Space Treaty.[55] In addition to astronauts, the Rescue Agreement also imposes specific obligations on States in relation to returned space objects:

(1) States are required to notify the launching authority and the Secretary-General of the United Nations of any information or discovery of any returned space object or its component parts;[56]
(2) if a space object or one of its component parts has returned to the territory of a State, that State is required to recover the object or component part, if requested, and to return it to the launching authority;[57]
(3) if a space object or one of its component parts has been found by a State beyond the territorial sovereignty of the launching authority, that State is required to recover it and to return it to the launching authority;[58] and

---

[51] Convention on the Law of the Sea, Article 98.

[52] Rescue Agreement, Article 1.

[53] Ibid., Articles 2 and 4.

[54] Ibid., Articles 3 and 4.

[55] Ibid., Article 6. In the event of the launch being conducted by an international organisation that has not accepted the provisions of the Rescue Agreement, its Member States would presumably be regarded as the "launching authorities" for the purposes of the Rescue Agreement.

[56] Ibid., Article 5(1).

[57] Ibid., Article 5(2) and (3).

[58] Ibid., Article 5(3).

3.2 Sources of Space Law   105

(4) if the returned space object or one of its component parts is of a hazardous
    or deleterious nature, the State must notify the launching authority and to
    immediately take effective steps to eliminate possible dangers of harm.[59]

The only reference to the reimbursement of costs by the launching authority
is found in Article 5(5) of the Rescue Agreement, which relates only to expenses
incurred in recovering and returning the space object or its component parts. There
is no specific provision in the Rescue Agreement that provides for the reimbursement of costs incurred in the rescue and repatriation of astronauts or the costs of any
cleanup of hazardous materials or risks. It appears from the language of the provision that the reimbursement of the necessary costs is to be dealt with separate to the
liability for damage caused by the space object.[60]

As at 1 January 2010, 91 States ratified the Rescue Agreement and a further
24 States have signed it.[61] Similar to the provisions of the Outer Space Treaty,
some commentators have asserted that some, if not all, of the provisions of the
Rescue Agreement have crystallised into customary international law.[62] However,
this view may be difficult to sustain considering the absence of substantial state
practice relating to these principles.

3.2.1.4 Liability Convention

It is not surprising that one of the more fundamental issues considered by COPUOS
in its early days was the codification of principles relating to liability for damage caused by space objects. In 1962, the United States submitted a draft text
on liability for space vehicle accidents,[63] followed by separate proposals from
Belgium[64] and Hungary.[65] In successive years, different amendments and alternative proposals were made by Argentina, Australia, Canada, India and Japan. It
was not until 1971, after years of discussions and negotiations within the Legal
Sub-Committee, that a draft treaty was finally submitted to COPUOS and, subsequently, to the General Assembly for their consideration. The Liability Convention
was subsequently adopted by the General Assembly on 29 November 1971.[66]

---

[59] Ibid., Article 5(4).

[60] The Principles Relevant to the Use of Nuclear Power Sources in Outer Space (the "Nuclear
Power Sources Principles") establishes a contrary position; see the Nuclear Power Sources
Principles, Principle 9(3).

[61] United Nations Office of Outer Space Affairs, supra note 41.

[62] Vladlen S. Vereshchetin and Gennady M. Danilenko, *Custom as a Source of International Law
of Outer Space* (1985) 13 J. SP. L. 22 and Gennady M. Danilenko, *Outer Space and the Multilateral
Treaty-Making Process* (1989) 4 BERKELEY TECH. L. J. 217.

[63] U.N.Doc. A/AC.105/C.2/L.8.

[64] U.N.Doc. A/AC.105/C.2/L.7.

[65] U.N.Doc. A/AC.105/C.2/L.10.

[66] General Assembly Resolution 2777 (XXVI).

The Liability Convention was the first United Nations space law instrument to introduce the concept of a "launching State". Article I defines a "launching State" as a State that:

(1) launches the space object;
(2) procures the launching of a space object;
(3) provides the territory for the launch; or
(4) provides the facility for the launch.

The Liability Convention deals with the compensation payable by the launching State for any damage caused by the launch or attempted launch of a space object, including its component parts and launch vehicle.[67] Where there is more than one launching State, a possibility clearly intended from the definition of "launching State", they are jointly and severally liable for any damage caused and a launching State that has paid compensation may present claims for contribution from the other launching States, or they may develop agreements to apportion their liability at any time.[68] The Liability Convention excludes from its scope any damage caused to the nationals of the launching State and to foreign nationals invited to participate in the launch.[69]

In the case of damage caused by space objects on the surface of the Earth or to aircraft in flight, the launching State is absolutely liable.[70] Provided the launch conforms to international law, then the absolute liability of the launching State would be exonerated to the extent that the damage was wholly or partly the result of gross negligence or an intentional act or omission by the victim State.[71] Where damage is caused by space objects in outer space or on celestial bodies, compensation is payable to the extent of the fault of the launching State.[72] If two or more space objects were involved in an accident that subsequently caused damage to a third State, then the liability of their launching States to the third State for the damage caused is joint and several.[73] In any event, no limit is placed on the liability of States under these provisions.[74]

The Liability Convention provides for any claims for compensation to be made within 1 year of being aware of the damage through diplomatic channels

---

[67] Liability Convention, Article I.

[68] Ibid., Article V.

[69] Ibid., Article VII.

[70] Ibid., Article II.

[71] Ibid., Articles II and VI.

[72] Ibid., Article III.

[73] Ibid., Article IV.

[74] This position should be contrasted with those found in maritime law in the 1976 Convention on Limitation of Liability for Maritime Claims, opened for signature on 19 November 1976, 1456 U.N.T.S. 221 (entered into force on 1 December 1986); and in international air law in the 1929 Warsaw Convention for the Unification of Certain Rules Relating to International Carriage by Air, opened for signature on 12 October 1929, 137 L.N.T.S. 11; 1933 U.K.T.S. 11 (entered into force on 13 February 1933), Chapter III.

3.2 Sources of Space Law

or the United Nations, without the need to first exhaust any remedies available to the victim State through the domestic courts of the launching State.[75] If no settlement is reached through negotiations, either party may request for the establishment of a Claims Commission to determine the merits and quantum of the compensation payable.[76] However, the determination of the Claims Commission is binding only where the parties have so agreed, for otherwise the determination is a recommendatory award that the parties must consider in good faith.[77]

As at 1 January 2010, 88 States ratified the Liability Convention and a further 23 States have signed it.[78] The practice of States in prescribing their liability in domestic legislation may constitute state practice in assessing the extent of its crystallisation into customary international law. For instance, Australia, Belgium, Brazil, France, Hong Kong, Japan, the Netherlands, Norway, Republic of Korea, Russia, South Africa, Sweden, the Ukraine, the United Kingdom and the United States have all enacted domestic legislation on private space activities.[79] In almost all of these States, they have legislated to transfer their liability under the Outer Space Treaty and the Liability Convention to private launch operators.[80]

3.2.1.5 Registration Convention

The Outer Space Treaty provides that the State of registry, defined as the State on whose registry a space object is registered, is to retain jurisdiction and control over the space object in outer space.[81] However, the Outer Space Treaty does not contain any specific stipulations concerning the nature and content of the registers to be kept

---

[75] Liability Convention, Articles IX, X and XI.

[76] Ibid., Article XIV.

[77] Ibid., Article XIX(2).

[78] United Nations Office of Outer Space Affairs, supra note 41.

[79] See Australia: *Space Activities Act 1998* (Cth.) and *Space Activities Regulations 2001* (Cth.); Belgium: *Law on the Activities of Launching, Flight Operations and Guidance of Space Objects 2005*; Brazil: *Resolution on Commercial Launching Activities from Brazilian Territory* (Res. No. 51 of 26 January 2001) and *Regulation on Procedures and on Definition of Necessary Requirements for the Request, Evaluation, Issuance, Follow-up and Supervision of Licences for Carrying out Launching Space Activities on Brazilian Territory* (No. 27); France: *Space Operations Act* (No. 2008-518 of 3 June 2008) and *Decree on the Authorisations Issued in Accordance with French Act No. 2008-518 of 3 June 2008 Relating to Space Operations* (No. 2009-643); Hong Kong: *Outer Space Ordinance 1997* (No. 65 of 1997); Japan: *Basic Space Law* (No. 43 of 28 May 2008); the Netherlands: *Space Activities Act* (13 June 2006); Norway: *Act on Launching Objects from Norwegian Territory into Outer Space* (No. 38 of 13 June 1969); Republic of Korea: *Space Liability Act* (No. 8852 of 21 December 2007); Russia: *Law on Space Activities 1993* (Decree 5663-1) and *Statute on Licensing Space Operations* (No. 104); South Africa: *Space Affairs Act 1993*; Sweden: *Act on Space Activities* (1982:963) and *Decree on Space Activities* (1982:1069); Ukraine: *Ordinance of the Supreme Soviet of Ukraine on Space Activity* (15 November 1996); United Kingdom: *Outer Space Act 1986*; and United States: *Commercial Space Launch Act 1994* (49 U.S.C. 701) and *Commercial Space Transportation Regulations* (14 C.F.R. chap. III).

[80] See, for example, *Space Activities Act 1998* (Australia), Part 4.

[81] Registration Convention, Article VI.

or any specific right or obligation concerning the act of registering space objects launched into outer space. Accordingly, the determination of which State would rightfully be the "State of registry" for the purposes of the Outer Space Treaty can be a difficult task at the best of times.

In 1968, a draft convention dealing with the registration of space objects was submitted by France to the Legal Sub-Committee.[82] In furtherance of this endeavour, the Legal Sub-Committee adopted a proposal from Canada in 1969 to request the Scientific and Technical Sub-Committee of COPUOS to study the technical aspects of registration appropriate for identifying and coordinating space objects.[83] In 1972, after the Scientific and Technical Sub-Committee provided the Legal Sub-Committee with some conclusions, Canada submitted a draft treaty,[84] which was later merged with the French proposal and jointly submitted.[85] After several years of deliberations, including consideration of an alternative text submitted by the United States,[86] the Legal Sub-Committee adopted a draft text for approval by COPUOS in 1974.[87] Subsequently, the Registration Convention was adopted on 12 November 1974 by the General Assembly of the United Nations.[88]

The Registration Convention continued with the use of the term "launching State" as defined in the Liability Convention.[89] When a space object is launched into Earth orbit or beyond, the launching State is required to make an appropriate entry in its register of space objects and to provide all relevant information, including the launching State(s), its registration number, the date and location of launch of the space object, its function and basic orbital parameters.[90] Where there is more than one launching State, the launching States must jointly determine which one of them is to be the State of registry, without prejudice to any separate agreement between the States concerning jurisdiction and control over the space object.[91] In other words, there can only be one State of registry even though there may be more than one launching State for a space object.[92]

---

[82] U.N.Doc. A/AC.105/C.2/L.45.

[83] U.N.Doc. A/AC.105/58.

[84] U.N.Doc. A/AC.105/C.2/L.82.

[85] U.N.Doc. A/AC.105/C.2/L.83.

[86] U.N.Doc. A/AC.105/C.2/85.

[87] U.N.Doc. A/AC.105/C.2/13.

[88] General Assembly Resolution 3235 (XXIX).

[89] Registration Convention, Article I.

[90] Ibid., Articles II(1) and IV. The basic orbital parameters required under the Registration Convention include the nodal period, inclination, apogee and perigee: Article IV.

[91] Ibid., Article II(2) and (3).

[92] The Registration Convention does not provide for the possibility of transfers of registrations, so the ability of non-launching States to exercise jurisdiction and control over a space object that has been sold or otherwise transferred remains difficult; see generally Marietta Benkö and Kai-Uwe Schrogl, *The 1998 European Initiative in the UNCOPUOS Legal Sub-Committee to Improve the Registration Convention* (1998) 41 PROC. COLL. L. OUTER SP. 58; Gabriel Lafferranderie, *L'application par l'Agence Spatial Européenne de la Convention sur l'immatriculation des objets*

3.2 Sources of Space Law

The Registration Convention was also innovative in being the first space law instrument to contain a review clause, providing for a scheduled review of its terms. Article X of the Registration Convention provides that the General Assembly is to consider a review of its provisions 10 years after it enters into force and, in any event, after 5 years of its entry into force, a third of the State Parties to the Registration Convention may request a conference of all State Parties to be convened at any time to review its provisions.

As at 1 January 2010, 53 States ratified the Registration Convention, with a further four States as signatories.[93] The 10th anniversary of the entry into force of the Registration Convention passed in 1986 without any proposals for amendments to be considered. However, the issue of the registration of space objects has increasingly become a topic of discussion in private and academic circles in recent times, as there are often multiple launching States involved in the launch of a single space object and there are increasing instances of ownership transfers of satellites.[94]

It should be noted that most States that have not ratified the Registration Convention nevertheless have been able to make entries on the United Nations register of space objects on a voluntary basis, pursuant to General Assembly Resolution 1721B of 1961.[95]

3.2.1.6 Moon Agreement

When the two superpowers of the Cold War were racing to reach the Moon, it was considered pertinent for the international community to adopt specific treaty provisions in relation to the Moon. In 1966, the Soviet Union submitted a draft treaty to the Legal Sub-Committee containing the legal principles governing the exploration and exploitation of the Moon and other celestial bodies, which was considered in the formulation of the Outer Space Treaty.[96] In 1969, Poland further submitted two proposals relating to the rules for human activities on the surface of the Moon and other celestial bodies,[97] and Argentina also submitted a proposal to regulate the legal status of substances and resources that originate from the Moon and other celestial bodies in the Solar System.[98]

---

lancés dans l'espace extra-atmospherique (1986) 11 ANN. AIR & SP. L. 229; and Ricky J. Lee, *Transferring Registration of Space Objects: The Interpretative Solution*, paper presented at the 47th Colloquium on the Law of Outer Space, 4–8 October 2004, in Vancouver, Canada.

[93] United Nations Office of Outer Space Affairs, supra note 41.

[94] See, for example, United Nations Office of Outer Space Affairs, PROCEEDINGS OF THE IISL / ECSL SYMPOSIUM: REINFORCING THE REGISTRATION CONVENTION (2003), *passim*.

[95] General Assembly Resolution A/16/1721B (1961). Paragraph 1 of the Resolution "*Calls upon* States launching objects into orbit or beyond to furnish information promptly to the Committee on the Peaceful Uses of Outer Space, through the Secretary-General, for the registration of launchings".

[96] U.N.Doc. A/AC.105/C.2/L.12.

[97] U.N.Doc. A/AC.105/C.2/L.53.

[98] U.N.Doc. A/AC.105/C.2/L.54.

In the following decade, in which there was consensus among its members that the legal issues concerning the Moon should be regulated by means of an international treaty, the Legal Sub-Committee failed, despite several attempts, to reach agreement on the provisions dealing with the use of lunar resources. In the end, the Legal Sub-Committee opted to defer the substantive issues relating to the regulation of the use of lunar resources to a later date and, consequently, the draft agreement as it stands today was submitted for adoption by COPUOS and the General Assembly. On 5 December 1979, the General Assembly resolved to adapt the proposed text of the Moon Agreement.[99]

The provisions of the Moon Agreement apply to the Moon and other celestial bodies in the Solar System, including orbits around, and trajectories to or from them.[100] The main principles include:

(1) the requirement that the Moon and other celestial bodies are to be used exclusively for peaceful purposes and threats or uses of force and all hostile acts are prohibited;[101]
(2) the exploration and use of the Moon and other celestial bodies are to be conducted without discrimination of any kind, including discrimination based on economic or scientific development;[102]
(3) disruptions to the lunar environment and that of other celestial bodies by adverse changes or harmful contamination are to be prevented;[103]
(4) the Moon, other celestial bodies and their natural resources are the "common heritage of mankind" (*res communis humanitatis*); and
(5) an international regime is to be established in the future to govern the exploration and exploitation of natural mineral resources on the Moon and other celestial bodies.[104]

It is this classification of the Moon, other celestial bodies and extraterrestrial natural resources as the "common heritage of mankind" that has proven to be one of the most controversial features of international space law. The same designation may be found in the United Nations Convention on the Law of the Sea in relation to the deep seabed and the mineral resources contained therein.[105] In the Moon Agreement, material expression is given to the common heritage of mankind principle as the "equitable sharing by all State Parties in the benefits derived from those resources, whereby the interests and needs of the developing countries, as well as the efforts of those countries which have contributed either directly or indirectly

---

[99] General Assembly Resolution A/34/68 (1979).

[100] Moon Agreement, Article 1.

[101] Ibid., Article 3.

[102] Ibid., Article 4.

[103] Ibid., Article 7.

[104] Ibid., Article 11.

[105] United Nations Convention on the Law of the Sea, Article 136.

3.2 Sources of Space Law

to the exploration of that space object, shall be given special consideration".[106] As the international community was unable to reach agreement on the terms of this "equitable sharing" of benefits, the Moon Agreement provides that the terms of this governing regime are to be contained in a future treaty when resource exploitation in space becomes feasible.[107]

The failure of the international community to reach agreement over the international legal regime to be created in order to implement Article 11 of the Moon Agreement, has resulted in the absence of any substantial acceptance of the Moon Agreement. As at 1 January 2010, only 13 States were parties to the Moon Agreement and, among them, only Australia would likely be considered to have a space launch capability.[108] Consequently, the legal status of the provisions contained in the Moon Agreement in the context of international space law remains a hotly debated issue.

3.2.2 General Assembly Declarations

3.2.2.1 Overview

When COPUOS and, in particular, the Legal Sub-Committee, began deliberations on the principles applicable to space activities, it was clear that a comprehensive legal code governing space activities would not be appropriate.[109] Instead, the Legal Sub-Committee opted to undertake a progressive approach to remain in step with the continuing development of space technology and applications.[110]

As suggested by Jericho and McCracken in 1985:

> The [space] treaties constitute the entirety of public international space law upon which further treaties will eventually be based. However, it is postulated that further developments in the international space law area will show a decreasing growth rate with a possible levelling of the curve in the next decade. Contemporaneously, international law developments likely will become more specific and tend toward the international commercial aspects of man's space endeavours.[111]

Accordingly, it was considered that, in relation to specific international commercial satellite applications, it was more appropriate to adopt an instrument containing

---

[106] Moon Agreement, Article 11(7).

[107] Ibid., Article 11(5).

[108] The 13 States that have ratified the Moon Agreement are Australia, Austria, Belgium, Chile, Kazakhstan, Lebanon, Mexico, Morocco, the Netherlands, Pakistan, Peru, the Philippines and Uruguay with France, Guatemala, India and Romania having signed but not ratified the Moon Agreement: United Nations Office of Outer Space Affairs, supra note 41.

[109] Vladímir Kopal, *The Role of United Nations Declarations of Principles in the Progressive Development of Space Law* (1988) 16 J. Sp. L. 5 at 6.

[110] United Nations, *Report of the Ad Hoc Committee on the Peaceful Uses of Outer Space to the United Nations General Assembly* (1959) U.N.Doc. A/4141, Part III.

[111] Eugene Jericho and David G. McCracken, *Space Law: Is it the Last Legal Frontier?* (1985) 51 J. Air L. & Com. 791 at 799–800.

legal principles in the form of a General Assembly declaration as contained in a resolution before completing the negotiations on the multilateral treaties.[112]

The first General Assembly declaration containing principles of space law was the Principles Declaration, which was adopted unanimously by the General Assembly on 13 December 1963.[113] The origins of most of the fundamental principles of space law that are found in the later treaties, especially the Outer Space Treaty, can be found in the nine operative paragraphs of the Principles Declaration. The first four paragraphs contain the positive principles later embodied in the Outer Space Treaty. For example, space activities are to be carried on:

(1) for the benefit and in the interest of all mankind;[114]
(2) in accordance with international law, including the Charter of the United Nations;[115] and
(3) in the interest of maintaining international peace and security and promoting international cooperation and understanding".[116]

Subsequent declarations, as discussed below, address specific issues arising from remote sensing of the Earth, direct television broadcasting, use of nuclear power sources and international cooperation in space activities. The precise legal force and effect of these resolutions continues to be the subject of intense academic debate. In addition to the relatively little state practice and *opinio juris*, a further complication is that the state practice or *opinio juris* made by a State that is already party to a relevant treaty may not be considered to be evidence of the existence of a customary principle. This is because the relevant state practice would have been done pursuant to its treaty obligations and not because the State considered itself bound by a relevant customary principle. Consequently, the widespread acceptance of the Outer Space Treaty by the international community may have a counterproductive effect in the context of creating customary norms of space law.

In the case of the Outer Space Treaty, the issue of the customary status of its provisions has been considered by several commentators.[117] It is generally accepted that the principles relating to freedom of use (Article I), non-appropriation (Article II), applicability of international law (Article III), State responsibility (Article VI), liability for damage caused by space objects (Article VII) and the retention of jurisdiction and control by States over space objects (Article VIII) may be considered part of customary law.[118] The basis for such a proposition includes the large number

---

[112] See Manfred Lachs, THE LAW OF OUTER SPACE: An EXPERIENCE IN CONTEMPORARY LAW MAKING (1972) at 27–41.

[113] General Assembly Resolution 1962 (XVIII).

[114] Principles Declaration, Paragraph 1.

[115] Ibid., Paragraph 4

[116] Ibid.

[117] See, for example, Bin Cheng, STUDIES IN INTERNATIONAL SPACE LAW (1998), ch. 7, *passim.*

[118] Ibid.

of States that have ratified the Outer Space Treaty, the support given by States not party to the Outer Space Treaty to General Assembly resolutions reiterating the contents of these provisions and the *opinio juris* of States given in speeches made during debates in intergovernmental organisations, such as COPUOS.[119] Although state practice may be considered to be in observance of a State's treaty obligation, it is unlikely that the same can be said for statements of *opinio juris*.

The Rescue Agreement, adopted for the elaboration and implementation of Article V of the Outer Space Treaty, is considered by some commentators to have crystallised into customary norms of international law, at the very least partially.[120] The main arguments made in support of this include the humanitarian nature of the astronaut assistance and rescue principles for which parallels may be found in the customary and treaty principles of international maritime law. Further, the compliance of States not party to the Rescue Agreement with the provisions of Article 5 of the Rescue Agreement concerning the duty to notify, recover and return discovered space objects may be considered as custom, though this must be distinguished with any state practice in compliance with the provisions of the Outer Space Treaty.[121]

It is very difficult to assert the provisions of the other space treaties, namely the Liability Convention, the Registration Convention and the Moon Agreement, as principles of customary international law. The lack of consistent, uniform and universal state practice or *opinio juris* for compliance with the liability provisions of the Liability Convention and the operative provisions of the Moon Agreement means that it is virtually impossible to consider the extent of their crystallisation into custom at this stage. The absence of widespread acceptance of the Moon Agreement makes it highly unlikely that any of its provisions, with the possible exception of those that repeat or affirm principles already contained in the Outer Space treaty, may be considered custom. In the case of the Registration Convention, some States not party to it have nevertheless notified the Secretary-General of the United Nations

---

[119] Ibid.

[120] See Stephen Gorove, *Interpreting Salient Provisions of the Agreement on the Rescue of Astronauts, the Return of Astronauts and Return of Objects Launched into Outer Space* (1968) 11 PROC. COLL. L. OUTER SP. 93; Vladímir Kopal, *Problems Arising from the Interpretation of Agreement on the Rescue of Astronauts, Return of Astronauts and Return of Objects Launched into Outer Space* (1968) 11 PROC. COLL. L. OUTER SP. 98; Maurice N. Andem, *The 1968 Rescue Agreement and the Commercialisation of Outer Space Activities During the 21st Century – Some Reflections* (1998) 41 PROC. COLL. L. OUTER SP. 75; and Robert C. Beckman, *1968 Rescue Agreement – An Overview*, in United Nations, PROCEEDINGS OF THE UNITED NATIONS / REPUBLIC OF KOREA WORKSHOP ON SPACE LAW (2003), 370–378.

[121] Recent examples of this include the discoveries of a U.S. launch vehicle component in Japan (U.N. Doc. A/AC.105/735), a component of the French Ariane rocket in Texas (U.N. Doc. A/AC.105/737), components of a U.S. launch vehicle in South Africa (U.N. Doc. A/AC.105/740) and remnants of a U.S. satellite in Saudi Arabia (U.N. Doc. A/AC.105/762) as referred to in Kenneth Hodgkins, *Procedures for Return of Space Objects under the Agreement on the Rescue of Astronauts, the Return of Astronauts and the Return of Objects Launched into Outer Space*, in United Nations, PROCEEDINGS OF THE UNITED NATIONS/INTERNATIONAL INSTITUTE OF AIR AND SPACE LAW WORKSHOP ON CAPACITY BUILDING IN SPACE LAW (2002), at 39.

of launched space objects explicitly pursuant to the Outer Space Treaty or General Assembly Resolution 1721B, which calls for the voluntary registration of space objects.[122]

3.2.2.2 Effect of the General Assembly Space Law Declarations in Customary International Law

Overview

In addition to the general principles contained in the space treaties, the international community through the COPUOS have chosen to adopt specific "legal" principles relating to particular space applications or activities such as remote sensing of the Earth and direct television broadcasting by satellite, international cooperation in space activities and the use of nuclear or radioisotopic power sources. The most important questions to be answered when considering these provisions is the binding effect they may have on States.

The same arguments favouring the creation of "instant custom" arising from the space treaties can also be made in the case of General Assembly declarations, in the sense that they have codified customary norms as agreed by States during discussions in COPUOS or became accepted and followed by the States as customary norms for them to be regarded as binding international law. In considering the customary effect of General Assembly resolutions, Brownlie suggested that the adoption of a resolution by the General Assembly may constitute evidence of the legal opinions of States, thus are *opinio juris* evidencing customary norms.[123] Unless the extreme position that only physical acts can constitute state practice is taken, there is no doubt that the adoption of such resolutions must in some way constitute practice. Akehurst, for example, defines state practice for these purposes as "any act or statement by a State from which views about customary law can be inferred; it includes physical acts, claims, declarations *in abstrato* (such as General Assembly resolutions), national laws, national judgments and omissions".[124]

The effect of such declarations on the obligations of States can be considered in three ways. On the one hand, Schwebel, when he was Deputy Legal Advisor to the Department of State, wrote that:

> As a statement of U.S. policy in this regard, I think it is fair to state that General Assembly resolutions are regarded as recommendations to Member States of the United Nations. To the extent, which is exceptional, that such resolutions, [which] are *meant to be* declaratory of

---

[122] See Petr Lála, *The United Nations Register of Objects Launched into Outer Space*, in United Nations, PROCEEDINGS OF THE UNITED NATIONS / INTERNATIONAL INSTITUTE OF AIR AND SPACE LAW WORKSHOP ON CAPACITY BUILDING IN SPACE LAW (2002), at 36.

[123] Ian Brownlie, PRINCIPLES OF PUBLIC INTERNATIONAL LAW (5th ed., 1998), at 14.

[124] Michael Akehurst, *Custom as a Source of International Law* (1974–1975) 47 BRIT. Y. B. INT'L. L. 1 at 53.

3.2 Sources of Space Law

international law, are adopted with the support of *all* members and are observed by the practice of States, such resolutions are evidence of customary international law on a particular subject matter.[125]

Consequently, it appears that a General Assembly declaration must have the explicit intention of creating customary law and at the time it was accepted as such by a true consensus, or sufficiently close to it, in order to be considered to reflect customary law. This calls into question the effect of abstentions as well as challenging the accepted view, since generality, rather than universality, is the requirement for the establishment of a customary principle.

As a middle view, in the *South West Africa Cases*, Judge Tanaka in his dissenting judgment held that:

> What is required for customary international law is the repetition of the same practice; accordingly, in this case resolutions, declarations, etc. ... must take place repeatedly ... This collective, cumulative and organic process of custom-generation can be characterised as the middle way between legislation by convention and the traditional process of custom making, and can be seen to have an important role from the viewpoint of the development of international law.[126]

It can be seen from this that, while Judge Tanaka did not appear to be as strict as Schwebel in prescribing a requirement of uniformity to be placed on custom-making resolutions, he did nevertheless require the resolutions or declarations to be repetitive in their statements in order to constitute evidence of customary norms of international law.

At the other end of the continuum, Sohn wrote that:

> There is wide consensus that these declarations actually establish new rules of international law binding upon all States. This is not treaty-making but a new method of creating customary international law. ... Thus the United Nations has made possible the creation of "instant international law" ... In a rapidly changing world the United Nations has found a method, albeit restricted by the rule of unanimity or quasi-unanimity, to adapt the principles of its Charter and the rules of customary international law to the changing times with an efficiency which even its most optimistic founders did not anticipate.[127]

In recent times, the debate over the value of General Assembly resolutions as customary norms has been somewhat overtaken by recent cases and commentaries. Sloan, for example, was of the view that General Assembly resolutions may be evidence of state practice when the following factors are taken into consideration:

(1) the legalistic nature of the language of its terms;
(2) the binding or mandatory nature of the language of its terms;
(3) the intention of the General Assembly as evidenced by statements of the States, especially those in the debate over the resolution;

---

[125] Stephen Schwebel, *Letter of U.S. State Department Deputy Legal Adviser Stephen Schwebel of 25 April 1975* [1975] U.S.D.I.L. 85.

[126] *South West Africa Case (Ethiopia and Liberia v. South Africa)* [1966] I.C.J. Rep. at 292.

[127] Louis B. Sohn, *The Development of the Charter of the United Nations: The Present State*, in Marteen Bos (ed.), THE PRESENT STATE OF INTERNATIONAL LAW AND OTHER ESSAYS (1973), 39–60 at 52–53.

116   3 State Responsibility and Liability for Compliance with International Space Law

(4) the voting record, in particular that of the States whose support may be essential for effective implementation, but nevertheless representing States from all economic, ideological and legal systems; and
(5) their repetition or recitation in subsequent resolutions, as Judge Tanaka suggested in his dissenting judgment in the *South West Africa Cases*.[128]

General Assembly resolutions are also increasingly gaining acceptance as evidence of state practice in cases of the International Court of Justice and various arbitral tribunals. In *Case Concerning Military and Paramilitary Activities in and against Nicaragua (Merits)*, the Court was confronted with the problem of having to formulate the customary norms in the unilateral use of force, in particular the *opinio juris* of States, and determined that:

> This *opinio juris* may, though with all due caution, be deduced from, *inter alia*, the attitude of the Parties and the attitude of States towards certain General Assembly resolutions .... The effect of consent to the text of such resolutions ... may be understood as an acceptance of the validity of the rule or set of rules declared by the resolution by themselves.[129]

This was further expanded by the Court in the *Advisory Opinion on the Legality of the Threat or Use of Nuclear Weapons*:

> The Court notes that General Assembly resolutions, even if they are not binding, may sometimes have normative value. They can, in certain circumstances, provide evidence important for establishing the existence of a rule or the emergence of an *opinio juris*. To establish whether this is true of a given General Assembly resolution, it is necessary to look at its content and the conditions for its adoption; it is also necessary to see whether an *opinio juris* exists as to its normative character. Or a series of resolutions may show the gradual evolution of the *opinio juris* required for the establishment of a new rule.[130]

It is apparent that the Court has moved progressively towards adopting the approach suggested by Sloan in considering the nature and language of the terms of a resolution as well as the intention of the States concerned in determining whether a General Assembly resolution contains customary norms. It is in this context that the General Assembly declarations concerning space activities are to be examined.

The Space Law Declarations

The five General Assembly declarations concerning space activities are:

(1) the Principles Declaration;
(2) the Principles Governing the Use by States of Artificial Earth Satellites for International Direct Television Broadcasting of 10 December 1982 (the "Broadcasting Principles");

---

[128] Blain Sloan, *General Assembly Resolutions Revisited* (1987) 58 BRIT. Y. B. INT'L. L. 39.

[129] *Case Concerning Military and Paramilitary Activities in and against Nicaragua (Merits) (Nicaragua v. United States)* [1986] I.C.J. Rep. at para. 188. See also Hilary C. M. Charlesworth, *Customary International Law and the Nicaragua Case* (1988) 11 AUST. Y. B. INT'L. L. 1; and Jonathan I. Charney, *Customary International Law in the Nicaragua Case Judgment on the Merits* (1988) 1 HAGUE Y. B. INT'L. L. 16.

[130] [1996] I.C.J. Rep. 226 at 254–255, para. 70.

3.2 Sources of Space Law

(3) the Principles Relating to Remote Sensing of the Earth from Outer Space of 3 December 1986 (the "Remote Sensing Principles");

(4) the Principles Relevant to the Use of Nuclear Power Sources in Outer Space of 14 December 1992 (the "Nuclear Power Sources Principles"); and

(5) the Declaration on International Cooperation in the Exploration and Use of Outer Space for the Benefit and in the Interest of All States, Taking into Particular Account the Needs of Developing Countries of 13 December 1996 (the "Cooperation Declaration").

In these General Assembly declarations it may be prudent, or at least convenient, to separate their provisions into four groups:

(1) operative provisions that merely repeat principles contained in existing treaties or declarations (the "Repeating Provisions");

(2) operative provisions that state, without more, the application of existing provisions of the Outer Space Treaty or the Rescue Agreement to specific situations (the "Applying Provisions");

(3) operative provisions that create new rights, duties or obligations of law that have not been previously stated in existing treaties or those that extend the applicability or content of the provisions of the Outer Space Treaty (the "New Provisions"); and

(4) declaratory or introductory provisions that appear to have no legal effect (the "Non-Legal Provisions").

In the case of Repeating Provisions, the extent of the original treaty provision's "migration" into a customary norm must first be considered. For example, when a provision of the Outer Space Treaty is considered by most States and commentators to have crystallised into customary international law, it would mean that a Repeating Provision embodying the same provision of the Outer Space Treaty would be restating the law as it already exists in custom. This is in addition to the weight, if any, that is to be given to the provisions in the declaration itself as evidence of customary international norms in the form of state practice or *opinio juris*.

Applying Provisions are, strictly speaking, not statements of legal principles but are instead "working examples" of specific applications of an existing treaty principle. Consequently, they may be considered to be lacking in the requisite legalistic language to be considered evidence of customary norms. In most cases the only relevant consideration when determining whether an Applying Provision is evidence of a customary norm is if the original treaty provision being applied is a customary norm of space law. Nevertheless, it is possible, though unlikely considering the lack of precedents, for a specific application of a legal principle to be considered by States to have binding legal effect while the general principle does not.

In the case of New Provisions, it will be necessary to consider whether they crystallise custom as agreed during negotiations or if they have been subsequently accepted and followed as custom. As an empirical exercise, the *travaux préparatoires* and, as a secondary source, the writings of legal scholars and commentators are examined to determine whether a New Provision has been accepted by States to be a principle of customary international law. As in the case of the provisions of the space treaties, it would be necessary to consider the relevant state practice and *opinio juris* of States to ascertain the extent to which the provisions may have crystallised into customary international law.

Principles Declaration

The Principles Declaration, adopted by the General Assembly in 1963, is the only resolution considered to contain principles of space law that was declared before the adoption of the Outer Space Treaty and the Liability Convention. Consequently, the Principles Declaration is unique in that, at the time it was adopted, its provisions were all New Provisions without an existing treaty precedent. Since the adoption of the 1967 Outer Space Treaty, however, it would be somewhat difficult to distinguish between the effects of the principles in the Principles Declaration vis-à-vis those of the Outer Space Treaty in the context of determining the customary norms of space law.

Between 1963 and 1967, the customary status of the provisions of the Principles Declaration would have required an evaluation of the *travaux préparatoires* to assess the level of their acceptance by States. Considering the positive statements made by almost all governmental representatives during the debates and negotiations, it is likely that the Principles Declaration is a codification of the customary principles accepted by States during the negotiations process. For example, those provisions relating to the freedom of movement in outer space may well have already represented customary international law by that time.[131] Jenks wrote in those intervening years that:

> The authority of the Declaration of Legal Principles may be expected to grow with the passage of years. While it is somewhat less than a treaty it must *already* be regarded as rather more than a statement of custom.[132]

Indeed, Fawcett valued the evidentiary weight of the Principles Declaration so highly that he saw the adoption of the Outer Space Treaty as a "retrograde step":

> In the Outer Space Treaty we have then a rigidly contractual instrument, in essence a bilateral arrangement between the principal space-users. Apart from its provision for partial demilitarisation of outer space, tracking and inspection, it does nothing to elaborate or secure the principles already set out in General Assembly Resolutions. It may even be that this ill-constructed and precarious instrument is a retrograde step. ... Though Resolution 1962 (XVIII) is for the most part a declaration, not of rules of international law, but of directive principles, it may like other similar General Assembly Resolutions, be regarded as forming part of an international *ordre public*, to which States should strive to make their policies conform.[133]

---

[131] Lachs, supra note 112, at 58.

[132] C. Wilfred Jenks, SPACE LAW (1965) at 185.

[133] James E. S. Fawcett, INTERNATIONAL LAW AND THE USES OF OUTER SPACE (1968) at 15–16.

3.2 Sources of Space Law

Table 3.1 Principles Declaration

Para.	Content	Type	Treaty provision
1	Space activities in the interest of all mankind	Repeating	Outer Space Treaty I
2	Freedom of exploration and use	Repeating	Outer Space Treaty I
3	Non-appropriation	Repeating	Outer Space Treaty II
4	Application of international law	Repeating	Outer Space Treaty III
5	International responsibility for space activities	Repeating	Outer Space Treaty VI
6	Due regard for the interests of other States	Repeating	Outer Space Treaty IX
7	Retention of jurisdiction and control over space objects on a State's registry	Repeating	Outer Space Treaty VIII
8	Liability for damage caused by space objects	Repeating	Outer Space Treaty VII
9	Assistance to astronauts in distress	Repeating	Outer Space Treaty V

Source: Adapted, with some variation, from Ricky J. Lee and Steven R. Freeland (2003, 126)

After 1967, the apparent acceptance of some of the Outer Space Treaty as customary law means that all that would be necessary is to compare the provisions of the Principles Declaration with those of the Outer Space Treaty. A comparison of the Principles Declaration and the Outer Space Treaty is contained in Table 3.1.

In this way, the extent of the crystallisation of the provisions of the Principles Declaration is linked to the legal effect of the crystallisation into custom of the provisions of the Outer Space Treaty and thus their legal effect from 1967 cannot be considered independently.

Broadcasting Principles

The Broadcasting Principles may be considered to be the first of three General Assembly declarations that relate to specific space activities (Table 3.2). The first three paragraphs provide for the purposes and objectives to be observed by States when conducting direct television broadcasting activities. Principle 6 requires States to make appropriate arrangements for international cooperation, taking into account the interests of developing States. Principle 11 requires States to cooperate on the protection of copyright and associated rights in the conduct of television broadcasting. A cursory glance over the *travaux préparatoires* of the Broadcasting Principles appears to show that there was substantial acceptance of the above provisions as being legal obligations, indicating their crystallisation into customary law. However, the substantial lack of state practice and *opinio juris* means these provisions would not be regarded as having crystallised into customary principles of international law.

Principle 10 provides that a broadcasting or receiving State has the right to request consultations with the other States on the satellite broadcasting service. Principles 13 and 14 provide that a State establishing a new broadcasting service must notify the proposed receiving States and enter into consultations if requested. Principle 15 states that the instruments of the International Telecommunication Union (the "ITU") are exclusively applicable to the signal overspill from the

120 3 State Responsibility and Liability for Compliance with International Space Law

Table 3.2 Broadcasting Principles

Para.	Content	Type	Treaty provision
1	Purposes and objectives	Non-Legal Provisions	
2			
3			
4	Applicability of international law	Applying	Outer Space Treaty III
5	Rights and benefits of States	Applying	Outer Space Treaty I
6	International cooperation in broadcasting	New Provision – Not Custom	
7	Peaceful settlement of disputes	Applying	Outer Space Treaty III
8	International responsibility for private acts	Applying	Outer Space Treaty VI
9	State responsibility for activities of international organisations	Applying	Outer Space Treaty VI
10	Duty and right of States to consult	New Provision – Not Custom	
11	Copyright in programs transmitted	New Provision – Not Custom	
12	Notification to the United Nations	Repeating	Outer Space Treaty XI
13	Consultations and agreements between States	New Provision – Not Custom	
14		New Provision – Not Custom	
15	Overspill of signals	Applying	I.T.U. instruments

Source: Adapted, with some variation, from Ricky J. Lee and Steven R. Freeland (2003, 127)

service. There does not appear to be widespread acceptance of these provisions by States during the debates and negotiations. For example, Argentina considered the spill-over issue not to be subject to the ITU, while Czechoslovakia and India both considered prior consent to be necessary instead of mere consultations.[134] Accordingly, it is likely that these particular provisions may not have the sufficient universality between States to be considered customary principles of law, even if there was sufficient evidence of a customary principle, which is not the case.

Remote Sensing Principles

Observations concerning the customary effects of the Broadcasting Principles can similarly be made about the Remote Sensing Principles. While most of the provisions are applications of the Outer Space Treaty, there is a set of provisions relating to international cooperation and another set of provisions relating to the duty and the right to consult that are not the subject of widespread international acceptance.

Principle II states that remote sensing activities are to be conducted for the benefit and in the interests of all States while "taking into particular consideration the needs of the developing countries". This is in effect a repetition of Article I of the Outer Space Treaty except that the original treaty provision did not require the particular needs of developing States to be taken into consideration. This reflects the growing influence of the developing States on the development of international law during

---

[134] See generally the *travaux préparatoire* of the Broadcasting Principles as reproduced in Nandasiri Jasentuliyana and Roy S. Lee, MANUAL ON SPACE LAW (1979), vol. III.

3.2 Sources of Space Law

this time. The fact that this "change" is repeated through the subsequent treaties and declarations, such as the Moon Agreement, suggests that this is now an additional requirement that is supported by acceptance of most States.[135] Consequently, a strong argument may be made supporting the crystallisation of this variation to the existing principle into customary international law.

Principles V to VIII, setting out the steps required of States in relation to mutual assistance and international cooperation, were not the subject of much controversy during the debates. Consequently, the uniformity requirement may be considered satisfied in terms of the creation of custom. Principle XI requires States to transmit relevant data and information to States affected or likely to be affected by natural disasters. Similarly, this may be considered to be a small extension of the obligations under Article IX of the Outer Space Treaty and the absence of controversy suggests that this was widely accepted by most States. Again, the problem here is the relative absence of state practice and *opinio juris*.

Principle XIII requires "sensing States" to enter into consultations upon request with "sensed States" in order to make available opportunities for participation. This infers, as with the Broadcasting Principles, that the sensing States are required to notify sensed States of their sensing activities. This is in contrast to the prior consent requirement that a significant number of States advocated during the negotiations as part of their permanent sovereignty over natural resources. In the absence of uniformity, it would be appropriate to suggest that the provision is not a customary norm, even if there is sufficient state practice on this subject.

Principle XII is a unique provision in that it requires sensed States to have access to primary and processed data relating to their territory on a non-discriminatory basis and on reasonable cost terms. This provision does not appear to be based on any existing principle of space law but is in fact a compromise reached between those States wanting free access and those that do not wish to provide any more rights to the sensed States beyond notification and consultation, arguing that the sensing of some States is an inevitable result of the orbit of the satellite. In one of the later reports of the Legal Sub-Committee Working Group on Remote Sensing, it was evident that there were divergent views concerning the basis of those access rights, the type of data involved and the timing of the access, and that the final wording was a compromise that did not satisfy a significant number of States.[136] Therefore, it would be difficult to support an argument that Principle XII is a legal obligation agreed to by all States (Table 3.3).

Nuclear Power Sources Principles

The Nuclear Power Sources Principles contain eleven provisions of which Principles 1, 6, 7, 8, 9 and 10 are Applying Provisions in the sense that they merely restate the application of provisions in existing space treaties to satellites containing nuclear

---

[135] Moon Agreement, Article 11(7).

[136] Report of the Chairman of the Working Group on Remote Sensing, U.N.Doc. A/AC.105/271.

122   3   State Responsibility and Liability for Compliance with International Space Law

Table 3.3   Remote Sensing Principles

Para.	Content	Type	Treaty provision
I	Definitions	Non-Legal Provision	
II	Rights and benefits of States	Repeating	Outer Space Treaty I
III	Application of international law	Applying	Outer Space Treaty III
IV	Rights and benefits of all mankind and permanent sovereignty over natural resources	Repeating	Outer Space Treaty I & IX
V	International cooperation	New Provision – Not Custom	
VI	Establishment of data processing facilities	Non-Legal Provision	
VII	Technical assistance to other States	New Provision – Not Custom	
VIII	Role of the United Nations in promoting international cooperation	Non-Legal Provision	
IX	Notification to the United Nations	Repeating	Outer Space Treaty XI
X	Protection of the natural environment	Applying	Outer Space Treaty IX
XI	Protection from natural disasters	New Provision – Not Custom	
XII	Access by sensed States to data on a non-discriminatory basis on reasonable cost terms	New Provision – Not Custom	
XIII	Right and duty to consult	New Provision – Not Custom	
XIV	International responsibility for activities	Applying	Outer Space Treaty VI
XV	Peaceful settlement of disputes	Applying	Outer Space Treaty III

Source: Adapted, with some variation, from Ricky J. Lee and Steven R. Freeland (2003, 128)

or radioisotopic power sources. In particular, Principle 6 merits some attention as it requires a right to consultations by States concerned with the re-entry of a nuclear power source. This may be considered to be a direct application of the right to consultations in the case of potential harmful interference as provided for in Article IX of the Outer Space Treaty. In contrast, it is distinguished from the consultation provisions of the Broadcasting Principles and the Remote Sensing Principles in that these rights are not dependent on the potential for harmful interference to the space activities of the affected States and therefore cannot find support in Article IX of the Outer Space Treaty (Table 3.4).

In the case of the New Provisions, namely Principles 3–5, it is pertinent to assess the legal basis of these provisions to ascertain their appropriate status in customary international law. Principles 3 and 4 are concerned with imposing a set of guidelines for the safe use of nuclear power sources, along with a requirement for a thorough safety assessment to be conducted prior to launch. This can be seen to be an elaboration of the obligations of States under Article IX of the Outer Space Treaty in that States are to have due regard to the interests of other States and to avoid harmful interference of the corresponding activities of other States.

Principle 5 requires States to inform other States and the United Nations if an object containing a nuclear power source is malfunctioning. The notification must include the orbital parameters of the spacecraft and its radiological risk. It is probable that Principle 5 constitutes customary international law, as the provision appears to have its origins from the following treaty provisions:

3.2 Sources of Space Law

Table 3.4 Nuclear Power Sources Principles

Para.	Content	Type	Treaty provision
1	Application of international law	Applying	Outer Space Treaty III
2	Definitional clause	Non-Legal Provision	
3	Guidelines and criteria for safe use	New Provision – Not Custom	
4	Safety assessment by a third party	New Provision – Not Custom	
5	Notification of re-entry of space objects containing a nuclear power source	New Provision – Not Custom	
6	Duty and right to consult	Applying	Outer Space Treaty IX
7	Assistance to States in case of re-entry	Applying	Rescue Agreement
8	International responsibility	Applying	Outer Space Treaty VI
9	Liability and determination of the amount of compensation payable	Applying	Outer Space Treaty VII Liability Convention XII
10	Peaceful settlement of disputes	Applying	Outer Space Treaty III
11	Review and revision of the Nuclear Power Sources Principles	Non-Legal Provision	

Source: Adapted, with some variation, from Ricky J. Lee and Steven R. Freeland (2003, 129)

- Article IX of the Outer Space Treaty, requiring States to avoid harmful contamination and adverse changes to the Earth's environment by introduction of "extraterrestrial" matter; and
- Article IV(3) of the Registration Convention, which imposes a duty on States to notify the Secretary-General of the United Nations if a registered spacecraft is no longer in Earth orbit.[137]

Cooperation Declaration

The Cooperation Declaration, adopted by the General Assembly without a vote in 1996, is special in that not all of its provisions contain legal obligations that would support a view that they are intended to have binding effect. For example, Paragraphs 7 and 8 merely suggest that States are to strengthen the role of COPUOS and to contribute to the United Nations space programs and initiatives. As such, they cannot reasonably be considered as legal obligations at all and certainly cannot be considered as part of customary law.

Paragraph 5 poses difficulties in its classification in that it requires States to keep in mind certain goals and objectives in framing activities involving international cooperation. It is unclear whether these goals are intended to be of a mandatory nature, as the words "promoting", "fostering" and "facilitating" are used. Even if the consideration of such issues is considered mandatory, the vagueness of the content of these goals makes it unlikely that States have intended for this to be a binding international legal obligation (Table 3.5).

---

[137] Article IV(2) of the Registration Convention allows States to provide "additional" information concerning a space object, although it does not impose a mandatory obligation.

124   3 State Responsibility and Liability for Compliance with International Space Law

Table 3.5 Cooperation declaration

Para.	Content	Type	Treaty provision
1	Application of international law	Applying	Outer Space Treaty III
2	Freedom of States to consider the extent of international cooperation	Applying	Outer Space Treaty I Outer Space Treaty IX
3	Contribution to promoting cooperation	Non-Legal Provision	
4	Considerations on the appropriate mode of international cooperation	Non-Legal Provision	
5	Priorities in international cooperation	Non-Legal Provision	
6	Consideration of development goals	Non-Legal Provision	
7	Role of the United Nations and the Committee on the Peaceful Uses of Outer Space	Non-Legal Provision	
8	Contribution to United Nations Program on Space Applications	Non-Legal Provision	

Source: Adapted, with some variation, from Ricky J. Lee and Steven R. Freeland (2003, 130)

Paragraphs 1 and 2 may be classified as Applying Provisions in that they are a specific application of existing treaty provisions in the context of international cooperation. Paragraph 2, for example, provides that States are free to determine all aspects of their participation in international space endeavours while ensuring compliance with legitimate interests of other States. In this context, the relevant provisions of the Outer Space Treaty that support these provisions as having some legal effect in customary international law include the following obligations requiring States to:

(1) undertake space activities with due regard to the corresponding interests of other States and Article I of the Outer Space Treaty in relation to freedom of exploration and use;[138]
(2) consider granting requests by other States to observe flights of space objects on the basis of equality;[139] and
(3) inform the Secretary-General of the United Nations, the public and the scientific community of the nature, conduct, locations and results of their space activities.[140]

3.2.3 Jus Cogens: Space Law Principles as Possible Peremptory Norms of International Law

At least one other possibility exists for the principles contained in the United Nations space treaties and General Assembly declarations to have a binding effect on States that are not party to the treaties, which occurs if some or all of them have attained

---

[138] Outer Space Treaty, Article IX.

[139] Ibid., Article X.

[140] Ibid., Article XI.

3.2 Sources of Space Law

the status of *jus cogens*, or a "peremptory norm of general international law". This is especially so considering the difficulty involved in finding a sufficient body of state practice to support their crystallisation into customary international law. Article 53 of the Vienna Convention on the Law of Treaties provides that:

> A treaty is void if, at the time of its conclusion, it conflicts with a peremptory norm of general international law. For the purposes of the present Convention, a peremptory norm of general international law is a norm accepted and recognised by the international community of States as a whole as a norm from which no derogation is permitted and which can be modified only by a subsequent norm of general international law having the same character.[141]

Commentators have suggested that Article 53 evidences "a *new* source, one that manifestly involves an intent of the community, as expressed in a community-wide forum, to create general norms directly".[142] It is argued that this new source of international law is created only when the international community assemble at a forum such as the General Assembly or at a universal international conference and without the need for state practice to have accumulated.[143] Danilenko, in a detailed study on the process of creating principles of *jus cogens*, concluded:

> It is generally recognised that in order to acquire the quality of *jus cogens* a norm must first pass the normative tests for rules of "general international law". It is also established that, secondly, such a norm must be "accepted and recognised" as a peremptory norm by "the international community of States as a whole". .... If the requirement of the acceptance and recognition by the international community of States "as a whole" is interpreted to mean the recognition by all the essential components of the international community, then the concept of *jus cogens* establishes a very strict threshold .... It follows that if there is an opposition to the proposed peremptory rule on the part of States comprising an important element of the international community, such a dissent would prevent the emergence of a rule of *jus cogens*.[144]

Generally, while it is unlikely that one single State can constitute an "important element of the international community", two observations must be made when considering the creation of *jus cogens* within the framework of international space law. Firstly, the relatively small number of States involved in space activities lends support to the view that, in space law, each spacefaring State may well constitute an "important element of the international community", along with the non-spacefaring States together as a bloc forming an important element in itself. Secondly, some of the principles concerning space law are contained in treaties that have not achieved a high number of ratifications or in General Assembly declarations. Subsequently,

---

[141] (1969) 1155 U.N.T.S. 331.

[142] Nicholas G. Onuf and Richard K. Birney, *Peremptory Norms of International Law: Their Source, Function and Future* (1974) 4 DENVER J. INT'L. L. & POL'Y. 187 at 195.

[143] José Joaquin Caicedo Perdomo, *La Teoría del Jus Cogens en Derecho Internacional a la Luz de la Convención de Viena Sobre el Derecho de los Tratados* (1975) 206–207 REV. ACA. COLOM. JURIS. 259. at 265.

[144] Gennady M. Danilenko, *International Jus Cogens: Issues of Law-Making* (1991) 2 EUR. J. INT'L. L. 42 at 55.

126   3 State Responsibility and Liability for Compliance with International Space Law

most States and commentators are of the view that they are not capable of creating custom, let alone *jus cogens*.[145]

Subject to these qualifications, it is arguable that some of the fundamental principles of space law contained in the Outer Space Treaty may have attained the status of *jus cogens*. Notable examples of this may include the principle of non-appropriation in Article II, the rescue and recovery obligations under Article V and the state responsibility provision in Article VI. Their overwhelming acceptance by States, including both parties to the Outer Space Treaty and otherwise, as well as the absence of clear objections during the negotiations lends support to that view. However, it is also clear that most of the other principles contained in other United Nations space treaties are unlikely to have been intended by the States to attain the status of *jus cogens*. Accordingly, an assessment of their status as principles of custom would nevertheless be necessary in order to determine their applicability to States that are not party to the treaties.

3.2.4 Other Space-Related Treaties

In addition to the treaties and General Assembly declarations as developed by the Committee on the Peaceful Uses of Outer Space, there are also other treaties and declarations that are applicable to space activities. This is the consequence of Article III of the Outer Space Treaty in extending the application of international law, particularly the Charter of the United Nations, to outer space. For example, in considering the law on the use of force in outer space, particular attention must be paid to the General Assembly resolutions on the use of military force on the Earth.[146]

There are other treaties and instruments on space law that are created by specialised agencies and international organisations for specific space applications. For example, the ITU regulates the use of radio frequencies in space and the allocation of orbital slots in the geostationary orbit through its Constitution and Convention and its subordinate Radio Regulations. The United Nations Educational, Scientific and Cultural Organisation ("UNESCO") and the World Intellectual Property Organisation ("WIPO") adopted the Brussels Convention Relating to the Distribution of Programme-Carrying Signals Transmitted by Satellite to provide international protection for the intellectual property rights in such programs.[147]

Significantly, a large number of bilateral and multilateral agreements have been used by States to provide a legal framework for specific institutions or activities.

---

[145] See discussion in Danilenko, supra note 144, at 62–63.

[146] See, for example, the Declaration on Principles of International Law Concerning Friendly Relations and Cooperation Among States in Accordance with the Charter of the United Nations, General Assembly Resolution 2625 (XXV) of 1970.

[147] Brussels Convention Relating to the Distribution of Programme-Carrying Signals Transmitted by Satellite, opened for signature on 21 May 1974, 1144 U.N.T.S. 3; 13 I.L.M. 1444 (entered into force on 25 August 1979).

The Convention for the Establishment of a European Space Agency (the "ESA Convention") and the specific treaties creating international institutions such as INTELSAT, INMARSAT and other regional satellite systems have played an important contribution to the development of space law.[148] Recently, States have also adopted agreements to deal with specific activities, such as the bilateral agreement between Australia and Russia in the field of cooperation on the regulation of launch activities,[149] and the Intergovernmental Agreement establishing the International Space Station.[150]

3.2.5 Space Law and the Lex Specialis Principle

The long-established principle *lex specialis derogat legi generali* provides that the special rules, the *lex specialis*, apply in preference to the general rules, the *lex generali*, even if the special rules derogate from the general rules. The *lex specialis* principle is not a new one in the development of public international law, for as early as in Roman civil law, the *Corpus Iuris Civilis* noted that "*in toto iure generi per speciem derogatur et illud potissimum habetur, quod ad speciem derectum est*".[151] Further, both Grotius and Vattel have expressed the view that rules that specifically regulate a subject matter are to apply in precedence to rules of a more general nature.[152]

It is an observation made by quite a number of commentators that the existing body of space law constitutes *lex specialis*.[153] Taken at its highest, a body of law that is *lex specialis* may apply as a legal code to the exclusion of other general principles of international law. The constituent instruments of the ITU, taken together, are

---

[148] See, for example, the Convention for the Establishment of a European Space Agency, opened for signature on 30 May 1975, 14 I.L.M. 864 (entered into force on 30 October 1980); the International Telecommunications Satellite Organization (INTELSAT) Agreement Between the United States of American and Other Governments and Operating Agreement, opened for signature on 20 August 1971, 23 U.S.T. 3813; T.I.A.S. 7532 (entered into force on 12 February 1973); and the Convention on the International Maritime Satellite Organisation (INMARSAT), opened for signature on 3 September 1976; 31 U.S.T. 1; T.I.A.S. 9605 (entered into force on 16 July 1979).

[149] Bilateral Agreement, supra note 8.

[150] International Space Station Intergovernmental Agreement, supra note 8.

[151] "In the whole of law, special takes precedence over genus, and anything that relates species is regarded as most important": Papinian, THE DIGEST OF JUSTINIAN, vol. IV (1985).

[152] Hugo Grotius, DE JURE BELLI AC PACIS. LIBRI TRES (1646), Book II, Ch. XVI, Sec. XXIX, at 428 and Emmerich de Vattel, LE DROIT DES GENS OU PRINCIPES DE LA LOI NATURELLE, APPLIQUES A LA CONDUITE ET AUX AFFAIRES DES NATIONS ET DES SOUVERAINS (1758), Vol. 1, Book II, Ch. XVII, at 511.

[153] See, for example, Hamilton DeSaussure and Peter P. C. Haanappel, *A Unified Multinational Approach to the Application of Tort and Contract Principles to Outer Space* (1978) 6 SYRACUSE J. INT'L. L. 1 at 10; Venkateswara S. Mani, *Development of Effective Mechanism(s) for Settlement of Disputes Arising in Relation to Space Commercialisation* (2001) 5 SING. J. INT'L. & COMP. L. 191 at 211; and Ricky J. Lee. *Reconciling Space Law and the Commercial Realities of the Twenty-First Century* (2000) 4 SING. J. INT'L. & COMP. L. 194.

an example of such a body of law. If applied to space law, the specific principles contained in the Outer Space Treaty and other United Nations space law instruments would apply to the exclusion of any other general principle of public international law. However, against this position are the express terms of Article III of the Outer Space Treaty, which state that:

> State Parties to the Treaty shall carry on activities in the exploration and use of outer space, including the Moon and other celestial bodies, in accordance with international law, including the Charter of the United Nations, in the interest of maintaining international peace and security and promoting international cooperation and understanding.

It is clear that, accordingly, general principles of international law would continue to have application to activities in outer space, although as a matter of practice the principles of space law as contained in the relevant treaties would have priority over such general rules.

3.3 State Responsibility and Jurisdiction

3.3.1 State Responsibility

3.3.1.1 Article VI of the Outer Space Treaty

In the increasingly private and multinational space sector, Article VI has become a provision of much concern to States whose private nationals are engaged in space activities. Article VI states that:

> State Parties to the Treaty shall bear international responsibility for national activities in outer space, including the Moon and other celestial bodies, whether such activities are carried on by governmental agencies or by non-governmental entities, and for assuring that national activities are carried out in conformity with the provisions set forth in the present Treaty. The activities of non-governmental entities in outer space, including the Moon and other celestial bodies, shall require authorisation and continuing supervision by the appropriate State Party to the Treaty. When activities are carried on in outer space, including the Moon and other celestial bodies, by an international organisation, responsibility for compliance with this Treaty shall be borne by the international organisation and by the State Parties to the Treaty participating in such organisation.

It is clear from this that Article VI imposes the following obligations on States:

(1) to bear responsibility for national activities in outer space regardless of whether such activities are carried out by public or private entities;
(2) to assure that national activities are conducted in conformity with the Outer Space Treaty and, through Article III, with international law;
(3) to authorise and continually supervise, where appropriate, the activities of non-governmental entities in outer space; and
(4) to share international responsibility for the activities of international organisations of which the State is a participant.

3.3 State Responsibility and Jurisdiction

3.3.1.2 Content of Responsibility

Presumably, the first issue to be determined is the content of this "international responsibility" to be borne by States. In the past, some commentators have sought to distinguish the "responsibility" prescribed under Article VI with "liability" as imposed under Article VII and the provisions of the Liability Convention.[154] From this perspective, Article VI would do no more than to prescribe a regulatory responsibility on States without the imposition of any liability on the State. In recent times, however, much emphasis has been placed by some commentators on the use of terms in other languages that are equally authentic for the purposes of interpreting the Outer Space Treaty.[155] In the French text, the term "responsabilité internationale" is used in both Articles VI and VII in place of both international "responsibility" and "liability" in the English text. Similarly, the Chinese term「国际责任」, the Russian term "международную ответственность" and the Spanish term "responsables internacionalmente" are used as the equivalent term for both "international responsibility" and "international liability" under Articles VI and VII. Accordingly, if there is to be no differentiation in meaning between "responsibility" under Article VI and "liability" under Article VII, then Article VI must be interpreted to mean that States are to be internationally liable for *national* space activities conducted by both public and private entities.[156] This is particular so considering Article 33(3) of the Vienna Convention on the Law of Treaties provides that the terms used in a treaty is presumed to have the same meaning in each authentic text.[157]

This is consistent with the principles of state responsibility in the body of customary international law in that liability for reparations must follow from a violation of international law. For example, in *Chorzów Factory (Indemnity) (Merits)*, the Permanent Court of International Justice held that:

> [It] is a principle of international law, and even a general conception of law, that any breach of an engagement involves an obligation to make reparation. ... the Court has already said

---

[154] See, for example, Ian Awford, *Commercial Space Activities: Legal Liability Issues*, in V. S. Mani, S. Bhatt and V. B. Reddy (eds.), RECENT TRENDS IN INTERNATIONAL SPACE LAW AND POLICY (1997) at 388.

[155] Article XVII of the Outer Space Treaty provides that the Chinese, English, French, Russian and Spanish texts are equally authentic. For example of recent commentaries that have made this observation, see Armel Kerrest, *Remarks on the Responsibility and Liability for Damage other than Those Caused by the Fall of a Space Object* (1997) 40 PROC. COLL. L. OUTER SP. 134.

[156] Some commentators have also noted that the *travaux préparatoires* and many domestic laws in civil law jurisdictions do not draw a distinction between "responsibility" and "liability". See, for example, Bin Cheng, *Article VI of the Outer Space Treaty Revisited: "International Responsibility", "National Activities" and "The Appropriate State"* (1998) 26 J. SP. L. 10 and Motoko Uchitomi, *State Responsibility / Liability for "National" Space Activities: Towards Safe and Fair Competition in Private Space Activities* (2001) 44 PROC. COLL. L. OUTER SP. 51.

[157] Article 33(4) of the Vienna Convention on the Law of Treaties further provides that, where there is a difference in meaning across different texts, the meaning that best reconciles the texts is to be adopted: Vienna Convention on the Law of Treaties, opened for signature on 23 May 1969, 1155 U.N.T.S. 331 (entered into force on 27 January 1980).

that reparation is the indispensable complement of a failure to apply a convention, and there is no necessity for this to be stated in the convention itself.[158]

Consequently, Article VI should be considered to have the effect of imposing liability on a State for activities in outer space that may be attributable to the State.

3.3.1.3 Imputability to the State and the Duty to Authorise and Continually Supervise Private Space Activities

Article VI requires for States to take international responsibility for "national" space activities conducted by both public and private entities. Generally, the space activities carried out by public entities are acts attributable to the State and, accordingly, they are activities for which the State must take international responsibility.[159] Further, if any damage or harm is caused to other States, the State is liable to pay reparations to the victim States to restore them as much as possible to their positions before the damage was inflicted (*restitutio in integrum*).[160] The issue, therefore, is whether Article VI is merely a restatement of the existing principle of state responsibility or if it expands the duty imposed on States.

Under general principles of international law, particularly with reference to the jurisprudence of the Iran-U.S. Claims Tribunal, the imputability of a "private" act to the State would appear to depend on an objective determination of any influence over, or benefit derived from, the activity that may be attributed to the State. If the acts are conducted by private persons or entities without the direction or influence of the State, then such acts are generally not imputable to the State.[161] In *Foremost Tehran Inc. v. Iran*, a company decided not to pay dividends to its shareholders, one of which was the claimant USA company. The Tribunal imputed that decision to the State because the company acted under the influence of some of its directors who were appointed by the Iranian Government and it was implementing government policy concerning the financial interests of foreigners.[162] In *Flexi-Van Leasing Inc. v. Iran*, it was held that, even if the entity was under the control of the State, it must be demonstrated that the specific conduct itself was directed or influenced by the State for it to be imputable to the State.[163]

The obligation under Article VI, with its qualification on "national" activities, may be seen as being no more than a restatement of the existing international law

---

[158] *Chorzów Factory (Indemnity) (Merits)* (1928) P.C.I.J. REP., Ser. A, No. 17, at 29.

[159] International Law Commission's Draft Articles on State Responsibility, Articles 5–7. See also *Massey (United States v. Mexico)* (1927) 4 R.I.A.A. 155.

[160] *Chorzów Factory*, supra note 158, at 46–48.

[161] International Law Commission's Draft Articles on State Responsibility, Article 11.

[162] *Foremost Tehran Inc. v. Iran* (1986) 10 Iran-U.S.C.T.R. 228.

[163] *Flexi-Van Leasing Inc. v. Iran* (1986) 12 Iran-U.S.C.T.R. 335. In that case, a private Iranian company under the control of the Iranian Government committed acts of expropriation, but the Tribunal held that they were not imputable to the State as it could not be demonstrated that the acts themselves were under the direction or influence of the State.

3.3 State Responsibility and Jurisdiction

before the adoption of the Outer Space Treaty itself.[164] In other words, States are to bear international responsibility for activities in outer space that are conducted under the State's direction or influence, regardless of whether the activities are conducted by public or private entities. To some extent, this can be seen as the logical interpretation of Article VI as States should not have to bear responsibility for acts beyond its control, direction or influence. For example, the acts and omissions of a Belgian national operating in Mexico ought not to be attributable to the Government of Belgium where:

(1) the Belgian national is not empowered by the law of Belgium to exercise elements of governmental authority;[165]
(2) the Belgian national is acting on the instructions of, or under the direction or control of, the Government of Belgium;[166]
(3) the acts have not been acknowledged or adopted by Belgium as its own.[167]

However, the analysis must not end there, as Article VI imposes a further requirement that the "appropriate" State is to authorise and continually supervise the space activities of private entities. This obligation is not qualified or confined by the use of the term "national" and, accordingly, is an obligation imposed on the State concerning all private activities, regardless of the existing degree of State control, direction or influence over the activity. Through the acts of authorisation and continuing supervision, the State would be asserting some degree of control, direction or influence over the private space activity, thus making it a "national" activity for which the State bears international responsibility. This produces the overall effect of requiring a State to bear international responsibility for all public or private space activities under its control, direction or influence, including those that it authorises and continually supervises as the "appropriate" State.

This conclusion has two ancillary effects. The first is that this international responsibility would apply to the State even if the relevant private space activity was conducted outside the territorial jurisdiction of the State, provided that the activity can be attributed to the State. This is consistent with the position taken under existing customary international law, as discussed by the International Court of Justice

---

[164] See, for example, Elisabeth Back-Impallomeni, *The Article VI of the Outer Space Treaty*, in United Nations, PROCEEDINGS OF THE UNITED NATIONS/REPUBLIC OF KOREA WORKSHOP ON SPACE LAW (2003), 348–351 and Ricky J. Lee, *Liability Arising from Article VI of the Outer Space Treaty: States, Domestic Law and Private Operators* (2005) 48 PROC. COLL. L. OUTER SP. 216.

[165] International Law Commission ("ILC") Articles on Responsibility of States for Internationally Wrongful Acts, Articles 5 and 7. See, for example, *Hyatt International Corporation v. Government of the Islamic Republic of Iran* (1985) 9 Iran-U.S.C.T.R. 72 at 88–94.

[166] Ibid., Article 8. See *Prosecutor v. Tadic* (1999) 38 I.L.M. 1518 and *Military and Paramilitary Activities*, supra note 129, at 14.

[167] Ibid., Article 11. See *U.S. Diplomatic and Consular Staff in Tehran* [1980] I.C.J. Rep. 3 and *Lighthouses* (1956) 12 R.I.A.A. 155.

STATE RESPONSIBILITY UNDER ARTICLE VI: SUMMARY OF ANALYSIS

States to bear international responsibility for "national" activities	→ "National" activities defined as those under State control, direction or influence	→ States bear international responsibility for private activities under State control, direction or influence
State to authorise continually supervise all space activities of its private entities	→ By authorising and supervising the activity, it comes under State direction and influence	States bear international responsibility for all private space activities

Fig. 3.2 State responsibility under Article VI: summary of analysis
Source: Lee, *Commentary Paper on Discussion Paper Titled "Commercial Use of Space, Including Launching" by Prof. Dr. Armel Kerrest*, in China Institute of Space Law, PROCEEDINGS OF THE 2004 SPACE LAW CONFERENCE (2004), at 220–231

and the International Criminal Tribunal for the Former Yugoslavia.[168] Further, this is also consistent with the approach adopted under the Liability Convention, which imposes liability on a State for the launch activities of its nationals, even if these activities take place outside the sovereign territory of that State (Fig. 3.2).[169]

3.3.1.4 The "Appropriate" State and the Duty to Authorise and Continually Supervise National Space Activities

The second ancillary effect is that the State of nationality may find itself in a situation where it is unable to supervise the space activities of its private nationals. For example, if the national, domiciled outside the territorial jurisdiction of his or her State of nationality, conducts his or her space activities within the territorial jurisdiction of another State, it would be difficult, if not impossible, for the State of nationality to authorise and continually supervise those space activities. Kerrest suggested that this difficulty is in itself a breach of international law:

> This is of course in total breach of international law. States have personal jurisdiction over their nationals. They must keep the capacity to implement international law in general and space law in particular and to make it applicable to their citizens whether they are natural or legal persons.[170]

When considered in the context of the duties imposed under Article VI being implemented through domestic legislation that carry criminal sanctions, this view

---

[168] See *Tadic*, supra note 166; and *Military and Paramilitary Activities*, supra note 129.

[169] Liability Convention, Article I.

[170] Armel Kerrest, *Commercial Use of Space, Including Launching* (2004), in China Institute of Space Law, 2004 SPACE LAW CONFERENCE: PAPER ASSEMBLE 199.

3.3 State Responsibility and Jurisdiction

does find some support in the existing body of international law. This is most notably the case in *S.S. Lotus*, in which the Permanent Court of International Justice held that Turkey was capable of extending its criminal jurisdiction over French nationals arrested in Turkey for committing a crime under Turkish law in the high seas.[171] As the Court stated:

> Though it is true that in all systems of law the principle of the territorial character of criminal law is fundamental, it is equally true that all or nearly all of these systems of law extend their action to offences committed outside the territory of the State which adopts them. . . . The territoriality of criminal law, therefore, is not an absolute principle of international law and by no means coincide with territorial sovereignty.[172]

Further in support of the ability of the "appropriate State" to implement its obligations imposed under Article VI of the Outer Space Treaty, Dickinson stated in his introductory comment on the Harvard Research Draft Convention on Jurisdiction with Respect to Crime that:

> The competence of the State to prosecute and punish its nationals on the sole basis of their nationality is based upon the allegiance which the person charged with crime owes to the State of which he is a national. . . . If international law permits a State to regard the accused as its national, its competence is not impaired or limited by the fact that he is also a national of another State.[173]

However, it must be noted that it is not in the legal competence of the State of nationality, but rather its physical inability to enforce its laws over nationals domiciled within the territorial sovereignty of another State that causes difficulties for the State of nationality to act as the "appropriate" State. Even in *Lotus*, which Brierly suggested to be "based on the highly contentious metaphysical proposition of the extreme positivist school",[174] the Court stated that the "exclusively territorial character of law relating to this domain constitutes a principle which, except as otherwise expressly provided, would *ipso facto*, prevent States from extending the criminal jurisdiction of their courts beyond their frontiers".[175] In other words, the personal jurisdiction that the State of nationality has over its nationals would allow for the later prosecution of any crimes committed by that national upon his or her return, but would not prevent that State from fulfilling its duty to authorise and continually supervise the space activities of that national.

In any event, if the "appropriate" State is intended to be the State of nationality, then it is doubtful that the drafters of the Outer Space Treaty would have found it necessary to invent a new term to describe it. Indeed, the term "appropriate" is best read with reference to the context in which it is placed, namely the act of authorising and continually supervising the space activities of private entities. In

---

[171] *S.S. Lotus (France v. Turkey)* (1927) P.C.I.J. REP., Ser. A, No. 10.

[172] Ibid., at 20.

[173] Edwin D. Dickinson, *Introductory Comment to the Harvard Research Draft Convention on Jurisdiction with Respect to Crime* (1935) 29 AM. J. INT'L. L. SUPP. 443 at 519 *et seq.*

[174] James L. Brierly, *The Lotus Case* (1928) 44 L. Q. REV. 154 at 155.

[175] *Lotus*, supra note 171, at 20.

134   3   State Responsibility and Liability for Compliance with International Space Law

the *travaux préparatoires* of the Principles Declaration, there was an U.S. proposal that contained the following provision:

> A state or international organisation from whose territory or with whose assistance or permission a space vehicle is launched bears international responsibility for the launching, and is internationally liable for personal injury, loss of life or property damage caused by such vehicle on the Earth or in air space.[176]

Therefore, it may be concluded that the "appropriate" State is better defined as the State in the best position to assert direct and immediate jurisdiction over the private entity to authorise and continually supervise its activities. In the case of a space activity conducted by a private entity within the territory of its State of nationality, the "appropriate" State is clearly that State. In the case of a private entity operating outside its State of nationality, however, the State in the best position to authorise and continually supervise is the State with territorial jurisdiction over the activities of that private entity. Consequently, in most circumstances, the territorial State may be designated as the "appropriate" State, a proposition that is supported by some eminent commentators of space law.[177]

In defining the term "appropriate State" as the territorial State, there are three implications that should be noted:

(1) this is a fairer outcome as the State of nationality should not be placed in a position where it must fulfil an impossible legal obligation;
(2) the problem of "double jeopardy" is avoided as the "appropriate" State and only that State is responsible under the Outer Space Treaty for authorising and continually supervising the space activities conducted by private entities within its territorial jurisdiction; and
(3) this definition does not affect or prejudice the effect of Article VII of the Outer Space Treaty or the provisions of the Liability Convention in imposing liability on the State of nationality, regardless of whether it had authorised and continually supervised the activity or otherwise.

3.3.2 *Jurisdiction*

On the issue of jurisdiction, Article VIII of the Outer Space Treaty provides that:

> A State Party to the Treaty on whose registry an object launched into outer space is carried shall retain jurisdiction and control over such object, and over any personnel thereof, while in outer space or on a celestial body. . . .

---

[176] U.N. Doc. A/C1/881 (14 October 1962).

[177] See, for example, Istvan Herczeg, *Interpretation of the Space Treaty of 1967 (Introductory Report)* (1967) 10 PROC. COLL. L. OUTER SP. 105 at 107; Stephen Gorove, *Liability in Space Law: An Overview* (1983) 8 ANN. AIR & SP. L. 373 at 377; and Michel Bourély, *Rules of International Law Governing the Commercialisation of Space Activities* (1986) 29 PROC. COLL. L. OUTER SP. 157 at 159.

3.3 State Responsibility and Jurisdiction

It is clear that the exercise of quasi-territorial jurisdiction over space objects depends on the carriage of that space object on the domestic registry of the State. In the case of where only one State is involved in the launch of a particular space object, this does not pose any difficulties. On the United Nations Register of Space Objects, almost all States have registered nearly all their space objects, either pursuant to the Registration Convention or on a voluntary basis under General Assembly Resolution 1721B (XVI).

However, where there is more than one launching State, these States must elect among themselves the State that would register the space object and, accordingly, exercise quasi-territorial jurisdiction and control over the space object pursuant to Article VIII of the Outer Space Treaty.

Article II(2) of the Registration Convention provides that:

> Where there are two or more launching States in respect of any such space object, they shall jointly determine which one of them shall register the object . . . , bearing in mind the provisions of Article VIII of the [Outer Space Treaty] and without prejudice to appropriate agreements concluded or to be concluded among the launching States on jurisdiction and control over the space object and over any personnel thereof.

For example, the United States adopted the position that it will only register all space objects that are owned by U.S. governmental or private entities, regardless of where they were launched and that non-U.S. payloads launched from the United States should be registered by the State whose nationals own the payload.[178] This is a positive development in terms of clarifying the practice of States in deciding on the appropriate State of registry between multiple launching States, as it reflects the commercial reality that it is the owner of the payload, rather than the other launching States, that would most desire the retention of jurisdiction and control.

What remains a subject of substantial academic debate is the problem of the transfer of registrations and, along with that transfer, the jurisdiction and control over the space object. Neither the Outer Space Treaty nor the Registration Convention contain any provisions for the transfer of the registration of a space object. This was seen as a particular issue during the privatisation of INMARSAT and the handover of the former British colony of Hong Kong to the People's Republic of China.

Several commentators have suggested that an amendment to the Registration Convention would ultimately be necessary in order to address this issue.[179] This

---

[178] See also Kenneth Hodgkins, *International Cooperation in the Peaceful Uses of Outer Space*, Remarks on Agenda Item 75 in the Fourth Committee of the United Nations General Assembly, 9 October 2002, at <http://www.state.gov/g/oes/rls/rm/2002/14362.htm>, last accessed on 9 April 2004.

[179] See, for example, Peter H. van Fenema, *The Registration Convention* (2002) in United Nations, PROCEEDINGS OF THE UNITED NATIONS WORKSHOP ON CAPACITY BUILDING IN SPACE LAW 33 and Kai-Uwe Hörl and Julian Hermida, *Change of Ownership, Change of Registry? Which Objects to Register, What Data to be Furnished, When, and Until When?* (2003) in United Nations, PROCEEDINGS OF THE I.I.S.L. / E.C.S.L. SYMPOSIUM: REINFORCING THE REGISTRATION CONVENTION 15 at 18.

position may be unnecessary when due consideration is given to existing state practice. Recently, both the Netherlands and the United Kingdom, parties to the Registration Convention, have sought to register space objects to which they were not launching States.[180] Particularly in the first case, the Netherlands asserted the view that it was not obliged to furnish any data pursuant to the Registration Convention as it was not a launching State but, having placed the satellites on its national register, was entitled to assert jurisdiction and control over the space object pursuant to Article VIII of the Outer Space Treaty.[181]

It is obvious that a *duty* to register the space object is imposed under the Registration Convention only on launching States and, accordingly, in order to register these space objects as non-launching States, both States must have found a *right* to register in the body of space law. As the Registration Convention clearly does not provide this right, the other logical candidates are Article VIII of the Outer Space Treaty with reference to domestic registries and General Assembly Resolution 1721B, with reference to the United Nations register. In this way, it is not necessary for the Registration Convention to be amended, but merely for the new State of registry to seek the registration of the transferred space object under Article VIII of the Outer Space Treaty and General Assembly Resolution 1721B to assert jurisdiction and control over the space object.

3.4 Liability

3.4.1 Overview

The fundamental concern of any enterprise in outer space would be liability. From the very beginnings of international space law, it has been recognised that States would have to accept international liability for any damage or injury they cause to third parties through the conduct of space activities. This is partly because space activities have been regarded by the international community as being inherently risky and dangerous and, consequently, third party States should be protected from any injury, loss or damage suffered resulting from the conduct of activities in outer space.

When the Outer Space Treaty was adopted by the United Nations General Assembly in 1968, space activities were the exclusive domain of the Soviet Union and the United States.[182] This remained the case when the Liability Convention was adopted in 1972.[183] At the time, there were no international joint efforts, even

---

[180] U.N.Doc. A/AC.105/806 (22 August 2003) for the purchase of satellites by NewSkies Satellites, a Dutch company; and U.N. Doc. ST/SG/Ser.E/417 (25 September 2002) for the registration of satellites of the privatised Inmarsat Limited, now a British company.

[181] U.N.Doc. A/AC.105/806 (22 August 2003).

[182] (1967) 610 U.N.T.S. 205; 6 I.L.M. 386.

[183] (1972) 961 U.N.T.S. 187; 10 I.L.M. 965.

without the participation of the private sector, in space activities. Three decades later, however, most space activities today are conducted by commercial concerns operating on a multinational level. The space treaties, in particular the Liability Convention, are proving to be inadequate in addressing the issues of third party liability, private space activities and the settlement of disputes.[184]

The privatisation and commercialisation of space activities in recent decades have prompted several States to pass on their international liability for private space activities to the launch operators. In order to assess comprehensively the liability regime applicable to private space activities, it is necessary not only to consider the international treaties but also the relevant domestic legislation concerning private space activities.[185]

3.4.2 International Liability

3.4.2.1 Development of the Liability Convention

One of the earliest issues debated among legal scholars with an interest in space law was the subject of responsibility and liability. As early as 1958, it was suggested that:

(1) the State launching a spacecraft accept full international responsibility for any possible damage;
(2) the State be entitled to make certain reservations as under the Warsaw Convention excluding, for example, liability in the case of force majeure; and
(3) an International Guaranty Fund be created to pay for damage caused by satellites except for intentional acts.[186]

In 1959, coinciding with this debate in academic circles, the Government of the United States circulated a proposal within the United Nations which suggested that, among other matters, the question of international liability for damage caused by the launching, flight and re-entry of payloads and associated launch vehicles

---

[184] See generally Karl-Heinz Böckstiegel, *Settlement of Disputes Regarding Space Activities* (1993) 21 J. Sp. L. 1; Philip D. Bostwick, *Going Private with the Judicial System: Making Creative Use of ADR Procedures to Resolve Commercial Space Disputes* (1995) 23 J. Sp. L. 1; Alexis Goh, *Coping with the Lack of a Mechanism for the Settlement of Disputes Arising in Relation to Space Commercialisation* (2001) 5 SINGAPORE J. INT'L. & COMP. L. 180; Patricia M. Sterns and Leslie I. Tennen, *Resolution of Disputes in the Corpus Juris Spatialis: Domestic Law Considerations* (1993) 36 PROC. COLL. L. OUTER SP. 172; and Hennake L. van Traa-Engelman, *Settlement of Space Law Disputes* (1990) 3 LEIDEN J. INT'L. L. 139.

[185] See Ricky J. Lee, *The Liability Convention and Private Space Launch Services – Domestic Regulatory Responses* (2006) 31 ANN. AIR & SP. L. 351.

[186] Isabella H. Ph. Rode-Verschoor, *The Responsibility of States for the Damage Caused by Launched Space-Bodies* (1958) 1 PROC. COLL. L. OUTER SP. 103.

as a priority issue.[187] During the first meeting of the Legal Sub-Committee to COPUOS in 1962, a set of substantive principles on liability was proposed by the United States.[188] Subsequently, it was agreed by all participating States to include a provision relating to liability in the Principles Declaration.[189]

The provision found in the Principles Declaration was substantially reproduced in Article VII of the Outer Space Treaty, stating:

> Each State Party to the Treaty that launches or procures the launching of an object into outer space, including the Moon and other celestial bodies, and each State Party from whose territory or facility an object is launched, is internationally liable for damage to another State Party to the Treaty or to its natural or juridical persons by such object or its component parts on the Earth, in air space or in outer space, including the Moon and other celestial bodies.

Considering the widespread acceptance of the Outer Space Treaty, it is likely that its terms or at least some of its essential provisions may be considered to have crystallised into customary law. The indicia of their crystallisation include the widespread acceptance of the Outer Space Treaty, the absence of objections by States and the repetition of its provisions in subsequent instruments.[190] The difficulty in declaring with any certainty the extent of the crystallisation of the provisions of the Outer Space Treaty, especially those dealing with responsibility and liability, is the continuing absence of state practice that remains a prerequisite for the formation of custom.[191]

The United States made a series of proposals for a set of principles on liability to the Legal Sub-Committee through the 1960s.[192] In its view, there must be four essential elements for a working international treaty on liability for activities in outer space:

(1) an explicit rule that the demonstration of fault cannot be a requirement of or prerequisite to liability;
(2) the standards to be applied to evaluate the damage suffered and the appropriate compensation payable;

---

[187] U.N. Doc. A/AC.105/C.2/L.4.

[188] U.N. Doc. A/AC.105/C.2/L.4.

[189] General Assembly Resolution 1962 (XVIII).

[190] See, for example, the discussion in Vladlen S. Vereshchetin and Gennady M. Danilenko, *Custom as a Source of International Law of Outer Space* (1985) 13 J. Sp. L. 22; D. Krstic, *Customary Law Rules in Regulating Outer Space Activities* (1977) 20 Proc. Coll. L. Outer Sp. 320; and Lee and Freeland (2003, 127).

[191] *Military and Paramilitary Activities in and against Nicaragua (Merits) (Nicaragua v. United States)* [1986] I.C.J. Rep. 14 at 94 and *North Sea Continental Shelf Cases (Germany v. Denmark; Germany v. the Netherlands)* [1969] I.C.J. Rep. 3. See, for example, Brownlie, supra note 123, at 4–11.

[192] U.N. Doc. A/AC.105/C.2/L.8.

3.4 Liability 139

(3) a denial of the traditional requirement for the claimant to exhaust all appropriate local remedies; and
(4) the imposition of specific time limits on negotiations for settlements and the establishment of a theoretically impartial claims commission to "advise" the parties.[193]

In addition to the proposals put forward by the United States, there were also several proposed texts from Belgium.[194] Reis, legal adviser to the United States Mission to the United Nations at the time, suggested that the Soviet Union did not pay serious attention to such proposals and "preferred ... not to put forward proposals under its own name but instead to rely upon Hungary".[195] Such a characterisation would appear to understate the contributions that the representatives of the Soviet Union made towards the formulation and discussion of the treaty.

While the Hungarian proposals did not differ from the United States proposals on the subject of absolute liability, they did suggest that the nature and amount of compensation payable should be determined by the law of the launching State.[196] The proposals also included a provision that rules of exception or exoneration from liability should have no application for "unlawful activities".[197]

After more than a decade of negotiations, the Liability Convention was adopted by the General Assembly in 1972. It was observed that the Liability Convention contained the fundamental elements sought by the United States through its proposals, while some less fundamental proposals were excluded in the interest of reaching a compromise. Reis suggested that it gave "maximum assurance that a launching State which has ratified the convention will pay a just claim" and encourages space powers not to "deal arrogantly with justified damage claims" from claimant States.[198] In order to scrutinise such claims, the provisions of the Liability Convention should be examined in detail.

3.4.2.2 Liability Provisions

The Liability Convention introduces the concept of a "launching State" which is subsequently used in the Convention on the Registration of Space Objects (the "Registration Convention") and other instruments on the law of outer space. Article I defines the terms "launching State" and "space object" with the cumulative effect that a "launching State" for the purposes of the Liability Convention includes:

---

[193] Herbert Reis, *Some Reflections on the Liability Convention for Outer Space* (1978) 6 J. Sp. L. 161.

[194] U.N. Doc. A/AC.105/C.2/L.7.

[195] Reis, supra note 193, at 126.

[196] U.N. Doc. A/AC.105/C.2/L.10/Rev.1, Article II.

[197] Ibid., Article V.

[198] Reis, supra note 193, at 128.

(1) a State that launches a space object, its component parts, its launch vehicle or parts thereof;

(2) a State that procures the launch of a space object, its component parts, its launch vehicle or parts thereof;

(3) a State from whose territory a space object, its component parts, its launch vehicle or parts thereof is launched; and

(4) a State from whose facility a space object, its component parts, its launch vehicle or parts thereof is launched.

From the above definition, it is clear that it is possible to have more than one launching State for each space object. For example, a satellite owned and to be operated by a French private concern to be launched by a German launch operator from a Russian facility located in Australia may result in France, Germany, Russia and Australia all being regarded as launching States. The Liability Convention imposes joint and several liability on the multiple launching States and each may present claims for indemnity or contribution from other launching States or to apportion their liability by agreement.[199] In the earlier days, when launch activities were the field of governmental agencies, this joint and several liability was much less of a concern than it is today with each segment of a launch operation being conducted by private multinational companies, making the imposition of international liability significantly more problematic.

Article II of the Liability Convention provides for absolute liability for any damage caused by space objects that is suffered on the surface of the Earth or in airspace. Whether the deep seabed would be considered part of the "surface of the Earth" is unclear. The provision states that:

> A launching State shall be absolutely liable to pay compensation for damage caused by its space object on the surface of the Earth or to aircraft in flight.

Provided that the space activity was conducted in accordance with international law, particularly the Charter of the United Nations and the Outer Space Treaty, the launching State would be exonerated from absolute liability to the extent that the damage resulted wholly or partly from the gross negligence or an intentional act or omission of the claimant State or its nationals and not from any unlawful act on the part of the launching State.[200] There is no explicit definition in the Liability Convention as to what would constitute gross negligence and this has been a matter of substantial academic discussion, though for present purposes a precise definition is probably unnecessary for the Liability Convention to apply to contemporary commercial applications.[201] For instance, Awford applied a more general meaning to the term "gross negligence" in such a context:

---

[199] Liability Convention, Article V.

[200] Ibid., Article VI.

[201] See, for example, Ronald E. Alexander, *Measuring Damages under the Convention on International Liability for Damage Caused by Space Objects* (1978) 6 J. SP. L. 151; Carl Q. Christol, *International Liability for Damage Caused by Space Objects* (1980) 74 AM. J. INT'L. L.

3.4 Liability

Exactly what is contemplated is not spelled out but if the claimant State caused or contributed to the damage it suffered by shooting down the Space Station of another State, using laser technology, presumably there will be complete exoneration. If, due to a lack of liaison, the communication or control system of the claimant State's satellite interferes with that of the satellite of another State, which then goes out of control and re-enters or collides with another satellite, then the facts may shown that the compensation for the damage needs to be apportioned in some way.[202]

Article III of the Liability Convention, on the other hand, provides that liability for damage caused in outer space by a space object will be determined on the basis of fault. Specifically, it states that:

> In the event of damage being caused elsewhere than on the surface of the Earth to a space object of one launching State or to persons or property on board such a space object by a space object of another launching State, the latter shall be liable only if the damage is due to its fault or the fault of persons for whom it is responsible.

It is uncertain at first glance what is meant by "another launching State" in Article III. However, considering the context of the provision as a whole, it is reasonable to assume that the appropriate meaning is "a launching State other than the launching State of the first object". In other words, Article III has application only where the damage caused is international and not domestic in nature.

Article IV (1) states:

> In the event of damage being caused elsewhere than on the surface of the Earth to a space object of one launching State or to persons or property on board such a space object by a space object of another launching State, and of damage thereby being caused to a third State or to its natural or juridical persons, the first two States shall be jointly and severally liable to the third State, to the extent indicated by the following:
>
> (a) if the damage has been caused to the third State on the surface of the Earth or to aircraft in flight, their liability to the third State shall be absolute;
> (b) if the damage has been caused to a space object of the third State or to persons or property on board that space object elsewhere than on the surface of the Earth, their liability to the third State shall be based on the fault of either of the first two States or on the fault of persons for whom either is responsible.

This provision deals with the situation where a collision occurs between two space objects in outer space and then causes damage to a third State, either on the surface of the Earth or in outer space. It should be noted that the provision deals only with the primary damage being caused in outer space and not if it is caused on the surface of the Earth, for in the latter case the liability would be absolute. In outer space, the liability of the two States is to be apportioned on the basis of the extent to

---

346; and Marc S. Firestone, *Problems in the Resolution of Disputes Concerning Damage Caused in Outer Space* (1985) 59 TUL. L. REV. 747.

[202] Ian Awford, *Legal Liability Arising from Commercial Activities in Outer Space*, paper presented at the Annual Conference of the International Bar Association, December 1990, in Paris, France.

which each one of them was at fault or, if the extent of fault cannot be established, proportioned equally between them.[203]

The Liability Convention does not apply to damage caused by a space object of a launching State to nationals of *that* launching State and to:

> Foreign nationals during such time as they are participating in the operation of that space object from the time of its launching or at any stage thereafter until its descent, or during such time as they are in the immediate vicinity of a planned launching or recovery area as the result of an invitation by that launching State.[204]

If the provision is to be interpreted with its prima facie meaning, the reference to *that* launching State instead of *a* launching State appears to indicate that, in the case of damage caused by a space object that has multiple launching States, a claim may be made by the nationals of one launching State against the other launching States. This ability of launching States to make claims against each other is questionable because if the claimant State presents a claim on behalf of its nationals against the other launching States, these States would have a right to indemnity or contribution from the claimant State, making the provision somewhat redundant. The appropriate interpretation thus appears to be the exclusion of claims made by nationals of launching States against any or all of the launching States from the scope of the Liability Convention.

The standard for determining the amount of compensation payable under the Liability Convention is found in Article XII, which provides that the quantum of damage is to be determined:

> ... in accordance with international law and the principles of justice and equity, in order to provide such reparation in respect of the damage as will restore the person, natural or juridical, State or international organisation on whose behalf the claim is presented to the condition which would have existed if the damage had not occurred.

3.4.3 Modern Liability Controversies

3.4.3.1 Launching State

Since the provisions of the Liability Convention have never been specifically invoked in anger (except in heated academic discussions), there are significant uncertainties in the interpretation of its provisions. The first and perhaps the most controversial today remains the definition of a "launching State" and its application to the multinational nature of the space industry today. This was particularly difficult in the context of the Sea Launch project, which involved a private joint venture of companies from Russia, the United States, Ukraine and Norway with the rockets launched from a converted oil drilling platform in the high seas.[205]

---

[203] Liability Convention, Article IV (2).

[204] Ibid., Article VII.

[205] Armel Kerrest, *Launching Spacecraft from the Sea and the Outer Space Treaty: The Sea Launch Project* (1997) 40 PROC. COLL. L. OUTER SP. 264.

Some scholars have suggested that this creates a lacuna in the application of the Liability Convention.[206] This may not necessarily be true as the launch operator in a launch from the high seas or in airspace above the high seas and the satellite operator who procured the launch would nevertheless be easily identifiable, and all launching States are jointly and severally liable. In the case of Sea Launch, for example, Norway, Russia, Ukraine and the United States would all be jointly and severally liable as launching States, though they may have agreed to apportion their potential liability by private contract.

In practice, States would generally prefer "launching States" to be defined as narrowly as possible, especially in the context of "procuring" the launch, as a broad definition may have the effect of stifling participation by some States in international endeavours or to approve the tangential involvement of their private concerns in order to avoid potential international liability.[207] Awford, for example, suggested that an expansive and "somewhat imaginative" interpretation of the term "launching State" may have the "remote possibility" of including the following States:

(1) the State that owns the launch facility;
(2) the State that manufactured and installed the launch facility;
(3) the State that supplied some of the components for the launch facility;
(4) the State that transported some of those components to the launch facility;
(5) the State that owns the territory on which the launch facility is built;
(6) the State that owns the satellite to be launched;
(7) the State that manufactured the satellite;
(8) the State that supplied components to the satellite;
(9) the State that transported the satellite to the launch facility;
(10) the State that arranged for the launch;
(11) the State that supplied the launch vehicle;
(12) the State that manufactured the component parts of the launch vehicle;
(13) the State that monitored the launch; and
(14) the State that provided the management of the overall project.[208]

One other commercial reality in the launch industry is that the launch operator is generally not the entity that will operate and control the satellite once it has been inserted into orbit. In such a case, it would be an injustice to continue to impose liability on the "launching States", namely the States responsible for the launch,

---

[206] Kai-Uwe Schrogl and Charles Davies, *A New Look at the "Launching State": The Results of the UNCOPUOS Legal Subcommittee Working Group "Review of the Concept of the 'Launching State'" 2000–2002* (2002) 45 PROC. COLL. L. OUTER SP. 286.

[207] William B. Wirin, *Practical Implications of Launching State – Appropriate State Definitions* (1994) 37 PROC. COLL. L. OUTER SP. 109 at 112.

[208] Awford, supra note 202, at 5–6. To this list one may add the State that leases the territory on which the launch facility is built; the State that financed the construction of the launch facility; the State that financed the manufacturing of the satellite; the State that financed the launch and the State that financed the construction of the launch vehicle.

when they no longer have any control or influence over the operation and control of the satellite. Some States today prepare bilateral agreements pursuant to Article V(2) of the Liability Convention to require the "operating States" to indemnify the "launching States" for any damage caused after orbital insertion. This is particularly important as the fault liability that forms the basis of liability for damage caused in outer space is based on the fault of the launching States collectively. Further, there is no treaty basis for the apportionment of liability between the launching States except by private intergovernmental agreement, as provided for under Article V of the Liability Convention. However, the practical enforcement of these agreements may be problematic considering that such agreements "shall be without prejudice to the right of a State sustaining damage to seek the entire compensation due under this Convention from any or all of the 'launching States' ".[209]

In addition to the conceptual difficulties associated with the definition of a "launching State", there are further interpretation problems associated with the wording of the definition itself. One such controversy is the issue of suborbital launches, which Gál suggested be excluded from the scope of the Liability Convention.[210] Böckstiegel advocated that not all suborbital launches would be excluded as the definition of "launch" includes attempted launches, though what would constitute an attempted launch was not clarified.[211] Gorove noted that, as with criminal law, an "attempt" must be intended and involved "perpetration" or "execution" of adequate means that have come close to success.[212] This approach does not appear to have met with widespread acceptance and, in any event, this would limit the scope of the applicability of the Liability Convention in a way that the drafters may not have intended at the time.[213]

The issue of "procuring" a launch for the purposes of the Liability Convention has also raised some questions, particularly in the context of private launch activities. Böckstiegel suggested that the mere link of nationality of a private launch operator is not sufficient to make that State a launching State – the State must actively request, initiate or promote the launching of the space object to have "procured" the launch.[214] This view is shared by Nesgos in the context of the "procuring" role of the State when one of its private enterprises provides a space object to be launched by a foreign State or a launch operator of a foreign State.[215] In

---

[209] Liability Convention, Article V.

[210] Gyula Gál, *Space Treaties and Space Technology: Questions of Interpretation* (1972) 15 PROC. COLL. L. OUTER SP. 105 at 106.

[211] Karl-Heinz Böckstiegel, *The Term "Launching State" in International Space Law* (1994) 37 PROC. COLL. L. OUTER SP. 80 at 81.

[212] Stephen Gorove, *Space Transportation Systems: Some International Legal Considerations* (1981) 24 PROC. COLL. L. OUTER SP. 117 at 118.

[213] See Reis, supra note 193.

[214] Böckstiegel, supra note 211, at 81.

[215] Peter D. Nesgos, *International and Domestic Law Applicable to Commercial Launch Vehicle Transportation* (1984) 27 PROC. COLL. L. OUTER SP. 98 at 102.

3.4 Liability

light of the obligation imposed on the "appropriate State" to authorise and continually supervise space activities of non-governmental entities and to take international responsibility for them under Article VI of the Outer Space Treaty, such an active role on the part of the State of nationality may be considered unnecessary for a State to be considered to have "procured" a launch.[216] In such a context, the suggested view by Wirin that "procurement" requires actual control over the launch or the payload in orbit is clearly met.[217]

3.4.3.2 Space Object

The Liability Convention defines a "space object" as including "component parts of a space object as well as its launch vehicle and parts thereof". It has often been noted that this is no more than a partial definition, or clarification, of a "space object", which in any event refers to itself.[218] Cheng outlined that "space object" covers "any object launched by humans into outer space, as well as any component part thereof, together with its launch vehicle and parts thereof" and so objects launched into the Earth's orbit and beyond are ipso facto regarded as space objects.[219] A similar legal definition for "space object" has been proposed by Kopal.[220]

This has particular relevance in the case of a space object launched by a rocket deployed from an aircraft in airspace. Böckstiegel suggested that, as the aircraft may be considered the first stage of the launch vehicle, the take-off of the aircraft would be considered the start of the launch procedure and therefore the State from whose territory the aircraft took off would be considered a launching State.[221] Gorove stated that it is more likely that the State in whose airspace the aircraft launched the rocket would be considered the launching State.[222] In order to resolve this conceptual impasse, it may be more appropriate to consider the aircraft as the "facility" for the launch and the airspace the "territory" from which the space object is launched. This is especially so considering one would be unlikely to consider the last port of call of the launch platform to be a launching State for a launch from the sea.

The definition of "space object" has specific relevance in the context of attributing liability for damage caused by space debris. Pieces, fragments and other substances of an object would generally be regarded as "parts" of that object. The

---

[216] This is partly suggested in Gorove, supra note 212, at 120.

[217] Wirin, supra note 207, at 113.

[218] See, for example, Stephen Gorove, *Definitional Issues Pertaining to "Space Object"* (1994) 37 PROC. COLL. L. OUTER SP. 87 at 88.

[219] Bin Cheng, *"Space Objects", "Astronauts" and Related Expressions* (1991) 34 PROC. COLL. L. OUTER SP. 17.

[220] Vladimír Kopal, *Some Remarks on Issues Relating to Legal Definitions of "Space Object", "Space Debris" and "Astronaut"* (1994) 37 PROC. COLL. L. OUTER SP. 99 at 101.

[221] Karl-Heinz Böckstiegel, *The Terms "Appropriate State" and "Launching State" in the Space Treaties: Indications of State Responsibility and Liability for State and Private Space Activities* (1992) 35 PROC. COLL. L. OUTER SP. 15.

[222] Gorove, supra note 218, at 91.

problem is that the partial definition in the Liability Convention refers to the inclusion of component parts of the space object and parts of its launch vehicle. As the term "component parts" has a clear meaning, the argument may therefore be forcefully made that the drafters of the Liability Convention intended for such a distinction to be maintained in the case of the "component parts" of a space object vis-à-vis the "parts" of a launch vehicle. However, as Gorove suggested, such a technical distinction does not appear to be maintained by state practice and, in practice, there does not appear to be a sound policy justification for such a distinction.[223]

One practical consequence of not maintaining a distinction between "component parts" and "parts" is that the launching States would be liable for damage caused by the orbital debris generated from their space objects to the space objects of other States. It is for this reason that Wirin proposed that the use of the term "component parts" was to specifically exclude small pieces and fragments that are not capable of surviving a re-entry into the atmosphere of the Earth.[224] On the other hand, Gorove was of the view that separating orbital debris from the definition of "space objects" would appear to run counter to the intention of the drafters of the Liability Convention.[225] Cheng further pointed out that "fragments of a space object that fall on the Earth are ... given the same status as the whole object ... [and] nothing suggests otherwise, or that shattered fuel tanks or flakes of paint from space objects in outer space should be treated any differently".[226] However, it is respectfully submitted that the drafters of the Outer Space Treaty, in referring to the "parts" of a launch vehicle and "component parts" of a space object, clearly intended to suggest an interpretation to "component parts" other than that proposed by Cheng.

3.4.3.3 Fault

The concept of "fault" as used in Article III of the Liability Convention, in which this term is not defined, has different meanings in different legal systems. In civil law systems, fault is generally interpreted by the courts on a case-by-case basis. In common law, fault is often associated with negligence in common law systems, thus necessitating considerations of the applicable duty and standard of care in its determination to be taken into account.[227]

In practice, this discrepancy in the legal notion of "fault" in different legal systems may not be of substantial consequence, as the facts of the circumstances in

---

[223] Stephen Gorove, *Toward a Clarification of the Term "Space Object": An International Legal and Policy Imperative?* (1993) 21 J. SP. L. 11 at 13–14.

[224] William B. Wirin, *Space Debris and Space Objects* (1991) 34 PROC. COLL. L. OUTER SP. 45. This was a view supported in He Qizhi, *Review of Definitional Issues in Space Law in the Light of Development of Space Activities* (1991) 34 PROC. COLL. L. OUTER SP. 32.

[225] Gorove, supra note 223, at 15.

[226] Cheng, supra note 219, at 24.

[227] Edward A. Frankle, *International Regulation of Orbital Debris* (2000) 43 PROC. COLL. L. OUTER SP. 369.

which damage was suffered may be *res ipsa loquitur*. For example, a satellite operator may be considered to be at fault if it placed the satellite in an orbit known to be already occupied by another satellite with which it is likely to collide or if the "victim" satellite operator failed to move its satellite out of the way of a known inert or "dead" satellite.

Consequently, one of the most noteworthy difficulties in the imposition of international liability for damage caused by orbital debris is not the identification of the origin of the debris, but rather the attribution of fault on the part of the launching States. In the context of common law notions of fault, it would be difficult to suggest that the launching States would be at fault because, although the risk of collisions with the generated debris is reasonably foreseeable, the launching States are unlikely to be able to take steps to prevent such a collision short of not launching the original space object at all or to use a substantial amount of fuel to take the satellite into either a sufficiently high "parking" orbit or to deorbit it back into the atmosphere of the Earth, although such a costly outcome is unlikely to be endorsed by most States.

3.4.3.4 Nuclear Power Sources

The requirements of the Outer Space Treaty and the liability provisions of the Liability Convention are repeated in the Principles Relevant to the Use of Nuclear Power Sources in Outer Space (the "Nuclear Power Sources Principles") as declared by the General Assembly in 1992 in the circumstance of space objects with nuclear and radioisotopic power sources onboard.[228] Similarly, the provision relating to the determination of the amount of compensation payable under Article XII of the Liability Convention can also be found in the Nuclear Power Sources Principles.[229]

In relation to the costs of the recovery and the clean-up, the Rescue Agreement and the Nuclear Power Sources Principles contain two substantially identical but procedurally different provisions. Under the Rescue Agreement, the expenses incurred for the recovery and return of the components of the space object are to be reimbursed by the States responsible for the launch.[230] The costs of the clean-up and other steps taken to eliminate the hazardous nature of the returned components are excluded from this reimbursement provision. Presumably this is because the costs of the recovery and return are technically not "damage", while the clean-up costs of eliminating hazardous materials are necessarily "damage". Consequently, it is appropriate to establish a head of liability for recovery costs that is separate to that of the liability for damage.

The Nuclear Power Sources Principles, on the other hand, provide that the compensation payable by the launching States in accordance with the Liability Convention and the Outer Space Treaty includes the reimbursement for "duly substantiated expenses for search, recovery and clean-up operations, including expenses

---

[228] Nuclear Power Sources Principles, Principle 9(1).

[229] Ibid., Principle 9(2).

[230] Rescue Agreement, Article 5(5).

148 3 State Responsibility and Liability for Compliance with International Space Law

for assistance received from third parties".[231] This means that, subject to the added requirement of "duly substantiating" the expenses, these costs are to be considered part of the "damage" to be compensated by the launching States. This is clearly inconsistent with the similar provision in the Rescue Agreement as discussed above. While this produces a procedural discrepancy, in practice it is doubtful that the relevant States concerned would make two separate claims relating to recovery costs and the damage arising from the return of a space object from outer space.

3.4.4 Calculation of Damages

3.4.4.1 Approach

Article I of the Liability Convention defines "damage" as being "loss of life, personal injury or other impairment of health; or loss of or damage to property of States or of persons, natural or juridical, or property of international intergovernmental organisations". Article XII further provides that the damages payable in compensation are to be determined "in accordance with international law and the principles of justice and equity" to the extent of restoring the injured parties to the condition prior to the damage occurring. The Liability Convention does not appear to include environmental damage as part of the "damage" potentially caused by space objects.[232] This approach has particular importance and relevance in considering clean-up costs of nuclear and radioisotopic power sources as discussed above.

3.4.4.2 Direct Damage

Article I of the Liability Convention refers to four specific heads of recoverable direct damage, namely, loss of life, personal injury, other impairment of health and loss of or damage to property. In the context of these damages, a claimant State would be required to demonstrate that the harm claimed flowed directly or immediately from and as the natural or probable result of the space object.[233] Some commentators have noted that "impairment of health" can result from both contamination as well as physical injury and that it is not necessary to have direct contact with the space object to suffer harm.[234] In this context, the radiation damage caused by the unexpected re-entry of *Cosmos-954* would be a recoverable damage, even without the need to rely on the Nuclear Power Sources Principles.[235]

Christol suggested that, in accordance with the United States view of the position in international law, compensation for the following items would also be appropriate and would be considered "direct" damages:

---

[231] Nuclear Power Sources Principles, Principle 9(3).

[232] See Christol, supra note 201.

[233] Christol, supra note 201, at 359.

[234] W. F. Foster, *The Convention on International Liability for Damage Caused by Space Objects* (1972) 10 CAN. Y. B. INT'L. L. 137 at 155 and Carl Q. Christol, *Protection of Space from Environmental Harms* (1979) 4 ANN. AIR & SP. L. 433.

[235] Paul G. Dembling, *Cosmos 954 and the Space Treaties* (1978) 6 J. SP. L. 129 at 133.

(1) lost time and earnings and impaired earning capacity;
(2) destruction or deprivation of the use of property, including where the property has been rendered unfit for its intended purposes;
(3) loss of profits resulting from business interruption;
(4) loss of rents;
(5) reasonable medical, hospital and nursing costs associated with injuries sustained by natural persons;
(6) physical and mental impairment;
(7) pain and suffering;
(8) humiliation;
(9) reasonable costs for the repair of property; and
(10) costs incurred in acts taken to mitigate the damage caused.[236]

3.4.4.3 Indirect Damage and Economic Loss

It is unclear from the Liability Convention whether it is intended to cover indirect or consequential damage. Articles II and III both refer to the damage being "caused" by the space object. Hungary and the Soviet Union opposed an interpretation that would allow recovery of indirect damage, while Italy and Japan both favoured it.[237] In the end, the question was left open, as "the word 'caused' should be interpreted as merely direct attention to the need for some causal connection between the accident and the damage, while leaving a broad discretion so that each claim can be determined purely on its merits".[238]

There appears to be some academic support for the proposition that, since "caused by" requires no more than a causal connection between the space object and the damage, the Liability Convention covers both direct and indirect damage.[239] In the situation of *Cosmos-954*, for example, Haanappel suggested that search and recovery costs incurred by Canada were incurred to mitigate probable damage and were recoverable indirect damage for the purposes of Article VII of the Outer Space Treaty and the provisions of the Liability Convention.[240]

3.4.4.4 Moral or Punitive Damages

In international law, moral damage is identified as the injury to the dignity or sovereignty of a State, such as a breach of a treaty obligation that does not produce a material injury and yet the violating State would be expected to pay adequate monetary penalties. Similarly, pain and suffering and the loss of capacity to enjoy life may also be considered to be moral damage to natural persons.

---

[236] Christol, supra note 201, at 359.

[237] Ibid.

[238] Foster, supra note 234, at 158. It should be noted that the same view can be found in Nicholas Mateesco Matte, AEROSPACE LAW (1977) at 157.

[239] Christol, supra note 201, at 362.

[240] Peter P. C. Haanappel, *Some Observations on the Crash of the Cosmos 954* (1978) 6 J. SP. L. 147 at 148.

150 3 State Responsibility and Liability for Compliance with International Space Law

The term "equity" has been noted as referring not to the common law concept of equity but rather to signify "moral justice".[241] In other words, the assessment of damages is a task to be undertaken with reference to the actual losses suffered rather than through the application of domestic or secondary international principles.

The United States has long expressed the view that moral damages are covered by the Liability Convention and that, if a claim is made in the future by the United States, such a claim would include a component for moral damages.[242] While the moral damage done to a natural person may establish a sufficient causal link with the space object, it is difficult to see how the moral damage suffered by a State would be recoverable if a causal connection cannot be made to the space object, considering the mere causation of damage by a space object is not prima facie a breach of an existing principle of treaty law.

Punitive damages have been considered by commentators to be both unnecessary and unrecoverable.[243] The reason why punitive damages is considered unnecessary is because of the provision for unlimited liability under the Liability Convention that allows the victims to recover sufficient compensation for their damages sustained. There appears to be three reasons why punitive damages may be unrecoverable:

(1) the provisions of the Liability Convention are very specific in tying the causation of the damage sustained to the space object and punitive damages cannot be included as they are not by their very nature compensation for damage actually sustained by the claimant State;
(2) punitive damages are generally assessed by tribunals only to punish the intentional acts of tortfeasors while the Liability Convention does not make any distinction on the liability of the launching States for intentional, reckless, negligent or accidental damage; and
(3) in the case of the launching State acting in breach of an existing legal principle, the appropriation "sanction" is the unavailability of any exoneration from absolute liability under Article VI of the Liability Convention and not the imposition of punitive damages.

3.5 Conclusions

From the above analysis, it is clear that the following implications arising from Articles VI and VII of the Outer Space Treaty and the provisions of the Liability Convention for a commercial mining venture in outer space must be considered:

---

[241] Alexander, supra note 201.

[242] U.S. Senate Committee on Foreign Relations, CONVENTION ON INTERNATIONAL LIABILITY FOR DAMAGE CAUSED BY SPACE OBJECTS (1972) S. EXEC. REP. 92–38, 92nd Cong., 2nd Sess. 9 at 7 and S. Neil Hosenball, *Space Law, Liability and Insurable Risks* (1976) 12 THE FORUM 141 at 151.

[243] See, for example, Christol, supra note 201, at 365–366.

3.5 Conclusions

(1) the "appropriate" State is required to authorise and continually supervise the space activities of private entities and ensure that the activities comply with the Outer Space Treaty;
(2) the State of registry of a space object can assert jurisdiction and control over that space object;
(3) States bear international responsibility for the "national" space activities of private entities; and
(4) the "launching States" are absolutely liable for any damage caused by their space objects on the surface of the Earth or to aircraft in flight and for any damage caused in outer space to the extent that the damage was caused by the fault of the launching States.

These implications have two important effects on the activities of a commercial space venture, particularly in the launch, transit and return segments of a commercial space mining venture. Firstly, it would be necessary for the venture to comply with the provisions of the Outer Space Treaty, as the "appropriate" State would presumably take steps in its "continuing supervision" of the activity to ensure this compliance. Secondly, the States are likely to take positive steps, such as the enactment of domestic laws and regulations, to pass on its international responsibility and liability under the Outer Space Treaty and the Liability Convention to the private entities engaged in space activities.[244]

With these in mind, it is evident that the legality of all aspects of a commercial space mining venture must be assessed, most notably, the effects of the freedom of exploration and use and the principle of non-appropriation on the exploration and extraction segments of the operation, as well as the "province of all mankind" and "common heritage of mankind" doctrines on the exploitation segment. These issues are explored in greater detail in the following chapters.

---

[244] Some States have already legislated to require private launch operators to indemnify and/or insure the State for any international liability incurred from the launch activity. See, for example, the *Space Activities Act 1998* (Cth) of Australia; the *Outer Space Act 1986* of the United Kingdom; and the *Commercial Space Launch Act 1984* of the United States. See Lee, supra note 185.

Chapter 4
Rights and Duties in the Commercial Exploration and Extraction of Mineral Resources on Celestial Bodies

4.1 Introduction

Along with all different types of space activities, the provisions of the United Nations space treaties all have important implications on the permissibility and confines of lawful activities in outer space and on celestial bodies. Although the treaties and their provisions have direct effect only on the public or governmental space activities of States, Article VI of the Outer Space Treaty extends their application to private space activities by requiring States to authorise, continually supervise and adopt international responsibility for private space activities. Accordingly, both governmental and private commercial space mining ventures, including their exploration and extraction segments, require compliance with the principles of international space law as contained in the United Nations space treaties and customary international law.

In particular, in the exploration and extraction segments of such commercial space mining ventures, there are three important provisions of the Treaty on Principles Governing the Activities of States in the Exploration and Use of Outer Space, including the Moon and other Celestial Bodies (the "Outer Space Treaty") that have the potential of imposing obligations that may affect the legality of the venture. They are:

(1) Article I, which prescribes the requirement that space activities are to be conducted for the benefit and in the interest of all States, the freedoms of exploration and use of outer space, and the freedom of access to all areas of celestial bodies;
(2) Article II, which embodies the principle of non-appropriation; and
(3) Article IX, which outlines the duty to have due regard to corresponding interests of other States when conducting space activities and to avoid harmful contamination of celestial bodies.

For all three provisions, there are implications that affect all commercial activities in outer space and implications that have direct and specific effect on space mining operations. For example, the principle of non-appropriation in Article II has

R.J. Lee, *Law and Regulation of Commercial Mining of Minerals in Outer Space*,
Space Regulations Library 7, DOI 10.1007/978-94-007-2039-8_4,
© Springer Science+Business Media B.V. 2012

implications for commercial activities in outer space *sensu stricto* and on celestial bodies generally as well as specific implications for the extraction segment of a mining operation. Therefore, it is pertinent to first consider the general content and effect of these provisions and, subsequently, to study their specific effect when applied to a commercial space mining venture. This would then allow for an analysis of the applicable space law principles to such exploration and extraction activities.

4.2 Commercial Use vs. Public Use

4.2.1 *Benefit and Interests of All Countries*

4.2.1.1 Overview

Article I of the Outer Space Treaty provides for three of the most fundamental principles of international space law, namely the freedoms of exploration, access and use by all States on a non-discriminatory basis and that space activities are to be carried out for the benefit and in the interest of all States. Specifically, Article I states:

> The exploration and use of outer space, including the Moon and other celestial bodies, shall be carried out for the benefit and in the interest of all countries, irrespective of their degree of economic or scientific development, and shall be the province of all mankind.
>
> Outer space, including the Moon and other celestial bodies, shall be free for exploration and use by all States without discrimination of any kind, on a basis of equality and in accordance with international law, and there shall be free access to all areas of celestial bodies.
>
> There shall be freedom of scientific investigation in outer space, including the Moon and other celestial bodies, and States shall facilitate and encourage international cooperation in such investigation.

Some commentators have suggested that these principles existed before the adoption of the Outer Space Treaty or that they had already become crystallised into customary law in any event.[1] In this context, it is pertinent to note that these provisions can be found in the 1963 United Nations General Assembly Declaration on Legal Principles Governing the Activities of States in the Exploration and Use of Outer Space (the "Principles Declaration"). This has led to some commentators asserting that these principles were part of existing customary international law at the time the Outer Space Treaty was adopted in 1967.[2] Of course, this would have meant an endorsement of the concept of "instant custom" as Cheng had advocated previously, whereby States adopt principles of customary international law by simultaneous and uniform state practice or *opinio juris*.[3] In turn, this had been

---

[1] See Bin Cheng, *United Nations Resolutions on Outer Space: "Instant" International Customary Law?* (1965) 5 INDIAN J. INT'L. L. 23 and Daniel Goedhuis, *Reflections on the Evolution of Space Law* (1966) 13 NEDERLANDS TIJDSCHRIFT 109.

[2] See, for example, Vladlen S. Vereshchetin and Gennady M. Danilenko, *Custom as a Source of International Law of Outer Space* (1985) 13 J. SP. L. 22 and He Qizhi, *The Outer Space Treaty in Perspective* (1997) 25 J. SP. L. 93.

[3] Cheng, supra note 1, at 36.

4.2 Commercial Use vs. Public Use

rejected by some scholars for the lack of constant and uniform practice.[4] Regardless, it is prudent to consider the terms of the Outer Space Treaty to be binding principles of international space law on all States from the adoption of the Outer Space Treaty in 1967, either through States ratifying the Outer Space Treaty or through the crystallisation of such terms into customary international law.

In the context of commercial space mining ventures, it is clear that the following two principles that are enshrined in Article I apply to such ventures:

(1) "exploration" and "use" of celestial bodies are to be carried out for the benefit and in the interest of all States; and
(2) all States have the freedom to access, explore and use all areas of celestial bodies on a basis of equality and without discrimination of any kind.

This is further reinforced by Article IX of the Outer Space Treaty, which states:

> In the exploration and use of outer space, including the Moon and other celestial bodies, States Parties to the Treaty shall be guided by the principle of co-operation and mutual assistance and shall conduct all their activities in outer space, including the Moon and other celestial bodies, with due regard to the corresponding interests of all other States Parties to the Treaty. States Parties to the Treaty shall pursue studies of outer space, including the moon and other celestial bodies, and conduct exploration of them so as to avoid their harmful contamination and also adverse changes in the environment of the Earth resulting from the introduction of extraterrestrial matter and, where necessary, shall adopt appropriate measures for this purpose. If a State Party to the Treaty has reason to believe that an activity or experiment planned by it or its nationals in outer space, including the moon and other celestial bodies, would cause potentially harmful interference with activities of other States Parties in the peaceful exploration and use of outer space, including the moon and other celestial bodies, it shall undertake appropriate international consultations before proceeding with any such activity or experiment. A State Party to the Treaty which has reason to believe that an activity or experiment planned by another State Party in outer space, including the moon and other celestial bodies, would cause potentially harmful interference with activities in the peaceful exploration and use of outer space, including the moon and other celestial bodies, may request consultation concerning the activity or experiment.

Within the context of the "benefit and interests of all countries" under Article I is the obligation imposed under Article IX that all activities must be conducted with "due regard to the corresponding interests of all other States" is of particular concern. As discussed below, this obligation in Article IX may have some influence over the appropriate interpretation to be applied to the requirement contained within Article I. It is important, however, to first analyse the potential interpretations that may be made of this requirement in Article I before discussing in detail the effect of the interaction between the two provisions on commercial space mining activities.

---

[4] See, for example, C. Wilfred Jenks, A NEW WORLD OF LAW? A STUDY OF THE CREATIVE IMAGINATION IN INTERNATIONAL LAW (1969) at 146; H. G. Darwin, *The Outer Space Treaty* (1967) 42 BRIT. Y. INT'L. L. 278 at 280; Hugh W. A. Thirlway, INTERNATIONAL CUSTOMARY LAW AND CODIFICATION: AN EXAMINATION OF THE CONTINUING ROLE OF CUSTOM IN THE PRESENT PERIOD OF CODIFICATION OF INTERNATIONAL LAW (1972), at 62–68; and Vereshchetin and Danilenko, supra note 2.

4.2.1.2 Article I of the Outer Space Treaty: "For the Benefit and in the Interest of All Countries"

The crucial determination to be made is whether the phrase "for the benefit and in the interest of all countries" imposes a positive and specific obligation "regarding the sharing of the benefits of space exploration and use" or merely an "expression of desire that the activities should be beneficial in a general sense".[5] Gorove, who analysed this provision in detail, argued for the latter and regarded most commercial space activities, such as telecommunications, broadcasting, remote sensing and power generation, as being beneficial in a general sense and were sufficient to satisfy the requirement.[6] In so doing, Gorove pointed to several factors that persuaded him to that view, which is shared, to some extent surprisingly, by some commentators from developing States.[7]

Firstly, the criteria for determining what is of benefit to a State is almost entirely subjective. What may be considered beneficial to one State may well be detrimental to another. Further, what may be considered beneficial today may be considered detrimental tomorrow with the aid of new information and the benefit of hindsight.[8] As there are no means for settling disputes between States over the definition of such terms in the Outer Space Treaty or otherwise, it is likely and foreseeable that each State would insist on determining the beneficial aspects of an activity based on its own subjective criteria without reference to the legitimate rights, interests and expectations of other States. This is unlikely to have been the intended outcome of the drafters of Article I of the Outer Space Treaty.

Secondly, the benefits and interests of all countries must include, by definition, the State conducting that particular exploration and use of outer space and/or the celestial bodies.[9] Accordingly, the interests of that State, presumably extending to commercial interests, would not be served if they were not taken into account in assessing the benefits derived from a particular activity in outer space, particularly with the need to provide some incentive or motivation for States to conduct space activities or at least to invest in them. In other words, even if the requirement imposed a specific duty to "share" the "benefits" among all States, such a requirement must be considered to some extent to be subject to the commercial interests, among other categories of interests, of the State conducting the relevant space activity in question.

---

[5] Stephen Gorove, *Implications of International Space Law for Private Enterprise* (1982) 7 ANN. AIR & SP. L. 319 at 321.

[6] Stephen Gorove, *Freedom of Exploration and Use in the Outer Space Treaty* (1971) 1 DENVER J. INT'L. L. & POL'Y. 93.

[7] See, for example, Silvia Maureen Williams, *Las Empresas Privadas en el Espacio Ultraterrestre* (1983) 8 REV. CEN. INV. DIF. AERO. ESP. at 39 and Luis F. Castillo Argañarás, *Benefits Arising From Space Activities and the Needs of Developing Countries* (2000) 43 PROC. COLL. L. OUTER SP. 50 at 57.

[8] Ibid., at 104.

[9] Gorove, supra note 5, at 321.

4.2 Commercial Use vs. Public Use

Thirdly, it is unclear from the provision whether it is the means of conducting the activity itself (*obligation de moyens*) or the results derived, or ends achieved, from such activity (*obligation de résultat*) that must be in the interest and for the benefit of all States.[10] If it is the results or ends derived from such activities, then it must be noted that the existing body of space law provides no mechanism for any sharing or distribution of such benefits in practice, even if the provisions of the Moon Agreement are taken into consideration in conjunction with those of the Outer Space Treaty. If it is the means of the activity itself, then the legal requirement would be no more than a negative prohibition on States conducting activities that are detrimental to the interests of other States. Monserrat Filho, in advocating the view that all space activities, public or private, must be subject to the "global public interest", suggested that this "does not admit any form of exploitation and use of the outer space [*sic*] capable of causing bad and damage [*sic*] to a country and to people, to the whole humankind or to part of it, as well as *hurting their legitimate interests*".[11]

The idea of Article I in practice as being no more than a moral obligation, instead of a legal one, is a view that is shared with Gorove by other commentators. Cheng, for example, observed that:

> Insofar as the preparatory work of the Treaty is concerned, the discussions which took place on several articles of the Treaty clearly showed that its draftsmen hardly intended this part of the Article I to be anything more than a declaration of principles from which no specific rights of a legal nature were to be derived, even though it may give rise to a moral obligation.[12]

Although this formulation may be considered the most favourable, especially in the context of promoting private and commercial space activities, it must be noted that there are two *indicia* to suggest that the requirement actually imposes a positive duty. The first is that the requirement in Article I utilises the plural form "interests" instead of the singular, which may indicate that this involves more than "just the vague, general 'interest' of all countries" and, instead, represents specific and identifiable interests.[13] This may be taken to mean that a particular set of interests of all States is to be taken into account in the conduct of space activities. The second is that while Article I may be considered to be "an aspiration couched in very general terms which could not be specifically implemented without further elaborations and guidelines", the provisions of the Agreement on the Activities of States on the Moon and other Celestial Bodies (the "Moon Agreement") may arguably constitute the further elaborations and guidelines to give effect to the "interests and benefits

---

[10] Such a distinction was made by Kerrest in the context of Article VI of the Outer Space Treaty. See Armel Kerrest, *Commercial Use of Space, Including Launching* (2004), in China Institute of Space Law, 2004 SPACE LAW CONFERENCE: PAPER ASSEMBLE 199 at 200.

[11] José Monserrat Filho, *Why and How to Define "Global Public Interest"* (2000) 43 PROC. COLL. L. OUTER SP. 22 at 24. Italics added.

[12] Bin Cheng, STUDIES IN INTERNATIONAL SPACE LAW (1998) at 234–235.

[13] Ibid.

of all countries" requirement.[14] Accordingly, even though the Moon Agreement has not won widespread acceptance as the *means* of implementing the requirement, this may not of itself prejudice the view that the requirement may nevertheless require implementation at a practical level.

The foregoing analysis may be condensed to produce at least three possible outcomes and the corresponding applications on the exploration and extraction segments of a commercial space mining operation:

(1) *Generalised mission statement rather than positive and specific duty.* If the requirement of "benefits and interests of all countries" is to be regarded as a generalised mission statement for all space activities instead of the imposition of a positive and specific duty, then clearly commercial mineral exploration and extraction activities on celestial bodies may be considered a positive development for all States, notwithstanding the absence of any sharing of financial or tangible benefits to other States.

(2) *Obligation imposed on the activity rather than the results derived thereof.* If the requirement does impose a specific and positive duty but such a duty is imposed on the activity itself instead of on the results and outcomes derived thereof, then the duty may be interpreted as no more than a negative duty of ensuring that the activity does not cause a detriment to any State. In such a case, commercial mineral exploration and extraction activities would not have much difficulty fulfilling such an obligation, as these activities or the means involved are unlikely to cause detriments for other States.

(3) *Positive duty to share the benefits derived from space activities.* If the requirement under Article I is to be interpreted as an actual obligation to share the resulting benefits derived from space activities, whether financial, tangible or both, then the Moon Agreement is an example, though an unacceptable one, of the practical means of fulfilling this obligation. However, it follows then that the obligation does not arise until the State or its private entities have gained a benefit that is capable of being shared on an equitable basis.[15] In the context of a commercial space mining venture, such a benefit would be produced only in the exploitation segment and thus the obligation would have no application on the exploration and extraction segments of the venture. Further, there is no suggestion that all States would be entitled to an equal share of such benefits. This is supported, for example, by the express stipulation in the Moon Agreement that there is to be an "equitable" sharing in the "benefits" derived from mineral resources extracted from celestial bodies, rather than an "equal" sharing in the "materials" or "profits" derived from such activities.[16]

---

[14] Ibid., at 322.

[15] Moon Agreement, Article 11.

[16] Ibid., Article 11(7)(d).

4.2 Commercial Use vs. Public Use

It can be seen from above that some doubts remain concerning the legal content and effect of the requirement that space activities be carried out "for the benefit and in the interest of all countries" and, correspondingly, the legality conducting commercial mineralogical prospecting and exploitation activities on celestial bodies. The existing body of state practice would tend to suggest that the provision is at least a generalised mission statement at the very least and at most an obligation on the nature of the activity rather than the benefits derived therefrom. However, the exploitation of mineral resources from celestial bodies may well be an activity that may attract an interpretation that involves a positive duty to share the derived benefits with other countries. This position is further supported by the existing formulation of the common heritage of mankind principle as contained in Article 11 of the Moon Agreement. Therefore, it is in the context of Article I and other relevant provisions of the Outer Space Treaty that the effect of the common heritage of mankind principle must be considered.

4.2.1.3 Article IX of the Outer Space Treaty: Due Regard to Corresponding Interests of Other States

As observed above, Article IX of the Outer Space Treaty requires States to conduct their activities in outer space and on celestial bodies with "due regard to the *corresponding* interests of all other States".[17] This requirement, as with all others in the Outer Space Treaty, extends to activities of private entities through the operation of Article VI. At first glance, it is apparent that any attempt at a definitive interpretation of this "due regard" requirement would suffer from the same defect as the "benefit and interests of all countries" requirement in Article I in that the "corresponding interests of other States" are not articulated or defined.

However, it is in the term "corresponding" that some legal salvation may be found for the commercial mining venture and its search for certainty in the applicable legal principles under international law. The adjective "corresponding", by definition, means something that is "equal" or "similar" to something else.[18] In this case, it is clear that the "corresponding interests" of other States must be "equal" or "similar" to the interests of the State undertaking the space activity, otherwise the meaning of "corresponding" would be defeated.

If Article I of the Outer Space Treaty imposes a positive and specific duty on States to share in the benefit of their space activities with other States in accordance with their interests, then there can be no "corresponding" interests. This is because the interest of the States undertaking space activities would collide with those of non-participating States that are receiving a share of the benefits derived from such activities due to the "equivalent" interests of the participating State conflict with those of the non-participating States. This lends support to the view that the term "interests" relates not to the financial interests of States but instead to a general

---

[17] Italics added.

[18] Shorter Oxford English Dictionary (5th ed., 2003).

interest of States and, accordingly, does not prescribe a positive and specific legal duty on States in their conduct of activities.[19] In this case, the corresponding interests of States in the context of Article IX are the freedoms provided for in Article I, namely the freedom of exploration and use of outer space and access to all areas of the Moon and other celestial bodies.

Consequently, it may be argued that Article I, taking into consideration Article IX, does no more than to impose a negative duty on States in their conduct of activities in outer space and on celestial bodies. This duty is to ensure that such activities do not interfere with the rights provided under the provisions of the Outer Space Treaty or cause any detriment to other States. It has been observed that the existing state practice already reflects this interpretation of the requirement under Article I.[20] As considered below, this conclusion poses as one of the legal obstacles to the extraction segment of a commercial mining venture in space.

4.2.2 Lawfulness of Commercial Use Generally

Kerrest noted that commercial uses of space, in general, pose legal problems mostly related to appropriation and that "sharing the common space resources, orbits and frequencies, establishing legal monopolies ... through patent laws ... may be in breach of space law".[21] However, Kerrest did not go on to explain what aspect of space law would be contravened by such commercial activities. Presumably, this is the result of the "for the benefit and in the interests of all countries" requirement in Article I of the Outer Space Treaty.[22] Arguably, the occupation of orbits and frequencies and the extension of intellectual property rights in outer space may certainly, to some extent, be in contravention of the requirements under the international law of outer space.[23]

It may be seen that commercial space activities may have some difficulty falling within the requirements of Article I. This is because commercial activities are, by definition, undertaken with a view to profit and such profits are to be shared only by the members of the private concern or the relevant governmental agency. Consequently, the benefit to be obtained by *all* States from commercial space

---

[19] See, for example, Bin Cheng, STUDIES IN INTERNATIONAL SPACE LAW (1997) at 234–235; James J. Trimble, *The International Law of Outer Space and Its Effect on Commercial Space Activity* (1983) 11 PEPP. L. REV. 521 at 546; and Ricky J. Lee, *Definitions of "Exploration" and "Scientific Investigation" with Focus on Mineralogical Prospecting and Exploration Activities* (2005), paper presented at the 56th International Astronautical Congress, 17–21 October 2005, in Fukuoka, Japan.

[20] Roger K. Hoover, *Law and Security in Outer Space from the Viewpoint of Private Industry* (1983) 11 J. SP. L. 115 at 123.

[21] Armel Kerrest, *Commercial Use of Space, Including Launching* (2004), in China Institute of Space Law, 2004 SPACE LAW CONFERENCE: PAPER ASSEMBLE 199 at 199.

[22] Ricky J. Lee, *Commentary Paper on Discussion Paper Titled "Commercial Use of Space, Including Launching" by Prof. Dr. Armel Kerrest* (2004), in China Institute of Space Law, 2004 SPACE LAW CONFERENCE: PAPER ASSEMBLE 220.

[23] Ibid.

4.2 Commercial Use vs. Public Use

activities must be assessed qualitatively and quantitatively in some manner in order to satisfy the requirements imposed under Article I of the Outer Space Treaty. There are at least four possible views on the legality of commercial space activities in such a context that may be placed on a continuum between two polarised positions:

(1) all commercial activities are, by definition, not for the benefit, nor in the interest of all States, though it may benefit one or a handful of States, and are thus unlawful under Article I of the Outer Space Treaty;

(2) commercial activities in space are lawful only to the extent that they provide, in conjunction to their commercial activities for profit, some element of "community service" to all States at no, or nominal cost, as is the case for some intergovernmental satellite organisations;[24]

(3) commercial activities in space are lawful only to the extent that the goods or services they provide may be purchased by any third party governmental or private consumer, regardless of national origin and on a non-discriminatory basis, as provided for in the case of remote sensing activities of the Earth;[25] or

(4) commercial activities in space are lawful provided that the activity does not, by its nature, structure or form, prevent any other commercial or non-commercial entity from undertaking the same activity in space.

From existing state practice and *opinio juris*, it is doubtful that commercial space activities would *per se* be unlawful or that some element of "community service", similar to those originally provided by INTELSAT and INMARSAT, would be required under international law.[26] This is because to do so would be to suggest that the requirement under Article I prescribes a positive duty on the sharing of "benefits" derived from activities in outer space. As discussed above, this is a view that was not widely accepted by either States or commentators when it was prescribed under Article 11 of the Moon Agreement in relation to mineral resource exploitation activities.[27]

It is arguable that, although there is no positive duty to share the derived benefits from non-exclusive commercial space activities, the fruits of such activities must be available for purchase by all potential customers on a non-discriminatory basis. This is because, even though the principle of non-discrimination contained in Article I of

---

[24] Amended Convention on the International Mobile Satellite Organisation, opened for signature on 24 April 1998 [2001] A.T.S. 11 (entered into force on 31 July 2001), Article 3 and International Telecommunications Satellite Organisation, *Agreement Relating to the International Telecommunications Satellite Organisation*, at <http://216.119.123.56/dyn4000/dyn/docs/ITSO/tpl1_itso.cfm?location=&id=5&link_src=HPL&lang=english>, last accessed on 13 January 2005, Article III.

[25] Principles Relating to Remote Sensing of the Earth from Outer Space (the "Remote Sensing Principles"), Principle XII.

[26] Convention on the International Mobile Satellite Organisation and Agreement Relating to the International Telecommunications Satellite Organisation.

[27] See, for example, Carl Q. Christol, *The American Bar Association and the 1979 Moon Treaty: The Search for a Position* (1981) 9 J. Sp. L. 77 and Martin Menter, *Commercial Space Activities Under the Moon Treaty* (1979) 7 Syracuse J. Int'l. L. & Com. 213 at 220.

the Outer Space Treaty relates to the freedom of exploration, use and access by States and not to the "benefit and interests of all countries" requirement, such an extension of the non-discrimination principle is not without precedent. For example, in the Principles Relating to Remote Sensing of the Earth from Outer Space (the "Remote Sensing Principles"), Principle XII provides that the sensed State shall have access to the primary and processed data "on a non-discriminatory basis on reasonable cost terms". Admittedly, the sensed State in the case of remote sensing is in a unique position but arguably this is remedied by the requirement that the access is to be given on reasonable cost terms. There is no such equivalent requirement for any other commercial space application, except for the non-discrimination requirement found in Article I of the Outer Space Treaty. Therefore, in the case of other commercial activities, or in considering the interests of other States, perhaps the commercial operator would be entitled to charge unreasonable costs provided that it is conducted in a non-discriminatory manner.

In terms of monopolistic and other exclusive practices, however, it is arguable that they would be lawful under Article I, subject to the caveat that such practices restrict only the *means* to access or use outer space but not to the *space* or the *use* itself. For example, a patent on a particular asteroid mining technology merely restricts one particular method of asteroid mining but does not inhibit the freedom of other States to mine asteroids by any other means. In the case of orbital slots and frequencies, it is arguable that the regulations of the International Telecommunication Union (the "ITU") may constitute a *lex specialis* that takes precedence over the requirements of Article I of the Outer Space Treaty, if not an expression or application thereof.

In the context of commercial mining operations on celestial bodies, therefore, it may be seen that the commercial nature may have the effect of enlivening causing two further obligations to be imposed under Article I:

(1) that the raw or processed ores are to be made available to other States on a non-discriminatory basis, though not necessarily on reasonable cost terms even though presumably they would be sold on the world market at listed prices as current at the time; and
(2) that any restrictive or monopolistic practices adopted by the venture relate only to the means of the activity and not to the general nature of the activity so that it does not prevent third parties from undertaking the same activity, albeit by different means.

4.3 Freedoms of Exploration and Use and the Principle of Non-appropriation

4.3.1 Freedoms Under Article I of the Outer Space Treaty

One of the further requirements imposed under Article I of the Outer Space Treaty may be referred to as the three fundamental freedoms in space law: the freedom

4.3 Freedoms of Exploration and Use and the Principle of Non-appropriation

of exploration, the freedom of use and the freedom of access to all areas of all celestial bodies, including the Moon. It is difficult to distinguish between "exploration" and "use" of outer space and celestial bodies, because although the two terms may, on first appearances, have very distinct meanings, it is very difficult in practice to determine their differences in applicability. As Böckstiegel observed:

> At first sight, the distinction between "exploration" and "use" may seem sufficiently clear. Indeed in connection with most space activities little doubt may come up which of these two terms is applicable. First doubts appear however, because the Outer Space Treaty speaks of exploration "of outer space". This wording could be interpreted to mean that space must be the object of the exploration. The consequent would be that the great part of research which has to take place "in space" in view of the specific physical conditions there, but which has as its object specific materials, would not be covered and might only be considered as "use" of space.[28]

It is conceivable that the distinction between "exploration" and "use" is the classical one as applied to the Polar Regions of the Earth, where "exploration" refers to scientific research while "use" relates to the practical implementation of this research, such as, in relation to the exploitation of natural resources.[29] This definition would nevertheless produce difficulties for the commercial space mining venture, since arguably mineral prospecting activities could fall into either "exploration", as merely research on the geology and mineralogy of a particular area of a celestial body, or "use", being a commercial operation driven by the motivation of financial gain.

In any event, such a delineation may not be sustainable considering the third paragraph of Article I refers to States having a "freedom of scientific investigation" and thus the term "exploration" must have a meaning other than scientific research. This is further supported by Article IX of the Outer Space Treaty, which provides, among other requirements, that States:

> ... shall pursue *studies* of outer space, including the Moon and other celestial bodies, and conduct *exploration* of them so as to avoid their harmful contamination and also adverse changes in the environment of the Earth resulting from the introduction of extraterrestrial matter . . . .[30]

This is clearly inconsistent with the proposition that "exploration" of outer space means scientific research in outer space, as the drafters of the Outer Space Treaty are unlikely to have repeated themselves in such blatant fashion. Consequently, other possible definitions for the word "exploration", vis-à-vis the meaning of the word "use", must be contemplated. If one chose to consider the results from activities

---

[28] Karl-Heinz Böckstiegel, *Reconsideration of the Legal Framework for Commercial Space Activities* (1990) 33 PROC. COLL. L. OUTER SP. 3. Böckstiegel also observed at 4 that the Outer Space Treaty refers only to exploration *of* outer space and not exploration *in* outer space, though this author is of the view that the difference between the two, in practical terms, would be subtle at best.

[29] Ibid.

[30] Italics added.

involving "exploration" and "use" rather than the means themselves, then a distinction may be drawn on the benefits to be derived from such activities. Specifically, "exploration" may be defined as activities in space that do not produce tangible benefits and "use" is in turn defined as activities that do produce tangible benefits. For example, the Apollo-Soyuz mission, where an U.S. Apollo craft docked with a Soviet Soyuz craft, was not intended to undertake scientific research nor did it produce any tangible benefits and thus may be classified as an "exploration" activity instead of a "use" of outer space.

In the context of Article I of the Outer Space Treaty and a commercial space mining venture, however, the distinction that may be made between "exploration" and "use" is not of much practical consequence. This is because in both cases the Outer Space Treaty provides for the prescription of a freedom for all States. Consequently, the important consideration is the content of these freedoms provided under Article I of the Outer Space Treaty. In accordance with the "benefit and interests of all countries" requirement in Article I, it may perhaps be presumed that the freedoms of exploration and use also require no more than a negative duty imposed on States not to inhibit the exploration and use of outer space as conducted by other States or their nationals.

With regard to the freedom of access, it is important to observe that this expressly relates to areas on celestial bodies only, in contrast to the freedoms of exploration and use which are applicable to both outer space sensu stricto and celestial bodies. Utilising the formulation outlined above for the freedoms of exploration and use, it is arguable that the freedom prescribes a corresponding duty on States not to undertake any activity that would exclude access of other States to a particular area on a celestial body. In other words, the obligations imposed under the freedoms of exploration and use are *logistical* ones prohibiting activities that exclude other States from undertaking the same activity. On the other hand, the obligation imposed under the freedom of access to all areas of celestial bodies is a *geographical* one, which prohibits activities that exclude other States from accessing the same area on a celestial body.

If these are the accepted effects of the freedoms provided under Article I, the practical question would be what level of activity would amount to an inhibition of another State's freedom to explore or use outer space or access an area on a celestial body. As discussed previously with respect to the exclusive and monopolistic commercial practices in space, it is arguable that any activity that purports to exclude other States from a particular type of activity or access to a specific area on a celestial body would contravene the freedoms contained in Article I.[31] Conversely, in order not to contravene the freedoms, a lawful activity must not prevent another State from undertaking a particular type of activity or for accessing a specific area on a particular celestial body.

The scope of such an obligation would be a question of degree. If the obligation is applied strictly, for example, then the occupation of an orbital position around

---

[31] See discussion on page 326.

4.3 Freedoms of Exploration and Use and the Principle of Non-appropriation

the Earth by any satellite, particularly on the geostationary orbit, would infringe the freedoms of exploration and use as this would prevent another satellite from providing coverage to a particular area for a particular activity.[32] On the other hand, if the prohibitions prescribed by the freedoms are applied too broadly, they may have very little legal effect, though their moral effects, if any, may be undiminished.

The other provisions of the Outer Space Treaty and present state practice appear to favour the latter position, or at least an approach that is closer to that particular end of the "stringency continuum". For example, Article XII of the Outer Space Treaty specifically provides that:

> All stations, installations, equipment and space vehicles on the Moon and other celestial bodies shall be open to representatives of other State Parties to the Treaty *on a basis of reciprocity*. Such representatives shall give reasonable *advance notice* of a projected visit, in order that the *appropriate consultations may be held* . . .[33]

If Article I provides an unconditional right not to be excluded from accessing any area on a celestial body, then there is no reason why access to facilities on celestial bodies as per Article XII of the Outer Space Treaty would be needed.[34] Further, on the issue of state practice, it must be noted that to date no protest or complaint has been lodged for the violation of any freedom prescribed under Article I, despite the large number of satellites that have been launched to date.

Even in a restricted form, the freedom of access to all areas of celestial bodies nevertheless poses a significant legal obstacle for a commercial space mining venture. This is because mining activities, especially in the extraction segment of the activity, necessarily require some degree of exclusivity over the area in which the mining activities are to take place. However, there is overlap between this issue and the non-appropriation principle contained in Article II of the Outer Space Treaty. Consequently, it is important to consider the implications of Article II in conjunction with the freedom of access to celestial bodies.

---

[32] The occupation of orbital slots on the geostationary orbit would not only occupy a particular orbital space to the exclusion of other satellites but also a particular radio frequency to be used for its transmissions: Yvon Henri, *Orbit/Spectrum Allocation Procedures Registration Mechanism*, paper presented at the ITU Biennial Seminar of the Radiocommunication Bureau, 15–19 November 2004, in Geneva, Switzerland. However, a persuasive argument may be made to support the view that the body of laws and regulations created by the ITU to regulate the use of the geostationary orbit and corresponding radio frequencies amount to a *lex specialis* to which Article I of the Outer Space Treaty has limited application.

[33] Italics added.

[34] Gorove made the same observation and suggested that this is an indication that the drafters of the Outer Space Treaty had not intended to fully abolish the extension of sovereignty of States into its facilities and installations on celestial bodies. See Stephen Gorove, *Sovereignty and the Law of Outer Space Re-examined* (1977) 2 ANN. AIR & SP. L. 311 at 316.

4.3.2 Article II of the Outer Space Treaty

4.3.2.1 Overview

It was clear from the beginning of space activities that the classical rules of international law on sovereignty, territory and delimitation cannot apply to outer space and celestial bodies. For example, in the modern world of rockets, missiles and interplanetary probes, the traditional "cannon-shot" rule of *potestas finitur ubi finitur armorum vis* cannot apply, regardless of whatever arbitrary limit is prescribed to be the boundary of sovereignty.[35] Article II of the Outer Space Treaty contains one of the most fundamental and universally acknowledged principles of space law, namely the principle of non-appropriation as stated in explicit terms:

> Outer space, including the Moon and other celestial bodies, is not subject to *national* appropriation by claim of sovereignty, by means of use or occupation, or *by any other means*.[36]

At first glance, two issues must be clarified in order to ascertain the precise content and effect of Article II. Firstly, the adjective "national" qualifies the principle that only "national" appropriation is prohibited and so the definition of the term "national appropriation" must be explored. Secondly, there are several possible interpretations concerning the scope of the phrase "by any other means" for the "national" appropriation of outer space and celestial bodies. Only then can the precise content and effect of Article II be distilled and applied in the context of commercial activities in outer space.

4.3.2.2 National Appropriation

The first question that needs to be addressed in the context of the scope, content and effect of Article II is its applicability to non-governmental and/or private entities. As Tennen noted, Article II does not refer explicitly to private entities even though the extension of the non-appropriation doctrine to private entities is "firmly established in space law".[37] As with the discussion in the context of Article VI of the Outer Space Treaty, any act of national appropriation in outer space and on celestial bodies conducted under the State's direction or influence, regardless of whether the act was undertaken by public or private entities, is prohibited. As Article VI requires the appropriate State to authorise and continually supervise the space activities of private entities, any act of national appropriation by private entities would be subject

---

[35] Virgiliu Pop, *A Celestial Body Is a Celestial Body Is a Celestial Body ...* (2001) 44 PROC. COLL. L. OUTER SP. 100 at 103; Ezra J. Reinstein, *Owning Outer Space* (1999) 20 Nw. J. INT'L. L. & BUS. 59; and Rosanna Sattler, *Transporting a Legal System for Property Rights: From the Earth to the Stars* (2005) 6 CHI. J. INT'L. L. 23.

[36] Italics added.

[37] Leslie I. Tennen, *Second Commentary on Emerging System of Property Rights in Outer Space* (2003) United Nations, PROCEEDINGS OF THE UNITED NATIONS/REPUBLIC OF KOREA WORKSHOP ON SPACE LAW 342 at 343.

4.3 Freedoms of Exploration and Use and the Principle of Non-appropriation

to the direction or influence of the State, thus contravening Article II of the Outer Space Treaty. Accordingly, it is clear that Article II must extend to private acts of national appropriation as well as those conducted directly by the State itself.

The second question arises because Article II does not purportedly prohibit all forms of appropriation but merely "national" appropriation. This must be considered as an issue of *scope* as distinct to the issue of whether Article II would have *application* to private and non-governmental entities, otherwise it may be possible for States to circumvent the prohibitions contained in the Outer Space Treaty simply by "privatising" the contravening activity.[38] There is a significant body of opinion among commentators that Article II also prohibits the creation of private property rights.[39] However, in considering the meaning of "national" appropriation, it is interesting to note that the French and Spanish texts both use similar wording to that of the English text.[40] The Chinese text, on the other hand, stipulates a different meaning that provides that "outer space, including the Moon and other celestial bodies, cannot, through the State by asserting sovereignty, use, occupation or any other means, be appropriated".[41] It is apparent that the Chinese text prohibits only appropriation of the Moon and other celestial bodies by the State and does not prohibit appropriation by private entities or, in the context of reconciling this with the other texts, that the meaning of "national" appropriation means appropriation by or for the State itself. Since Article XVII of the Outer Space Treaty makes the Chinese text equally authentic with the English, French, Russian and Spanish texts, the construction that is contained in the Chinese text must be given some degree of weight in determining the content and effect of Article II. This is particularly so considering the Vienna Convention on the Law of Treaties determined that the terms used in each authentic text is presumed to have the same meaning and, where there is a difference, the meaning that best reconciles the texts is to be adopted.[42]

Further, it may be useful to consider the relevant provisions of the Moon Agreement, because although it has not received widespread acceptance in the international community, its provisions may provide some guidance in the interpretation

---

[38] See discussion in Tennen, supra note 37, at 344 and Patricia M. Sterns and Leslie I. Tennen, *Privateering and Profiteering on the Moon and Other Celestial Bodies: Debunking the Myth of Property Rights in Space* (2003) 31 ADV. SPACE RES. 2433.

[39] See, for example, Leslie I. Tennen, *Outer Space: A Preserve for All Humankind* (1979) 2 HOUS. J. INT'L. L. 145 at 149.

[40] The French text of Article II provides that "L'espace extra-atmosphérique, y compris la Lune et les autres corps célestes, ne peut faire l'objet d'appropriation nationale par proclamation de souveraineté, ni par voie d'utilisation ou d'occupation, ni par aucun autre moyen." Similarly, the Spanish text provides that "El espacio ultraterrestre, incluso la Luna y otros cuerpos celestes, no podrá ser objeto de apropiación nacional por reivindicación de soberanía, uso u ocupación, ni de ninguna otra manera."

[41] Translated by the author. The Chinese text of Article II states:
「外层空间、包括月球与其他天体 在内、不得由国家通过提出主权主张、通过使用或占领、或以任何其他方法、据为己有。」

[42] Vienna Convention on the Law of Treaties, opened for signature on 23 May 1969, 1155 U.N.T.S. 331 (entered into force on 27 January 1980), Article 33.

of Article II of the Outer Space Treaty, to which the Moon Agreement is intended to be an extension and thus complementary.[43] Indeed, the Vienna Convention on the Law of Treaties mandates that subsequent treaties that interpret or apply the provisions of an earlier treaty are to be taken into account when interpreting the terms of the earlier treaty.[44] To that end, Article 11 of the Moon Agreement provides that:

1. ...
2. The Moon is not subject to national appropriation by any claim of sovereignty, by means of use or occupation, or by any other means.
3. Neither the surface nor the subsurface of the Moon, nor any part thereof or natural resources in place, shall become property of any State, international intergovernmental or non-governmental organisation, national organisation or non-governmental entity or of any natural person. The placement of personnel, space vehicles, equipment, facilities, stations and installations on or below the surface of the Moon, including structures connected with its surface or subsurface, shall not create a right of ownership over the surface or the subsurface of the Moon or any areas thereof.

   ...

If "national" appropriation as contained in Article II of the Outer Space Treaty and Article 11(2) of the Moon Agreement means appropriation by both the State and private entities, then the first provision of Article 11(3) is redundant, at least to the extent that it applies to the surface of the Moon. One further noteworthy observation that may be made from this is that Article 11(3) of the Moon Agreement states that the Moon cannot become the "property" of any State, even though this would apparently be the existing effect of Article 11(2) by prohibiting the national appropriation of the Moon.

It appears from the above discussion that, if Article 11(3) of the Moon Agreement is to have a meaning distinct to that of Article 11(2) and, therefore, Article II of the Outer Space Treaty, then "national appropriation", as a term, must have a meaning different to that of attaining property rights by the State. Some commentators support this narrow approach to the interpretation of Article II, rather than a broader one that includes exclusive property rights.[45] To that end, it may be prudent to contrast these provisions with the terms of Article 137 of the United Nations Convention on the Law of the Sea, which states that:

No State shall claim or exercise sovereignty or sovereign rights over any part of the Area or its resources, nor shall any State or natural or juridical person appropriate any part thereof.

---

[43] Eilene M. Galloway, *Agreement Governing the Activities of States on the Moon and Other Celestial Bodies* (1980) 5 ANN. AIR & SP. L. 481 at 498–499.

[44] Vienna Convention on the Law of Treaties, Article 31(3).

[45] See, for example, Carl Q. Christol, *The Common Heritage of Mankind Provision in the 1979 Agreement Governing the Activities of States on the Moon and Other Celestial Bodies* (1980) 14 INT'L. LAWYER 429 at 448 and Stephen Gorove, *Interpreting Article II of the Outer Space Treaty* (1969) 37 FORDHAM L. REV. 349 at 351.

4.3 Freedoms of Exploration and Use and the Principle of Non-appropriation

No such claim or exercise of sovereignty or sovereign rights nor such appropriation shall be recognised.[46]

It is clear from the above that Article 137(1) of the Convention on the Law of the Sea expressly prohibits the following acts:

(1) claim of sovereignty over any part of the Area by a State;
(2) exercise of sovereignty over any part of the Area by a State;
(3) appropriation of any part of the Area by a State; and
(4) appropriation of any part of the Area by a natural or juridical person.

It is apparent from Article 137(1) that the Convention on the Law of the Sea does not prohibit the exercise of sovereignty by natural or juridical persons. From this it may be suggested that the Convention envisaged that only States can assert or exercise sovereignty over territory whereas both States and nationals can appropriate land. This is consistent with the distinction drawn in customary international law, which:

(1) considers sovereignty and the ability to assert jurisdiction, to be the exclusive province of States; and
(2) appropriation or title and the ability to obtain exclusive possession, to be capable of assertion by both States and private nationals.[47]

When read in light of this distinction, "national appropriation" in Article II of the Outer Space Treaty may mean no more than the "exercise of sovereignty". Accordingly, Articles II does not prescribe any rights or duties concerning the assertion of title by private nationals, as long as they do not amount to an exercise of sovereignty by the State as the British East India Company once did for Great Britain in earlier centuries.[48] Similarly, Article 11(2) of the Moon Agreement would now be consistent and complementary with Article 11(3), the former dealing with the exercise of sovereignty by States and the latter with the ability to assert title by both States and private nationals. This is considered in detail below.

4.3.2.3 Prohibition on Property Rights as a Customary Norm?

As Article II may not apply to prohibit the actual creation of private property rights on celestial bodies expressly, but merely the assertion of state sovereignty,

---

[46] The "Area" is defined in the Convention on the Law of the Sea, Article 1(1) as "the seabed and ocean floor and subsoil thereof, beyond the limits of national jurisdiction".

[47] Gorove, supra note 45, at 351 and Wayne N. White, *Real Property Rights in Outer Space* (1997) 40 PROC. COLL. L. OUTER SP. 370 at 372.

[48] See, for example, Stephen D. Krasner, *Think Again: Sovereignty* (2001) 122 FOREIGN POLICY 20.

it is necessary to consider the possibility that such a prohibition is a norm of customary international law. This is not a question of a treaty provision crystallising into customary international law, but rather the existence of a customary principle notwithstanding the express terms of Article II to prohibit private property rights on celestial bodies.

As early as 1961, the formulation of Article II focused only on States and not on natural or juridical persons. As the United States submitted, "man should be free to venture into space without any restraints except those imposed by the laws of his own nation and by international law".[49] This may be seen as an implicit recognition that nationals, not being subjects of international law, would be bound not to exercise property rights on celestial bodies in any event.

Alternatively, the provision prescribes that state sovereignty cannot be asserted by the acts of private nationals, though the language appears to suggest the former view.[50] This uncertainty was further emphasised by Australia as, after several drafts that did not include express language concerning property rights, its representative said that draft Article II "did not make it clear that outer space was not subject to national sovereignty and that no one could acquire property rights in outer space, including the Moon and other celestial bodies".[51]

There was a significant number of observations and statements made by several participating States that either affirmed or denied the application of Article II to property rights. For example, Belgium took the view that no one has yet denied that the term "appropriation" included "both the establishment of sovereignty and the creation of titles to property in private law".[52] Further, in the First Committee proceedings on the draft of Article II, France noted that the provision prohibited claims to both "sovereignty and property rights in space".[53] On the other hand, the statements made by Brazil,[54] Chile,[55] Japan,[56] the Netherlands,[57] and the Philippines,[58] in which they referred to the effect of the non-appropriation provision in preventing colonialism, international rivalries and internationalisation of outer space, would suggest that they were of the view that the provision related to the prohibition of

---

[49] Submission by Australia, Canada, Italy and the United States of America to the First Committee of the General Assembly, 4 December 1961, U.N.Doc. A/C.1/L.301 and A/C.1/SR.1210 at 245.

[50] Ibid.

[51] (1966) U.N.Doc. A/AC.105/C.2/SR.71 and Add. 1, at 15.

[52] Ibid., at 7.

[53] (1966) U.N.Doc. A/C.1/PV.1492 at 429.

[54] Ibid., at 432.

[55] Ibid.

[56] Ibid., at 439.

[57] Ibid., at 440.

[58] (1966) U.N.Doc. A/C.1/SR.4393 at 444.

4.3 Freedoms of Exploration and Use and the Principle of Non-appropriation

state sovereignty only. It is clear, however, on a detailed review of the *travaux préparatoires* that no State stated positively that Article II of the Outer Space Treaty does not and should not extend to prohibit property rights on celestial bodies.[59]

In light of there being somewhat widespread acceptance by States that there is a prohibition on the claim and exercise of property rights on celestial bodies and, in the absence of any contrary *opinio juris* from States, it is respectfully submitted that the potential for the existence of such a customary norm must be recognised. Consequently, it may be prudent to consider that, regardless of the appropriate interpretation to be given to Article II of the Outer Space Treaty, States and private nationals are bound by customary international law not to claim or exercise exclusive property rights on celestial bodies.

4.3.2.4 The Bogotá Declaration as an Example of Potential Breach of Article II of the Outer Space Treaty

Overview

On 3 December 1975, eight equatorial States, namely Brazil, Colombia, Ecuador, Indonesia, Congo, Kenya, Uganda and Zaïre,[60] adopted the Declaration of the First Meeting of Equatorial Countries at Bogotá, Colombia (the "Bogotá Declaration").[61] The Bogotá Declaration asserts that segments of the geostationary orbit ("GEO") form an integral part of the territory of the subjacent States and thus the subjacent State was able to exercise sovereignty. Goedhuis suggested that three of the reasons that supported this prima facie contravention of the non-appropriation principle are:

(1) the existence of the geostationary orbit relies exclusively on the gravity of the Earth and therefore cannot be considered part of outer space;
(2) the principles of the Outer Space Treaty were formulated without providing sufficient scientific advice to the developing States and were thus drafted for the benefit of the industrialised States only; and
(3) the lack of delimitation between airspace and outer space meant that the non-appropriation principle does not apply to the geostationary orbit.[62]

The first argument is answered easily by the observation that the existence of *all* orbits around the Earth relies exclusively on the gravity of the Earth and, accordingly, this is no justification for the singular treatment for the geostationary orbit. If this is accepted, then no orbits around the Earth would be considered part of outer space and therefore subject to the provisions of the Outer Space Treaty. This would

---

[59] See generally Nandasiri Jasentuliyana and Roy S. Lee (eds.), MANUAL ON SPACE LAW (1979), vol. 1.

[60] Zaïre is now the Democratic Republic of the Congo.

[61] (1976) 16 I.L.M. 1.

[62] Daniel Goedhuis, *Influence of the Conquest of Outer Space on National Sovereignty: Some Observations* (1978) 6 J. SP. L. 36 at 38–39.

be contrary to the express wording of Article IV, which prohibits the placement of any object carrying nuclear weapons or any other weapon of mass destruction "in orbit around the Earth". The second and third arguments, on the other hand, merit more discussion and analysis.

Power Inequality of Industrialised States

During debates in the United Nations Committee on the Peaceful Uses of Outer Space ("COPUOS"), Colombia pointed out that she had not ratified the Outer Space Treaty and that the prescribed freedoms and the principle of non-appropriation did not constitute customary international law or *jus cogens*.[63] Since the launch of *Sputnik-1* in October 1957, there has been no claim of sovereignty over any part of outer space or celestial bodies until the Bogotá Declaration. On the contrary, statements and international instruments in the intervening years have acknowledged and emphasised the binding character of the fundamental principles of space law, even outside the confines of the Outer Space Treaty. Accordingly, it may be argued that the prescribed freedoms and the principle of non-appropriation have the requisite state practice and *opinio juris* to attain the status of being principles of customary international law. Even where the eight equatorial States may arguably be able to contract out of customary principles through the Bogotá Declaration, they can only do so to affect themselves and not States that are not party to the Declaration.[64]

In any event, as discussed earlier, a principle may be considered a peremptory norm of international law, or *jus cogens*, if it has received the acceptance of "all important elements of the international community".[65] Even if the eight equatorial States that were party to the Bogotá Declaration were to be considered collectively to be an important element of the international community, the fact that three of the eight had ratified the Outer Space Treaty at the time of the Bogotá Declaration and a further three had signed it seriously undercuts such an argument.[66] Accordingly, it may be said that the prescribed freedoms and the principle of non-appropriation may have gained the support of all important elements of the international community to attain the status of *jus cogens*.

---

[63] U.N. Doc. A/AC.105/PV.173 at 56 (1977).

[64] Article 34 of the Vienna Convention on the Law of Treaties provides that "A treaty does not create either rights or obligations for a third State without its consent".

[65] See discussion at 242 et seq.

[66] Brazil signed the Outer Space Treaty on 30 January 1967 and ratified it on 5 March 1969; Colombia signed it on 27 January 1967; Ecuador signed it on 27 January 1967 and ratified it on 7 March 1969; Indonesia and the Democratic Republic of the Congo (formerly Zaïre) signed it on 27 January 1967; Kenya and Uganda acceded to the Outer Space Treaty on 19 January 1984 and 24 April 1968, respectively; and Congo never signed nor ratified the Outer Space Treaty: U.S. Department of State, *Outer Space Treaty*, <http://www.state.gov/t/ac/trt/5181.htm>, last accessed on 18 April 2004.

4.3 Freedoms of Exploration and Use and the Principle of Non-appropriation

Absence of Delimitation Between Airspace and Outer Space

Resulting from the politically sensitive nature of the issue of delimitation, there has never been an agreed international boundary between airspace and outer space.[67] Recently, however, there has been a growing number of theoretical suggestions that have been advanced on this subject, particularly the methods by which the delimitation of airspace and outer space can be prescribed by international law.[68] The two predominant theories are the "spatial" approach and the "functionalist" approach.

Spatial Approach

The approach that has gained increasing support from States is the spatial approach, which simply seeks to draw a clear boundary between airspace and outer space at a particular altitude.[69] At least eight possible altitudes have been suggested for such a demarcation by the 1970 and 1977 background papers prepared by the Legal Sub-Committee of COPUOS.[70] Some spatial theorists have sought to establish this boundary between airspace and outer space on a geophysical basis, namely at the upper limit of the atmosphere of the Earth.[71]

Accordingly, the boundary would be at the limits of the "meteorological atmosphere", which is defined as the altitude beyond which physical phenomena would have negligible effect on the surface of the Earth, at about 80–85 km from the surface of the Earth.[72] This approach has received support from a number of international conventions, such as Article 1 of the Convention on the Regulation of Aerial Navigation, the Spanish-American Convention on Aerial Navigation and the Convention on International Civil Aviation, all of which have held that the expression "airspace" should be the region where air exists.[73] Presumably, this means the body of air surrounding the Earth, for otherwise air law would have application on all celestial bodies with an atmosphere, such as Mars or even the Europa, a satellite of Jupiter.

One other suggestion that has been advanced concerning the appropriate altitude for a boundary between airspace and outer space is based on the aerodynamic characteristics of flight instruments. This is referred to as the "von Karman line", at which aerodynamic lift is exceeded by ascensional pressure.[74] Accordingly, the aeronautical ceiling theorists propose that air sovereignty extend up to 100 km

---

[67] Henneke L. van Traa-Engelman, COMMERCIAL UTILIZATION OF OUTER SPACE (1993), at 47.

[68] Damodar Wadegaonkar, ORBIT OF SPACE LAW (1984), at 38.

[69] Ibid., at 46.

[70] Ibid.

[71] Ibid., at 39.

[72] Ibid.

[73] E. R. C. van Bogaert, ASPECTS OF SPACE LAW (1986), at 12.

[74] Marietta Benkö, Willem de Graaff and Gijsbertha C. M. Reijnen, SPACE LAW IN THE UNITED NATIONS (1985), at 12.

above the Earth.[75] This is in effect the average of the maximum altitude for an aircraft, namely about 80 km, and the minimum altitude for space activities, or lowest perigee, which presently ranges between 70 and 160 km above the surface of the Earth.[76]

There are theorists that have adopted a multilevel boundary whereby demarcation is based upon the division of airspace and outer space into zones.[77] Representatives of the Committee on Space Research ("COSPAR") have suggested that the division of the space above the Earth in three zones, the first up to the 50 or 60 km over which the underlying State could exercise full sovereignty as its "airspace". The second zones lies between the "airspace" and an altitude of around 100 km above the Earth to be designated as "mesospace", subject to a new specific legal regime.[78] The third zone is, of course, outer space, which lies beyond 100 km above the surface of the Earth.[79] Similarly, some have proposed the "effective control" test, which decides the boundary on the basis of the altitude at which the States would be able to exercise effective control over its territory.[80] The problems with these approaches are that, firstly, this depends greatly on a subjective determination of the capacity of the subjacent States to exercise sovereignty upwards from its territories and, secondly, the ability of States collectively to effectively control its airspace must increase over time with the continuing advances of economic and technological development.

The most common suggestion advocated by States that favour a spatial delimitation is a demarcation based on the lowest perigee of a satellite orbiting the Earth, which is presently 110 km above the surface of the Earth.[81] Several studies, such as those conducted by COSPAR, have indicated that below the altitude of 100 km above sea level, satellites would not be able to continue in orbit and would fall to the Earth.[82] It has been contended that this approach satisfies the demands of the freedom for space flight and space exploration for it defines outer space as the region in which, as a minimum, a satellite can still prescribe a full orbit around the Earth.[83]

This also satisfies the needs of international air law, as the boundary lies well above the maximum altitude that a civilian aircraft can operate, currently being no higher than 60 km above sea level.[84] Even when State and military aircrafts are also

---

[75] Wadegaonkan, supra note 68, at 46.

[76] Gbenga Oduntan, *The Never Ending Dispute: Legal Theories on the Spatial Demarcation Boundary Plane Between Airspace and Outer Space* (2003) 1 HERTS. L. J. 64 at 79–80.

[77] Ibid.

[78] C. de Jager and Giljbertha C. M. Reijnen, *Mesospace: The Region Between Airspace and Outer Space* (1975) 18 PROC. COLL. L. OUTER SP. 107 and Peter P. C. Haanappel, *Airspace, Outer Space and Mesospace* (1976) 19 PROC. COLL. L. OUTER SP. 160.

[79] Gennady P. Zhukov, *Delimitation of Outer Space* (1980) 23 PROC. COLL. L. OUTER SP. 221 at 227.

[80] Wadegaonkar, supra note 68.

[81] Benkö, supra note 74, at 136 and Wadegaonkar, supra note 68, at 47.

[82] Ibid.

[83] Benkö, supra note 74, at 128.

[84] Ibid., at 128.

taken into consideration, the von Karman line nevertheless lies below this proposed physical boundary between airspace and outer space.[85]

This approach also appears to be the one supported by state practice, with the notable exception of the United States, discussed below. It must be observed that no State has yet protested against satellites passing over its territories at altitudes above 100 km as being an infringement of its sovereign airspace.[86] Further, some States have imposed a regulatory demarcation between airspace and outer space to confine the applicability of its domestic legislation regulating space launch activities. For example, the *Space Activities Act 1998* (Cth) of Australia expressly defines "outer space" as being beyond 100 km above the surface of the Earth.[87] Nevertheless, this falls short of being accepted as a customary boundary between airspace and outer space, as the requirements of universality would not be satisfied, even among spacefaring States, for the acceptance of a physical delimitation between airspace and outer space in the absence of express treaty provisions.[88]

Functionalist Approach

The failure of the spatial approach to agree to a scientific or technical criteria for the physical delimitation of air space and outer space has, to some extent, caused the functionalist approach, which provides for a non-physical delimitation of airspace and outer space, to gain increasing acceptance. The functionalist approach concentrates on the character and nature of the activities and their objectives rather than the physical locale of the activities themselves at a given time.[89]

Proponents of this approach argue that within the same space above the Earth, there simultaneously operate two international legal orders regulating aeronautical and astronautical activities of States respectively.[90] Accordingly, the issue of delimitation should be based primarily on the nature and the type of the particular activity, being either aeronautical and astronautical, wherein the latter should be subject to space law irrespective of the altitude at which they are carried out.[91] Consequently, space activities would be governed by space law, regardless of the precise location where the activities are conducted at the time, such as the location of the launch facility for the launch vehicle of the space object.[92] Effectively,

---

[85] Ibid.

[86] Ibid.

[87] *Space Activities Act 1998* (Cth), s. 8.

[88] Ian Brownlie, PRINCIPLES OF PUBLIC INTERNATIONAL LAW (5th ed., 1998), at 5–11.

[89] Manfred Lachs, THE LAW OF OUTER SPACE: AN EXPERIENCE IN CONTEMPORARY LAW MAKING (1972), at 56.

[90] Gennady P. Zhukov and Yuri M. Kolosov, INTERNATIONAL SPACE LAW (1984), at 154.

[91] Benkö, supra note 74, at 129.

[92] Isabella H. Ph. Diederiks-Verschoor, AN INTRODUCTION TO SPACE LAW (2nd ed., 1999), at 20.

176   4 Rights and Duties in the Commercial Exploration and Extraction of Mineral ...

this proposes different frontiers for the different types of space activities, which are conditional on the degree of tolerance accorded to them by subjacent States.[93]

This is by no means an unworkable approach, as any access to outer space from the surface of the Earth must transverse airspace, in some cases that of other States or the high seas, in order to reach outer space. For example, an equatorial launch from Christmas Island, an Australian external territory, would necessarily fly over the territories of Indonesia, Papua New Guinea and, in some cases, East Timor. Similarly, a launch by Sea Launch near the territorial waters of Kiribati may fly over various States in the South Pacific, such as Tuvalu and Samoa, as well as French Polynesia of France and Tokelau and the Cook Islands of New Zealand.[94] It would be surprising if such a launch vehicle would be subject to the provisions of international and domestic air law. The *Commercial Space Launch Act 1984* of the United States, for example, defines its applicability by reference to the launch vehicle, being the involvement of a rocket of 200,000 pounds per second of impulse and a ballistic coefficient of 12 psi, rather than by a physical delimitation between airspace and outer space at a specified altitude.[95]

Effect on the Bogotá Declaration

While it is true that there is no settled international delimitation between airspace and outer space, it is clear that all States intend for any activities involving the orbiting of a satellite around the Earth to be considered space activities.[96] If activities involving satellites in low Earth orbit are considered to be space activities, then the deployment of satellites in the geostationary orbit must accordingly be considered space activities as well. Consequently, the argument that the geostationary orbit may be considered to be part of airspace rather than outer space is an interesting one that has failed to gain the widespread acceptance of States.

In conferences of the ITU, which has the primary responsibility of regulating the orbital slot allocations on the geostationary orbit, some delegations have expressly stated that the claims of sovereignty under the Bogotá Declaration cannot be recognised.[97] However, even though these claims have been widely rejected, it demonstrates a deep sentiment among some States that the space law framework

---

[93] Lachs, supra note 89.

[94] Sea Launch Company LLC, *Sea Launch User's Guide* (2003), located at <http://www.sea-launch.com/customers_webpage/sluw/>, last accessed on 19 January 2005.

[95] *Commercial Space Launch Regulations* (U.S.) 14 C.F.R. 400.2.

[96] Oduntan, supra note 76.

[97] See, for example, commentary in Australian Department of Foreign Affairs and Trade, *Final Protocol and Partial Revision of the 1998 Radio Regulations, as incorporated in the Final Acts of the World Radiocommunication Conference (WRC-2000), Done at Istanbul on 2 June 2000* [2001] A.T.N.I.A. 32, <http://www.austlii.edu.au/au/other/dfat/nia/2001/32.html>, last accessed on 18 April 2004.

4.3 Freedoms of Exploration and Use and the Principle of Non-appropriation

under the Outer Space Treaty does not adequately protect the interests of the developing States, a problem made far more intractable later in the negotiations for the Moon Agreement.[98]

4.3.2.5 "By Any Other Means"

Lachs, who held the chair of the Legal Sub-Committee during the debates on the Outer Space Treaty, emphasised the prohibition of appropriation based on "use" and "occupation", as he was of the view that Article II prevented the creation of "titles".[99] As discussed previously, the use of the term "title" in the context of "national appropriation" is clearly meant to indicate claims of national sovereignty by States rather than proprietary or private ownership rights.[100] In any event, having reached such a conclusion, Lachs noted the phrase "by any other means" and asked: "What other means are there?"[101]

Some commentators suggested that the phrase "by any other means" was not meant to refer to specific means but that it includes "whatever residue of international law applies to national appropriation, and has no limitation".[102] Lachs lent further support to this view by asserting that all other means were discussed "precisely to illustrate the unreality of their application to it. It was *ex abundante cautela* that these titles were indicated and at once discarded".[103] Lachs went on to suggest three possible "other means", namely discovery, contiguity and parts of outer space bordering airspace, and considered them all inadequate in asserting a claim of national appropriation.[104]

The difficulty with the approach adopted by Lachs is that it assumed that the phrase "by any other means" was subject *ejusdem generis* to the means already enumerated. Christol, on the other hand, was of the view that the phrase "by any other means" has a life of its own.[105] This is because the provision "by claims of sovereignty, by means of use or occupation" is all encompassing and thus the phrase "by any other means" would not add anything to its legal effect. Christol suggested

---

[98] Gbenga Oduntan, *Legality of the Common Heritage of Mankind Principle in Space Law: Reconciliation of the Views from the North and South*, paper presented at the International Institute of Space Law Symposium, 6 May 2003, in Sydney, Australia.

[99] Lachs, supra note 89, at 43. The British delegation was of the same view, in that "no State is able to establish an exclusive title to any part of outer space": H. G. Darwin, *The Outer Space Treaty* (1967) 42 BRIT. Y. B. INT'L. L. 282.

[100] Ivan A. Vlasic, *The Space Treaty: A Preliminary Evaluation* (1967) 5 CAL. L. REV. 512.

[101] Lachs, supra note 89, at 43.

[102] S. Bhatt, *Legal Control of the Exploration and Use of the Moon and Celestial Bodies* (1968) 8 INDIAN J. INT'L. L. 38 and E. Brooks, *Control and Use of Planetary Resources* (1969) 11 PROC. COLL. L. OUTER SP. 342.

[103] Lachs, supra note 89, at 43–44.

[104] Ibid., at 43.

[105] Carl Q. Christol, *Article 2 of the 1967 Principles Treaty Revisited* (1984) 9 ANN. AIR & SP. L. 217 at 241.

that the negotiating history of Article II, as evidenced by the *travaux préparatoires* of the Outer Space Treaty, implied the phrase "by any other means" was designed to impose the same restrictions on individuals and private entities.[106] If this interpretation is accepted, then "by any other means" would include the exercise of sovereign rights by States through private use, private occupation and assertions of private exclusive rights. This interpretation, though creative, is nevertheless consistent with the idea that Article II relates only to the exercise of state sovereignty or "national appropriation" and, in that context, refers only to a State exercising sovereign rights through private use or occupation of celestial bodies.

4.3.2.6 Précis: Content and Effect of Article II of the Outer Space Treaty on Commercial Space Activities

Setting aside the controversy concerning the legal validity of the claims made by States under the Bogotá Declaration, there remains a significant degree of disagreement among commentators even on the effect of Article II on exclusive claims of title asserted by non-governmental entities, such as private individuals or companies. Gorove, for example, adopted the "literalist" approach and was of the view that individuals could lawfully appropriate any part of outer space, including the Moon and other celestial bodies.[107] This position has found support among some commentators, especially in the context of the allocation and use of the geostationary orbit by private entities.[108]

According to Christol, the more commonly accepted views on the effects of the non-appropriation principle in Article II would include:

(1) prohibition on the appropriation of States of areas, or parts of areas of the space environment;
(2) prohibition on the appropriation of intergovernmental organisations of areas, or parts of areas of the space environment;
(3) prohibition on a State to grant to its nationals or private entities exclusive rights to the space environment; and
(4) prohibition on an intergovernmental organisation from exercising or granting exclusive rights to the space environment.[109]

In the context of private and commercial entities, this effectively means that Article II operates to prohibit the appropriation or assertion of exclusive rights by States, their nationals and private entities. In other words, States and private entities would not have the legal authority to assert any exclusivity over any area of

---

[106] Ibid., at 263.

[107] Gorove, supra note 47, at 351.

[108] See, for example, Clyde E. Rankin III, *Utilization of the Geostationary Orbit – A Need for Orbital Allocation* (1974) 13 COLUM. J. TRANS. L. 101.

[109] Christol, supra note 105, at 263.

space. For example, while a State or private entity can have a satellite occupying a particular orbital position around the Earth, it would not be able to assert exclusive use and occupation of that orbital position without a satellite. Similarly, States and private entities are free to build facilities and installations on the Moon and other celestial bodies and sell those facilities, but they cannot exclusively occupy or sell the underlying "land" or other vacant "land". As discussed, however, this may not be correct in light of what "national appropriation" best meaning the exercise of sovereign rights. Accordingly, it may be prudent to suggest that Article II of the Outer Space Treaty is in fact silent on the issue of exclusive property rights but does have the effect of prohibiting the exercise of sovereign rights, which is prohibited whether by claim, use or occupation by the State or its nationals.

4.3.3 Relevant Provisions of the Moon Agreement

4.3.3.1 Non-appropriation: Article 11(2)

Article 11 of the Moon Agreement, in seeking to repeat the provisions of Articles I and II of the Outer Space Treaty, has presented in itself some issues of interpretation that it would be prudent to investigate. To begin with, it should be noted that the Moon Agreement applies not only to the Moon, but also to other celestial bodies in the Solar System and orbits and trajectories around them.[110] Accordingly, the provisions of the Moon Agreement would be applicable to the Moon, the other planets and their natural satellites as well as asteroids.

In an identical manner to Article II of the Outer Space Treaty, Article 11(2) of the Moon Agreement prohibits "national appropriation" by any claim of sovereignty, by means of use or occupation or by any other means. From the discussion in Section 4.3.2, "national appropriation" would mean no more than exercise of state sovereignty so that Article 11(2), as is the case with Article II of the Outer Space Treaty, prohibits only the exercise of state sovereignty but has no effect on the creation of exclusive property rights by States or their private nationals.

4.3.3.2 Freedom of Exploration and Use: Articles 11(4) and 6

The three freedoms provided for under Article I of the Outer Space Treaty, namely the freedom of exploration, freedom of use and freedom of scientific investigation, find expression in Articles 11(4) and 6 of the Moon Agreement. Article 11(4) of the Moon Agreement provides that:

---

[110] Article 1(1) of the Moon Agreement provides that: "The provisions of this Agreement relating to the Moon shall also apply to other celestial bodies within the solar system, other than the Earth, except insofar as specific legal norms enter into force with respect to any of these celestial bodies." Article 1(2) further provides that "For the purposes of this Agreement reference to the Moon shall include orbits around and other trajectories to or around it."

180   4 Rights and Duties in the Commercial Exploration and Extraction of Mineral ...

> State Parties have the right to exploration and use of the Moon without discrimination of any kind, on the basis of equality and in accordance with international law and the terms of this Agreement.

It is clear that Article 11(4) is simply a reproduction of the language contained in Article I of the Outer Space Treaty, except that the Moon Agreement does not provide for "free access to all areas of celestial bodies". This may be considered to be of no significance in light of the fact that the assertion and maintenance of exclusionary title on the surface and subsurface of the Moon is prohibited under Article 11(3) of the Moon Agreement. In any event, the full force and effect of Article I of the Outer Space Treaty would continue to apply as it is not inconsistent with Article 11(4) of the Moon Agreement.

Similarly, Article 6(1) of the Moon Agreement provides that:

> There shall be freedom of scientific investigation on the Moon by all State Parties without discrimination of any kind, on the basis of equality and in accordance with international law.

The requirements that scientific investigations on the Moon be conducted on the basis of equality and without discrimination of any kind are not found in Article I of the Outer Space Treaty. This also may not necessarily be of great significance in the context of mineral exploration and extraction activities for at least two reasons:

(1) the activities involved in scientific investigations may well encompass the exploration and/or use of outer space and celestial bodies and, consequently, would be subject to the equality and non-discrimination requirements under Article I of the Outer Space Treaty and Article 11(4) of the Moon Agreement; and
(2) Article 6(2) of the Moon Agreement, as discussed below, provides specific rights and duties concerning the collection of mineral samples from celestial bodies, thus giving specific content to the limitations on the freedom of scientific investigation in this context.

4.3.3.3 Prohibition of Private Title: Article 11(3)

Article 11(3) of the Moon Agreement contains the following specific prohibitions:

(1) the surface of a celestial body or any part thereof cannot become "property" of any State, intergovernmental or non-governmental organisation, domestic governmental or non-governmental organisation and natural persons;
(2) the subsurface of a celestial body or any part thereof cannot become "property" of any State, intergovernmental or non-governmental organisation, domestic governmental or non-governmental organisation and natural persons;
(3) natural resources in place on the surface or subsurface of a celestial body cannot become "property" of any State, intergovernmental or non-governmental organisation, domestic governmental or non-governmental organisation and natural persons; and

4.3 Freedoms of Exploration and Use and the Principle of Non-appropriation

(4) placement of personnel, vehicles, equipment, facilities, stations and installations on the surface or subsurface of a celestial body cannot create a right of "ownership" over that surface or subsurface.

There is little doubt that "property" in this case means having title, especially when taking into account the wording of the other authentic texts.[111] This is because, although the French word "*propriété*" and the Spanish word "*propiedad*" basically translates to "property", the Chinese term「财产」can be translated as both "asset" and "property".[112] This is further reinforced by the reference to "ownership" in the last provision of Article 11(3), indicating that "property" in this context must be the exercise of some form of title or property right over the surface or subsurface of the Moon or other celestial bodies, including its natural resources.

This effectively means that, although there is a significant number of commentators who are of the view that Article II of the Outer Space Treaty prohibits the creation of property rights on celestial bodies, this prohibition did not in fact come into existence until the adoption of Article 11(3) of the Moon Agreement. In any event, these prohibitions clearly impose a severe constraint on the ability of States and private entities to engage in the extraction of mineral resources from the surface or subsurface of celestial bodies. This is because the extraction of mineral resources would require the ownership, or at least some form of title thereof, in the surface of the celestial body, in its subjacent subsurface and in the mineral resources, both before and after their extraction. As the Outer Space Treaty does not contain equivalent prohibitions and the Moon Agreement has not received widespread acceptance

---

[111] Article 21 of the Moon Agreement provides that the Arabic, Chinese, English, French, Russian and Spanish texts are equally authentic.

[112] Commercial Press, *A New English-Chinese Dictionary* (2nd ed., 1984), at 75. The French text of Article 11(3) provides that "Ni la surface ni le sous-sol de la Lune, ni une partie quelconque de celle-ci ou les ressources naturelles qui s'y trouvent, ne peuvent devenir la propriété d'États, d'organisations internationales intergouvernementales ou non gouvernementales, d'organisations nationales ou d'entités gouvernementales, ou de personnes physiques. L'installation à la surface ou sous la surface de la Lune de personnel ou de véhicules, matériel, stations, installations ou équipements spatiaux, y compris d'ouvrages reliés à sa surface ou à son sous-sol, ne crée pas de droits de propriété sur la surface ou le sous-sol de la Lune ou sur une partie quelconque de celle-ci ...". The Spanish text states that "Ni la superficie ni la subsuperficie de la Luna, ni ninguna de sus partes o recursos naturales podrán ser propiedad de ningún Estado, organización internacional intergubernamental o no gubernamental, organización nacional o entidad no gubernamental ni de ninguna persona física. El emplazamiento de personal, vehículos espaciales, equipo, material, estaciones e instalaciones sobre o bajo la superficie de la Luna, incluidas las estructuras unidas a su superficie o la subsuperficie, no creará derechos de propiedad sobre la superficie o la subsuperficie de la Luna o parte alguna de ellas ...". The Chinese text provides that「月球的表面或表面下层或其任何部分或其中的自然资源均不应成为任何国家、政府间或非政府国际组织国家组织、或非政府实体或任何自然人的财产。在月球表面或表面下层，包括与月球表面或表面下层相连接的构造物在内，安置人员、外空运载器、装备设施、站所和装置、不应视为对月球或其任何领域的表面或表面下层取得所有权。...」

in the international community, the ability of a mining venture to lawfully extract mineral resources remains somewhat uncertain.

4.3.3.4 Extraction of Mineral Samples: Article 6(2)

The Moon Agreement also contains a specific provision for the extraction of mineral samples from the surface or subsurface of the Moon and other celestial bodies in the Solar System. Specifically, Article 6(2) of the Moon Agreement provides that:

> In carrying out scientific investigations and in furtherance of the provisions of this Agreement, the State Parties shall have the right to collect on and remove from the Moon samples of its mineral and other substances. Such samples shall remain at the disposal of those State Parties which caused them to be collected and may be used by them for scientific purposes. State Parties shall have regard to the desirability of making a portion of such samples available to other interested State Parties and the international scientific community for scientific investigation. State Parties may in the course of the scientific investigations also use mineral and other substances of the Moon in quantities appropriate for the support of their missions.

In this context, Article 6(2) of the Moon Agreement may be seen as an exception to the prohibition on the exercise of title and property rights over celestial bodies and their natural resources, as provided under Article 11(3). Specifically, it appears that Article 6(2) provides several rights to States:

(1) the right to collect on and remove from celestial bodies samples of its mineral and other substances in carrying out scientific investigations and in furtherance of the provisions of the Moon Agreement;
(2) the right to retain the samples collected and used for scientific purposes;
(3) the right to share a portion of a sample collected with the international scientific community for scientific investigations; and
(4) the right to use mineral and other substances on the Moon in quantities appropriate for the support of missions of scientific investigations.

In each case, the right provided under Article 6(2) is confined to the purposes of scientific investigations. In the context of the present study, the obvious next step would be to consider whether, as a matter of international law, exploration and prospecting of mineral resources can be considered "scientific investigation". To some degree, the geological and mineralogical study of celestial bodies is a necessary part of the "scientific investigation" of the Solar System. For example, the scientific community has continually reaffirmed the value of such mineralogical study to discovering the origins of the Solar System.[113] At the same time,

---

[113] See, for example, A. G. W. Cameron, *Origin of the Solar System* (1988) 26 ANN. REV. ASTRON. & ASTROP. 441 and Hans E. Suess, *Chemical Evidence Bearing on the Origin of the Solar System* (1965) 3 ANN. REV. ASTRON. & ASTROP. 217.

the conduct of mineralogical studies of celestial bodies for the purposes of commercial prospecting and profit may not appropriately be considered to be part of humankind's "scientific investigation" of the Solar System.

The difficulty in maintaining this delineation lies in the practical reality that the scientific and commercial sectors of the space community are intertwined and greatly interdependent. In practice, for any mineralogical study of a celestial body or part thereof, no matter how scientific in nature and how broad the coverage of the study, the data obtained may give rise to mining operations in a specific part of the surface or subsurface of the celestial body with a view to commercial gain.[114] Similarly, a commercial mineral prospecting activity on a celestial body, no matter how capitalist oriented the mission or how confined the geographical scope of the prospecting, may give rise to the production of valuable data that may be shared to assist in the scientific investigation of that celestial body.

Since it is difficult to reach a conclusion by practical differentiation between commercial prospecting and scientific investigation, it may be prudent to compare and contrast Article 6(2) of the Moon Agreement with similar provisions in other international resource development regimes. For example:

(1) in the Convention on the Law of the Sea, "marine scientific research" in the deep seabed is regulated pursuant to Article 143 and Part XIII of the Convention, while the exploration of mineral resources is regulated by Article 153 and Part XI of the Convention;[115]
(2) in the Madrid Protocol on Environmental Protection to the Antarctic Treaty, Article 7 prohibits "any activity relating to mineral resources, other than scientific research"; and
(3) the scope of what constitutes scientific research in relation to mineral resources in Antarctica was further clarified in Article 1 of the failed Wellington Convention on the Regulation of Antarctic Mineral Resource Activities to exclude prospecting and exploration, where:

(a) "prospecting" means activities aimed at identifying areas of mineral resource potential for possible exploration and development, including geological, geochemical and geophysical investigations and field observations, the use of remote sensing techniques and collection of surface, seafloor and sub-ice samples, but not including dredging and excavations, except for the purpose of obtaining small-scale samples, or drilling, except shallow drilling into rock and sediment to depths not exceeding 25 m; and

---

[114] See, for example, Predictive Mineral Discovery Cooperative Research Centre, *Utilisation and Application of the Research: Commercialisation and Links with Users* (2002), located at <http://www.pmdcrc.com.au/repspubs/annrep.html>, last accessed on 20 January 2005.

[115] The effect of these provisions are not affected by the adoption of the Agreement relating to the Implementation of Part XI of the United Nations Convention on the Law of the Sea (1995) U.N.Doc. A/RES.48/263.

(b) "exploration" means activities aimed at identifying and evaluating specific mineral resource occurrences or deposits, including exploratory drilling, dredging and other surface or subsurface excavations required to determine the nature and size of mineral resource deposits and the feasibility of their development, but excluding pilot projects or commercial production.[116]

If the analogies between celestial bodies and the deep seabed and/or between celestial bodies and Antarctica may be maintained, then it would be reasonable to assume that the term "scientific investigation" in the Moon Agreement would also exclude mineral exploration and prospecting with a view of assessing the feasibility of mineral resource development for commercial gain. With this in mind, it is then important to consider the force and effect of the Moon Agreement in light of its limited acceptance in the international community.

4.3.3.5 Force and Effect of the Moon Agreement Generally

Overview of the Issues

As of 1 January 2010, the Moon Agreement is in force but has been ratified by only 13 States, of which only Australia is likely to be considered to have a space capability, and signed by a further four States, of which only France is likely to be considered to have an advanced space capability.[117] Of course, this does not prevent a State without space capability from contracting with a third State or its nationals or private entities to launch space objects on its behalf or to provide it with space capability.[118] In any event, it is important to consider the legal force and effect of the Moon Agreement on third States that are not parties to the Moon Agreement.

Generally, without focusing on specific provisions, there are at least three circumstances that are to be considered in this discussion:

(1) the force and effect of a right or obligation contained in the Moon Agreement for a State that has ratified it;
(2) the force and effect of a right or obligation contained in the Moon Agreement for a State that has signed it but not ratified it; and

---

[116] Article 1 of the Wellington Convention on the Regulation of Antarctic Mineral Resource Activities (1989) 27 I.L.M. 868.

[117] The States that have ratified the Moon Agreement are Australia, Austria, Belgium, Chile, Kazakhstan, Lebanon, Mexico, Morocco, the Netherlands, Pakistan, Peru, the Philippines and Uruguay and the States that have signed but not ratified it are France, Guatemala, India and Romania: United Nations Office of Outer Space Affairs, *Status of International Agreements Relating to Activities in Outer Space* (2008) United Nations, 1 January 2008, at <http://www.unoosa.org/pdf/publications/ST_SPACE_11_Rev1_Add1E.pdf>, last accessed on 16 July 2006.

[118] This is, of course, subject to international regulation on the transfer of arms and advanced military technologies under the Wassenaar Agreement and the Missile Technology Control Regime: see, for example, discussion in Ricky J. Lee and Steven R. Freeland, *The Impact of Arms Limitation Agreements and Export Control Regulations on International Commercial Launch Activities* (2002) 45 Proc. Coll. L. Outer Sp. 321.

4.3 Freedoms of Exploration and Use and the Principle of Non-appropriation

(3) the force and effect of a *right* or *obligation* contained in the Moon Agreement for a third State.

Rights Under the Moon Agreement

Prima facie, the force and effect of a right in each of the circumstances outlined above arising from the Moon Agreement should also be considered. However, in the case of the present discussion, only the force and effect of *obligations* should actually be considered as there are no rights arising from the Moon Agreement. The reasons being that:

(1) the prohibitions on national appropriation and property rights under Articles 11(2) and 11(3) are clearly obligations and not rights;
(2) the freedoms of exploration, use and scientific investigation in Articles 6 and 11(4) are rights that are already contained in the Moon Agreement and the only substantial difference between them lies in the additional obligation that scientific investigations must be conducted on the basis of equality and without discrimination of any kind; and
(3) the right to collect and use mineral samples in carrying out scientific investigations as contained in Article 6(2) is merely a specific example of the freedom of scientific investigation, with the requirement to consider the desirability of sharing the samples with other States being an additional obligation under the Moon Agreement, while noting that the obligation extends only to the need to consider, not to actually share any such mineral samples collected.

In any event, a right contained in the Moon Agreement can only be exercised by a third State that is not party to it if the treaty intended for third States to have that right and be able to exercise it against parties and other non-parties of the Moon Agreement. This is because Article 36 of the Vienna Convention on the Law of Treaties provides that:

> A right arises for a third State from a provision of a treaty if the parties to the treaty intend the provision to accord that right either to the third State, or to a group of States to which it belongs, or to all States, and the third State assents thereto. Its assent shall be presumed so long as the contrary is not indicated, unless the treaty otherwise provides. . . .

In the case of the Moon Agreement, it should be noted that each provision containing a right refers to a right to be exercised by "States Parties". For example, Article 9(1) of the Moon Agreement provides that "*States Parties* may establish manned and unmanned stations on the Moon".[119] Accordingly, it is clear that the rights contained in the Moon Agreement were not intended to extend to third States without them being party to it and, therefore, cannot be exercised by them.[120]

---

[119] Italics added.

[120] See generally Ian Robertson Sinclair, THE VIENNA CONVENTION ON THE LAW OF TREATIES (2nd ed., 1984) and Ian Brownlie, PRINCIPLES OF PUBLIC INTERNATIONAL LAW (1990) at 603–635.

Obligations Under the Moon Agreement

It is generally accepted law that a State party to a treaty is bound by its terms and they must be performed by that State in good faith, a principle commonly referred to as *pacta sunt servanda*.[121] Accordingly, the 13 States that have ratified the Moon Agreement are clearly bound by any obligations arising from it and must perform them in good faith. Similarly, a State cannot be held to be bound by an obligation arising from a treaty to which the State is not a party, unless the party has expressly consented or if the obligation is one that exists also as a customary norm of international law.[122] As for the five States that have signed but not ratified the Moon Agreement, Article 18 of the Vienna Convention on the Law of Treaties relevantly provides that:

> A State is obliged to refrain from acts which would defeat the object and purpose of a treaty when it has signed the treaty or has exchanged instruments constituting the treaty subject to ratification, acceptance or approval, until it shall have made its intention clear not to become a party to the treaty ...

It should be noted that Article 18 of the Vienna Convention on the Law of Treaties refers to acts that defeat the "object" and "purpose" of a treaty rather than the breach or contravention of a treaty. This is appropriate, because otherwise there would be no need for consent by ratification if a State is capable of being bound by terms of a treaty that it has signed but not ratified. However, considering the number of years that have passed since the five States have signed the Moon Agreement and yet refrained from ratifying it, these States may well be considered to have "made their intention clear not to become a party to the treaty". In any event, it would be prudent to consider whether the action or conduct of a State defeats the objects and purposes of the treaty rather than the mere contravention of an obligation.

In the case of the Moon Agreement, its objects and purposes may be identified from its preamble as follows:

(1) promote the further development of cooperation among States in the exploration and use of the Moon and other celestial bodies in the Solar System on the basis of equality;
(2) prevent the Moon from becoming an area of international conflict; and
(3) define and develop the provisions of existing space law instruments in relation to the Moon and other celestial bodies.

In the case of the prohibition on property rights under Article 11(3) of the Moon Agreement, it is arguable that the creation of such rights by States that have signed but not ratified the treaty would be contrary to the object of promoting the further development of cooperation among States in the exploration and use of the Moon.

---

[121] Vienna Convention on the Law of Treaties (1969) 1155 U.N.T.S. 311, Article 26.

[122] Ibid., Articles 35 and 38.

4.3 Freedoms of Exploration and Use and the Principle of Non-appropriation

Similarly, the same can be said of the obligation to conduct scientific investigations on the basis of equality and without discrimination of any kind and the need to consider the desirability to share mineral samples collected with other States.

4.3.4 Defining a Celestial Body

4.3.4.1 The Problem

One of the first problems identified during the early stages of the debate among commentators on the Outer Space Treaty and the Moon Agreement related to the definition of "celestial body".[123] This is because objects in the Solar System including the planets, their natural satellites, as well as asteroids and meteorites, differ markedly in size and physical conditions. They range from small solid objects of varying shapes, such as asteroids, to large liquid or gaseous objects that are wholly unsuitable for landings, such as Jupiter and the comets. In one of the earliest attempts to resolve this issue, Zhukov differentiated between bodies that may be objects of exploration and/or exploitation and others that are inappropriate for human activities due to their dimension, nature or substance.[124] In his view, "celestial bodies" includes the "planets and their natural satellites, asteroids and large meteorites" but excludes "micrometeorites, smaller meteorites and comets".[125] In addition to the problem of introducing more terms that may require legal definition, this has the conceptual result of excluding objects that may well be the subject of human interest. In any event, because the treaties, especially the Moon Agreement, refer repeatedly to "celestial bodies", a more precise legal definition should be advocated for the further development of space law.[126]

4.3.4.2 Definition Based on Potential Human Interest

It has been argued that two distinct legal regimes should be created to regulate space objects depending on their dimensions and the composition of their surfaces. For some, including Zhukov, a distinction based on the existence of a human economic value was seen as important, while others would criticise this as being contrary to the common interest principle already established in space law.[127] Gál suggests that

---

[123] See, for example, discussion in E. Brooks, *National Control of Natural Planetary Bodies: Preliminary Considerations* (1966) 32 J. AIR L. & COM. 315.

[124] Gennady P. Zhukov, KOSMICHESKOYE PRAVO (1966) at 270–275. See also Marko G. Markoff, *La Lune et le Droit International* (1964) 68 REV. GEV. DR. INT'L. PUB. 248.

[125] Gennady P. Zhukov, *The Problem of the Definition of Outer Space* (1967) 10 PROC. COLL. L. OUTER SP. 271 at 273.

[126] Andrzej S. Górbiel, *Remarques sur la définition de l'espace extra-atmosphérique* (1978) 21 PROC. COLL. L. OUTER SP. 89.

[127] See F. G. Rusconi, *An Essay on the Lawful Concept of Heavenly Bodies* (1966) 9 PROC. COLL. L. OUTER SP. 55.

188   4   Rights and Duties in the Commercial Exploration and Extraction of Mineral ...

the existence of a solid surface for the landing of space vehicles is an important factor to be considered.[128] Specifically, he stated:

> Under the aspect of space law (*sic*) celestial bodies are the Moon, and the planets, moons, asteroids (or planetoids) of our solar system which are suitable for landing of manned or unmanned spacecraft, are of natural origin, and cannot be deviated from their celestial orbit. In the astronomical sense the concept of celestial bodies is much wider; the lawyer, however, is not interested in those which cannot become the scene of legally relevant actions, like the sun, our solar systems, comets, etc.[129]

The adoption of this definition has the effect of excluding comets and the large gaseous planets, namely Jupiter, Saturn, Uranus and Neptune, from the definition of "celestial bodies", even though they may well be objects of human exploration and possible exploitation. Assuming that the definition of celestial bodies may be expanded to include any solid mass in space, thus including comets and tiny asteroids as well as some of the gas giants, Fasan posed an interesting scenario.[130]

First, if a comet or asteroid is discovered to cross the Earth's orbit, it poses a threat to the Earth environment as well as to the safety of astronauts. Although this would undoubtedly be "a phenomenon which could endanger human life" and would "reach the surface of the Earth by natural means", the Moon Agreement would apply to it as long as it has not reached the surface of the Earth.[131] The use of any technology to deflect such an asteroid or comet from its natural orbit and guide it towards the sun would raise few questions with respect to the necessity of the action. However, if one assumes that the destruction is the ultimate form of appropriation, any state that destroys an asteroid or comet would, strictly speaking, contravene the provisions of the Moon Agreement.

4.3.4.3 Definition Based on Potential Removal by Humans

Fasan also presented another example in which the definition of celestial body may be controversial, more pertinent to the present study, where an asteroid, with a diameter of a few hundred meters, is removed from its orbit and moved to an orbit around the Earth, high above the geostationary orbit. The question would be whether the asteroid remains a natural celestial body even if it is then hollowed out and its surface and interior covered with artificial installations and structures.[132] Fasan suggested that this asteroid would then cease to be a "celestial body" as it is transformed

---

[128] Gyula Gál, SPACE LAW (1969) at 186.

[129] Ibid.

[130] Ernst Fasan, *Asteroids and Other Celestial Bodies – Some Legal Differences* (1998) 26 J. SP. L. 33. See also Myres S. McDougal, Harold D. Lasswell and Ivan A. Vlasic, LAW AND PUBLIC ORDER IN SPACE (1963) at 767.

[131] See Articles 1(3) and 5(3) of the Moon Agreement.

[132] Ernst Fasan, *Large Space Structures and Celestial Bodies* (1984) 27 PROC. COLL. L. OUTER SP. 243.

4.3 Freedoms of Exploration and Use and the Principle of Non-appropriation

into an artificial "space object" to which notions involving ownership, control, registration and liability are applied differently in international law.[133] Accordingly, the legal definition of "celestial bodies" ought to be limited to objects that cannot be transported through space by human intervention, thus excluding smaller and less massive objects from any international legal regime that may be imposed under the Moon Agreement.[134] In particular, Smirnoff suggested that "celestial bodies":

> ... in the sense of the treaties and agreements on outer space are natural objects in outer space including their eventual gaseous coronas which cannot be artificially moved from their natural orbits.[135]

However, there are a number of commentators who have opposed this view and suggested that the Moon Agreement and the other treaties would apply to any kind of object in outer space as "celestial bodies".[136] Indeed, if any object that can be moved from their natural orbits would not be considered a "celestial body" for the purposes of the international treaties, then it would have the undesirable consequences that the scope of applicability of any international regulatory regime would be at the mercy of technological developments, particularly in the field of propulsion, and it would not prevent the wholesale consumption of such bodies in outer space for mineral exploitation and other purposes.

Indeed, if the suggestion made by Working Group Three of the International Institute of Space Law in 1964 was accepted, "celestial bodies" would be defined as those "natural objects in outer space ... which cannot be artificially removed from their natural orbits".[137] As it is envisaged that asteroids will be moved in the near future in order to assist in the exploitation of their mineral resources, this definition would assist in transforming such asteroids into "space objects" once moved and thus outside the regulatory ambit of some aspects of international law, particularly the provisions of the Moon Agreement, a result unpalatable for some commentators. Hosenball had stated the contrary view that if an asteroid was moved into an orbit around the Earth for exploitation of its mineral resources, it would nevertheless remain a celestial body within the meaning of that term and thus would not change its character by its artificial movement.[138]

4.3.4.4 Definition Based on Size

The question then is whether such a distinction can be maintained on the basis of size. For example, what appropriate designation is applicable to "meteorites", which

---

[133] Fasan, supra note 130, at 40.

[134] Ibid., at 37.

[135] Michael Smirnoff, *Report from Working Group Three on the Law of Outer Space* (1964) 7 PROC. COLL. L. OUTER SP. 352.

[136] See Neil S. Hosenball, *Current Issues of Space Law Before the United Nations* (1974) 2 J. SP. L. 8 and Gennady P. Zhukov, WELTRAUMRECHT (1968) at 272.

[137] Smirnoff, supra note 135.

[138] Marian Nash Leich, DIGEST OF UNITED STATES PRACTICE IN INTERNATIONAL LAW (1980).

are technically asteroids or comets, or fragments thereof, that either have no fixed or discernable orbit or have been captured by the gravity of the Earth or other celestial bodies? Sztucki, for example, was of the belief that they are not included in the definition of "celestial bodies":

> [Meteorites are] celestial bodies in the astronomical sense but certainly cannot be subjected to [the] legal regime envisaged for celestial bodies and, e.g. excluded from appropriation. There is, however, an essential difference between meteorites and asteroids. Freedom of exploration and use of outer space, naturally, presupposes taking samples of meteorites, etc., which because of the unaccountable number of meteorites and no fixed trajectory, does not impair possibilities of other States to do exactly the same.[139]

Williams, in turn, refers to a definition that describes a meteorite as "a solid object moving in outer space, of considerably smaller proportions than an asteroid but considerably larger than an atom or molecule".[140] This is the definition accepted by Commission 22 of the International Astronomical Union in 1961.[141] For a lawyer, however, this presents little value as there is a very large range of dimensions between atoms and asteroids, with no assistance in the definition of the terms "considerably larger" and "considerably smaller".[142] In any event, it would be difficult to sustain the proposition that the Sun, which of itself has little value for exploration or exploitation, would be a celestial body while a meteoroid, which may be of much human interest despite its size, would not be designated as such.

4.3.4.5 Search for a Commercially and Legally Viable Definition

It is apparent from the above discussion that the following factors would have to be considered in seeking a commercially and legally viable definition of "celestial bodies", especially in light of their potential value for human exploitation:

(1) it is clearly the intention of the framers of the Outer Space Treaty, the Moon Agreement and of the States that have ratified them for human exploitation of mineral resources on "celestial bodies" to be regulated by the terms and provisions of those treaties;[143]

(2) to exclude natural objects of potential human interest for exploration and/or exploitation from the definition of "celestial bodies" would appear to be contrary to the intent, if not the spirit, of declaring them the "province of all mankind" and may, in any event, render meaningless some provisions in the

---

[139] Jerzy Sztucki, *Remarks During the Discussion on the Introductory Report* (1966) 9 Proc. Coll. L. Outer Sp. 64 at 64.

[140] Sylvia Maureen Williams, *Utilisation of Meteorites and Celestial Products* (1969) 12 Proc. Coll. L. Outer Sp. 271.

[141] See, for example, American Meteor Society, *Definition of Terms by the IAU Commission 22, 1961,* at <http://www.amsmeteors.org/define.html>, last accessed on 27 December 2004.

[142] See, for example, discussion in Pop, supra note 35, at 105.

[143] See Outer Space Treaty, Article I and Moon Agreement, Article 11.

4.3 Freedoms of Exploration and Use and the Principle of Non-appropriation

Moon Agreement relating specifically to the mineral resources on the Moon and other celestial bodies in the Solar System;[144]

(3) conversely, the definition of "celestial bodies" cannot refer only to natural objects of human interest as they cannot be ascertained as a class of objects at any given point in time, giving rise to unnecessary uncertainty in the application of international space law; and

(4) similarly, the definition cannot be based on the ability of humans to remove such bodies from their natural orbits, as such bodies would only be moved if it was the subject of human interest and, along with advancements in propulsion technology, does not create a defined class of natural objects to be depicted as "celestial bodies".[145]

It is apparent that the main reason for requiring a definition of "celestial bodies" is to exclude, for most intents and purposes, objects that may be removed from their natural orbits from the scope of the definition. The reason for this is so that the resources contained on such objects may be exploited without the need to refer to the terms and provisions of the Outer Space Treaty and the Moon Agreement concerning the appropriation of celestial bodies and the exploitation of resources.[146] However, there is no real reason other than commercial interests why this should be allowed in the context of outer space and celestial bodies being the "province of mankind". This is because the desire and the ability to move even the largest of asteroids and comets would eventually enable the exploitation of their resources without the need to have regard to the interests of other States.

Accordingly, it may be unnecessary to restrict the scope of the definition of "celestial bodies" as all natural objects within the Solar System and to define the scope of what would be defined as "appropriation" or "extraction". Specifically, the removal of a celestial body from its natural orbit or location should amount to both "appropriation" and "extraction", if it is so removed for the purpose of resource exploitation. On the other hand, if the removal of an object from its orbit is effected for reasons other than resource exploitation, such as necessity or science, then such removal should not be considered "appropriation" or "extraction" within the scope of the space treaties.

Of course, the extraction of mineral resources in substantial amounts from a celestial body may result in variations to the natural orbit of the object as a result of changes in the mass of the object due to mass being removed from the object and the relatively small mass of the object itself. However, as the extraction of mineral resources *per se* would invoke the relevant provisions of the treaties, it is unnecessary to prescribe a minimum amount of variation to the natural orbit of the object, even if it was to be relative to the mass of the object itself. In this way, the real problem behind the definition of "celestial bodies" may be resolved without the

---

[144] Ibid.

[145] Ibid.

[146] Ibid.

4.4 Environmental Protection of Celestial Bodies

4.4.1 Article IX of the Outer Space Treaty

The provision of the Outer Space Treaty that is generally cited with reference to the protection of the space environment is its Article IX, which requires States to take appropriate action in order to avoid the harmful contamination of outer space. Specifically, Article IX requires all States to conduct explorations of outer space in a manner that avoids the harmful contamination of outer space and celestial bodies, in addition to any adverse changes to the environment of the Earth through the introduction of extraterrestrial matter.

There is no definition as to what would constitute "harmful contamination" of outer space and celestial bodies, though it would appear to be quite specific and narrow in scope.[147] For instance, the avoidance of harmful contamination is an obligation restricted to the "study" and "exploration" of outer space and not to the "use" of outer space.[148] It is somewhat surprising that the "use" of outer space and celestial bodies would not be subject to the requirement to avoid harmful contamination of the space environment and celestial bodies. In any event, the lack of a precise definition of "harmful contamination" nevertheless renders the analysis of the obligations imposed on States and their private entities somewhat difficult. Although some commentators have suggested that "harmful contamination" means any activity or residue thereof that may be deemed harmful in any way, it is difficult to see how this can be sustained as all activities conducted in outer space would have an impact in some way on the environment, however minor, regardless of the nature of the activities involved.[149]

Further, States are required to engage in appropriate international consultations if it has a reasonable belief that its planned space activity, or that of another State, could cause potential "harmful interference" with the peaceful exploration and use of outer space.[150] This requirement has severe limitations, the first of which is that this provision is not retrospective in nature and relates only to activities proposed and not to activities already completed.[151] The second is that there is no definition

---

[147] Delbert D. Smith, *The Technical, Legal and Business Risks of Orbital Debris* (1997) 6 N. Y. U. ENVT'L. L. J. 50 at 56.

[148] Outer Space Treaty, Article IX.

[149] See, for example, M. Miklody, *Some Remarks to the Legal Status of Celestial Bodies and Protection of the Outer Space Environment* (1983) 25 PROC. COLL. L. OUTER SP. 13.

[150] Outer Space Treaty, Article IX.

[151] Smith, supra note 147, at 57.

4.4 Environmental Protection of Celestial Bodies

of "harmful interference" and, if the similar provision relating to the use of radio frequencies in space in the Constitution of the ITU is any guide, harmful interference would mean an impairment or total restriction to the ability of the other State to conduct its space activities.[152] The third is that, even if such a proposed activity results in harmful interference with the activities of other States, the obligation of the State proposing the activity is limited to conducting consultations with the affected States, without any obligation to modify its proposed activity, subject to its obligation to have due regard to the corresponding interests of those States under Article IX of the Outer Space Treaty.

It is prudent also to note that Article IX of the Outer Space Treaty requires States to avoid "adverse changes in the environment of the Earth resulting from the introduction of extraterrestrial matter", such as materials extracted from celestial bodies, and are to "adopt appropriate measures for this purpose".[153] Consequently, it will be necessary to adopt international regulations for the quarantine of extraterrestrial materials and for the States to undertake the continuing domestic supervision of private actors in the implementation of such regulations.

4.4.2 Article 7 of the Moon Agreement

Article 7(1) of the Moon Agreement does not provide much assistance in the interpretation of Article IX of the outer space Treaty with respect to harmful contamination of celestial bodies. Specifically, Article 7(1) provides that:

> In exploring and using the Moon, States Parties shall take measures to prevent the disruption of the existing balance of its environment, whether by introducing adverse changes in that environment, by its harmful contamination through the introduction of extra-environmental matter or otherwise.

It should be noted that Article IX of the Outer Space Treaty prohibits the harmful contamination of outer space and celestial bodies, while Article 7 of the Moon Agreement prohibits the disruption of the existing balance of celestial bodies. This disruption of the existing balance is defined as:

(1) introducing adverse changes in that environment;
(2) harmful contamination of that environment through the introduction of extra-environmental matter; and
(3) other means.

---

[152] Constitution and Convention of the International Telecommunication Union, opened for signature on 22 December 1992, 1825 U.N.T.S. 3; 28 U.S.T. 7645 (entered into force on 1 July 1994), Article 45.

[153] See Philip McGarrigle, *Hazardous Biological Activities in Outer Space* (1984) 18 AKRON L. REV. 103 and George S. Robinson III, *Earth Exposure to Martian Matter: Back Contamination Procedures and International Quarantine Regulations* (1976) 15 COLUM. J. TRANSNAT'L. L. 17.

194   4 Rights and Duties in the Commercial Exploration and Extraction of Mineral ...

Further, Article 7 of the Moon Agreement applies not only to "exploration", as is the case for Article IX of the Outer Space Treaty, but also to "use" and "scientific investigation" as provided for under the Moon Agreement.

It is clear that Article 7 of the Moon Agreement does not extend or interpret Article IX of the Outer Space Treaty but rather creates new and additional obligations concerning the preservation of the "balance" of celestial bodies. Accordingly, it is probable that Article 7 would only have force and effect on States that have ratified the Moon Agreement but not to States that have merely signed it or third States, as provided for in the customary norms applicable to the law of treaties.[154]

In this context it may be useful to consider the specificity of the provisions in Article 145 of the Convention on the Law of the Sea concerning the protection of the environment in the deep seabed, although one must be mindful that, unlike the deep seabed, most celestial bodies are unlikely to contain life or an ecological balance. In particular, Article 145 stipulates that:

> Necessary measures shall be taken in accordance with this Convention with respect to activities in the Area to ensure effective protection for the marine environment from harmful effects which may arise from such activities. To this end, the Authority shall adopt appropriate rules, regulations and procedures for *inter alia*:
>
> ...

(e) the prevention, reduction and control of pollution and other hazards to the marine environment, including the coastline, and of interference with the ecological balance of the marine environment, particular attention being paid to the need for protection from harmful effects of such activities as drilling, dredging, excavation, disposal of waste, construction and operation or maintenance of installations, pipelines and other devices related to such activities;

(f) the protection and conservation of the natural resources of the Area and the prevention of damage to the flora and fauna of the marine environment.

4.5 Legal Implications for a Regulatory Regime for Exploration of Mineral Resources on Celestial Bodies

4.5.1 Two Types of Exploration

In the interest of simplicity and convenience, the types of activities that the exploration segment may involve can be grouped into two categories: "prospecting" and "exploration". These two terms are defined in the Wellington Convention on the Regulation of Antarctic Mineral Resource Activities as:

(1) "prospecting" means all activities directly or indirectly associated with the aim of identifying areas of mineral resource potential for possible exploration and

---

[154] For the principles relating to the effect of a treaty on States that have signed but not ratified it, see the Vienna Convention on the Law of Treaties, Article 18; for the principles relating to the effect of a treaty on third States, see the Vienna Convention on the Law of Treaties, Article 34.

extraction, including geological, geochemical and geophysical investigations and field observations, the use of remote sensing techniques and the collection of surface samples and small scale subsurface samples with drilling to depths not exceeding 25 m; and

(2) "exploration" means all activities directly or indirectly connected with the aim of identifying and evaluating specific mineral resource deposits or occurrences, including exploratory drilling and other surface or subsurface excavations required to determine the nature and size of mineral resource deposits and the feasibility of their development.

For both prospecting and exploration activities, it will be necessary for any future international regulatory regime for such activities to address the legal issues that arise generally and specifically.

4.5.2 The Exploration Segment Generally

Article VI of the Outer Space Treaty requires the State to authorise and continually supervise the activities of its private entities. Consequently, it would be necessary for any private entity conducting space activities to comply with the obligations imposed on the State under the Outer Space Treaty, the Moon Agreement and other relevant international instruments.

Further, it would be necessary for the State or a private "explorer" to consider the following general obligations under the Outer Space Treaty when undertaking prospecting or exploration on celestial bodies:

(1) the need not to interfere with the rights provided to other States under the Outer Space Treaty;[155]
(2) the need to have due regard to the corresponding interests of other States in the conduct of space activities;[156]
(3) the obligation to avoid harmful interference with the activities of other States in outer space;[157]
(4) the avoidance of harmful contamination of the space environment in the exploration of outer space and celestial bodies;[158] and
(5) for States that are party to the Moon Agreement, there is an addition duty to avoid conducting activities that disrupt the existing balance on celestial bodies in their exploration and use.[159]

---

[155] Outer Space Treaty, Articles I and IX.

[156] Ibid., Article IX.

[157] Ibid.

[158] Ibid.

[159] Moon Agreement, Article 7(1).

4.5.3 Prospecting Activities

Prospecting activities may well be considered "exploration" or, at the very least, "use", for the purposes of Article I of the Outer Space Treaty, and thus subject to the freedoms provided for under Article I. It is likely that prospecting activities may also be considered "scientific investigations". In any event, States and their private entities are clearly free to undertake prospecting activities conditional upon the obligations imposed under international law. Further, in addition to the general obligations detailed above, prospecting activities conducted by States Parties to the Moon Agreement are bound by the requirement under Article 6(2) of the Moon Agreement that States are to consider the desirability of sharing the samples collected with other States.

It should be noted that, in the case of prospecting by means of remote sensing technology, the provisions in the Principles Relating to Remote Sensing of the Earth from Outer Space would not apply as those Principles relate only to the remote sensing of Earth from outer space and not of other celestial bodies.

4.5.4 Exploration Activities

As is the case with prospecting activities, exploration activities would fall within the scope of "exploration" or "use" of outer space and celestial bodies, though in this case it is doubtful whether exploration activities can be considered "scientific investigation" as such activities are clearly conducted with a view to commercial exploitation and profit. As in the case of prospecting activities, States that are parties to the Moon Agreement, as well as their nationals, must comply with Article 6(2) of the Moon Agreement and consider the desirability of sharing the samples collected with other States.

The predominant issue with exploration of specific areas on celestial bodies is exclusivity over that area for exploration purposes. If Article II of the Outer Space Treaty is taken to mean the prohibition on the creation of exclusive property rights on celestial bodies, then it may be arguable that a State or one of its private entity would not be able to lawfully create exclusive exploration rights over a specific area of a celestial body.

This is particularly so in light of the prohibition on private property rights in Article 11(3) of the Moon Agreement. However, this may be more of a practical than conceptual issue, as the answer may lie in the manner in which any exclusive right is asserted, especially in the absence of an intergovernmental institution for the provision of such rights as envisaged by Article 11(5) of the Moon Agreement. If the State or private entity merely asserted an exclusive right to explore a particular area without any restrictions placed on access to third parties, the primary issue would be one of practical enforcement of such asserted right. If this was possible, then this may nevertheless be a breach of Articles I and IX of the Outer Space Treaty in denying freedom of use to other States and failing to have due regard to the corresponding interests of other States. If the State or private entity fenced off the area or impeded access to the area for others by any means, then it may be seen to have asserted property rights over the area. In addition to possible breaches of Articles I and IX of the Outer Space Treaty, this would contravene Article II of the Outer Space Treaty and, if applicable, Article 11(3) of the Moon Agreement.

4.6 Legal Implications for a Regulatory Regime for Extraction of Mineral Resources on Celestial Bodies

4.6.1 Sovereignty over Mineral Resources

The first question that requires consideration is whether state sovereignty can exist over mineral resources as distinct to state sovereignty over territory. The reason why this issue is significant is that, if sovereignty can exist over mineral resources, then the extraction of resources from celestial bodies may constitute a "national appropriation" for the purposes of Article II of the Outer Space Treaty. This would effectively prohibit any extraction of mineral resources of celestial bodies. Even though there is some uncertainty in the legality of extracting mineral resources from celestial bodies, it is clear that the express terms of the Outer Space Treaty and the other United Nations space treaties do not embody a specific legal prohibition.

Article 11 of the Moon Agreement appears to attempt to put the issue beyond doubt. Article 11(2) reaffirms that national appropriation of celestial bodies, whether by any claim of sovereignty, means of use or occupation or by any other means, is prohibited. Article 11(3) goes further by declaring that neither the surface or the subsurface of the celestial bodies "nor any part thereof or natural resources in place, shall become property of any State, ... national organisations or non-governmental entity or of any natural person". Article 11(1) also declares natural resources on celestial bodies to be the "common heritage of mankind".

In such an analysis, it is prudent to keep in mind that the Moon Agreement is binding only on States that are party to it. Consequently, it may be argued that States that are not party to the Moon Agreement may refuse to consider mineral resources on celestial bodies to be the "common heritage of mankind" or that the extraction of such resources is prohibited. This is because neither provision exists specifically in the other United Nations space treaties and the provisions of the Moon Agreement have not been expressed to reflect customary international law or have crystallised into custom since their adoption.

As for States that are party to the Moon Agreement, it may be noted that neither Article 11(1) nor Article 11(3) are expressed in absolute terms. Both provisions refer to, and may be subject to, the express undertaking in Article 11(5) that States are to establish an international regime to govern exploitation of natural resources on celestial bodies when such exploitation is about to become feasible. Accordingly, it may be said that not only does the Moon Agreement not prohibit the extraction of mineral resources, it actually permits such an activity and mandates that an international regulatory regime is to be created to govern the "exploitation" of natural resources on celestial bodies.

4.6.2 Extraction Methods

Assuming that the extraction of mineral resources from celestial bodies is not unlawful per se, it is then appropriate to consider the lawfulness of the methods that may be used to extract such mineral resources. There are at least four possible means of extracting ores from a celestial body such as a Near Earth Asteroid. In each of these scenarios, different provisions of the Outer Space Treaty and the Moon Agreement have varying effects on the ability of a State or its private entity to extract mineral resources from a celestial body. Consequently, it is necessary to consider each scenario separately.

4.6.2.1 Commercial Extraction of Natural Mineral Resources In Situ on Celestial Bodies

In analysing the legality of the commercial extraction of mineral resources in situ on celestial bodies, it is prudent to consider some of the international legal principles that have been previously discussed, as adumbrated below:

(1) the requirement that exploration and use of outer space be "carried out for the benefit and in the interest of all countries" in Article I of the Outer Space Treaty and Article 11 of the Moon Agreement;
(2) the principle of non-appropriation in Article II of the Outer Space Treaty and Article 11 of the Moon Agreement;
(3) the prohibition of private property rights on celestial bodies in Article 11 of the Moon Agreement or as exists in customary international law; and
(4) the need for some authorisation from the creation of a regulatory regime as anticipated under Article 11 of the Moon Agreement.

As discussed above, whether the commercial extraction of resources from celestial bodies can be categorised as being "carried out for the benefit and in the interest of all countries" within the meaning of Article I of the Outer Space Treaty would depend on the interpretation of the content and scope of the provision. As mentioned above, Article I and Article IX may be interpreted in this context as imposing a negative duty on States not to interfere with the rights of, or to cause any detriment to, other States. This may pose a particular difficulty for the extraction of mineral resources from celestial bodies, though in the case of extraction in situ it may be suggested that the rights and interests of other States are not significantly affected. This is because the *in situ* extraction activities of one State would not, in most cases, prevent another State from undertaking the same activities on the same celestial body for the same mineral resources, though not from the same area.

There is a risk that such an interpretation constitutes too liberal a view of the duties and obligations imposed under Article I of the Outer Space Treaty. This is because it may be said that the commercial nature of the activity and the deprivation of the particular mineral ores of a celestial body by one State may be perceived to be carried out for the benefit of the State or States involved rather than "for the benefit and in the interest of all countries". As outlined in Section 4.2.2 and in the context of existing state practice in other fields such as remote sensing of the Earth by satellite, the following additional legal obligations may be imposed on States by Article I of the Outer Space Treaty on the commercial extraction of mineral resources:

(1) to make available the raw or processed ores to other States on a non-discriminatory basis, presumably at market value; and
(2) that nothing in the activity performed should prevent other States from undertaking the same activity.

In any event, the principle of non-appropriation in Article II of the Outer Space Treaty and the corresponding provisions of the Moon Agreement presents further legal difficulties in the commercial extraction of mineral resources from celestial bodies. This is due to the nature of the activity itself probably involving the assertion of exclusive private property rights over the surface and subsurface of the relevant celestial body. Even if there is no technical need for exclusivity in the use of the mining area on the celestial body, the commercial need would necessitate the assertion of exclusivity as financiers and investors would never be willing to provide the funds necessary to finance the mining venture without assurances that their investment is protected. The most effective means of protecting their investment is by being able to preclude competitors from accessing the same area of the celestial body for mining and other incidental purposes.

Although it may be prudent to suggest that Article II of the Outer Space Treaty is in fact silent on the issue of exclusive property rights, this is not an interpretation that has won widespread acceptance in the international community, for it is generally considered that the existence of private property rights requires the existence of state sovereignty, which is expressly prohibited by Article II of the Outer Space Treaty.[160] Further, it is clear that Article 11 of the Moon Agreement, by its express terms, prohibits the creation and assertion of private property rights. The absence or illegality of exclusive private property rights necessitates changes to the existing body of space law or the creation of a new international regulatory regime that would allow for the creation of such rights for the purpose of exploring and extracting mineral resources on celestial bodies.

---

[160] See Tennen, supra note 39; Prevost, supra note 39; and Michael E. Davis and Ricky J. Lee, *Twenty Years After: The Moon Agreement and Its Legal Controversies* [1999] AUST. INT'L. L. J. 9.

4.6.2.2 Extraction of Resources In Situ on the Objects That Results in Substantial Depletion of the Mass of the Object

One of the unique features of the mining of celestial bodies is that, unlike mining on Earth, the materials extracted from the surface or sub-surface of a celestial body is removed from that celestial body and sent to the Earth, causing some depletion of the mass of that body. In addition to the legal issues that relate to the in situ extraction of mineral resources generally as discussed above, activities leading to the substantial depletion of the mass of a celestial body may be unlawful as a matter of international space law, in particular the provisions of the Outer Space Treaty. This is because:

(1) destruction can be considered to be the ultimate form of appropriation, thus contravening Article II of the Outer Space Treaty;[161]
(2) other States are deprived of the ability to exercise similar rights over the same celestial body, thus possibly violating Articles I and IX of the Outer Space Treaty;[162] and
(3) the disruption of the environment of the celestial body is in breach of Article 7 of the Moon Agreement.

From a policy and legal perspective, taking into account the legal uncertainty over the definition of "celestial body" in the Outer Space Treaty and the Moon Agreement, several options exist in addressing the unique issues arising from such an activity:

(1) determine that, despite the substantial depletion of mass of a celestial body, such a mode of ore extraction would not contravene the provisions of the Outer Space Treaty and the Moon Agreement;
(2) define a minimum mass for an object to be classified as a "celestial body" and then set a maximum limit on materials, defined as a percentage of the total mass of the celestial body, that may be extracted from it;
(3) define a maximum limit on materials, expressed as a percentage of the total mass of a celestial body, that may be extracted from a celestial body regardless of its mass or size;
(4) require some level of processing of the ores extracted to be conducted in situ so that the bulk of the extracted ores would remain on the celestial body; and/or
(5) the prohibition of mining activities on celestial bodies that have less than a prescribed minimum mass.

It will be necessary for the above options to be considered and, from a policy perspective, for the right balanced approach to be adopted in any future legal or regulatory regime. This is particularly the case considering the depletion of mass of

---

[161] See also Article 11(2) of the Moon Agreement.

[162] See also Article 15 of the Moon Agreement.

celestial bodies through mining activities may affect the orbital mechanics of that celestial body as well as those of nearby celestial bodies that may be subject to its gravitational effects.[163]

4.6.2.3 Changing the Orbit of Objects or Moving Them to Lunar Orbit or Earth Orbit for the Extraction of Resources

For some smaller objects, such as small asteroids, it may be more cost effective to move the orbit of the object closer to the orbit of the Earth or even to remove the object to an orbit around the Earth or the Moon. As discussed above, it is questionable as to whether an object in the Solar System that orbits around the Sun or a planet would be classified as a "celestial body" if one is capable of moving the object from its orbit. It is difficult to exclude objects that are capable of being moved from the definition of "celestial bodies", as this would become a diminishing classification as propulsion technology improves, enabling larger objects to be moved from their orbits over time. Regardless, in addition to the legal issues noted above in relation to ore extraction generally and the substantial depletion of mass from a celestial body, it is necessary to consider the specific issue of changing the orbital parameters of a celestial body or removing it from its natural orbit.

There is no specific provision in the Outer Space Treaty or the Moon Agreement for changing the orbital parameters of, or removing celestial bodies, except for the legal principles that would apply to the activity as described above. In practice, as discussed in Chapter 6, the international community would need to make a policy determination as to whether extraction of mineral resources from celestial bodies involving such acts would be lawful and, if so, what limits would apply to these acts. This is particularly so given the possible effect such activities would have on the orbital mechanics of nearby celestial bodies that may even increase the possibility of such objects impacting the Earth in the future.[164]

4.6.2.4 Moving the Object to Earth Orbit for a Controlled Descent to the Surface of the Earth for Extraction of Resources

If it is technologically feasible to move a celestial body to the Earth's orbit for the purpose of extracting mineral resources, then it may become desirable for the object to be placed in a controlled descent to the surface of the Earth, either as a whole or broken up in parts. This would allow for the processing of the materials from the

---

[163] See, for example, Stanley G. Love and Thomas J. Ahrens, *Catastrophic Impacts on Gravity Dominated Asteroids* (1996) 124 ICARUS 141; Andrea Carusi, Giovanni B. Valsecchi, Germano D'Abramo and Andrea Boattini, *Deflecting NEOs in Route of Collision with the Earth* (2002) 159 ICARUS 417; and Thomas J. Ahrens and Alan W. Harris, *Deflection and Fragmentation of Near-Earth Asteroids*, in Tom Gehrels (ed.), HAZARDS DUE TO COMETS AND ASTEROIDS (1994), at 897–927.

[164] See, for example, Carusi, Valsecchi, D'Abramo and Boattini, supra note 163 and Ahrens and Harris, supra note 163.

space object to be done entirely on the surface of the Earth. In addition to the legal issues outlined above, one would need to consider the desirability of such activities in the implementation of any international regulatory framework, considering the risk of damage to the surface of the Earth and the risk of contamination by the introduction of extraterrestrial matter.

4.7 Conclusions

The mineralogical exploration of celestial bodies and the consequential physical extraction of mineral resources from their surface and/or subsurface raises specific legal and policy issues, in addition to the general issues of space law that are applicable to all space activities. It is clear from the above discussion that an international regulatory framework for exploration and extraction activities on celestial bodies will need to consider the legal and policy issues as adumbrated below:

(1) the lawfulness of commercial prospecting, exploration and extraction activities in the context of the need to be conducted for the benefit and in the interest of all countries under Article I of the Outer Space Treaty;
(2) the granting of exclusive property rights on celestial bodies for exploration and extraction activities;
(3) the lawfulness of physically removing materials from celestial bodies for the purpose of commercial exploration and extraction of mineral resources;
(4) what obligations, if any, are placed on the dissemination of the prospecting and exploration data and the materials extracted from celestial bodies;
(5) the appropriate limits, if any, on the physical extraction of materials that may substantially deplete the mass of the celestial body;
(6) the appropriate limits, if any, on the size or mass of a celestial body that may be the subject of commercial mining activities;
(7) the legality of changes to the orbital parameters of a celestial body or the removal of a celestial body from its natural orbit; and
(8) the measures that may be prescribed or adopted to avoid adverse effects on the Earth through contamination by the introduction of extraterrestrial materials to the environment of the Earth.

It is evident from the issues identified above on the exploration and extraction segments of a commercial space mining venture that such activities pose significant challenges for the existing body of space law. Before considering the appropriate international regulatory framework for such activities, it is prudent, if not necessary, to analyse the legal and policy issues arising from the exploitation segment of a commercial space mining venture. In particular, this will entail a discussion of the declaration in the Moon Agreement that celestial bodies are the "common heritage of mankind", generally considered to be the most controversial concept in the existing body of space law. It is in resolving this controversy that the key to creating an appropriate international regulatory framework may be found.

Chapter 5
Exploitation Rights: Evolving from the "Province of Mankind" to the "Common Heritage of Mankind"

5.1 Introduction

The most controversial international legal impasse associated with the commercial exploitation of mineral resources in outer space is the concept that the Moon and other celestial bodies are the "common heritage of mankind". The common heritage of mankind is a relatively new concept of international law which provides to some extent for the non-appropriation, equitable benefit distribution, peaceful use, preservation and shared management of certain spatial areas that are currently not subject to territorial or sovereign control of any State.[1] So far, the only multilateral treaties to formally incorporate the common heritage of mankind doctrine have been:

(1) the Agreement Governing the Activities of States on the Moon and Other Celestial Bodies (the "Moon Agreement") in relation to the Moon and other celestial bodies;[2] and
(2) the United Nations Convention on the Law of the Sea, in relation to mineral resource exploitation of the deep seabed.[3]

In the Moon Agreement, there are some problems that have emerged regarding various interpretations of the common heritage of mankind doctrine. From the experience of the debates over the Convention on the Law of the Sea, the prospect of mandatory transfers of benefits to non-participants, an indeterminate international

---

[1] These are the five tenets of the "common heritage of mankind" enunciated by most commentators. See, for example, Kemal Baslar, THE CONCEPT OF THE COMMON HERITAGE OF MANKIND IN INTERNATIONAL LAW (1998) at 80 and Kevin V. Cook, *The Discovery of Lunar Water: An Opportunity to Develop a Workable Moon Treaty* (1999) 11 GEORGETOWN INT'L. ENVT'L. L. REV. 647 at 656–659.

[2] Agreement Governing the Activities of States on the Moon and Other Celestial Bodies (the "*Moon Agreement*"), opened for signature on 18 December 1979, 1363 U.N.T.S. 3; 18 I.L.M. 1434 (entered into force on 11 July 1984).

[3] United Nations Convention on the Law of the Sea, opened for signature on 10 December 1982, 1833 U.N.T.S. 3; 21 I.L.M. 1261 (entered into force on 28 July 1994).

R.J. Lee, *Law and Regulation of Commercial Mining of Minerals in Outer Space*,
Space Regulations Library 7, DOI 10.1007/978-94-007-2039-8_5,
© Springer Science+Business Media B.V. 2012

governance regime and insecure private property rights have caused many States to shy away from the Moon Agreement. The impasse over the Moon Agreement has led at least one scholar to suggest that there is now created a de facto moratorium on mining activities in outer space.[4] This absence of certainty over the legal framework for celestial bodies has hindered moves towards commercial utilisation and exploitation of the Moon and has thereby inhibited possible lunar and celestial resource developments.

In order for a solution to be found, the ideological, financial and practical implications the common heritage of mankind doctrine in the Moon Agreement must be analysed in the context of comparable international regimes, such as those governing Antarctica and the deep seabed under the Convention on the Law of the Sea, especially the failures in those negotiations to reach a settlement that was agreeable to both the industrialised States and the developing States. Further, it is prudent also to consider the origins of the doctrine in the New International Economic Order and its evolution in space law from the concepts and provisions contained in the Moon Agreement and other United Nations space treaties.

5.2 Antarctica and the 1988 Wellington Convention

5.2.1 The Antarctic Treaty Framework

In the nineteenth century, France, Great Britain, Russia the United States and other States undertook a series of exploratory expeditions to Antarctica, previously *terra australis nondum cognita*, which is now the earliest model of the international management of a geographical land area.[5] By the early Twentieth Century, whaling from mainland shore stations commenced and led to the establishment of temporary settlements along the coast.[6] After the Second World War, scientific research became the predominant activity in Antarctica, and newly developed transportation and communication technologies enabled the establishment of large-scale, permanent research stations on the continent.[7] Despite this new scientific focus, seven States had made territorial claims at that time over the continent, specifically Argentina,

---

[4] See, for example, Martin Menter, *Commercial Space Activities Under the Moon Treaty* (1979) 7 Syracuse J. Int'l. L. & Com. 213 at 221.

[5] Douglas M. Zang, *Frozen in Time: The Antarctic Mineral Resource Convention* (1991) 76 Cornell L. Rev. 722 at 724–726. See also John Hanessian, *The Antarctic Treaty 1959* (1960) 9 Int'l. & Comp. L. Q. 436.

[6] J. Peter A. Bernhardt, *Sovereignty in Antarctica* (1975) 5 Cal. W. Int'l. L. J. 297 and Kurt M. Shusterich, *The Antarctic Treaty System: History, Substance and Speculation* (1984) 39 Int'l. J. 800.

[7] Patrick T. Bergin, *Antarctica, the Antarctic Treaty Regime and Legal and Geopolitical Implications of Natural Resource Exploration and Exploitation* (1988) 4 Fl. Int'l. L. J. 1, at 20.

Australia, Chile, France, New Zealand, Norway and the United Kingdom.[8] In the 1970s, the possibility of mineral resource exploitation began to be considered, with speculation that Antarctica contained large deposits of iron, copper, chromium, platinum, nickel, zinc, tin, silver and gold.[9] There was also speculation of significant oil reserves in offshore Antarctica.[10]

Comprehensive co-operative efforts to explore and regulate resource recovery activities in Antarctica started at the same time as regulatory developments in outer space; i.e., during the International Geophysical Year of 1957–1958.[11] During this period, over 30,000 scientists representing 70 countries undertook coordinated research studies of Antarctica and its environment,[12] and established 60 staffed bases to gather scientific data.[13] The international legal response spawned by this co-operative venture culminated in 1959 in the signing of the Antarctic Treaty by 12 countries with established interests in Antarctica, primarily to defer further territorial claims over the continent.[14] Currently, the Antarctic Treaty System ("*ATS*") is comprised of the 1959 Antarctic Treaty in conjunction with numerous subsequent agreements adopted by parties to the original treaty.[15] Over time, the Antarctic

---

[8] Dagmar Butte, *International Norms in the Antarctic Treaty* (1992) 3 INT'L. LEG. PERSP. 1. See also Alfred van der Essen, *The Origin of the Antarctic System* (trans. Susan Fisher), in Francesco Francioni and Tullio Scovazzi (eds.), INTERNATIONAL LAW FOR ANTARCTICA (2nd ed., 1996), 17–30, at 17–18. The United States did not make territorial claims to Antarctica: Todd Jay Parriott, *Territorial Claims in Antarctica: Will the United States Be Left Out in the Cold* (1986) 22 STANFORD J. INT'L. L. 67 and D. Michael Hinkley, *Protecting American Interests in Antarctica: The Territorial Claims Dilemma* (1990) 39 NAVAL L. REV. 43.

[9] Richard W. Bentham, *Antarctica: A Minerals Regime* (1990) 8 J. ENERGY NAT. RES. L. 120 at 123; Jonathan D. Weiss, *The Balance of Nature and Human Needs in Antarctica: The Legality of Mining* (1995) 9 TEMPLE INT'L. & COMP. L. J. 387 at 398–400; Ellen B. Heim, *Exploring the Last Frontier for Mineral Resources: A Comparison of International Law Regarding the Deep Seabed, Outer Space and Antarctica* (1990) 23 VAND. J. TRANSNAT'L. L. 819, at 836–837; and Zang, supra note 5, at 724–726.

[10] See James E. Carroll, *Of Icebergs, Oil Wells and Treaties: Hydrocarbon Exploitation Offshore Antarctica* (1983) 19 STAN. J. INT'L. L. 207.

[11] Paul Lincoln Stoller, *Protecting the White Continent: Is the Antarctic Protocol Mere Words or Real Action?* (1995) 12 AZ. J. INT'L. & COMP. L. 335 at 344–346; Grier C. Raclin, *From Ice to Ether: The Adoption of a Regime to Govern Resource Exploitation in Outer Space* (1986) 7 J. INT'L. L. & BUS. 727 at 730–732; and Nicholas M. Matte, *Legal Principles Relating to the Moon*, in Jasentuliyana and Lee (eds.), MANUAL ON SPACE LAW (1979), at 317.

[12] Raclin, supra note 11, at 730–732.

[13] Zang, supra note 5, at 724–726 and Stoller, supra note 11, at 344–346.

[14] The Antarctic Treaty, opened for signature on 1 December 1959, 402 U.N.T.S. 71; 1961 U.K.T.S. 97 (entered into force on 23 June 1961). For further discussion on the Antarctic Treaty see Raclin, supra note 11, at 730–732; Stacey L. Lowder, *A State's International Legal Role: From the Earth to the Moon* (1999) 7 TULSA J. COMP. & INT'L. L. 253 at 265–266; Zang, supra note 5, at 726–728; and Stoller, supra note 11, at 346–348.

[15] Agreed Measures for the Conservation of Antarctic Flora and Fauna, opened for signature on 13 June 1964, 17 U.S.T. 991; 1998 A.T.S. 6 (entered into force on 1 November 1982) (the "*Agreed Measures*"); Convention on the Conservation of Antarctic Seals, opened for signature on 1 June 1972, 29 U.S.T. 441 (entered into force on 11 March 1978) (the "*CCAS*"); Convention

Treaty has proven to be remarkably resilient to external pressure and changing international focus on the use of Antarctica.[16] As Stokke and Østreng noted:

> While the shift of emphasis from peaceful use to scientific freedom to environmental protection largely reflects changing priorities among the Consultative Parties themselves, it was also spurred by the fact that external criticism was increasingly targeting environmental matters: in some measure, the Consultative Parties seem to have sought to adapt the ATS to the substance of the external pressure.[17]

5.2.2 Operation of the Antarctic Treaty System

Given that the Antarctic Treaty does not displace the claims to title and sovereignty made by seven of the original Consultative Parties, the treatment of Antarctica under the Antarctica Treaty cannot in any manner be consistent with any form or concept of universal ownership, such as the common heritage of mankind.[18] The common heritage of mankind doctrine is not articulated in any of the treaties of the ATS,[19] even though Balch suggested as early as 1910 that Antarctica should become "the possession of all members of the family of nations".[20] There are some provisions of the Antarctic Treaty that nevertheless resemble elements of the common heritage of mankind doctrine. For example, the Antarctic Treaty:

---

on the Conservation of Antarctic Marine Living Resources, opened for signature on 20 May 1980, 33 U.S.T. 3476 (entered into force on 7 April 1982) (the "*CCAMLR*"); Convention on the Regulation of Antarctic Mineral Resource Activities, opened for signature on 2 June 1988, 27 I.L.M. 859 (not in force) (the "*Wellington Convention*"); and Protocol on Environmental Protection to the Antarctic Treaty, opened for signature on 4 October 1991, 30 I.L.M. 1461 (entered into force on 14 January 1998) (the "*Madrid Protocol*"). See John Vogler, THE GLOBAL COMMONS: A REGIME ANALYSIS (1995) at 79; Heim, supra note 9, at 839–840; Zang, supra note 5, at 722–723; George N. Barrie, *The Antarctic Treaty Forty Years On* (1999) 116 S. AFR. L. J. 173; Karen Scott, *Institutional Developments Within the Antarctic Treaty System* (2003) 52 INT'L. & COMP. L. Q. 473.

[16] See, for example, Gilliam Triggs, *The Antarctic Treaty Regime: A Workable Compromise or a "Purgatory of Ambiguity"?* (1985) 17 CASE W. RES. J. INT'L. L. 195.

[17] Olav Shram Stokke and Willy Østreng, *The Effectiveness of ATS Regimes: Introduction*, in Olav Shram Stokke and Davor Vidas (eds.), GOVERNING THE ANTARCTIC: THE EFFECTIVENESS AND LEGITIMACY OF THE ANTARCTIC TREATY SYSTEM (1996), 113–119, at 115.

[18] See Benedetto Conforti, *Territorial Claims in Antarctica: A Modern Way to Deal with an Old Problem* (1986) 19 CORNELL INT'L. L. J. 249 and Ellen S. Tenenbaum, *A World Park in Antarctica: The Common Heritage of Mankind* (1991) 10 VA. ENVT'L. L. J. 109.

[19] See Vogler, supra note 15, at 93–94 and Zang, supra note 5, at 765–766.

[20] Thomas Willing Balch, *The Arctic and Antarctic Regions and the Law of Nations* (1910) 4 AM. J. INT'L. L. 265. Balch is quoted and discussed in David C. Marko, *A Kinder, Gentler Moon Treaty: A Critical Review of the Current Moon Treaty and a Proposed Alternative* (1992) 8 J. NAT. RES. & ENVT'L. L. 293, at 310–313; Raclin, supra note 11, at 737–738; and Zang, supra note 5, at 726.

5.2 Antarctica and the 1988 Wellington Convention

(1) is made in "the interest of all mankind";[21]
(2) specifies that "Antarctica shall be used for peaceful purposes only";[22]
(3) contains an elaborate system for the preservation of the Antarctic environment for future generations.[23]

However, in sharp contrast to common heritage of mankind provisions contained in the Moon Agreement and the Convention on the Law of the Sea, the Antarctic Treaty specifies that it does not nullify or affirm various pre-existing territorial claims over portions of Antarctica and it also does not assert the common ownership of Antarctica for all mankind. [24] Instead, the ATS has the effect of deferring the issue of claims to title and sovereignty over Antarctica to a later time and nothing contained in the Antarctic Treaty or during the life of the ATS was to disturb the *status quo ante* in relation to sovereignty.[25] Scott noted that:

> The agreement to disagree regarding sovereignty has been reconfirmed in agreements concluded subsequent to the Antarctic Treaty. Despite a Third World push in the 1980s and early 1990s to have the Common Heritage of Mankind principle accepted as the basis for a new international Antarctic regime, the unresolved question of national sovereignty continues to underpin debate on all Antarctic issues, including current concerns such as the management of tourism and establishment of a secretariat.[26]

In any event, the fact that Antarctica is a terrestrial land mass means that it is, by its very nature, subject to the possibility of assertion of title and sovereignty by States. While this may mean that the Arctic may be a better case study than Antarctica for present purposes, there is at present no regulatory regime applicable

---

[21] Antarctic Treaty, Preamble.

[22] Ibid., Preamble and Article I.

[23] Ibid., Article IX(1)(f) and see Agreed Measures, CCAS, CCAMLR, Wellington Convention and Madrid Protocol. See also discussion by Thomas M. Franck and Dennis M. Sughrue, *Symposium: The International Role of Equity-as-Fairness* (1993) 81 GEORGETOWN L. J. 563 at 590–594.

[24] Antarctic Treaty, Article IV.

[25] Ibid. See Sir Arthur Watts, INTERNATIONAL LAW AND THE ANTARCTIC TREATY SYSTEM (1992), at 136–140.

[26] Shirley V. Scott, *Universalism and Title to Territory in Antarctica* (1997) 66 NORDIC J. INT'L. L. 33 at 39. See also Peter J. Beck, *Regulating One of the Last Tourism Frontiers: Antarctica* (1990) 10 APP. GEOG. 343; Colin Michael Hall and Simon McArthur, *Ecotourism in Antarctica and Adjacent Sub-Antarctic Islands: Development, Impacts, Management and Prospects for the Future* (1993) 14:2 TOURISM MAN. 117; Debra J. Enzenbacher, *Antarctic Tourism: An Overview of 1992/93 Season Activity, Recent Developments and Emerging Issues* (1994) 30 POLAR REC. 105; and Debra J. Enzanbacher, *The Regulation of Antarctic Tourism*, in Colin Michael Hall and Margaret E. Johnson (eds.), POLAR TOURISM: TOURISM IN THE ARCTIC AND ANTARCTIC REGIONS (1995).

to the Arctic Ocean and its ice sheet besides the Convention on the Law of the Sea.[27] As Chopra had observed:

> Since Antarctica had been conceived of as a land mass and only later was found to be a frozen continent, it was readily accepted as *terra nullius*. A large part of Antarctica lies above the sea level and can easily be identified as *terra* (land). However, a significant portion of the Antarctic continent does not fall clearly under the category of *terra nullius* – let alone the category of *terra* (land) – because it is below sea level. These areas are more like frozen seas than frozen land and could be considered identical to the shallow, frozen Arctic Sea. ... Taking these factors into account, it appears that one-third of the Antarctic continental area is nothing but frozen sea. The only difference is that Antarctic ice is described as *terra firma*, i.e., firmly attached to the land mass, whereas Arctic ice is not firmly attached to the land beneath it in all places.[28]

With this background, the regulation of activities in Antarctica provides both a comparison and contrast to the laws governing the deep seabed and outer space, particularly in relation to the common heritage of mankind. In contrast to both the deep seabed and outer space, the Consultative Parties to the Antarctic Treaty, on the other hand, have adopted an oligarchic formula to manage the area, in preference to the democratic formula that was chosen for the Moon Agreement.[29] Ultimately, the oligarchic regime reserves management of Antarctica and its environment for States that have proven their technological capacities and financial investments in Antarctica under the "activity criterion" of the Antarctic Treaty.[30] By this criterion, States must demonstrate sufficient "interest in Antarctica by conducting substantial

---

[27] See Brent Carpenter, *Warm Is the New Cold: Global Warming, Oil, UNCLOS Article 76 and How an Arctic Treaty Might Stop a New Cold War* (2009) 39 ENVT'L. L. 215; Barnaby J. Feder, *A Legal Regime for the Arctic* (1978) 6 ECOLOGY L. Q. 785; Scott G. Borgerson, *Arctic Meltdown: The Economic and Security Implications of Global Warming* (2008) 87:2 FOREIGN AFF. 63; Bo Johnson Theutenberg, *The Arctic Law of the Sea* (1983) 52 NORDISK TIDS. INT'L. RET 3; David Vanderzwaag, John Donihee and Mads Faegteborg, *Towards Regional Ocean Management in the Arctic: From Coexistence to Cooperation* (1988) 37 U. N. B. L. J. 1; Lincoln P. Bloomfield, *The Arctic: Last Unmanaged Frontier* (1982) 60 FOREIGN AFF. 87; Christopher C. Joyner, *Ice-Covered Regions in International Law* (1991) 31 NAT. RES. J. 213; Barry Hart Dubner, *On the Basis for Creation of a New Method of Defining International Jurisdiction in the Arctic Ocean* (2005) 13 Mo. ENVT'L. L. & POL'Y. REV. 1; Melissa A. Verhaag, *It Is Not Too Late: The Need for a Comprehensive International Treaty to Protect the Arctic Environment* (2003) 15 GEO. INT'L. ENVT'L. L. REV. 555; and Parker Clote, *Implications of Global Warming on State Sovereignty and Arctic Resources Under the United Nations Convention on the Law of the Sea: How the Arctic Is No Longer Communis Omnium Naturali Jure* (2008) 8 RICH. J. GLOBAL L. & BUS. 195.

[28] Sudhir K. Chopra, *Antarctica as a Commons Regime: A Conceptual Framework for Cooperation and Coexistence*, in Christopher C. Joyner and Sudhir K. Chopra (eds.), THE ANTARCTIC LEGAL REGIME (1988), 163–186 at 165–166.

[29] See René-Jean Dupuy, *The Notion of the Common Heritage of Mankind Applied to the Seabed* (1983) 18 ANN. AIR & SP. L. 347, at 349–350.

[30] See Cook, supra note 1, at 677–679; Raclin, supra note 11, at 745–747; Vogler, supra note 15, at 82–84; and Stoller, supra note 11, at 346–348.

5.2 Antarctica and the 1988 Wellington Convention

research activity there, such as the establishment of a scientific station or the dispatch of a scientific expedition" in order to gain consultative status in the meetings of the Consultative Parties.[31]

Currently, full participation in decision-making under the Antarctic Treaty is restricted to the 12 original members and 16 specially-selected additional members since 1961 that have satisfied the "activity criteria", along with 19 non-consultative parties that have recently been granted "observer status" as a concession to the General Assembly of the United Nations.[32] As Weiss commented:

> This split of voting privileges has resulted in a two-tiered system among the thirty-nine nations who have agreed to be bound by the [Antarctic] Treaty, with all the power in the hands of the [Consultative Parties] So far no party has been denied [Consultative Party] status that has agreed to the Treaty and has built a scientific base and conducted research on the continent, but very few countries can afford to build a scientific base in Antarctica.[33]

Furthermore, decisions of the Consultative Parties of the Antarctic Treaty are made by consensus, rather than by a majority vote.[34] This custom ensures that States engaged in Antarctic activities are only bound by amendments and legal developments that they explicitly agree to.[35] This also means that the number of Consultative Parties to the Antarctic Treaty cannot be increased dramatically, for such a step may cause the process of decision-making to be frustrated by dissensions.

Consensus decision-making was also adopted in the Agreement Relating to the Implementation of Part XI of the United Nations Convention on the Law of the Sea of 10 December 1982 (the "1994 Agreement") governing deep seabed mining.[36] In redressing the imbalance of power which resulted from its initial one nation-one

---

[31] See Antarctic Treaty, Article IX.

[32] See Vogler, supra note 15, at 82–84. The 12 original Consultative Parties are Argentina, Australia, Belgium, Chile, France, Japan, New Zealand, Norway, Russia, South Africa, the United Kingdom and the United States. The 16 additional Consultative Parties are Brazil, Bulgaria, China, Ecuador, Finland, Germany, India, Italy, the Netherlands, Peru, Poland, Republic of Korea, Spain, Sweden, Ukraine and Uruguay. The 19 non-consultative parties are Austria, Belarus, Canada, Colombia, Cuba, Czech Republic, Denmark, Estonia, Greece, Guatemala, Hungary, Democratic People's Republic of Korea, Monaco, Papua New Guinea, Romania, Slovakia, Switzerland, Turkey and Venezuela: Australian Antarctic Division, *Antarctic Law & Treaty: Treaty Parties*, Department of the Environment, Water, Heritage and the Arts, <http://www.aad.gov.au/default.asp?casid=80>, 19 August 2009, last accessed on 14 November 2009.

[33] Weiss, supra note 9, at 394–395.

[34] Arnfinn Jørgensen-Dahl and Willy Østreng, THE ANTARCTIC TREATY SYSTEM IN WORLD POLITICS (1991), at 95; Christopher C. Joyner and Ethel R. Theis, EAGLE OVER THE ICE: THE U.S. IN THE ANTARCTIC (1997) at 174; and Christopher C. Joyner and Ethel R. Theis, *The United States and Antarctica: Rethinking the Interplay of Law and Interests* (1987) 20 CORNELL INT'L. L. J. 65.

[35] Cook, supra note 1, at 677–679; Raclin, supra note 11, at 753.

[36] Agreement Relating to the Implementation of Part XI of the United Nations Convention on the Law of the Sea of 10 December 1982 (the "*1994 Agreement*"), opened for signature on 28 July 1994, 1836 U.N.T.S. 3; 33 I.L.M. 1309 (entered into force on 28 July 1996) and Cook, supra note 1, at 682–685.

vote system, the 1994 Agreement also amended the composition of the governing Council to more equitably balance the interests of developing countries, geographical regions, consumers, importers, investors and exporters of mineral products, and appropriately allocating proportional representation for these interest areas.[37] The cumulative effect of these management provisions is to prevent legal or economic progress in Antarctica or the deep seabed being dominated by large numbers of developing and non-participating states, and to consolidate a degree of stability in the governance of these areas. The Moon Agreement would benefit from incorporating comparable provisions in any revised amendments to its initial regime provisions, which are ultimately more conducive to the development of common heritage resources.

On 29 September 1982, Malaysia proposed to the General Assembly of the United Nations that, as Antarctica belonged to the international community, the United Nations ought to administer the continent or to require the "present occupants" to act as trustees for the international community.[38] Supporting by a significant number of developing States, Malaysia proposed to place the issue on the agenda of the General Assembly, a move opposed by the Consultative Parties of the Antarctic Treaty.[39] During the debates, a number of challenges were made by various developing States in relation to the ATS, in particular:

(1) the exclusivity of decision-making by the Consultative Parties under the Antarctic Treaty, a charge justified by the Consultative Parties on the basis of the practical, financial and legal responsibilities assumed by the Consultative Parties as well as their unique knowledge and expertise in their involvement in activities in Antarctica;
(2) the meetings and negotiations between the Consultative Parties are confidential and held behind closed doors; and
(3) the interest of the international community in the preservation of the Antarctic environment.[40]

---

[37] 1994 Agreement, Annex, Section 3; see Cook, supra note 1, at 682–685.

[38] U.N. G.A.O.R. 37th Sess. (10th Meeting), U.N. Doc. A/37/PV.10 (1982), at 17. See Christopher C. Joyner and Blair G. Ewing, Jr., *Antarctica and the Latin American States: The Interplay of Law, Geopolitics and Environmental Priorities* (1992) 4 GEORGETOWN INT'L. ENVT'L. L. REV. 1.

[39] The countries that supported Malaysia in placing the issue on the agenda of the General Assembly included Antigua and Barbuda, Singapore, the Philippines, Thailand, Pakistan, Algeria, Sierra Leone and Guyana, while at that time the Consultative Parties of the Antarctic Treaty were Argentina, Australia, Belgium, Brazil, Chile, China, France, India, Japan, New Zealand, Norway, South Africa, the Soviet Union, the United Kingdom, the United States and Uruguay: see Moritaka Hayashi, *The Antarctica Question in the United Nations* (1986) 19 CORNELL INT'L. L. J. 275 at 276–277.

[40] Ibid., at 282–286. See also Sudhir K. Chopra, *Antarctica in the United Nations: Rethinking the Problems and Prospects* (1986) 80 AM. SOC. INT'L. L. PROC. 269; Bruno Simma, *The Antarctic Treaty as a Treaty Providing for an "Objective Regime"* (1986) 19 CORNELL INT'L. L. J. 189; Christopher C. Joyner, *Japan and the Antarctic Treaty System* (1989) 16 ECOLOGY L. Q. 155; Boleslaw Adam Boczek, *The Soviet Union and the Antarctic Regime* (1984) 78 AM. J. INT'L. L.

The greatest challenge to the legitimacy of the Antarctic Treaty in the context of the international community, however, is the prospect of the exploitation of mineral resources in Antarctica and its continental shelf. It was apparent that there is a need to balance the regulation of activities relating to the exploitation of mineral resources and the concerns of the international community over the preservation of the Antarctic environment and the desire of the developing States to share in the anticipated mineral bounty of Antarctica.[41] The compromise was the Convention on the Regulation of Antarctic Mineral Resource Activities (the "Wellington Convention"), which was ultimately rejected by the international community and never entered into force.[42]

5.2.3 The Wellington Convention

5.2.3.1 Origins of the Wellington Convention

The Wellington Convention was negotiated during a time when a policy tightrope was attempted by the Consultative Parties to the Antarctic Treaty, namely the balancing act between the need to preserve the environment and to regulate activities of mineral exploitation in Antarctica.[43] As suggested by the Consultative Parties themselves at the time:

> The first is an *internal* accommodation between those Consultative Parties claiming sovereignty in Antarctica and those which neither make nor recognise such claims. The second is an *external* accommodation between the Consultative Parties who have assumed

---

834; David A. Colson, *The United States Position on Antarctica* (1986) 19 CORNELL INT'L. L. J. 291; Robert Friedheim and Tsuneo Akaha, *Antarctic Resources and International Law: Japan, the United States and the Future of Antarctica* (1989) 16 ECOLOGY L. Q. 119; and Roderic Alley, *New Zealand and Antarctica* (1984) 39 INT'L. J. 911.

[41] See Roland Rich, *A Minerals Regime for Antarctica* (1982) 31 INT'L. & COMP. L. Q. 709; Brian Robert Murphy, *Antarctic Treaty System – Does the Minerals Regime Signal the Beginning of the End?* (1991) 14 SUFFOLK TRANSNAT'L. L. J. 523; and Francesco Francioni, *Legal Aspects of Mineral Exploitation in Antarctica* (1986) 19 CORNELL INT'L. L. J. 163.

[42] Wellington Convention on the Regulation of Antarctic Mineral Resource Activities (the "*Wellington Convention*"), opened for signature on 25 November 1988, 27 I.L.M. 868 (not in force).

[43] See Steven J. Burton, *New Stresses on the Antarctic Treaty: Toward International Legal Institutions Governing Antarctic Resources* (1979) 65 VA. L. REV. 421; Frank C. Alexander, Jr., *Legal Aspects: Exploitation of Antarctic Resources: A Recommended Approach to the Antarctic Resource Problem* (1978) 33 U. MIAMI L. REV. 371; Patrick T. Bergin, *Antarctica, the Antarctic Treaty Regime and Legal and Geopolitical Implications of Natural Resource Exploration and Exploitation* (1988) 4 FL. INT'L. L. J. 1; Minturn T. Wright, *The Ownership of Antarctica, Its Living and Mineral Resources* (1987) 4 J. L. & ENV'T. 49; Ronald W. Scott, *Protecting United States Interests in Antarctica* (1989) 26 SAN DIEGO L. REV. 575; and Christopher C. Joyner, *The Antarctic Minerals Negotiating Process* (1987) 81 AM. J. INT'L. L. 888.

212   5 Exploitation Rights: Evolving from the "Province of Mankind" to the "Common . . .

that initiative for taking measures relating to Antarctica and the rest of the international community, viewed either as individual states or collectively.[44]

If the Wellington Convention had entered into force, an Antarctic Mineral Resources Commission would be created to oversee development in certain zones of Antarctica.[45] Private ventures would have been required to pay fees and taxes on the mineral resources they extract but there was no sharing of benefits or mandatory technology transfer. It should be noted, however, that the Wellington Convention never came into force and has been shelved for 50 years.[46]

5.2.3.2 Creating a Legal Regime for Mineral Exploitation in Antarctica and Its Continental Shelf

The Wellington Convention was adopted to regulate the mineral resource exploration and exploitation activities of States in Antarctica and the surrounding ice shelves and continental shelves, the latter being a necessary intersection between the ATS and the Convention on the Law of the Sea.[47] Specifically, the Wellington Convention was adopted to regulate:

---

[44] Christopher D. Beeby, *An Overview of the Problems Which Should Be Addressed in the Preparation of a Regime Governing the Mineral Resources of Antarctica*, in Francisco Orrego-Vicuna (ed.), ANTARCTIC RESOURCES POLICY: SCIENTIFIC, LEGAL AND POLITICAL ISSUES (1983), at 194. See also Gregory J. Lohmeier, *Keeping Cool Amidst the Ice: Addressing the Challenge of Antarctic Mineral Resources* (1988) 2 EMORY J. INT'L. DISP. RESOL. 141 and E. Paul Newman, *The Antarctica Mineral Resources Convention: Developments from the October 1986 Tokyo Meeting of the Antarctic Treaty Consultative Parties* (1987) 15 DENVER J. INT'L. L. & POL'Y. 421.

[45] Wellington Convention, Article 21.

[46] Edith Brown Weiss, *International Environmental Law: Contemporary Issues and the Emergence of a New World Order* (1993) 81 GEORGETOWN L. J. 675 at 704.

[47] Joan E. Moore, *The Polar Regions and the Law of the Sea* (1976) 8 CASE W. RES. J. INT'L. L. 204; Christopher C. Joyner and Peter J. Lipperman, *Conflicting Jurisdictions in the Southern Ocean: The Case of an Antarctic Minerals Regime* (1987) 27 VA. J. INT'L. L. 1; Liesbeth Peeters, *Square Peg, Round Hole: Jurisdiction over Minerals Mining Offshore Antarctica* (2004) 1 MQ. J. INT'L. & COMP. ENVT'L. L. 217; Allan Young, *Antarctic Resource Jurisdiction and the Law of the Sea: A Question of Compromise* (1985) 11 BROOKLYN J. INT'L. L. 45; Christopher C. Joyner, *Antarctica and the Law of the Sea: An Introductory Overview* (1983) 13 OCEAN DEV. & INT'L. L. 277; John Warren Kindt, *Ice-Covered Areas and the Law of the Sea: Issues Involving Resource Exploitation and the Antarctic Environment* (1988) 14 BROOKLYN J. INT'L. L. 27; Christopher C. Joyner, *The Antarctic Treaty System and the Law of the Sea – Competing Regimes in the Southern Ocean?* (1995) 10 INT'L. J. MARINE & COASTAL L. 301; Donald R. Rothwell, *A Maritime Analysis of Conflicting International Law Regimes in Antarctica and the Southern Ocean* (1994) 15 AUST. Y. B. INT'L. L. 155; and Bernard H. Oxman, *Antarctica and the New Law of the Sea* (1986) 19 CORNELL INT'L. L. J. 211; and Christopher C. Joyner, *Antarctica and the Law of the Sea: Rethinking the Current Legal Dilemmas* (1981) 18 SAN DIEGO L. REV. 415.

5.2 Antarctica and the 1988 Wellington Convention

> ... mineral resource activities which take place on the continent of Antarctica and all Antarctic islands, including all ice shelves, south of 60 south latitude and in the seabed and subsoil of adjacent offshore areas up to the deep seabed.[48]

The Wellington Convention provides for the establishment of the Antarctic Mineral Resources Commission, a Special Meeting of the Parties, an Advisory Committee, Regulatory Committees and a Secretariat. The Antarctic Mineral Resources Commission was to comprise the Consultative Parties of the Antarctic Treaty as on 25 November 1988 and decision-making was to be by three-quarters majority,[49] except for a number reserved matters on which the "absence of a formal objection" or consensus is required.[50] The most important of these matters was the role to identify possible areas for mineral exploration and development, a decision that must also be agreed with by the Special Meeting of the Parties.[51]

One notable feature of the Wellington Convention is that prospecting activities may be undertaken by any "Sponsoring State" without the need to obtain consent from either the Antarctic Mineral Resources Commission or any other institution created under the Wellington Convention.[52] Before any exploration and extraction activities can take place, however, approval by the Antarctic Mineral Resources Commission is required after an assessment of the financial feasibility of the venture and the environmental impact of the project.[53] Once the Antarctic Mineral Resource Commission issues an exploration permit, the applicant operator would have the exclusive right to explore for the mineral resources until a development permit is then obtained to extract and exploit the mineral resources in the designated area.[54]

The problem arising from this process is that the freedom to undertake prospecting activities, which can have a significant impact on the local environment, was provided for expressly under the Wellington Convention. Although it was possible for a State who was a member of the Antarctic Mineral Resources Commission to continually and repeatedly frustrate any resource development in Antarctica and Antarctic waters by vetoing any application for an exploration permit, let alone a development permit, significant damage may already be done by the prospecting activities of States.

---

[48] Wellington Convention, Article 5(2). See also Helena M. Tetzeli, *Allocation of Mineral Resources in Antarctica: Problems and a Possible Solution* (1987) 10 HASTINGS INT'L. & COMP. L. REV. 525 and Frank C. Alexander, Jr., *Legal Aspects: Exploitation of Antarctic Resources* (1979) 33 U. MIAMI L. REV. 371 at 381–382.

[49] Wellington Convention, Article 22(1).

[50] Ibid., Article 22(2) and (5).

[51] Ibid., Article 28.

[52] Ibid., Article 37.

[53] Ibid., Article 44.

[54] Ibid., Articles 53 and 54.

5.2.3.3 Failure of the Wellington Convention

The Wellington Convention was open for signature by the States that participated in the negotiations for a year.[55] However, for Australia and France, the need to promote the adoption of a comprehensive agreement on the protection of the Antarctic environment was paramount and, accordingly, they decided not to sign the Wellington Convention.[56] Coincidentally, this was followed by a number of oil spills in 1989 that highlighted the concern for environmental damage to polar regions and led other countries, particularly Belgium, India and Italy, to decide against signing the Wellington Convention.[57] These oil spills include:

(1) the *Bahia Paraiso* of Argentina, which ran aground 2 miles off Palmer Station of Antarctica on 28 January 2009;[58]
(2) the *Humboldt* of Peru, which ran aground near King George Island in Antarctic waters on 27 February 1989;[59] and
(3) the *Exxon Valdez* of the United States, which was grounded on Bligh Reef in Prince William Sound off Alaska on 24 March 1989.[60]

Considering the conservation of the Antarctic environment was already an important concern of the international community before the adoption of the Wellington Convention, it cannot be said that the reaction of the relevant States to the international attention drawn to the oil spills was entirely unpredictable.[61] These concerns

---

[55] Ibid., Article 60.

[56] James Crawford and Donald R. Rothwell, *Legal Issues Confronting Australia's Antarctica* (1991) 13 AUST. Y. B. INT'L. L. 53; Francesco Francioni, *Resource Sharing in Antarctica: For Whose Benefit?* (1990) 1 EUR. J. INT'L. L. 258 at 259. See also Rodney R. McColloch, *Protocol on Environmental Protection to the Antarctic Treaty – The Antarctic Treaty – Antarctic Minerals Convention – Wellington Convention – Convention on the Regulation of Antarctic Mineral Resource Activities* (1992) 22 GA. J. INT'L. & COMP. L. 211.

[57] Ibid.

[58] Mary Lynn Canmann, *Antarctic Oil Spills of 1989: A Review of the Application of the Antarctic Treaty and the New Law of the Sea to the Antarctic Environment* (1990) 1 COLO. J. INT'L. ENVT'L. L. & POL'Y. 211 at 211–212.

[59] Ibid.

[60] Richard T. Carson, Robert C. Mitchell, Michael Hanemann, Raymond J. Kopp, Stanley Presser and Paul A. Ruud, *Contingent Valuation and Lost Passive Use: Damages from the Exxon Valdez Oil Spill* (2003) 25 ENVR'L. & RES. EC. 257. See also Alex T. Leonhard, *Ixtoc I: A Test for the Emerging Concept of the Patrimonial Sea* (1980) 17 SAN DIEGO L. REV. 617.

[61] See Christopher C. Joyner, *Oceanic Pollution and the Southern Ocean: Rethinking the International Legal Implications for Antarctica* (1984) 24 NAT. RES. J. 1; Nigel D. Bankes, *Environmental Protection in Antarctica: A Comment on the Convention on the Conservation of Antarctic Marine Living Resources* (1981) 19 CAN. Y. B. INT'L. L. 303; Christopher C. Joyner, *The Southern Ocean and Marine Pollution: Problems and Prospects* (1985) 17 CASE W. RES. J. INT'L. L. 165; Stephen A. Seach, *Conflicting Interests in Antarctica: People or Nature? Who Decides?* (1991) 5 TEMPLE INT'L. & COMP. L. J. 109; William M. Welch, *The Antarctic Treaty System: Is It Adequate to Regulate or Eliminate the Environmental Exploitation of the Globe's Last Wilderness?* (1992) 14 HOUSTON J. INT'L. L. 597; Michael Koch, *The Antarctic Challenge:*

5.2 Antarctica and the 1988 Wellington Convention

were later addressed in 1991 in the Madrid Protocol on Environmental Protection to the Antarctic Treaty.[62] As Rothwell noted about the Madrid Protocol:

> It was negotiated at a time when there was considerable debate over whether mining should be permitted in Antarctica and not long after the Treaty parties had concluded negotiations for a specific Antarctic minerals regime. That the parties could so quickly about-turn and adopt a new instrument which not only sought to prohibit mining but also comprehensively protect the Antarctic environment is a testament to their goodwill to cooperatively manage Antarctica and the robustness of the Treaty system.[63]

The failure of the Wellington Convention and the attitude of a significant number of the Consultative Parties of the Antarctic Treaty would make the mineral exploitation of Antarctica, Antarctic islands and Antarctic waters somewhat inconceivable in the present socio-legal climate.[64] Meanwhile, some of the attention of the advocates of mineral exploitation has shifted to the potential for the exploitation of icebergs that seasonally drift from Arctic and Antarctic waters to warmer waters.[65]

Although there are a number of similarities between the conceptual treatment of Antarctica and other geographical areas of international regulation, such as the deep seabed and celestial bodies, there are several reasons why the Antarctic regime under the Wellington Convention would not be applicable to celestial bodies. First, the sharing of benefits is required on an equitable basis under the Moon Agreement, whereas there is no such sharing requirement under the Wellington Convention. Second, the Antarctic system is dominated by its Consultative Parties, whereas any international regime created for celestial bodies must involve substantial participation by developing States. Third, the States under the Wellington Convention and the Antarctic Treaty mostly operate by consensus, which would be impractical for any

---

Conflicting Interests, Cooperation, Environmental Protection and Economic Development (1984) 15 J. MARIT. L. & COM. 117; and John Warren Kindt and Todd Jay Parriott, *Ice-Covered Areas: The Competing Interests of Conservation and Resource Exploitation* (1984) 21 SAN DIEGO L. REV. 941.

[62] Madrid Protocol. See Steve T. Madsen, *A Certain False Security: The Madrid Protocol to the Antarctic Treaty* (1993) 4 COLO. J. INT'L. ENVT'L. L. & POL'Y. 458.

[63] Donald R. Rothwell, *Polar Environmental Protection and International Law: The 1991 Antarctic Protocol* (2000) 11 EUR. J. INT'L. L. 591 at 592.

[64] See Andrew N. Davis, *Protecting Antarctica: Will a Minerals Agreement Guard the Door or Open the Door to Commercial Exploitation?* (1990) 23 GEO. WASH. J. INT'L. L. & EC. 733; David W. Floren, *Antarctic Mining Regimes: An Appreciation of the Attainable* (2001) 16 J. ENVT'L. L. & LIT. 467; Jem M. Spectar, *Saving the Ice Princess: NGOs, Antarctica and International Law in the New Millennium* (2000) 23 SUFFOLK TRANSNAT'L. L. REV. 57; Colin Deihl, *Antarctica: An International Laboratory* (1991) 18 B. C. ENVT'L. AFF. L. REV. 423; Joseph J. Ward, *Black Gold in a White Wilderness – Antarctic Oil: The Past, Present and Potential of a Region in Need of Sovereign Environmental Stewardship* (1998) 13 J. LAND USE & ENVT'L. L. 363; and Robert E. Money, Jr., *The Protocol on Environmental Protection to the Antarctic Treaty: Maintaining a Legal Regime* (1993) 7 EMORY INT'L. L. REV. 163.

[65] Bryan S. Geon, *A Right to Ice? The Application of International and National Water Laws to the Acquisition of Iceberg Rights* (1998) 19 MICH. J. INT'L. L. 277; and Thomas R. Lundquist, *The Iceberg Cometh? International Law Relating to Antarctic Iceberg Exploitation* (1977) 17 NAT. RES. J. 1.

216   5   Exploitation Rights: Evolving from the "Province of Mankind" to the "Common ...

decisions in space due to the number of States that would necessarily be involved in the relevant decision-making process.[66]

5.3 Outer Space as the Province of All Mankind

5.3.1 Interpreting Article I of the Outer Space Treaty

Article I of the Outer Space Treaty states that:

> The exploration and use of outer space, including the Moon and other celestial bodies, shall be carried out for the benefit and in the interest of all countries, irrespective of their degree of economic or scientific development, and shall be the province of all mankind.
>
> ...

The concept of the "province of all mankind" is not defined in the Outer Space Treaty, though some commentators have made observations on whether the word "mankind" is meant to designate:

(1) all States;
(2) all States, particularly developing States;
(3) all nations;
(4) all living human beings; or
(5) all living and future human beings.[67]

In an effort to resolve the question over the meaning of "mankind" in Article I of the Outer Space Treaty, Gorove had suggested that:

> Mankind as a concept should be distinguished from that of man in general. The former refers to a collective body of people, whereas the latter stands for individuals making up that body. Therefore, the rights of mankind should be distinguished, for instance, from the so-called human rights. Human rights are rights to which individuals are entitled on the basis of their belonging to the human race, whereas the rights of mankind relate to the rights of the collective entity and would not be analogous with the rights of the individuals making up that entity.[68]

In other words, the term "mankind" is intended to refer to a collective or commune of human beings and not to States. Williams also noted further that:

---

[66] Raclin, supra note 11, at 753. See also Gerald Staub, *The Antarctic Treaty as Precedent to the Outer Space Treaty* (1974) 17 PROC. COLL. L. OUTER SP. 282 and Patricia Minola, *The Moon Treaty and the Law of the Sea* (1981) 18 SAN DIEGO L. REV. 455.

[67] See, for example, David Tan, *Towards a New Regime for the Protection of Outer Space as the "Province of All Mankind"* (2000) 25 YALE J. INT'L. L. 145 at 162; Ernst Fasan, *The Meaning of the Term "Mankind" in Space Legal Language* (1974) 2 J. SP. L. 125 at 131; and Carl Q. Christol, SPACE LAW: PAST, PRESENT AND FUTURE (1991) at 389.

[68] Stephen Gorove, *The Concept of "Common Heritage of Mankind": A Political, Moral or Legal Innovation?* (1972) 9 SAN DIEGO L. REV. 390 at 393.

5.3 Outer Space as the Province of All Mankind

A growing trend of the international community is to take account of the positions and interests of medium and non-space powers in the exploitation of these new areas of human activity. Mankind, in the words of René-Jean Dupuy, is an interspatial and intertemporal concept and includes not only those who are present but also those who are to come.[69]

It appears that, according to Gorove and Williams, the term "mankind" is intended to designate humanity as a whole, both present and future, rather than States or a collection of individuals. If so, then the term "province of mankind" would denote some practical form of collective or communal sovereignty and ownership on the one hand or merely an idealistic and declaratory statement intended to negate any possible exercise of sovereignty or appropriation on the other. Some commentators have adopted the former position and some of those suggest further that the application of the "common heritage of mankind" doctrine to outer space substantiates and clarifies, if not replaces, the abstract concept of "province of all mankind".[70] However, it is the latter position that appears to have wider acceptance. After all, the Moon Agreement declares only celestial bodies and their natural resources, rather than outer space *sensu lato*, as the common heritage of mankind.[71] Further, the limited acceptance of the Moon Agreement among the international community further lends support to the position that the "province of mankind" merely negates any exercise of sovereignty or appropriation in outer space.[72]

In any event, both the Outer Space Treaty and the Moon Agreement refers to the "exploration and use of the Moon" and other celestial bodies, rather than the Moon and the other celestial bodies themselves, as the "province of all mankind".[73] Accordingly, it may not be appropriate to consider the abstract "province of all mankind" concept in the Outer Space Treaty to have any practical implications on the status of celestial bodies in international law.

5.3.2 Correlation with Article II of the Outer Space Treaty

Suffice to note, however, that Article II of the Outer Space Treaty prohibits any national appropriation of outer space, the Moon and other celestial bodies, whether by claim of sovereignty, by means of use or occupation, or by any other means. It was noted by Adams that eliminating state sovereignty from outer space and celestial bodies have practical implications:

---

[69] Sylvia Maureen Williams, *The Law of Outer Space and Natural Resources* (1987) 36 INT'L. & COMP. L. Q. 142 at 150–151.

[70] See, for example, Baslar, supra note 1, at 91–111; Aldo Armando Cocca, *The Advances in International Law Through the Law of Outer Space* (1981) 9 J. SP. L. 13 at 16; and Nicholas Mateesco Matte, AEROSPACE LAW: TELECOMMUNICATIONS SATELLITES (1982) at 77.

[71] Moon Agreement, Article 11(2).

[72] Nicolas Mateesco Matte, *The Draft Treaty on the Moon, Eight Years Later* (1978) 3 ANN. AIR & SP. L. 511 at 531 and Tan, supra note 67, at 163.

[73] Moon Agreement, Article 4(1).

The fundamental difficulty with eliminating sovereignty, though, is that another basis must be provided for performing the general functions that may now be premised on sovereignty. While states apparently recognise international interests in space exploration, they must still protect their legitimate interests in defence and safety. Without sovereignty, some basis must be established for creating and enforcing a regulatory regime. ... How successful this provision will be depends on how effectively the remaining provisions in the treaty establish legal relationships, rights and duties to replace some of those ordinarily flowing from sovereignty.[74]

To that end, a distinction is to be drawn between the characterisation of outer space and celestial bodies as *res nullius* or *res extra commercium* on the one hand, being the traditional characterisations of terrestrial areas not subject to national sovereignty, and the new concept of *res communis* on the other.[75] Outer space, including the Moon and other celestial bodies, are not subject to assertions of national sovereignty or unconditional and unrestricted exploitation or use of outer space but instead are "commonly needed by humanity as a condition of survival and are to be used for the common benefit" and thus "cannot be subject to private ownership or state sovereignty".[76] Gál has taken the concept one step further and suggested that, as space exploration and use are considered the venture of all humankind, outer space and celestial bodies must be deemed to be *res communis omnium*.[77]

The concept of *res communis* or *res communis omnium* is based on the ideological assumption that States have a common interest in the exploration, use and exploitation of the global commons and this concept has found expression in the provisions of Articles I and II of the Outer Space Treaty.[78] The one central problem with such an ideology is that each individual enjoys the benefit of exploitation and use of the resources to its maximum extent but spreads the cost of such exploitation across all users or even all humankind, thus providing a powerful incentive for individual over-exploitation.[79] As Hardin illustrated:

Picture a pasture open to all. ... The positive component is the benefit to the individual peasant from grazing one additional animal. The negative component is the reduction of

---

[74] Thomas R. Adams, *The Outer Space Treaty: An Interpretation in Light of the No-Sovereignty Provision* (1968) 9 HARV. INT'L. L. J. 140 at 143.

[75] For a discussion of the concepts of *res nullius* and *res extra commercium* in international law, see Robert Y. Jennings, THE ACQUISITION OF TERRITORY IN INTERNATIONAL LAW (1963); Pacifico A. Ortiz, LEGAL ASPECTS OF THE NORTH BORNEO QUESTION (1964); and Malcolm N. Shaw, TITLE TO TERRITORY IN AFRICA (1971); and Surya P. Sharma, TERRITORIAL ACQUISITION, DISPUTES AND INTERNATIONAL LAW (1997).

[76] See Andronico O. Adede, *The System for the Exploitation of the "Common Heritage of Mankind" at the Caracas Conference* (1975) 69 AM. J. INT'L. L. 31 and Tan, supra note 67, at 161.

[77] Gyula Gál, SPACE LAW (1969) at 189–190. Lachs had made similar comments on the issue: Manfred Lachs, THE LAW OF OUTER SPACE (1972) at 23.

[78] See Tan, supra note 67, at 161.

[79] Erin A. Clancy, *The Tragedy of the Global Commons* (1998) 5 IND. J. GLOBAL LEGAL STUD. 601 at 604.

grass available to feed his other animals. But since the effects of overgrazing are shared by all the herdsmen, the negative component is measured by any given herdsman is overshadowed by the positive benefit to him of grazing an additional animal. Therein lies the tragedy. Each man is locked into a system that compels him to increase his herd without limit – in a world that is limited. Ruin is the destination towards which all men rush, each pursuing his own best interest in a society that believes in the freedom of the commons.[80]

When this hypothetical common pasture is extrapolated to outer space and celestial bodies, States "are free to make maximum use of resources because no outside mechanism exists to force their acceptance of external costs, either the cost of resource degradation or the cost of resource depletion".[81] With the impact of the New International Economic Order (the "NIEO") in the 1970s, States were soon confronted with a methodology devised by the international community, particularly the developing world, with the external mechanism needed to enforce the restrictions on the exploitation and use by States of the *res communis* of outer space, particularly in the form of the doctrine of the common heritage of mankind.

5.4 The New International Economic Order

5.4.1 Origins of the New International Economic Order

The concept of the NIEO has much of its origins in the early decades of the Cold War, though some commentators suggest that the seeds for such a conceptual and philosophical development were planted much earlier.[82] In particular, the completion of the reconstruction of Europe after the Second World War shifted the focus from post-war rebuilding to the economic disparity between the industrialised States and the developing world.[83] In any event, the focus on reconstruction of Europe has led partly to the progressive decolonisation in Africa and Asia and, rightly or wrongly, the newly decolonised and independent States had expectations of a more equitable international order.[84] As Rozental suggested:

---

[80] Garrett Hardin, *The Tragedy of the Commons* (1968) 162 SCIENCE 1243.

[81] Joan Eltman, *A Peace Zone on the High Seas: Managing the Commons for Equitable Use* (1993) 5 INT'L. LEGAL PERSP. 47 at 64.

[82] See, for example, Richard B. Lillich, *Economic Coercion and the "New International Economic Order": A Second Look at Some First Impressions* (1976) 16 VA. J. INT'L. L. 233 at 234–237 and Kirsten Borgsmidt, *The Generalised System of Preferences in Favour of Developing Countries against the Historical Background in the Light of Public International Trade Law and the New International Economic Order* (1985) 54 NORDISK TIDS. INT'L. RET. 33 at 34–41.

[83] D. H. N. Johnson, *The New International Economic Order* [1983] Y.B. WORLD AFF. 204 at 209. See also Thomas W. Wälde, *A Requiem for the "New International Economic Order": The Rise and Fall in International Economic Law and a Post-Mortem with Timeless Significance*, in Gerhard L. G. Hafner and Gerhard Loibl (eds.), LIBER AMICORUM PROFESSOR SIEDL-HOHENVELDERN (1998) at 771–804.

[84] See, for example, Norbert Horn, *Normative Problems of a New International Economic Order* (1982) 16 J. WORLD TRADE L. 338 at 348; Eisuke Suzuki, *Self-Determination and World Public*

The international order which emerged in the postwar era has been unable to satisfy the most basic aspirations of the majority of the world's peoples. The accession to independence of a multitude of nations in Africa, Asia and the Caribbean has created an entirely new set of problems and demands affecting the international community, especially in the field of economic relations. .... This new state of affairs has not benefitted those countries whose needs are the greatest.[85]

It was observed at the time that the industrialised States continued to dominate the global economy, based on almost constant full employment and rapid technological evolution.[86] In the meantime, the developing States continued to fulfil their old colonial roles of supplying the industrialised States with natural resources and raw materials, lacking the technology and capital necessary to process or refine such materials into more valuable finished products.[87] It was also suggested that the conduct of transnational corporations amounted to the imposition of an inequitable and exploitative terms of trade on developing States.[88] As Ellis observed:

Frequently, these entities employed their economic power and monopolistic positions with respect to technology and manufactured products, to favourably control the terms of trade and investment in bargaining with developing nations. Likewise, transnational corporations often controlled the commodity markets in which the developing nations sold their raw materials. These entities used this control to impose inequitable terms of trade and exploit the developing nations. The developing nations increasingly viewed such business practices as coercive and restrictive.[89]

Since the pace of decolonisation quickened during the 1950s and 1960s, the developing States found themselves increasingly able to control the agenda and voting patterns at the United Nations. This was particularly the case with the

---

Order: Community Responses to Territorial Separation (1976) 16 VA. J. INT'L. L. 779; and Rudolf von Albertini, DEKOLONISATION (1996).

[85] Andres Rozental, *The Charter of Economic Rights and Duties of States and the New International Economic Order* (1976) 16 VA. J. INT'L. L. 309 at 309. See also N. J. Udombana, *The Third World and the Right to Development: Agenda for the Next Millennium* (2000) 22 HUMAN RIGHTS Q. 753.

[86] G. W. Haight, *The New International Economic Order and the Charter of Economic Rights and Duties of States* (1975) 9 INT'L. LAWYER 591 at 592 and ibid. at 310.

[87] Haight, supra note 86, at 592–593 and Rozental, supra note 85, at 310.

[88] See K. Ventata Raman, *Transnational Corporations, International Law and the New International Economic Order* (1979) 6 SYRACUSE J. INT'L. L. & COM. 17; Joel Davidow and Lisa Chiles, *The United States and the Issue of the Binding or Voluntary Nature of International Codes of Conduct Regarding Restrictive Business Practices* (1978) 72 AM. J. INT'L. L. 247; and Timothy W. Stanley, *International Codes of Conduct for MNCs: A Skeptical View of the Process* (1981) 30 AM. U. L. REV. 973.

[89] Mark E. Ellis, *The New International Economic Order and General Assembly Resolutions: The Debate over the Legal Effects of General Assembly Resolutions Revisited* (1985) 15 CAL. W. INT'L. L. J. 647 at 652–653.

5.4 The New International Economic Order

General Assembly and the Economic and Social Council, as well as some of the international economic institutions such as the United Nations Conference on Trade and Development (the "UNCTAD"), which was established on 1964.[90]

In the 1970s, the world saw an end to an international economic framework dominated by the United States. Two successive devaluations of the U.S. dollar and the ending of its convertibility to gold and the advent of floating currencies coincided with the oil crisis of the 1970s when energy prices increased rapidly in the face of limited oil supply from the Middle East.[91] This unravelling of the international order occurred at the same time as the number of developing States increased dramatically with the acquisition of independence by former colonies and the prevailing belief throughout the newly-independent developing world that political independence and non-alignment with superpowers would be the beginning to the solution of all of their other socio-economic problems.[92] The reality, of course, was somewhat different, as Ferguson observed:

> As some not quite facetiously used to say, everyone thought that when the tri-colour flag or the Union Jack came down, and you got a new flag in the light of independence, a new seat in the United Nations, a new national anthem and a new national airline, then you had arrived. The arrival never came, however. The lines of trade and communications still ran to the old metropol. Banking lines ran the same way. Even cultural lines ran the same way. Today in Africa, the easiest way to go from the east coast to the west coast is to go to Rome, Paris or London and then come back. I once made a telephone call between two capitals 110 miles apart, Abidjan in the Ivory Coast and Accra in Ghana, and the call was routed from Abidjan to Paris, Paris to London, and London back to Accra. The Ivory Coast had been French and the lines still ran to Paris. Ghana had been the Gold Coast under the English and the lines still ran to London.[93]

Ryan had suggested that there were two principal causes of dissatisfaction among developing States with the pre-existing international economic order:

(1) the Bretton Woods institutions, namely the International Bank for Reconstruction and Development (the "IBRD" or the "World Bank"), the International Monetary Fund (the "IMF") and the General Agreement on Tariffs and Trade

---

[90] Ian Taylor and Karen Smith, UNITED NATIONS CONFERENCE ON TRADE AND DEVELOPMENT (UNCTAD) (2007), at 14 and Iqbal Haji, *Finance, Money, Developing Countries and UNCTAD*, in Michael Zammit Cutajar and Waldek R. Malinowski, UNCTAD AND THE SOUTH-NORTH DIALOGUE: THE FIRST TWENTY YEARS (1985) at 138.

[91] Robert Mundell, *Dollar Standards in the Dollar Era* (2007) 29 J. POLICY MODELLING 677 at 681–686; Karel Jansen, MONETARISM, ECONOMIC CRISIS AND THE THIRD WORLD (1983), at 45; C. Clyde Ferguson, *The New International Economic Order* [1980] U. ILL. L. F. 693 at 694; Francisco Orrego Vicuna, *From the Energy Crisis to the Concept of an Economic Heritage of Mankind: Guidelines for Reorganising the International Economic System* (1976) 1 INT'L. TRADE L. J. 87; and Ibrahim F. I. Shihata, *Arab Oil Policies and the New International Economic Order* (1976) 16 VA. J. INT'L. L. 261.

[92] See, for example, Karen A. Hudes, *Towards a New International Economic Order* (1976) 2 YALE STUD. WORLD PUB. ORD. 88.

[93] Ferguson, supra note 91, at 695.

("GATT"), were created in the 1940s in a world that was vastly different to the world that existed with the global economic needs in the 1970s; and

(2) the Bretton Woods institutions and the international economic framework of the time were devised to serve the interests of the industrialised States and satisfaction of the urgent needs of the developing States will require a fundamental reshaping of these institutions.[94]

With the realisation that the problems of the developing world were fundamentally economic and not political in nature, the developing States banded together into the so-called Group of 77 and established the Decade for Development to promote a greater flow of aid from the developed to the underdeveloped world.[95] However, with the world gripped in the throes of an oil crisis, it was clear that the resources of the developed world were to be devoted to domestic needs. As the developing world had been persuaded by then that a complete overhaul of the international economic structure was required to accelerate their economic development, they began to call for the establishment of a new economic order. The NIEO that they envisaged was a new world of which many industrialised States at the time considered to be "frightening and unimaginable".[96] As Scali, the U.S. Permanent Representative to the United Nations at that time, stated during the debates of the General Assembly concerning the adoption of the NIEO principles in resolutions:

> The document in question contains elements supported by all United Nations Members. It also contain elements which many Members . . . – large and small, and on every continent – do not endorse. The United states delegation, like many others, strongly disapproves of some provisions in the document and has in no sense endorsed them. The document we have produced is a significant political document, but it does not represent unanimity of opinion in this Assembly. To label some of these highly controversial conclusions as agreed is not only idle; it is self-deceiving. In this house, the steamroller is not the vehicle for solving vital, complex problems.[97]

---

[94] Kevin W. Ryan, *Towards a New International Economic Order* (1975) 9 U. QLD. L. J. 135 at 136.

[95] See Rose D'Sa, *The "Right to Development" and the New International Economic Order with Special Reference to Africa* [1984] THIRD WORLD LEG. STUD. 140; Ferguson, supra note 91, at 696; Ellis, supra note 89, at 651–654; and Alan G. Friedman and Cynthia A. Williams, *The Group of 77 at the United Nations: An Emergent Force in the Law of the Sea* (1979) 16 SAN DIEGO L. REV. 555.

[96] Ferguson, supra note 91, at 696.

[97] U.N. Doc. A/PV.2229 (1 May 1974) at 41–50, as reproduced in United Nations, *Other Documents: United Nations General Assembly Sixth Special Session – Reservations Entered by the United States* (1974) 13 I.L.M. 744 at 744–745.

5.4 The New International Economic Order

5.4.2 Sixth Special Session of the General Assembly and the Principles of the New International Economic Order

5.4.2.1 Adoption of the NIEO Resolutions

By the 1960s, the developing States began realising that the United Nations presented itself as an accessible forum for them to challenge the existing international economic system that existed at the time.[98] The increasing number of developing States in the international community meant that the Third World began to enjoy a numerical superiority in many of the United Nations organs in which they were able to control the agenda and voting on many economic and developmental issues, particularly in the General Assembly, the UNCTAD and the Economic and Social Council.[99] Crucially, the United Nations was perceived by the developing world as having a quasi-legislative or "limited legislative" role.[100] For example, the following resolutions were adopted by the General Assembly and other intergovernmental organisations to create favourable quasi-legal principles of public international law:

- Declaration on the Granting of Independence to Colonial Countries and People of the General Assembly;[101]
- Resolution of the General Assembly on Permanent Sovereignty over Natural Resources;[102]
- Declaration of a United Nations Development Decade by the General Assembly;[103] and
- the amendment to GATT to create non-reciprocal preferential trade benefits for the developing States.[104]

---

[98] Christopher C. Joyner, *U.N. General Assembly Resolutions and International Law: Rethinking the Contemporary Dynamics of Norm Creation* (1981) 11 CAL. W. INT'L. L. J. 445.

[99] Edward McWhinney Q.C., THE WORLD COURT AND THE CONTEMPORARY LAW-MAKING PROCESS (1979) at 132 and James W. Skelton, Jr., *UNCTAD's Draft Code of Conduct on the Transfer of Technology: A Critique* (1981) 14 VAND. J. TRANSNAT'L. L. 381.

[100] Ram Prakash Anand, NEW STATES AND INTERNATIONAL LAW (1972), at 74.

[101] General Assembly Resolution 1514 (1960), U.N. G.A.O.R. Supp. (No. 16) at 66, U.N. Doc. A/4684.

[102] General Assembly Resolution 1803 (1962), U.N. G.A.O.R. Supp. (No. 17) at 14, U.N. Doc. A/5217. See Emeka Duruigbo, *Permanent Sovereignty and Peoples' Ownership of Natural Resources in International Law* (2006) 38 GEO. WASH. INT'L. L. REV. 33.

[103] General Assembly Resolution 1710 (1961), U.N. G.A.O.R. Supp. (No. 17) at 17, U.N. Doc. A/5100.

[104] General Agreement on Tariffs and Trade, opened for signature on 30 October 1947, 55 U.N.T.S. 187; 1948 A.T.S. 23 (entered into force on 1 January 1948) and Protocol Amending the General Agreement on Tariffs and Trade, opened for signature on 8 February 1965, 572 U.N.T.S. 320 (entered into force on 27 June 1966). See also Philip E. Jacobs, Alexine L. Atherton and Arthur M. Wallenstein, THE DYNAMICS OF INTERNATIONAL ORGANISATIONS (1972), at 415.

224   5 Exploitation Rights: Evolving from the "Province of Mankind" to the "Common ...

In 1974, the Sixth Special Session of the United Nations General Assembly was convened to address issues in the international economy.[105] At this session, two resolutions were adopted at the urging of the developing States, namely the Declaration on the Establishment of a New International Economic Order (the "NIEO Declaration"),[106] and the Programme of Action on the Establishment of a New International Economic Order (the "NIEO Programme of Action").[107] The principles and concepts contained in the NIEO Declaration and the NIEO Programme of Action were later codified and elaborated in the Charter of Economic Rights and Duties of States (the "Charter of Economic Rights") that was also adopted by the General Assembly in 1974.[108]

5.4.2.2 Objects and Goals of the NIEO Resolutions

The aims and objects of the NIEO Declaration, the NIEO Programme of Action and the Charter of Economic Rights were controversial not because they were new but because some, if not all, of them were opposed by the industrialised States.[109] The industrialised States also feared and opposed the perceived ultimate objective of the NIEO, which was a "revolutionary restructuring of the existing international economic status quo" that would create a global "welfare state".[110]

The major objectives to be achieved in the implementation of the NIEO, to which the industrialised States oppose as being steps towards the creation of a global "welfare state", include:

(1) the reduction of debt by the developing States;
(2) recognition of the economic sovereignty of the developing States;
(3) increasing the international purchasing power of exports of raw materials and commodities;
(4) increasing control by the developing States over the degree and nature of foreign investment and developmental aid;
(5) increasing foreign access to markets of industrialised States;
(6) promoting and reducing the costs of technology transfers; and

---

[105] See Robin C. A. White, *A New International Economic Order* (1975) 24 INT'L. & COMP. L. Q. 542.

[106] General Assembly Resolution 3201 (S-VI), U.N. G.A.O.R. Supp. (No. 1)at 3, U.N. Doc. A/9559 (1974) (the "*NIEO Declaration*").

[107] General Assembly Resolution 3202 (S-VI), U.N. G.A.O.R. Supp. (No. 1) at 5, U.N. Doc. A/9559 (1974) (the "*NIEO Programme of Action*").

[108] General Assembly Resolution 3281 (XXIX), U.N. G.A.O.R. Supp. (No. 30) at 50, U.N. Doc. A/9631 (1974) (the "*Charter of Economic Rights*").

[109] See James C. Ingram, INTERNATIONAL ECONOMIC PROBLEMS (2nd ed., 1970) at 73; Sartaj Aziz, *The New International Order: Search for Common Ground* in Pradip K. Ghosh (ed.), NEW INTERNATIONAL ECONOMIC ORDER: A THIRD WORLD PERSPECTIVE (1984), at 48; Davidow and Chiles, supra note 88, at 247–248; and Horn, supra note 84, at 348.

[110] Ellis, supra note 89, at 648. See also Adeoye Akinsanya and Arthur Davies, *Third World Quest for a New International Economic Order: An Overview* (1984) 33 INT'L. COMP. L. Q. 208.

5.4 The New International Economic Order

(7) increasing the power and influence of the developing States in the Bretton Woods institutions, namely the World Bank, the IMF and the GATT.[111]

The belief was that these goals, if implemented and achieved, would allow for the establishment of a new international economic order "based on equity, sovereign equality, interdependence, common interest and cooperation among all States".[112] These goals were embodied in the terms and provisions of the NIEO Declaration, the NIEO Programme of Action and the Charter of Economic Rights.

5.4.2.3 Principles Contained in the NIEO Resolutions

The NIEO Declaration, the NIEO Programme of Action and the Charter of Economic Rights provide for the following principles that have been considered to be the principal manifestations of the NIEO:

(1) the right of every State to sovereignty over its domestic economy and their natural resources;[113]
(2) the right of all States to "restitution and full compensation" for past exploitation of their territory and resources;[114]
(3) the right to increase domestic controls over foreign investment, including the activities of multinational corporations;[115]
(4) increased influence and decision-making power in the World Bank and the IMF;[116]
(5) reducing the cost and procedural difficulties in technology transfers from industrialised States to developing States;[117]
(6) non-reciprocal and non-discriminatory preferences in trade and tariffs;[118]
(7) increased flow of developmental aid;[119]

---

[111] Jeffrey A. Hart, THE NEW INTERNATIONAL ECONOMIC ORDER: CONFLICT AND COOPERATION IN NORTH-SOUTH ECONOMIC RELATIONS 1974–1977 (1983) at 33.

[112] Ellis, supra note 89, at 658. See also Charles N. Brower and John B. Tepe, Jr., *The Charter of Economic Rights and Duties of States: A Reflection or Rejection of International Law?* (1975) 9 INT'L. LAWYER 295 and Charles N. Brower, *The Charter of Economic Rights and Duties of Rights and the American Constitutional Tradition: A Bicentennial Perspective on the "New International Economic Order"* (1976) 10 INT'L. LAWYER 701.

[113] Charter of Economic Rights, Articles 1 and 2(1). See also the Resolution on Permanent Sovereignty over Natural Resources, General Assembly Resolution 1803.

[114] NIEO Declaration, paragraph 4(f).

[115] Charter of Economic Rights, Article 2(2). See M. Sornarajah, *The New International Economic Order, Investment Treaties and Foreign Investment Laws in ASEAN* (1985) 27 MALAYA L. REV. 440.

[116] Charter of Economic Rights, Article 10.

[117] Ibid., Article 13.

[118] Ibid., Articles 18 and 19.

[119] Ibid., Article 22.

(8) the maintenance of the international purchasing power of exports of raw materials and commodities;[120]

(9) the right to nationalise and expropriate foreign interests in accordance with the domestic law of the developing State;[121] and

(10) the "seabed and ocean floor and the subsoil thereof, beyond the limits of national jurisdiction, as well as the resources of the area are the common heritage of mankind".[122]

Unlike the various General Assembly resolutions and declarations dealing with other areas of international law, such as those concerning space activities, the provisions of the NIEO Declaration, the NIEO Programme of Action or the Charter of Economic Rights are neither codifications of existing principles of customary international law nor the elaborations or specific applications of existing treaty provisions or customary provisions. Similar to the 1963 Declaration on Legal Principles Governing the Activities of States in the Exploration and Use of Outer Space,[123] which preceded the 1967 Treaty on Principles Governing the Activities of States in the Exploration and Use of Outer Space, including the Moon and other Celestial Bodies,[124] it is unlikely that the NIEO Declaration, the NIEO Programme of Action and the Charter of Economic Rights have any legally binding effect without their further implementation by way of an international treaty.[125] As White had observed:

---

[120] Ibid., Article 28.

[121] Ibid., Article 2(2).

[122] Ibid., Article 29. See Kilaparti Ramakrishna, *North-South Issues, Common Heritage of Mankind and Global Climate Change* (1990) 19 MILL. J. INT'L. STUD. 429; Bradley Larschan and Bonnie C. Brennan, *The Common Heritage of Mankind Principle in International Law* (1983) 21 COLUM. J. TRANSNAT'L. L. 305; and Edward Guntrip, *The Common Heritage of Mankind: An Adequate Regime for Managing the Deep Seabed?* (2003) 4 MELB. J. INT'L. L. 376.

[123] General Assembly Resolution 1962 (XVIII).

[124] Treaty on Principles Governing the Activities of States in the Exploration and Use of Outer Space, including the Moon and other Celestial Bodies, opened for signature on 27 January 1967, 610 U.N.T.S. 205; 18 U.S.T. 2410; T.I.A.S. 6347; 6 I.L.M. 386 (entered into force on 10 October 1967).

[125] See John King Gamble, Jr. and Maria Frankowska, *International Law's Response to the New International Economic Order: An Overview* (1986) 9 B. C. INT'L. & COMP. L. REV. 257; Mitchell Wigdor, *Canada and the New International Economic Order: Some Legal Implications* (1982) 20 CAN. Y. B. INT'L. L. 161; Christian Tomuschat, *The Charter of Economic Rights and Duties of States: Some Thoughts on the Significance of Declarations of the United Nations General Assembly* (1976) ZEIT. AUS. RECHT. VÖLK. 36444; Guy Feuer, *La Charte des Droits et Devoirs Economiques des Etats* (1975) 79 REVUE GEN. DR. INT'L. PUB. 2272; Edward McWhinney Q.C., *The International Law-Making Process and the New International Economic Order* (1976) 14 CAN. Y. B. INT'L. L. 57; and generally Jonathan I. Charney, *Universal International Law* (1993) 87 AM. J. INT'L. L. 529. However, it ought be noted that the Soviet Union arguably took a different view: see William E. Butler, *Socialist International Institutions and the New International Economic Order* (1984) 3 PUB. L. FORUM 152.

5.4 The New International Economic Order

> While most States realise that the Charter [of Economic Rights] and the related U.N. documents and conventions have no legally binding weight, they nevertheless have had an important impact on the international economic community. No developed country, of course, has rushed to transfer resources to an underdeveloped neighbour. Rather, the virtue of these instruments lies in the fact that they have brought both the plight and the power of the developing countries to the attention of the Western industrial States. The developed States now realise that the frustration which the Third World countries have faced is capable of creating a serious rift between developed and developing nations in the international economic community.[126]

However, some of the principles of the NIEO have found limited expression in various human rights instruments, particularly in relation to the provision of economic and social rights.[127] Nevertheless, the absence of a binding treaty has been a continuing criticism of the ability of international law to address the revolutionary changes proposed within the NIEO context.[128]

Other than the concept of the common heritage of mankind, the principles of the NIEO have achieved remarkable success in their application and implementation in three particular areas of the regulation of the global economy. First, various instruments implementing preferential treatment for developing States in various trade and tariff arrangements had been adopted. In the GATT, for example, decisions made on 25 June 1971 and 28 November 1979 introduced a generalised system of preferences to accord preferential tariff treatment to developing States without the usual application of the most favoured nation principle to such preferences.[129] The most expansive trade agreement that has been perceived to implement most of the elements of the NIEO has been the trade treaties between Europe and the African, Caribbean and Pacific Group of States, which contained provisions dealing with, *inter alia*, trade preferences, stabilisation of purchasing power for mineral and commodity exports and developmental aid.[130]

---

[126] Gillian White, *A New International Economic Order?* (1976) 16 VA. J. INT'L. L. 323 at 341.

[127] Seymour J. Rubin, *Economic and Social Human Rights and the New International Economic Order* (1986) 1 AM. U. J. INT'L. L. & POL'Y. 67 and Noel Dias, *The NIEO Revisited* (1996) 8 SRI LANKA J. INT'L. L. 27.

[128] Eric Allen Engle, *The Failure of the Nation State and the New International Economic Order: Multiple Converging Crises Present Opportunity to Elaborate a New Jus Gentium* (2004) 16 ST. THOM. L. REV. 187 and David P. Fidler, *Revolt Against or from Within the West? TWAIL, the Developing World and the Future Direction of International Law* [2003] CHINESE J. INT'L. L. 29.

[129] General Agreement on Tariffs and Trade, BASIC INSTRUMENTS AND SELECTED DOCUMENTS, 18th Supp. (1971), at 24–26 and General Agreement on Tariffs and Trade, BASIC INSTRUMENTS AND SELECTED DOCUMENTS, 26th Supp. (1979), at 203–225. See Bela Balassa, *The Tokyo Round and the Developing Countries* (1980) 14 J. WORLD TRADE L. 93 and Borgsmidt, supra note 82.

[130] ACP-EEC Convention of Lomé, opened for signature on 28 February 1975 [1976] O.J. L25/2 (entered into force on 1 April 1976); Second ACP-EEC Convention of Lomé, opened for signature on 31 October 1979 [1980] O.J. L347/1 (entered into force on 1 January 1981); Third ACP-EEC Convention of Lomé, opened for signature on 31 March 1986 [1986] O.J. L86/3 (entered into force on 1 May 1986); Fourth ACP-EEC Convention of Lomé, opened for signature on 15 December 1989 [1991] O.J. L229/3 (entered into force on 1 March 1990); Partnership Agreement between the

Second, the voting quotas of the developing States in the World Bank and the IMF have increased steadily since the adoption of the NIEO instruments.[131] This is due in part to the regular reviews of quotas and in another part to the pressure brought upon the industrialised States by the developing States, most notably in the IMF, which was seen correctly by those States to have a regulatory function as well as a financial one.[132]

Third, the actions by some developing States in nationalising or expropriating public and private commercial interests and assets of industrialised States, have been met with opposition and hostility from industrialised States. The developing States have sought to justify their actions, particularly those assets relating to mining interests, as the Charter of Economic Rights provided for developing States with permanent sovereignty over their natural resources and the right to expropriate and nationalise foreign assets and interests in accordance with domestic law.[133] The actions of these developing States, mostly notably of Iran and Libya, have led to a substantial number of international arbitrations to determine the liability and quantum of compensation that ought to be paid for the expropriated assets and business interests.[134] Whether the actions of the developing States in expropriating foreign interests in mineral resources, and the subsequent political and legal consequences that followed, amount to either a vindication or repudiation of the NIEO principles remain a matter of conjecture for the international community.[135]

Members of the African, Caribbean and Pacific Group of States, of the One Part, and the European Community and its Member States, of the Other Part, opened for signature on 15 December 2000 [2000] O.J. L317/3 (entered into force on 1 April 2003). See Isaac K. Minta, *The Lome Convention and the New International Economic Order* (1984) 27 HOWARD L. J. 953; Wolfgang Benedek, *The Lomé Convention and the International Law of Development: A Concretisation of the New International Economic Order?* (1982) 26 J. AFR. L. 74; and David L. Perrott, *EEC Attitudes and Responses to the New International Economic Order* (1984) 3 PUB. L. FORUM 115.

[131] Richard W. Edwards, Jr., *Responses of the International Monetary Fund and the World Bank to the Call for a "New International Economic Order": Separating Substance from Rhetoric* (1984) 3 PUB. L. FORUM 89 at 96–98.

[132] Ibid.

[133] Charter of Economic Rights, Article 2.

[134] See, for example, *Mobil Oil Iran, Inc. v. Iran* (1987) 16 IRAN-U.S.C.T.R. 3; *Texaco Overseas Petroleum Co. v. Libya* (1978) 53 I.L.R. 389; *BP Exploration Company (Libya) Ltd. v. Libya* (1978) 53 I.L.R. 297; *Kuwait v. The American Independent Oil Co.* (1982) 66 I.L.R. 518; and *Libyan American Oil Co. v. Libya* (1982) 62 I.L.R. 140. See also Robert B. von Mehren and P. Nicholas Kourides, *International Arbitrations Between States and Foreign Private Parties: The Libyan Nationalisation Cases* (1981) 75 AM. J. INT'L. L. 476; Grant Hanessian, *General Principles of Law in the Iran-U.S. Claims Tribunal* (1989) 27 COLUM. J. TRANSNAT'L. L. 309; Robin C. A. White, *Expropriation of the Libyan Oil Concessions: Two Conflicting International Arbitrations* (1981) 30 INT'L. & COMP. L. Q. 1; Burns H. Weston, *The Charter of Economic Rights and Duties of States and the Deprivation of Foreign-Owned Wealth* (1981) 75 AM. J. INT'L. L. 437; and Fernando R. Teson, *State Contracts and Oil Expropriations: The Aminoil-Kuwait Arbitration* (1984) 24 VA. J. INT'L. L. 323.

[135] See, for example, Gabrielle Marceau, *Some Evidence of a New International Economic Order in Place* (1991) 22 REVUE GÉN. DR. 397; Emily Carasco, *A Nationalisation Compensation Framework in the New International Economic Order* [1983] THIRD WORLD LEG. STUD. 49; and

The most controversial of the principles of the NIEO and the issue that has the most long-term impact on international relations and the international political economy is the designation of the seabed and its mineral resources to be the common heritage of mankind. The implementation of this concept in the negotiations over the Convention on the Law of the Sea has led to deadlocks during both the negotiations and the implementation of the treaty over a number of decades.[136] Further, the extension of the same concept to celestial bodies and their mineral resources has led to a similar deadlock in the negotiations and the acceptance of the Moon Agreement.

5.5 Origins of the Common Heritage of Mankind Concept in the Context of the Deep Seabed

5.5.1 The Old Law of the High Seas

The historical development of the law relating to the high seas can be described as a series of combats between those States that favoured closed seas and those that favoured the freedom of the seas.[137] With the dominance of the maritime powers by the seventeenth century, the freedom of the seas had emerged victorious over the short-lived claims of Britain over the other waters surrounding the British Isles after the accession of James I,[138] and even the partition of the oceans along with the non-Christian lands of the world between Spain and Portugal by Pope Julius II under the Papal Bull *Ea quae* of 1506 that confirmed the Treaty of Tordesillas of 1494.[139] The high seas has long been recognised to be *res extra commercium* and,

---

Francisco V. García-Amador, *The Proposed New International Economic Order: A New Approach to the Law Governing Nationalisation and Compensation* (1980) 12 LAWYER AM. 2.

[136] See Martti Koskenniemi and Marja Lehto, *The Privilege of Universality: International Law, Economic Ideology and Seabed Resources* (1996) 65 NORDIC J. INT'L. L. 533; Elisabeth Mann Borgese, *The New International Economic Order and the Law of the Sea* (1977) 14 SAN DIEGO L. REV. 584; John Norton Moore, *The Law of the Sea and the New International Economic Order* (1984) 3 PUB. L. FORUM 13; Robert Friedheim, *UNCLOS and the New International Economic Order* (1987) 4 J. L. & ENV'T. 17; John King Gamble, Jr., *The Third United Nations Conference on the Law of the Sea and the New International Economic Order* (1983) 6 LOYOLA L. A. INT'L. & COMP. L. J. 65; and Tommy T. B. Koh, *Negotiating a New World Order for the Sea* (1984) 24 VA. J. INT'L. L. 761.

[137] See Thomas Wemyss Fulton, THE SOVEREIGNTY OF THE SEA (1911).

[138] David Armitage, *Making the Empire British: Scotland in the Atlantic World 1542–1707* (1997) 155 PAST & PRESENT 34; David Sandler Berkowitz, JOHN SELDEN'S FORMATIVE YEARS: POLITICS AND SOCIETY IN EARLY SEVENTEENTH CENTURY (1988) at 53; and René-Jean Dupuy and Daniel Vignes, A HANDBOOK ON THE NEW LAW OF THE SEA (Volume 2, 1991) at 387.

[139] Frances Gardiner Davenport, EUROPEAN TREATIES BEARING ON THE HISTORY OF THE UNITED STATES AND ITS DEPENDENCIES TO 1648 (1917) and Charles Gibson, SPAIN IN AMERICA (1966).

230   5 Exploitation Rights: Evolving from the "Province of Mankind" to the "Common ...

as such, are free from the sovereignty and appropriation of individual States, but has indeed been regarded since Roman times as *omnium communia*.[140]

Historically, the high seas were considered to be held on international "public trust", in the sense that private property rights were excluded and vested trusteeship rights and duties in the international community to ensure public access and public benefit.[141] The concept of the high seas being held on public trust can trace its origins to Roman law, as Marcianus had declared that the seas and its fishes to be "communis omnium naturali jure".[142] Similarly, Domitius Ulpianus had written as early as the second century and later adopted by Justinian I:

> Naturali iure communia sunt omnium haec: aer et aqua profluens et mare et per hoc litoria maris. Nemo igitur ad litus maris accedere prohibetur, dum tamen villis et monumenti et aedificiis abstineat, quia non sunt iuris gentium, sicut est mare.[143]

In 1169, Pope Alexander III confirmed to the Consul of Genoa that freedom was to reign on the seas.[144] When Spain claimed a monopoly of commerce in the West Indies and protested an expedition by Sir Francis Drake, on both occasions Elizabeth I of England asserted the freedom of the seas in response.[145] Similarly, Russia had asserted the same principle in 1587 in its diplomatic correspondence

---

[140] Ibid., at 123.

[141] Matthew Hale, DE JURE MARIS ET BRACHIORUM EJUSDEM (1667) and Robert Gream Hall, ESSAY ON THE RIGHTS OF THE CROWN AND THE PRIVILEGES OF THE SUBJECT IN THE SEA SHORES OF THE REALM (2nd ed., 1875). See also the English cases of *Ward v. Creswell* (1741) 125 E.R. 1165; and *Gann v. Free Fishers of Whitstable* (1865) 11 E.R. 1305; as well as the U.S. case of *Martin v. Lessee of Waddell* (1842) 41 U.S. 367 at 410–411. It has been suggested that the concept is being revived in the context of the modern law of the high seas: see Peter H. Sand, *Public Trusteeship for the Oceans*, in Tafsir Malick Ndiaye and Rüdiger Wolfrum (eds.), LAW OF THE SEA, ENVIRONMENTAL LAW AND SETTLEMENT OF DISPUTES: LIBER AMICORUM JUDGE THOMAS A. MENSAH (2007), at 521–544.

[142] W. Paul Gormley, *The Development and Subsequent Influence of the Roman Legal Norm of "Freedom of the Seas"* (1963) 40 U. DET. L. J. 561.

[143] "By natural law, these things are common in mankind: the air, running water, the sea and the shores of the sea. Thus no one is forbidden to approach the seashore, provided he respects the homes, monuments and buildings, which are subject not only to the laws of nations, like the sea.": Thomas Collett Sanders, THE INSTITUTES OF JUSTINIAN (4th ed., 1903) at 90. Johnson had suggested that "On the third day, God created the oceans; and of course with the creation of the oceans, he must have created the public trust doctrine to protect the oceans": Ralph W. Johnson, *The Public Trust Doctrine*, in Douglas J. Canning and James Scott (eds.), THE PUBLIC TRUST DOCTRINE IN WASHINGTON STATE: PROCEEDINGS OF THE SYMPOSIUM, NOVEMBER 18, 1992 (1993).

[144] André-Louis Sanguin, *Geopolitical Scenarios, From the Mare Liberum to the Mare Clausum: The High Seas and the Case of the Mediterranean Basin* (1997) 2 GEOADRIA 51 at 52.

[145] Louis B. Sohn, *Managing the Law of the Sea: Ambassador Pardo's Forgotten Second Idea* (1998) 36 COLUM. J. TRANSNAT'L. L. 285 at 286; Ian Brownlie, PRINCIPLES OF PUBLIC INTERNATIONAL LAW (6th ed., 2003), at 224; and Tommy T. B. Koh, *The Origins of the 1982 Convention on the Law of the Sea* (1987) 29 MALAYA L. REV. 1.

5.5 Origins of the Common Heritage of Mankind Concept in the Context of . . .

with England.[146] Grotius, in his famous work *Mare Liberum* published in 1609, opined that States cannot unilaterally or collectively attain sovereignty or title to the high seas by occupation or otherwise because they are *res communis* or *res extra commercium*.[147] Butler observed that, since Grotius:

> Naval power and commercial shipping interests in the nineteenth century ensured European and American support for, indeed insistence upon, the principle of the freedom of the seas. In the twentieth century the freedom of the seas has come to be accepted as a "general", "basic" or "fundamental" principle of international law: some are even prepared to treat it as *jus cogens*.[148]

The concept of the freedom of the seas is negative or prohibitive in nature, in the sense that it is a restrictive doctrine that is asserted to counter attempts to interfere in the exercise of that freedom. As such, the exercise of such a freedom is nevertheless subject to limits that apply to the extent that they prevent a State from using the high seas in a manner that interfere with or even deny the use of the high seas by another State. As Gidel observed:

> La liberté de la haute mer, essentiellement négative, ne peut pas cependant ne pas comporter des conséquences positives. Dirigée contre l'exclusivité d'usage elle se résout nécessaire-ment en une idée d'égalité d'usage. . . . Tous les pavillons maritimes ont un droit égal à tirer de la haute mer les diverses utilités qu'elle peut comporter. Mais l'idée d'égalité d'usage ne vient qu'en second lieu. L'idée essentielle contenue dans le principe de liberté de la haute mer est l'idée d'interdiction d'interférence de tout pavillon dans la navigation en temps de paix de tout autre pavillon.[149]

The Convention on the Law of the Sea embodies this conflict between the guarantee of the freedom of the high seas and the prevention of interference with the

---

[146] Vladimir E. Grabar (trans. William E. Butler), THE HISTORY OF INTERNATIONAL LAW IN RUSSIA 1647–1917 (1991).

[147] William E. Butler, *Grotius and the Law of the Sea*, in Hedley Bull, Benedict Kingsbury and Adam Roberts (eds.), HUGO GROTIUS AND INTERNATIONAL RELATIONS (1992), at 229–220. See also discussion in Arvid Pardo, *The Law of the Sea: Its Past and Its Future* (1984) 63 OR. L. REV. 7; Arcangelo Travaglini, *Reconciling Natural Law and Legal Positivism in the Deep Seabed Mining Provisions of the Convention on the Law of the Sea* (2001) 15 TEMPLE INT'L. & COMP. L. J. 313; Joseph W. Dellapenna, *Treaties as Instruments for Managing Internationally-Shared Water Resources: Restricted Sovereignty vs. Community of Property* (1994) 26 CASE W. RES. J. INT'L. L. 27; and Jan Schneider, *Something Old, Something New: Some Thoughts on Grotius and the Marine Environment* (1978) 18 VA. J. INT'L. L. 147.

[148] Ibid., at 212.

[149] "The freedom of the high seas, which is essentially a negative freedom, cannot – however – be without positive consequences. Directed against the idea of an exclusive use, the freedom nec-essarily solves itself in an idea of equal right of use. . . . All maritime flags have the same right to draw from the high seas the different uses it may offer. But the idea of an equal right to use comes only in a second step. The essential idea enshrined in the principle of the freedom of the high seas is the idea to prohibit any interference of any flag in the navigation of any other flag in times of peace." (extract translated for the author by Dr. Ulrike M. Bohlmann with thanks): Gilbert Charles Gidel, LE DROIT INTERNATIONAL PUBLIC DE LA MER (1932). See also Ram Prakash Anand, ORIGIN AND DEVELOPMENT OF THE LAW OF THE SEA (1983), at 232–233.

232   5 Exploitation Rights: Evolving from the "Province of Mankind" to the "Common ...

exercise of that freedom by others.[150] While all States have the freedoms to navigation, over flight, lay undersea cables and pipelines, construction of artificial islands, fishing and scientific research, these freedoms must be exercised "with due regard for the interests of other States in their exercise of the freedom of the high seas.[151]

5.5.2 The Old Law of the Deep Seabed

The deep seabed, as with the high seas, is not subject to sovereignty or appropriation by States and it has been suggested that the same freedom of the high seas apply to the deep seabed beneath the high seas.[152] The 1958 Convention on the High Seas provided for freedom of the high seas and that States may lay submarine cables and pipelines on the deep seabed.[153] However, as Oda stated the deep seabed and the high seas must be considered separately in terms of the legal principles applicable to them:

> It is quite defensible to maintain that the submarine areas have always been tacitly regarded as an international realm which can never be possessed by any State. The author does not believe that the submarine areas, merely by reason of their being beneath the superjacent waters, thereby become a part of them.[154]

Although the deep seabed is not subject to sovereignty or appropriation by any State, the use of resources from the deep seabed is a different matter. The use of certain sedentary fisheries in the seabed, such as pearl, oyster and sponge fisheries, has been allowed to States on the basis of historical title or prescription.[155] To some

---

[150] Convention on the Law of the Sea, Article 87. See Jon van Dyke and Christopher Yuen, *"Common Heritage" v. "Freedom of the High Seas": Which Governs the Seabed?* (1982) 19 SAN DIEGO L. REV. 493; Lea Brilmayer and Natalie Klein, *Land and Sea: Two Sovereignty Regimes in Search of a Common Denominator* (2001) 33 N. Y. U. J. INT'L. L. & POL'Y. 703; John Temple Swing, *Who Will Own the Oceans?* (1976) 54 FOREIGN AFF. 527; Bernard H. Oxman, *The Territorial Temptation: A Siren Song at Sea* (2006) 100 AM. J. INT'L. L. 830; and Steven J. Burton, *Freedom of the Seas: International Law Applicable to Deep Seabed Mining Claims* (1977) 29 STANFORD L. REV. 1135.

[151] Ibid.

[152] Francis T. Christy, Jr., *Property Rights in the World Ocean* (1975) 15 NAT. RES. J. 695; Brownlie, supra note 145, at 241; Dennis W. Arrow, *Seabeds, Sovereignty and Objective Regimes* (1984) 7 FORDHAM INT'L. L. J. 169; and A. Vaughan Lowe, *Reflections on the Waters: Changing Conceptions of Property Rights in the Law of the Sea* (1986) 1 INT'L. J. ESTUARINE & COASTAL L. 1.

[153] Convention on the High Seas, opened for signature on 29 April 1958, 450 U.N.T.S. 11 (entered into force on 30 September 1962), Articles 2 and 26. See Shigeru Oda, *Some Reflections on Recent Developments in the Law of the Sea* (2002) 27 YALE J. INT'L. L. 217; Sir Kenneth Bailey, *Australia and the Law of the Sea* (1962) 1 ADEL. L. REV. 1; Michael J. McCabe, Ignazio J. Ruvolo and M. Howard Wayne, *Recent Developments in the Law of the Seas II: A Synopsis* (1971) 8 SAN DIEGO L. REV. 658; and Jon Gregory Jackson, *Deepsea Ventures: Exclusive Mining Rights to the Deep Seabed as a Freedom of the Sea* (1976) 28 BAYLOR L. REV. 170.

[154] Shigeru Oda, INTERNATIONAL CONTROL OF SEA RESOURCES (1989), at 155.

[155] Brownlie, supra note 145, at 241.

5.5 Origins of the Common Heritage of Mankind Concept in the Context of ...

extent, this is no different to the regulation of the use of fisheries in the high seas.[156] As early as 1857, it was asserted in Great Britain that the right to own and operate mines on the deep seabed beyond the low tide mark and title to the mineral resources vested in the Crown.[157] For example, Oda stated a somewhat balanced view that:

> The principle of non-appropriation of the deep ocean floor does not lead us to conclude that the exploration or the exploitation of this area should be suspended. On the contrary, the most effect exploitation of the resources should be encouraged and the incentives for this should not be removed. Free access to the resources of the deep ocean floor should be the right of all nations, however, not merely the right of those possessing advanced technologies.[158]

In the early debates on the regulation of the exploitation of resources of the deep seabed, there were those of one school who advocated the idea of international control of deep seabed resources, exercised by a specialised agency of the United Nations.[159] There were those of another school who were of the view that international regulation was not necessary until actual exploitation began and conflicts of interests between States arose and, until then, the deep seabed was free to all for use and exploitation.[160] For example, Ely went so far as to suggest that:

> In this writer's opinion, the most constructive accomplishment that the Law of the Sea conference could present to the mineral consumers of the world would be to forever refrain from meeting again, now or hereafter, either in this world or the next.[161]

Although the exploitation of mineral resources from the deep seabed remained a distant possibility at the time, the international community began turning their minds to the idea of international regulation for such activities under the auspices

---

[156] Shigeru Oda, FIFTY YEARS OF THE LAW OF THE SEA (2003), at 80–86.

[157] *Cornwall Submarine Mines Act 1858* (U.K.).

[158] Oda, supra note 156, at 322.

[159] See, for example, Clark M. Eichelberger and Francis T. Christy, Jr., *Comments on International Control of the Sea's Resources*, in Lewis M. Alexander (ed.), THE LAW OF THE SEA (1967) at 299–309; Clark M. Eichelberger, *The United Nations and the Bed of the Sea* (1969) 6 SAN DIEGO L. REV. 339; Francis T. Christy, Jr., *Property Rights in the World Ocean* (1975) 15 NAT. RES. J. 695; Francis M. Auburn, *The International Seabed Area* (1971) 20 INT'L. & COMP. L. Q. 173; and Gonzalo Biggs, *Deepsea's Adventures: Grotius Revisited* (1975) 9 INT'L. LAWYER 271.

[160] See, for example, Northcutt Ely, *United States Seabed Minerals Policy* (1971) 4 NAT. RES. LAWYER 597; Northcutt Ely, *The Draft United Nations Convention on the International Seabed Area – American Bar Association Position* (1971) 4 NAT. RES. LAWYER 60; L. Frederick E. Goldie, *Mining Rights and the General International Law Regime of the Deep Ocean Floor* (1975) 2 BROOKLYN J. INT'L. L. 1; L. Frederick E. Goldie, *A General International Law Doctrine for Seabed Régimes* (1973) 7 INT'L. LAWYER 796; and Luke W. Finlay and Maxwell S. McKnight, *Law of the Sea: Its Impact on the International Energy Crisis* (1974) 6 L. & POL'Y. INT'L. BUS. 639; and David L. Larson, *The United States Position on the Deep Seabed* (1979) 3 SUFFOLK TRANSNAT'L. L. J. 1.

[161] Northcutt Ely, *Potential Regimes for Deep Seabed Mining* (1978) 6 INT'L. BUS. LAWYER 93 at 104.

234   5   Exploitation Rights: Evolving from the "Province of Mankind" to the "Common . . .

of the United Nations or new specialised agencies.[162] Unfortunately, progress on such a multilateral development was arrested by the coincidental timing of such an effort with the movement to establish and implement the NIEO and designate the deep seabed as the common heritage of mankind, although the proposal for such a designation actually originated from Malta.

5.5.3   Proposal from Malta that "Detonated the Time Bomb"

On 17 August 1967, the Maltese Ambassador to the United Nations proposed to the General Assembly a "Declaration and Treaty concerning the reservation exclusively for peaceful purposes of the seabed and ocean floor underlying the seas beyond the limits of present national jurisdiction and the use of their resources in the interests of mankind", described by one commentator as having "detonated the time bomb".[163] At the time, a significant number of States, some being industrialised States and some being developing States, were receptive of the proposal.[164] Most of the industrialised States preferred a much more cautious approach to the proposal, with the Soviet Union unwilling to agree to anything beyond a study of present regulatory activities and the United States adopting an attitude of vehement opposition.[165] Most of the opposition stemmed from his idea that of making the deep seabed and the ocean floor a common heritage of mankind and to draft a multilateral treaty containing the following legal principles:

(1) the seabed and the ocean floor are not subject to national appropriation in any manner whatsoever and are reserved for peaceful purposes;[166]

---

[162] See Michael A. Wagner, *International Regulation of the Oceans and Their Resources* (1971) 37 BROOKLYN L. REV. 402.

[163] U.N. Doc. A/6695 (1967). See J. Henry Glazer, *The Maltese Initiatives Within the United Nations – A Blue Planet Blueprint for Transnational Space* (1975) 4 ECOLOGY L. Q. 279.

[164] See, for example, Cyprus: U.N. Doc. A/C.1/PV.1530 at 5–7; Ghana: U.N. Doc. A/C.1/PV.1526 at 7–8; and Sweden: U.N. Doc. A/C.1/PV.1527 at 12–14. It was noted that Ceylon (Sri Lanka), Libya, Nigeria and Somalia also supported the Maltese proposal: Günter Weissberg, *International Law Meets the Short-Term National Interest: The Maltese Proposal on the Sea-Bed and Ocean Floor – Its Fate in Two Cities* (1969) 18 INT'L. & COMP. L. Q. 41 at 53.

[165] See, for example, Soviet Union: U.N. Doc. A/C.1/PV.1525 at 3–4; United Kingdom: U.N. Doc. A/C.1/PV.1524 at 2–4; and United States: U.N. Doc. A/C.1/PV.1524 at 4–5. See also Sevinc Carlson, *Soviet Policy on the Seabed and the Ocean Floor* (1973) 1 SYRACUSE J. INT'L. L. & COM. 104. The position of the Soviet Union was supported by Bolivia, Bulgaria, Belarus, Hungary, Iceland, India, Iran, Poland and Yugoslavia: Weissberg, supra note 164, at 56; Christopher C. Joyner, *Towards a Legal Regime for the International Seabed: The Soviet Union's Evolving Perspective* (1975) 15 VA. J. INT'L. L. 871; Artemy A. Saguirian, *The USSR and the New Law of the Sea Convention: In Search of Practical Solutions* (1990) 84 AM. SOC'Y. INT'L. L. PROC. 295; and Serguei Karev, *The Russian Federation and the UN Conference on the Law of the Sea* (1995) 89 AM. SOC'Y. INT'L. L. PROC. 455.

[166] See, for example, Tullio Treves, *Military Installations, Structures and Devices on the Seabed* (1980) 74 AM. J. INT'L. L. 808.

5.5 Origins of the Common Heritage of Mankind Concept in the Context of ...

(2) use of the seabed and the ocean floor and their economic exploitation are to be undertaken with the aim of "safeguarding the interests of mankind";

(3) the "net financial benefits derived from the use and exploitation of the seabed and the ocean floor shall be used primarily to promote the development of poor countries"; and

(4) the creation of an international agency to assume jurisdiction over the seabed and the ocean floor "as trustee for all countries" and to regulate all activities involving their use and economic exploitation.[167]

The United States reacted to the Maltese proposal rapidly, with a series of over 20 resolutions by the U.S. Congress aimed at deterring any progress on the proposal by the U.S. or the international community and refusing to vest jurisdiction and control over the deep seabed and the ocean floor with the United Nations.[168] The U.S. was of the view that it has a right of access to seabed minerals and this right was afforded it by customary international law as part of the freedom of the high seas, a position later legislated into U.S. domestic law.[169] When reflecting on this opposition on the part of the industrialised States, particularly the United States, Pardo wrote that:

> I have a feeling that many may consider me a prophet of doom and gloom and somewhat of a utopian because I have predicted that the present uncertain status of the seabed may lead to a competitive scramble by a few countries to appropriate for national purposes the land under the world's seas and oceans; that this would cause an escalation of the arms race, increased political tensions and progressive impairment of the marine environment as a whole; and that the establishment of an effective international regime for the seabed and ocean floor beyond national jurisdiction is the only way to avoid the incalculable dangers to all of us that appear to me inevitable if the present situation is allowed to continue much longer.[170]

It is the practical application of the concept of the common heritage of mankind to the deep seabed and the regulation of its use and economic exploitation by an international organisation with exclusive jurisdiction that became the focus of the

---

[167] U.N. Doc. A/6695 (1967). See Arvid Pardo, *Who Will Control the Seabed?* (1969) 47 FOREIGN AFF. 123; Arvid Pardo, *Development of Ocean Space – An International Dilemma* (1971) 31 LA. L. REV. 45; and Arvid Pardo, *Before and After* (1983) 46 L. & CONTEMP. PROBS. 95.

[168] U.S. House Res. 816–824, 828–830, 834–835, 837, 839–840, 843–844, 854–857, 865, 876, 881 and 916, 90th Cong., 1st Sess. (1967). See Alan V. Lowe, *The International Seabed and the Single Negotiating Text* (1976) 13 SAN DIEGO L. REV. 489 at 490; and Weissberg, supra note 164.

[169] *Deep Seabed Hard Mineral Resources Act 1980* (U.S.); James B. Morell, THE LAW OF THE SEA: AN HISTORICAL ANALYSIS OF THE 1982 TREATY AND ITS REJECTION BY THE UNITED STATES (1992); Theodore G. Kronmiller, THE LAWFULNESS OF DEEP SEABED MINING (1981); Louis Henkin, Oscar Schachter, Richard C. Pugh and Hans Smit, INTERNATIONAL LAW: CASES AND MATERIALS (3rd ed., 1993) at 1314; Kenneth Mwenda, *Deep Sea-Bed Mining Under Customary International Law* (2000) 7 MURDOCH U. ELEC. J. L. 2; Wolfgang Friedmann, *Selden Redivivus – Towards a Partition of the Seas?* (1971) 65 AM. J. INT'L. L. 757.

[170] Arvid Pardo, *An International Regime for the Deep Seabed: Developing Law or Developing Anarchy?* (1969) 5 TEX. INT'L. L. F. 204 at 204.

debate in the negotiations over the seabed issues and, for that matter, the entire Convention on the Law of the Sea negotiations during the Third United Nations Conference on the Law of the Sea ("UNCLOS III"). The developing States, by then organised together as the Group of 77, were determined that the mineral resources of the deep seabed were to be administered by a United Nations authority and exploited by an "Enterprise" controlled by the authority with the proceeds of exploitation going to an international fund for distribution primarily to the developing States.[171] For the industrialised States, the argument was that commercial exploitation of resources would take place only if sufficient incentives are given to the States with the technological capability and the financial capacity to take on such a venture, particularly the European States, Japan, the Soviet Union and the United States.[172] This formed the impasse that was to continue throughout the negotiations in UNCLOS over the terms of the Convention on the Law of the Sea.

5.5.4 Negotiating History on the Deep Seabed

5.5.4.1 Early Negotiations

During the course of UNCLOS III, a combination of the improved scientific understanding and new economic prospects had initiated a set of novel issues to the governance of the ocean commons, particularly the deep seabed.[173] In particular, this included the discovery of rich mineral deposits on the deep seabed in the form of polymetallic or manganese nodules, consisting primarily of nickel, copper, manganese and cobalt, which are found on the floor of all oceans.[174] This is in addition to

---

[171] See Richard W. Bentham and P. C. Quine, *The Status of the Law of the Sea Negotiations* (1978) 6 INT'L. BUS. LAWYER 76 at 83; Elisabeth Mann Borgese, *Boom, Doom and Gloom over the Oceans: The Economic Zone, the Developing Nations and the Conference on the Law of the Sea* (1974) 11 SAN DIEGO L. REV. 541 at 552; Milenko Milic, *Third United Nations Conference on the Law of the Sea* (1976) 8 CASE W. RES. J. INT'L. L. 168; Arthur D. Martinez, *The Third United Nations Conference on the Law of the Sea: Prospects, Expectations and Realities* (1976) 7 J. MARIT. L. & COM. 253; Clark M. Eichelberger, *The Seabed Question in Context: One of the Many Issues Massing for the 1973 Conference* (1971) 8 SAN DIEGO L. REV. 653; and Rajeev Amarasuriya, *The Third United Nations Conference on the Law of the Sea: An Introductory Insight* (2001) 13 SRI LANKA J. INT'L. L. 137.

[172] Bentham and Quine, supra note 171, at 84.

[173] Vogler, supra note 15, at 47–50; John Vogler, THE GLOBAL COMMONS: ENVIRONMENTAL AND TECHNOLOGICAL GOVERNANCE (2000); John Briscoe and Jo Lynn Lambert, *Seabed Mineral Discoveries Within National Jurisdiction and the Future of the Law of the Sea* (1984) 18 U. S. F. L. REV. 433; and Craig H. Allen, *Protecting the Oceanic Gardens of Eden: International Law Issues in Deep-Sea Vent Resource Conservation and Management* (2001) 13 GEO. INT'L. ENVT'L. L. REV. 563.

[174] See Brian T. Chu, *The United States and UNCLOS III in the New Decade: Is It Time for a Compromise?* (1992) 4 J. CONTEMP. LEG. ISSUES 253 at 254; Heim, supra note 9, at 822–823; Cook, supra note 1, at 679–682; and Frederick W. Kosmo, Jr., *The Commercialisation of Space: A Regulatory Scheme that Promotes Commercial Ventures and International Responsibility* (1988) 61

5.5 Origins of the Common Heritage of Mankind Concept in the Context of ...

gold, platinum, titanium, chromium, tin, diamonds and other rare minerals that can be found as placer deposits in the deep seabed.[175] This coincided with the increasing demands of the developing States for the NIEO and the Maltese proposal to have the deep seabed and the ocean floor designated the common heritage of mankind and for their use and exploitation to be regulated by an international agency.[176] The natural result of these diametrically opposed positions was deadlock.

Ironically, it was the United States that first presented a "Draft United Nations Convention on the International Seabed Area" in 1970.[177] The negotiations over the new Convention on the Law of the Sea, being the first working session of UNCLOS III, took place in Caracas, Venezuela, in 1974.[178] By the end of that session, the industrialised States had already won a significant victory by attaining support for a 200-mile exclusive economic zone that would give the coastal States all of the resources found within 200 miles of their coasts, thus capturing most of the resource

---

S. CAL. L. REV. 1055, at 1075–1077. See also John Pinna Craven, *Technology and the Law of the Sea: The Effect of Prediction and Misprediction* (1985) 45 LA. L. REV. 1143; John M. Murphy, *The Politics of Manganese Nodules: International Considerations and Domestic Legislation* (1979) 16 SAN DIEGO L. REV. 531; Richard B. Frank and Bruce W. Jenett, *Murky Waters: Private Claims to Deep Ocean Seabed Minerals* (1975) 7 L. & POL'Y. INT'L. BUS. 1237; Frank L. La Que, *Different Approaches to International Regulation of Exploitation of Deep-Ocean Ferromanganese Nodules* (1978) 15 SAN DIEGO L. REV. 477; and Theodore G. Kronmiller, *Reconciling Public and Private Interests in Multilateral Negotiations: The Law of the Sea Conference* (1981) 1 PUB. L. F. 159.

[175] See Peter M. Bartlett and W. C. Jillian van Rensburg, *Technical, Economic and Institutional Constraints on the Production of Minerals from the Deep Seabed* [1986] ACTA JURIDICA 69 and John M. Murphy, *Deep Ocean Mining: Beginning of a New Era* (1976) 8 CASE W. RES. J. INT'L. L. 46.

[176] See Boleslaw Adam Boczek, *Ideology and the Law of the Sea: The Challenge of the New International Economic Order* (1984) 7 B. C. INT'L. & COMP. L. REV. 1; and William J. Martin, Jr., *Legal Aspects of Seabed Mineral Exploitation* (1975) 3 INT'L. BUS. LAWYER 148.

[177] Oliver L. Stone, *The United States Draft Convention on the International Seabed Area* (1971) 45 TUL. L. REV. 527; Sir Robert Y. Jennings Q.C., *The United States Draft Treaty on the International Seabed Area – Basic Principles* (1971) 20 INT'L. & COMP. L. Q. 433; William Palmer, *The United States Draft United Nations Conference on the International Seabed Area and the Accommodation of Ocean Uses* (1973) 1 SYRACUSE J. INT'L. L. & COM. 110; Edward J. Brunet, *Musing on the Bottom: Economic and Legal Implications of the United States' Proposed Draft United Nations Convention on the International Seabed* [1974] U. ILL. L. F. 251; and H. Gary Knight, *The Draft United Nations Conventions on the International Seabed Area: Background, Description and Some Preliminary Thoughts* (1971) 8 SAN DIEGO L. REV. 459.

[178] See David P. Stang, *Political Cobwebs Beneath the Sea* (1973) 7 INT'L. LAWYER 1; George P. Smith II, *Apostrophe to a Troubled Ocean* (1972) 5 IND. LEG. F. 267; Katherine W. Schoonover, *The History of Negotiations Concerning the System of Exploitation of the International Seabed* (1977) 9 N. Y. U. J. INT'L. L. & POL'Y. 483; Barry Buzan, *Seabed Issues at the Law of the Sea Conference: The Caracas Session* (1974) 12 CAN. Y. B. INT'L. L. 222; and Shabtai Rosenne, *The Third United Nations Conference on the Law of the Sea* (1976) 11 ISR. L. REV. 1; Leigh S. Ratiner and Rebecca L. Wright, *United States Ocean Mineral Resource Interests and the United Nations Conference on the Law of the Sea* (1973) 6 NAT. RES. LAWYER 1; Claiborne Pell, *A New Era in Ocean Policy* (1997) 12 INT'L. J. MARINE & COASTAL L. 1; and Richard J. Greenwald, *Problems of Legal Security of the World Hard Minerals Industry in the International Ocean* (1971) 4 NAT. RES. LAWYER 639.

238   5   Exploitation Rights: Evolving from the "Province of Mankind" to the "Common ...

wealth in the oceans for the wealthy coastal States.[179] Danzig observed that, instead of compelling the industrialised States to limit their exclusive economic zones or to share the resources contained in them, as the United States had offered to do:

> The developing countries have joined a stampede to divide the best part of the ocean treasure *colonial style*. ... This means that roughly ninety percent of the oil lying in the seabed would fall under national, as distinguished from international, jurisdiction and control. ... It is one of the greatest give-aways in history. ... I can attribute the stupid position adopted by so many of the developing countries only to: (a) the developing countries' distrust of anything proposed by an imperial power; ... (b) a conflict of interest among the developing states that hasn't come out into the open.[180]

This outcome is also at variance with the position that the developing States had adopted or thought to have adopted prior to the convening of UNCLOS III in 1974.[181] Despite this concession, the industrialised States continued their opposition to the internationalisation of the deep seabed and its mineral resources.

---

[179] See Lewis M. Alexander and Robert D. Hodgson, *The Impact of the 200-Mile Economic Zone on the Law of the Sea* (1975) 12 SAN DIEGO L. REV. 570; Richard Hudson, THREE SCENARIOS: THE LAW OF THE SEA, OCEAN MINING AND THE NEW INTERNATIONAL ECONOMIC ORDER (1977) at 14; Dean Rusk and Milner S. Ball, *Sea Changes and the American Republic* (1979) 9 GA. J. INT'L. & COMP. L. 1; L. G. Weeks, *Subsea Petroleum Resources*, U.N. Doc. A/AC.138/87 (1973); Milan Thamsborg and Jørgen Lilje-Jensen, *Demarcation of the Area* (1998) 67 NORDIC J. INT'L. L. 215; William Turbeville, *American Ocean Policy Adrift: An Exclusive Economic Zone as an Alternative to the Law of the Sea Treaty* (1983) 35 U. FLA. L. REV. 492; Ted L. McDorman, *The Entry into Force of the 1982 LOS Convention and the Article 76 Outer Continental Shelf Regime* (1995) 10 INT'L. J. MARINE & COASTAL. L. 165; Gea Tung, *Jurisdictional Issues in International Law: Kelp Farming Beyond the Territorial Sea* (1982) 31 BUFF. L. REV. 885; John R. Stevenson and Bernard H. Oxman, *The Third United Nations Conference on the Law of the Sea: The 1974 Caracas Session* (1975) 69 AM. J. INT'L. L. 1; and Barry Hart Dubner, *A Proposal for Accommodating the Interests of Archipelagic and Maritime States* (1976) 8 N. Y. U. J. INT'L. L. & POL'Y. 39.

[180] Aaron L. Danzig, *A Funny Thing Happened to the Common Heritage on the Way to the Sea* (1975) 12 SAN DIEGO L. REV. 655 at 656–658 and Luke W. Finlay, *United States Policy with Respect to High Seas Fisheries and Deep Seabed Minerals – A Study in Contrasts* (1976) 9 NAT. RES. LAWYER 629. It ought to be noted that the People's Republic of China, though it aspired to lead the Third World and has asserted the need for internationalisation of the deep seabed, it had not been willing to share the resources within its exclusive economic zone: see Charles Douglas Bethill, *People's China and the Law of the Sea* (1974) 8 INT'L. LAWYER 724.

[181] See, for example, Maurice Hope-Thompson, *The Third World and the Law of the Sea: The Attitude of the Group of 77 Toward the Continental Shelf* (1980) 1 B. C. THIRD WORLD L. J. 37; Derry J. Devine, *Southern Africa and the Law of the Sea: Problems Common, Uncommon and Unique* [1986] ACTA JURIDICA 29; Melvin A. Conant and Christa G. Conant, *Resource Development and the Seabed Regime of UNCLOS III: A Suggestion for Compromise* (1978) 18 VA. J. INT'L. L. 61; A. Chauncey Newlin, *An Alternative Legal Mechanism for Deep Sea Mining* (1980) 20 VA. J. INT'L. L. 257; Roy S. Lee, *Deep Seabed Mining and Developing Countries* (1979) 6 SYRACUSE J. INT'L. L. & COM. 213; John Breaux, *The Diminishing Prospects for an Acceptable Law of the Sea Treaty* (1979) 19 VA. J. INT'L. L. 257; Woodfin L. Butte, *The Law of the Sea – Breakers Ahead* (1972) 6 INT'L. LAWYER 237; and Jon L. Jacobson and Thomas A. Hanlon, *Regulation of Hard Mineral Mining on the Continental Shelf* (1971) 50 OR. L. REV. 425.

5.5.4.2 Unilateral Approaches

By the time the Informal Composite Negotiating Text was produced in 1977, it was clear that no compromise over the issue of the treatment of the deep seabed was within sight.[182] A significant number of commentators was already writing off the prospect of reaching a compromise agreement over the terms of the Informal Composite Negotiating Text.[183] In order to break the impasse, the United States began to move towards a unilateral approach as U.S. Ambassador Elliot Richardson sought passage of legislation through the U.S. Congress that would enable authorisation of private mining of the deep seabed.[184] After a few years of deliberations, Congress in 1980 duly enacted the *Deep Seabed Hard Mineral Resources Act*.[185]

---

[182] See Alexander Yankov, *The Law of the Sea Conference at the Crossroads* (1978) 18 VA. J. INT'L. L. 31; Frederick Arnold, *Toward a Principled Approach to the Distribution of Global Wealth: An Impartial Solution to the Dispute over Seabed Manganese Nodules* (1980) 17 SAN DIEGO L. REV. 557; Elisabeth Mann Borgese, *A Constitution for the Oceans: Comments and Suggestions Regarding Part XI of the Informal Composite Negotiating Text* (1978) 15 SAN DIEGO L. REV. 371; James R. Silkenat, *Solving the Problem of the Deep Seabed: The Informal Composite Negotiating Text for the First Committee of UNCLOS III* (1977) 9 N. Y. U. J. INT'L. L. & POL'Y. 177; Theodore M. Beuttler, *The Composite Text and Nodule Mining – Over-regulation as a Threat to the "Common Heritage of Mankind"* (1978) 1 HASTINGS INT'L. & COMP. L. REV. 167; Robert B. Krueger, *Policy Options in the Law of the Sea Negotiations* (1978) 6 INT'L. BUS. LAWYER 89; and Robert F. Pietrowski, Jr., *Hard Minerals on the Deep Ocean Floor: Implications for American Law and Policy* (1978) 19 WM. & MARY L. REV. 43.

[183] See Arthur J. Goldberg, *The State of the Negotiations on the Law of the Sea* (1980) 31 HASTINGS L. J. 1091; John Thomas Smith II, *The Seabed Negotiation and the Law of the Sea Conference – Ready for a Divorce?* (1978) 18 VA. J. INT'L. L. 43; Margaret L. Dickey, *Should the Law of the Sea Conference be Saved?* (1978) 12 INT'L. LAWYER 1; Jonathan I. Charney, *United States Interests in a Convention on the Law of the Sea: The Case for Continued Efforts* (1978) 11 VAND. J. TRANSNAT'L. L. 39; Lewis Alexander, Francis Cameron and Dennis Nixon, *The Costs of Failure at the Third Law of the Sea Conference* (1978) 9 J. MARIT. L. & COM. 1; Geoffrey Hornsey, *International Law – All at Sea* (1978) 29 N. IR. LEG. Q. 250; and Henry C. Byrum, Jr., *An International Seabed Authority: The Impossible Dream?* (1978) 10 CASE W. RES. J. INT'L. L. 621.

[184] See Elliot L. Richardson, *United States Interests and the Law of the Sea* (1978) 10 LAWYER AM. 651; Robert Everett Bostrom, *The United States' Legislative Response to the Third United Nations Conference on the Law of the Sea Deadlock* (1979) 2 B. C. INT'L. & COMP. L. J. 409; Paul N. McCloskey, Jr. and Ronald K. Losch, *The U.N. Law of the Sea Conference and the U.S. Congress: Will Pending U.S. Unilateral Action on Deep Seabed Mining Destroy Hope for a Treaty?* (1979) 1 NW. J. INT'L. L. & BUS. 240; Charles E. Biblowit, *Deep Seabed Mining: The United States and the United Nations Convention on the Law of the Sea* (1984) 58 ST. JOHN'S L. REV. 267; Kathryn Surace-Smith, *United States Activity Outside of the Law of the Sea Convention: Deep Seabed Mining and Transit Passage* (1984) 84 COLUM. L. REV. 1032; Scott C. Whitney, *Environmental Regulation of United States Deep Seabed Mining* (1978) 19 WM. & MARY L. REV. 77; Charles Douglas Oliver, *Interim Deep Seabed Mining Legislation: An International Environmental Perspective* (1981) 8 J. LEGIS. 73; and Cheryl Hein Johnston, *Deep Seabed Mineral Resources Act* (1980) 20 NAT. RES. J. 163.

[185] See Elliot L. Richardson, *Law in the Making: A Universal Regime for Deep Seabed Mining?* (1981) 8 J. LEGIS. 199; Dana B. Ott, *An Analysis of Deep Seabed Mining Legislation* (1978) 10 NAT. RES. LAWYER 591; David W. Proudfoot, *Guarding the Treasures of the Deep: The Deep*

240   5 Exploitation Rights: Evolving from the "Province of Mankind" to the "Common ...

This was followed by similar legislation enacted by other industrialised States, including France, Germany and the United Kingdom.[186] In justification for the unilateral approach, Ambassador Richardson had argued that:

> Far from jeopardising the Conference, sea-bed mining legislation should facilitate the early conclusion of a general acceptable treaty by dispelling any impression that the Governments of the countries preparing to engage in such mining could be induced to acquiesce in an otherwise unacceptable treaty as the only means of obtaining the minerals of the seabed beyond national jurisdiction.[187]

Ambassador Satya Nandan of Fiji, who at the time was chairman of the Group of 77, retorted that such action on the part of the United States amounted to an exercise of sovereignty in violation of international law:

> ... unilateral legislation relating to sea-bed resources beyond national jurisdiction has no validity in international law, and activities conducted thereunder had no legal status. ... There could be no substitute for a universally agreed treaty for a rational and

---

Seabed Hard Mineral Resources Act (1973) 10 HARV. J. ON LEGIS. 596; Dennis W. Arrow, *The Proposed Regime for the Unilateral Exploitation of Deep Seabed Mineral Resources by the United States* (1980) 21 HARV. INT'L. L. J. 337; J. P. Lawless, *Implementation of the Deep Seabed Hard Mineral Resources Act*, paper presented at the Offshore Technology Conference, 3–6 May 1982, in Houston, TX, USA; Francis M. Auburn, *The Deep Seabed Hard Mineral Resources Bill* (1972) 9 SAN DIEGO L. REV. 491; Betsy Cox and Frank Brogan, *Law of the Sea – Proposed Deep Seabed Hard Mineral Resources Act* (1979) 9 GA. J. INT'L. & COMP. L. 641; Harry M. Collins, *Deep Seabed Hard Mineral Resources Act – Matrix for United States Deep Seabed Mining* (1981) 13 NAT. RES. LAWYER 571; Thomas M. Franck and Evan R. Chesler, *An International Regime for the Seabed Beyond National Jurisdiction* (1975) 13 OSGOODE HALL L. J. 579; and Porter Hoagland III, *The Conservation and Disposal of Ocean Hard Minerals: A Comparison of Ocean Mining Codes in the United States* (1988) 28 NAT. RES. J. 451.

[186] *Deep Sea Mining Act 1981* (U.K.); *Law on the Exploration and Exploitation of Mineral Resources of the Deep Seabed* (81–1135) (France); and *Act of Interim Regulation of Deep Seabed Mining 1981* (Federal Republic of Germany). See James C. F. Wang, HANDBOOK ON OCEAN POLITICS & LAW (1992) at 284–286; David D. Caron, *Municipal Legislation for Exploitation of the Deep Seabed* (1980) 8 OCEAN DEV. & INT'L. L. J. 259; Francisco Orrego-Vicuña, *National Laws on Seabed Exploitation: Problems of International Law* (1981) 13 LAWYER AM. 139; Mark S. Bergman, *The Regulation of Seabed Mining Under the Reciprocating States Regime* (1981) 30 AM. U. L. REV. 477; Paul N. McCloskey, Jr., *Domestic Legislation and the Law of the Sea Conference* (1979) 6 SYRACUSE J. INT'L. L. & COM. 225; Richard Todd Luoma, *A Comparative Study of National Legislation Concerning the Deep Sea Mining of Manganese Nodules* (1983) 14 J. MARIT. L. & COM. 243; and Barry Hart Dubner, *The Caspian: Is It a Lake, a Sea or an Ocean and Does It Matter? The Danger of Utilising Unilateral Approaches to Resolving Regional/International Issues* (2000) 18 DICK. J. INT'L. L. 253.

[187] U.N. Doc. A/CONF.62/BUR/SR.41 (1978) at 8–9. See also David Lawrence Treat, *The United States' Claims of Customary Legal Rights Under the Law of the Sea Convention* (1984) 41 WASH. & LEE L. REV. 253; Suzette D. Stenersen, *Mining the Seabed: International v. National Control* (1993) 4 U.S.A.F.A. J. LEG. STUD. 103; Jack N. Barkenbus, *Seabed Negotiations: The Failure of United States Policy* (1977) 14 SAN DIEGO L. REV. 623; and Elliot L. Richardson, *The United States Posture Toward the Law of the Sea Convention: Awkward but not Irreparable* (1983) 20 SAN DIEGO L. REV. 505.

5.5 Origins of the Common Heritage of Mankind Concept in the Context of ... 241

equitable development of the resources of the deep sea-bed area in the interests of the world community as a whole. Over-all agreement should not be jeopardised through hasty and short-sighted actions.[188]

Rather than breaking the deadlock, it is apparent that the opposite result was achieved, as the unilateral act on the part of the United States and other industrialised States only increased intransigence of the developing States.[189] The developing States considered such unilateral actions to be in breach of existing resolutions of the General Assembly.[190] Accordingly, the unilateral approach did no more than to contribute towards entrenching the impasse between the industrialised States and developing States in UNCLOS III.

5.5.4.3 Proposal for a Common Heritage Fund

Meanwhile, the developing States attempted to break the impasse with an approach, opposite to the unilateral approach of the industrialised States in ideology and practice. On 19 May 1978, Nepal introduced a proposal to establish a Common Heritage Fund during the Seventh Session of UNCLOS III in Geneva, Switzerland.[191] In an attempt to reverse the advantage obtained by the coastal States in the agreement over the exclusive economic zones, the proposal called for the substantial income from the use and resource exploitation of both the deep seabed and the exclusive economic zones to be used for developmental aid, combat marine pollution and to finance the United Nations and its peacekeeping operations.[192] It was argued by Nepal that the proposal is an attempt to reconcile the competing common heritage of mankind and the exclusive economic zone concepts in UNCLOS III:

The concept of the Common Heritage of Mankind has been damaged by those who contend that there is a necessary incompatibility between the idea of the Common Heritage and the idea of the economic zone. We believe that both ideas are essential and we believe that they are necessarily intermixed, *i.e.* the economic zone can and should make a substantial contribution to the implementation of the concept of the Common Heritage.[193]

---

[188] U.N. Doc. A/CONF.62/BUR/SR.41 (1978) at 7–8.

[189] See H. Gary Knight, *The Deep Seabed Hard Mineral Resources Act – A Negative View* (1973) 10 SAN DIEGO L. REV. 446; Jeffrey D. Wilson, *Mining the Deep Seabed: Domestic Regulation, International Law and UNCLOS III* (1983) 18 TULSA L. J. 207; Steven J. Molitor, *The Provisional Understanding Regarding Deep Seabed Matters: An Ill-Conceived Regime for U.S. Deep Seabed Mining* (1987) 20 CORNELL INT'L. L. J. 223; Michael R. Molitor, *The U.S. Deep Seabed Mining Regulations: The Legal Basis for an Alternative Regime* (1982) 19 SAN DIEGO L. REV. 599; Virginia A. Pruitt, *Unilateral Deep Seabed Mining and Environmental Standards: A Risky Venture* (1982) 8 BROOKLYN J. INT'L. L. 345; Richard G. Darman, *The Law of the Sea: Rethinking U.S. Interests* (1978) 56 FOREIGN AFF. 373; and Jonathan I. Charney, *Law of the Sea: Breaking the Deadlock* (1977) 55 FOREIGN AFF. 598.

[190] General Assembly Resolutions 2749 (XXV) and 2574D (XXIV).

[191] U.N. Doc. A/CONF.62/65 (1978).

[192] Ibid.

[193] Ibid., at 3.

242   5   Exploitation Rights: Evolving from the "Province of Mankind" to the "Common ...

However, the timing of the proposal and its requirement that contributions be made by coastal States from income generated from their exclusive economic zones ensured that the proposal was bound for failure.[194] Norway, for example, made clear the day after the proposal was made that there could be no sharing of any mineral revenues from the exclusive economic zones and access for landlocked and other States with special geographic characteristics must be restricted to living resources only.[195] In any event, there could not have been any realistic prospects of acceptance and success when the proposal calls for an increase in international regulation and financial contributions towards developing States when the industrialised States, particularly the United States, were seeking to reduce or even eliminate them. Such a proposal probably only added to the element of fear held by the industrialised States that these were steps towards the creation of a global welfare state.

5.5.4.4 Conclusion of UNCLOS III

The negotiations in UNCLOS III had operated on the basis of consensus for the majority of its time, as it was believed that a comprehensive legal framework for the law of the sea required an absence of dissent during its negotiations for it to be acceptable and effective in practice.[196] However, the rules of procedure allowed

---

[194] See discussion in John J. Logue, *The Nepal Proposal for a Common Heritage Fund* (1979) 9 CAL. W. INT'L. L. J. 598; and William C. Lynch, *The Nepal Proposal for a Common Heritage Fund: Panacea or Pipedream?* (1980) 10 CAL. W. INT'L. L. J. 25.

[195] U.N. Doc. A/CONF.62/SR.102 (1978) at 11. The 44 landlocked States of the world are Afghanistan, Andorra, Armenia, Austria, Azerbaijan, Belarus, Bhutan, Bolivia, Botswana, Burkina Faso, Burundi, Central African Republic, Chad, Czech Republic, Ethiopia, Hungary, Kazakhstan, Kosovo, Kyrgyzstan, Laos, Lesotho, Liechtenstein, Luxembourg, Macedonia, Malawi, Mali, Moldova, Mongolia, Nepal, Niger, Paraguay, Rwanda, San Marino, Serbia, Slovakia, Swaziland, Switzerland, Tajikistan, Turkmenistan, Uganda, Uzbekistan, Vatican City, Zambia and Zimbabwe. See Thomas M. Franck, Mohamed El Baradei and George Aron, *The New Poor: Land-Locked, Shelf-Locked and Other Geographically Disadvantaged States* (1974) 7 N. Y. U. J. INT'L. L. & POL. 33; Tiyanjana Maluwa, *Southern African Landlocked States and Rights of Access under the New Law of the Sea* (1995) 10 INT'L. J. MARINE & COASTAL L. 529; Ibrahim J. Wani, *An Evaluation of the Convention on the Law of the Sea from the Perspective of the Landlocked States* (1982) 22 VA. J. INT'L. L. 627; Glenn McGowan, *Geographic Disadvantage as a Basis for Marine Resource Sharing Between States* (1987) 13 MONASH U. L. REV. 209; Samuel Pyeatt Menefee, *"The Oar of Odysseus": Landlocked and "Geographically Disadvantaged" States in Historical Perspective* (1993) 23 CAL. W. INT'L. L. J. 1; Susan Ferguson, *UNCLOS III: Last Chance for Landlocked States?* (1977) 14 SAN DIEGO L. REV. 637; Martin Ira Glassner, *The Status of Developing Landlocked States Since 1965* (1973) 5 LAWYER AM. 480; Tariq Hassan, *Third Law of the Sea Conference Fishing Rights of Landlocked States* (1976) 8 LAWYER AM. 686; and Martin Ira Glassner, *Developing Landlocked States and the Resources of the Seabed* (1974) 11 SAN DIEGO L. REV. 633.

[196] See Anthony D'Amato, *An Alternative to the Law of the Sea Convention* (1983) 77 AM. J. INT'L. L. 281; E. D. Brown, *The UN Convention on the Law of the Sea 1982* (1984) 2 J. ENERGY NAT. RES. L. 259 at 260–261. See also Harry M. Collins, Jr., *Mineral Exploitation of the Seabed: Problems, Progress and Alternatives* (1979) 12 NAT. RES. LAWYER 599; G. David Robertson and Gaylene Vasaturo, *Recent Developments in the Law of the Sea 1981–1982* (1983) 20 SAN DIEGO

5.5 Origins of the Common Heritage of Mankind Concept in the Context of ...

for formal voting in UNCLOS III after the failure of all attempts to achieve consensus.[197] At the final negotiating session on 23 April 1982, the United States called for a vote and the Convention on the Law of the Sea was adopted by 130 votes in favour with 4 votes against and 17 abstentions.[198] Of the four States that voted against its adoption, only the United States expressed firm objections to the terms of the provisions in Part XI of the Convention on the Law of the Sea, as Israel cast its negative vote due to the standing given to the Palestine Liberation Organisation under the Convention while Turkey and Venezuela voted against the Convention due to maritime delimitation disputes with neighbouring States.[199] However, the most telling statement was in the States that abstained, namely Belgium, Bulgaria, Belarus, Czechoslovakia, East Germany, West Germany, Hungary, Italy, Luxembourg, Mongolia, the Netherlands, Poland, Spain, the Soviet Union, Thailand, Ukraine and the United Kingdom, as most of them did so because they had the same objections as those of the United States. Rosenne noted that:

> This includes all the industrialised States of both Western and Eastern Europe (except Canada, France and Japan), three of the permanent members of the Security Council, and all (except the three mentioned) of the leading maritime powers, including those with the longest coastline, the USSR (as it then was) and the United States of America. ... This boded ill for a major international convention intended to be universal in time and in space.[200]

The Convention on the Law of the Sea was opened for signature on 10 December 1982.[201] From the time it was opened for signature in 1982 to its eventual entry into force in 1994, none of the States that abstained or voted against it had reversed their position. Further, no industrialised States, even those that voted in favour of the Convention on the Law of the Sea, proceeded to sign and ratify it, with the exception

---

L. REV. 679; Philip Allott, *Power Sharing in the Law of the Sea* (1983) 77 AM. J. INT'L. L. 1; Dennis W. Arrow, *The "Alternative" Seabed Mining Regime: 1981* (1982) 5 FORDHAM INT'L. L. J. 1; and Robert L. Brooke, *The Current Status of Deep Seabed Mining* (1984) 24 VA. J. INT'L. L. 361.

[197] U.N. Doc. A/CONF.62/30/Rev. 3, Rule 37.

[198] UNCLOS III, *Official Records*, vol. XVII (U.N. Doc. A/CONF.62/122); 1833 U.N.T.S. 3. See also Brown, supra note 196, at 260; and Luke T. Lee, *The Law of the Sea Convention and Third States* (1983) 77 AM. J. INT'L. L. 541.

[199] UNCLOS III, *Official Records*, vol. XVI, at 155.

[200] Shabtai Rosenne, *The United Nations Convention on the Law of the Sea 1982: The Application of Part XI: An Element of Background*, in Shabtai Rosenne, ESSAYS ON INTERNATIONAL LAW AND PRACTICE (2007) at 457–458. See also Shigeru Oda, *Sharing of Ocean Resources – Unresolved Issues in the Law of the Sea* (1981) 3 N. Y. J. INT'L. & COMP. L. 1; and John King Gamble, Jr. and Maria Frankowska, *The 1982 Convention and Customary Law of the Sea: Observations, a Framework and a Warning* (1984) 21 SAN DIEGO L. REV. 491.

[201] See Bernardo Zuleta, *The Law of the Sea After Montego Bay* (1983) 20 SAN DIEGO L. REV. 475 at 475; Ted L. McDorman, *The 1982 Law of the Sea Convention: The First Year* (1984) 15 J. MARIT. L. & COM. 211; James L. Malone, *The United States and the Law of the Sea After UNCLOS III* (1983) 46 L. & CONTEMP. PROBS. 29; and Leslie M. Malone, *Customary International Law and the United Nations' Law of the Sea Treaty* (1983) 13 CAL. W. INT'L. L. J. 181.

of Cyprus, Malta and Iceland, by the time the 60th instrument of ratification was deposited on 16 November 1993.[202] It is noteworthy that China, as one of the largest consumers of mineral resources in the world and though it usually professes itself to be a developing State, had not ratified the Convention on the Law of the Sea at the time of its entry into force.[203] This was the case even though it had been warmly embraced by a significant number of African and Latin American States.[204] In any event, the deposit of the 60th instrument of ratification precipitated the Convention on the Law of the Sea entering into force 12 months later, on 16 November 1994.[205]

5.6 Part XI of the Convention on the Law of the Sea and the Effects of the Common Heritage of Mankind

5.6.1 Overview and the Features of Part XI

The concept of the common heritage of mankind concept was first enunciated in relation to outer space. However, within months of the concept being discussed in COPUOS as being applicable to the outer space, the Moon and celestial bodies in

---

[202] The 60 States that had ratified the Convention on the Law of the Sea by 16 November 1993 were Angola, Antigua and Barbuda, Bahamas, Bahrain, Barbados, Belize, Botswana, Brazil, Cameroon, Cape Verde, Costa Rica, Côte d'Ivoire, Cuba, Cyprus, Djibouti, Dominica, Egypt, Fiji, Gambia, Ghana, Grenada, Guinea, Guinea-Bissau, Guyana, Honduras, Iceland, Indonesia, Iraq, Jamaica, Kenya, Kuwait, Mali, Malta, Marshall Islands, Mexico, Micronesia, Namibia, Nigeria, Oman, Paraguay, the Philippines, Saint Kitts and Nevis, Saint Lucia, Saint Vincent and the Grenadines, São Tomé and Principe, Senegal, Seychelles, Somalia, Sudan, Tanzania, Togo, Trinidad and Tobago, Tunisia, Uganda, Uruguay, Yemen, Yugoslavia, Zaïre (now the Democratic Republic of the Congo), Zambia and Zimbabwe: United Nations, *Circular Letter C.N. 418.1993.TREATIES-7 (Depositary Notification)* (14 January 1994). See also Ted L. McDorman, *Will Canada Ratify the Law of the Sea Convention?* (1988) 25 SAN DIEGO L. REV. 535; Hilton Staniland, *A Sea-Change: The United Nations Convention on the Law of the Sea* (1983) 100 S. AFR. L. J. 700; J. Kevin McCall, *A New Combination to Davy Jones' Locker: Melee over Marine Minerals* (1978) 9 LOYOLA U. CHI. L. J. 935; and Rob Huebert, *Canada and the Law of the Sea Convention* (1997) 52 INT'L. J. 69.

[203] See Zou Keyuan, *China's Efforts in Deep Seabed Mining: Law and Practice* (2003) 18 INT'L. J. MARINE & COASTAL L. 481; Paul C. Yuan, *The United Nations Convention on the Law of the Sea from a Chinese Perspective* (1984) 19 TX. INT'L. L. J. 415; and Victor H. Li, *Sovereignty at Sea: China and the Law of the Sea Conference* (1979) 15 STANFORD J. INT'L. STUD. 225.

[204] See Francisco Orrego-Vicuña, *The Regime for the Exploitation of the Seabed Mineral Resources and the Quest for a New International Economic Order of the Oceans: A Latin-American View* (1978) 10 LAWYER AM. 774; Jane Gilliland Dalton, *The Chilean Mar Presencial: A Harmless Concept or a Dangerous Precedent?* (1993) 8 INT'L. J. MARINE & COASTAL L. 397; Walter van Overbeek, *Article 121(3) LOSC in Mexican State Practice in the Pacific* (1989) 4 INT'L. J. ESTUARINE & COASTAL L. 252; Eduardo Ferrero Costa, *Pacific Resources and Ocean Law: A Latin American Perspective* (1989) 16 ECOLOGY L. Q. 245; and Kaldone G. Nweihed, *Venezuela's Contribution to the Contemporary Law of the Sea* (1974) 11 SAN DIEGO L. REV. 603.

[205] Convention on the Law of the Sea, Article 308. See Jonathan I. Charney, *Entry into Force of the 1982 Convention on the Law of the Sea* (1995) 35 VA. J. INT'L. L. 381.

1967, Malta advocated the acceptance on the part of the international community that the high seas and the deep seabed and its mineral resources were also the common heritage of mankind.[206] The concept of the common heritage of mankind was ultimately incorporated into the Declaration of Principles by the United Nations General Assembly in 1970,[207] and was formalised into the Convention on the Law of the Sea.[208] To date, the Convention on the Law of the Sea represents the most elaborated application of the concept of the common heritage of mankind in any instrument of international law.[209]

Under Part XI of the Convention on the Law of the Sea, the deep seabed and its mineral resources are deemed to be the common heritage of mankind and can only be exploited in accordance with the terms of Part XI of the Convention on the Law of the Sea.[210] The ideological conflict between the industrialised States and developing States, as well as the politics of the Cold War, meant that a balancing act was tried and failed between the call for the implementation of the NIEO and the need to maintain sufficient economic incentives for developed States who held the necessary capital and technology to develop seabed resources.[211] Further, parallel negotiations at UNCLOS III and in COPUOS created a situation in which negotiations in one field often influenced the position of States on corresponding issues in the other,

---

[206] See discussion in, for example, Anastasia Strati, *Deep Seabed Cultural Property and the Common Heritage of Mankind* (1991) 40 INT'L. & COMP. L. Q. 859; Jan van Ettinger, Alexander King and Peter Payoyo, *The Common Heritage of Mankind and Four Problem Areas*, the United Nations University, <http://www.unu.edu/unupress/unupbooks/uu15oe/uu15oe0q.htm>, last accessed on 12 February 2001; Baslar, supra note 1, at 80–82; Nandasiri Jasentuliyana, INTERNATIONAL SPACE LAW AND THE UNITED NATIONS (1999), at 139–140; Nandasiri Jasentuliyana, *The Role of Developing Countries in the Formulation of Space Law* (1995) 20:2 ANN. AIR & SP. L. 95, at 105–106; and Gennady M. Danilenko, *The Concept of the "Common Heritage of Mankind" in International Law* (1988) 13 ANN. AIR & SP. L. 247, at 249–250.

[207] General Assembly Resolution 2749 (XXV). See discussion in René-Jean Dupuy, *The Notion of the Common Heritage of Mankind Applied to the Seabed* (1983) 18 ANN. AIR & SP. L. 347; Paul Lawrence Saffo, *The Common Heritage of Mankind: Has the General Assembly Created a Law to Govern Seabed Mining?* (1979) 53 TUL. L. REV. 492; L. Frederick E. Goldie, *A Note on Some Diverse Meanings of "The Common Heritage of Mankind"* (1983) 10 SYRACUSE J. INT'L. L. & COM. 69; Martin A. Harry, *The Deep Seabed: The Common Heritage of Mankind or Arena for Unilateral Exploitation?* (1992) 40 NAVAL L. REV. 207; Steven Kotz, *"The Common Heritage of Mankind": Resource Management of the International Seabed* (1978) 6 ECOLOGY L. Q. 65; Alexandre Kiss, *Conserving the Common Heritage of Mankind* (1990) 59 REV. JUR. U. P. R. 773; Douglas M. Johnston, *The New Equity in the Law of the Sea* (1976) 31 INT'L. J. 79.

[208] Convention on the Law of the Sea, Article 136. See discussion in Jasentuliyana (1999), supra note 206, at 139–140; and Jasentuliyana (1995), supra note 206, at 105–106.

[209] See Stephen Vasciannie, *Part XI of the Law of the Sea Convention and Third States: Some General Observations* (1989) 48 CAM. L. J. 85; Shabtai Rosenne, *The United Nations Convention on the Law of the Sea, 1982: The Application of Part XI: An Element of Background* (1995) 29 ISR. L. REV. 491; and Bernard H. Oxman, *The High Seas and the International Seabed Area* (1989) 10 MICH. J. INT'L. L. 526.

[210] Convention on the Law of the Sea, Articles 136 and 137.

[211] Cook, supra note 1, at 679–682.

thus creating a high degree of inertia for any compromise to be offered by any State in either set of negotiations.[212]

The main features of Part XI of the Convention on the Law of the Sea are:

(1) the deep seabed and its resources are declared the common heritage of mankind and no State or any entity can acquire rights to the mineral resources except pursuant to the Convention on the Law of the Sea;[213]
(2) the International Seabed Authority (the "*ISA*") is empowered to licence mining operations and to undertake them for itself through the "Enterprise" and licence applicants are required to identify two areas of estimated equal value, one for the applicant and the other for the Enterprise or to developing States;[214]
(3) determination of applications is to be done by a 36-member Council of the ISA, elected through a complex mechanism;[215]
(4) the Enterprise is to be initially funded by a complex financing system;[216]
(5) licensees are required to transfer mining technologies to the Enterprise on a compulsory basis;[217]
(6) production is controlled under an elaborate system to protect land-based miners so that seabed miners;[218] and
(7) the equitable sharing of financial and other economic benefits derived from the exploitation of mineral resources in the deep seabed, as well as payments into the compensation fund.[219]

The industrialised States, such as the United States, were particularly antagonistic to the final formulation to Part XI of the Convention on the Law of the Sea as it is now formulated. Many industrialised States argued that Part XI excessively favoured the special interests of developing States and land-based mineral exporters and, further, it imposed undue regulations and taxes, which were considered to be inconsistent with the *laissez-faire* philosophy of international trade.[220]

---

[212] Danilenko, supra note 206, at 252–254.

[213] Convention on the Law of the Sea, Articles 136 and 137.

[214] Ibid., Annex III.

[215] Ibid., Article 161.

[216] Ibid., Article 171.

[217] Ibid., Article 144.

[218] Ibid., Article 151.

[219] Ibid., Articles 140, 151 and 160. See Eugene P. Miller and Joseph H. Delehant, *Deep Seabed Mining: Government Guaranteed Financing under the Maritime Aids of the Merchant Marine Act 1936 as an Alternative to Treaty-Related Loss Compensation* (1980) 11 J. MARIT. L. & COM. 453 and Jonathan I. Charney, *The Equitable Sharing of Revenues from Seabed Mining* (1975) 8 STUD. TRANSNAT'L. LEG. POL'Y. 53.

[220] For further discussion of specific criticisms levelled at Part XI of the Convention on the Law of the Sea by developed States, see Lilliana Torreh-Bayouth, *UNCLOS III: The Remaining Obstacles to Consensus on the Deepsea Mining Regime* (1981) 16 TX. INT'L. L. J. 79 and Marlene Dubow, *The Third United Nations Conference on the Law of the Sea: Questions of Equity for American Business* (1982) 4 NW. J. INT'L. L. & BUS. 172.

The unilateral approaches of the industrialised States in rejecting the Convention on the Law of the Sea and opting instead to enact domestic legislation for the authorisation of private mining activities in the deep seabed meant that the Convention on the Law of the Sea never achieved the universality that it had set out to attain.

It was apparent that there were at least six areas of concern for the industrialised States in the manifestation of the common heritage of mankind concept for the deep seabed that prevented their acceptance of the Convention on the Law of the Sea. These areas of concern are:

(1) costs of the proposed International Seabed Authority ("ISA") on its Member States as well as its decision-making process;
(2) the nature and operations of the "Enterprise";
(3) compulsory transfer of technology;
(4) limitations on production;
(5) the financial terms of contracts; and
(6) the compensation fund.

It is noteworthy, however, that it is not only the industrialised States that have expressed concern or objection to the provisions of the Convention on the Law of the Sea in relation to the deep seabed. Significant concern has also been expressed by the developing States and those in favour of international regulation and other elements of the common heritage of mankind doctrine. Pardo, for instance, wrote that:

> ... the Seabed Authority to be established under the Convention, while possessing very detailed rulemaking authority, is substantially so weak as to be unlikely to be viable and structurally so complex as to be unworkable. ... The composition and decision-making procedures in the key organ of the future Authority – the Council – are so complex as to make timely and effective decisions on important matters very difficult. Unrealistic production limitations, heavy bureaucratic controls, substantial fees, production charges and other payments payable by those who have obtained production authorisations further weaken the Authority.[221]

5.6.2 Decision-Making in the International Seabed Authority

While the Assembly of the ISA has competence in deciding the administrative issues, policies and operations of the ISA, it is the Council that has direct supervisory control over the development of deep seabed resources.[222] It is the Council and not the Assembly that has the power to make final decisions for the ISA on applications for mineral resource development in the deep seabed, as well as decisions concerning rules, regulations and procedures of the ISA, the election of its

---

[221] Pardo, supra note 167, at 103.
[222] Convention on the Law of the Sea, Article 162.

248   5 Exploitation Rights: Evolving from the "Province of Mankind" to the "Common ...

Secretary-General, the production limits to be imposed as well as the annual budget of the ISA.[223]

While all Member States of the Convention on the Law of the Sea are members of the Assembly of the ISA, only 36 Member States would be represented on the Council.[224] These States are supposed to represent the largest mineral consumer States, the deep seabed mining States, the major mineral exporting States, special interest States and the various geographical regional groupings used in the practice of the United Nations.[225] The rules regarding the composition of the Council have been criticised as being too complex and, in any event, does not change the fact that developing States would have a majority on the Council.[226] Specifically, however, the Convention on the Law of the Sea requires the States of the socialist bloc, as they then were, be represented in both groups of the consumer and producer States and that the "largest consumer" is to be represented in the consumer group of States, thus guaranteeing a seat on the Council for the United States if it ratified the Convention on the Law of the Sea.[227] In fact, the United States and other developed States would qualify as the largest consumers, producers and miners of deep seabed resources and, as such, significant representation of their interests in the Council of the ISA is almost guaranteed, particularly if the United States became a Member State.[228] Further, as the adoption of rules, regulations and procedures relating to the equitable sharing of the benefits derived from mineral exploitation activities in the deep seabed are to be adopted by consensus, the United States and other industrialised States are further assured by the terms of the Convention on the Law of the Sea that such decisions cannot be made in a manner that does not have their agreement and acceptance.

5.6.3 The Enterprise

The Convention on the Law of the Sea establishes the ISA's own international business venture called the "Enterprise". Administered by a Governing Board and a Director-General, the Enterprise is vested with a wide range of powers to engage

---

[223] Ibid. See also Elizabeth Riddell-Dixon, *The Preparatory Commission on the International Seabed Authority: "New Realism"?* (1992) 7 INT'L. J. ESTUARINE & COASTAL L. 195.

[224] Ibid., Article 161.

[225] Ibid.

[226] See Kathryn E. Yost, *The International Seabed Authority Decision-Making Process: Does It Give a Proportionate Voice to the Participant's Interests in Deep Sea Mining?* (1983) 20 SAN DIEGO L. REV. 659 and Andronico O. Adede, *Law of the Sea – Developing Countries' Contribution to the Development of the Institutional Arrangements for the International Seabed Authority* (1978) 4 BROOKLYN J. INT'L. L. 1.

[227] Convention on the Law of the Sea, Article 161.

[228] See Jonathan I. Charney, *The Law of the Deep Seabed Post UNCLOS III* (1984) 63 OR. L. REV. 19.

in mineral resource development of the deep seabed in direct competition with any commercial venture that has been granted a licence by the ISA.[229] The start-up costs of the Enterprise were to be met by loans, grants and other subsidies, including limited direct contributions from the Member States and some access to the funds of the ISA. The profits derived from the operations of the Enterprise are to be used to contribute to the budget of the ISA as well as for sharing with the international community, with particular attention to the developing States.[230]

The idea that the international regulator was to go into business in competition with the entities that it was to regulate has caused concern to the industrialised States, fearing that this conflict of interest would led to the Authority favouring the needs of the Enterprise over that of any commercial venture.[231] Further, the ISA may be tempted to adopt rules and regulations that favour the operations of the Enterprise at the expense of commercial ventures.[232] Not surprisingly, these concerns have made the Enterprise one of the major stumbling blocks to the acceptance of the Convention on the Law of the Sea by the industrialised States.

5.6.4 Compulsory Transfer of Technology

One of the principal tenets of the NIEO is the transfer of relevant technology from the industrialised States to the developing States. The Convention on the Law of the Sea provides for transfer of technology that would enable the Enterprise to exploit mineral resources on the deep seabed and to facilitate the redressing of the economic imbalance between North and South. This takes place at two levels.

First, the Convention on the Law of the Sea requires the transfer of technology to the ISA, not merely as a *quid pro quo* for applying for a licence for the exploitation of mineral resources in the deep seabed.[233] In other words, an industrialised State "with no aspirations to mine the seabed might still incur an obligation to transfer technology to the Enterprise and to other parties".[234] Second, every commercial venture for deep seabed mining is required to make available to the Enterprise

---

[229] Convention on the Law of the Sea, Article 170.

[230] Convention on the Law of the Sea, Article 173.

[231] See Doug Bandow, *UNCLOS III: A Flawed Treaty* (1982) 19 SAN DIEGO L. REV. 475 at 484.

[232] Ibid.

[233] Convention on the Law of the Sea, Article 144. See Douglas Yarn, *The Transfer of Technology and UNCLOS III* (1984) 14 GA. J. INT'L. & COMP. L. 121 and George D. Haimbaugh, Jr., *Technological Disparity and the United Nations Seabed Debates* (1973) 6 IND. L. REV. 690.

[234] See John King Gamble, Jr., *Assessing the Reality of the Deep Seabed Regime* (1985) 22 SAN DIEGO L. REV. 779 at 784–785; David Silverstein, *Proprietary Protection for Deepsea Mining Technology in Return for Technology Transfer: New Approach to the Seabeds Controversy* (1978) 60 J. PAT. OFF. SOC'Y. 135; and Colin M. Alberts, *Technology Transfer and Its Role in International Environmental Law: A Structural Dilemma* (1992) 6 HARV. J. L. & TECH. 63.

the technology that it used to carry out its activities under its contract with the ISA.[235] The Convention on the Law of the Sea makes some ineffective attempt at compensating the proprietor of the technology, as Gamble noted:

> Far more details are contained in Annex III; many of these attempt to provide fair payment for the technology transferred. Yet it remains evident that such provisions would cause industry representatives to cringe. They would respond that such technology is invaluable. If in the process they exaggerate the cost, then the Enterprise could evoke compulsory settlement of the dispute in accordance with Part XI.[236]

5.6.5 Limitations on Production

Article 151 of the Convention on the Law of the Sea prescribes a mechanism for determining the limits on the production of mineral resources from the deep seabed during the "interim period", which begins 5 years before the year in which the earliest commercial production is planned to begin and lasts for 25 years.[237] Further, during the interim period, the levels of production for other metals, including copper, cobalt and manganese cannot exceed what had been authorised as the maximum allowed production level for nickel.[238] The Convention on the Law of the Sea expressly states that rights and obligations relating to anti-competitive and unfair economic practices would apply to exploration and exploitation activities in the deep seabed, but only to the extent that they are applicable under relevant multilateral trade agreements, such as the GATT.[239]

It is apparent that such practices and measures are, in theory, incompatible with the concepts of an international free market and its economic practices. The industrialised States, in particular, have suggested that such measures, designed to protect land-based mining activities in developing States, would impair the appeal of mining the deep seabed to a commercial venture.[240] In practice, however, such measures have existed for some time in domestic contexts, including in the regulation of the production of fossil fuels in the United States.[241]

---

[235] Convention on the Law of the Sea, Annex III, Article 5–3.

[236] Gamble, supra note 234, at 785.

[237] The "production ceiling" for any year during the interim period is the sum of the difference between the trend line values for nickel consumption for the first five years of the interim period and 60% of the difference between the trend line values for nickel consumption of the year immediately before the earliest commercial production and the year the production authorisation is applied for: Convention on the Law of the Sea, Article 151(4).

[238] Ibid., Article 151(7).

[239] Ibid., Article 151(8).

[240] See Bandow, supra note 231.

[241] *Burford v Sun Oil Co.* (1943) 319 U.S. 315.

5.6.6 Financial Terms of the Contracts

There was no historical precedent to the financial costs of deep seabed mining at the time the Convention on the Law of the Sea was adopted. Consequently, the financial arrangements for the regulation of such commercial ventures was based on an economic and financial model for the future seabed mining industry formulated by the Massachusetts Institute of Technology.[242] This model produced a serious of different rates of return based on various assumptions, with an average internal rate of return on investment ("ROI") of around 15% based on this model.[243]

Commercial mining ventures are subject to four separate charges payable to the ISA: an application processing fee, a fixed annual fee, an annual production charge and a share of the net proceeds from the mineral exploitation.[244] The share of the net proceeds payable to the ISA was reducible at the option of the venture in exchange for an increase in the annual production charge, which was seen to be a concession to the socialist States, as they then were.[245] The processing fee and some of these payments are comparable in magnitude and kind to the "premium payments" made to the developing States in return for the mining concessions.[246] Further, the share of the proceeds that is payable by the commercial mining venture to the ISA ranges from 35% of the net proceeds for an ROI that is less than 10% to 50% of the net proceeds for an ROI of greater than 20%.[247] These rates of payments increase to 40% and 70% of the net proceeds, respectively, once the commercial venture has recovered the full costs of its investment.[248] While the subject of much objection from the industrialised States, as Katz noted:

> These shares of net proceeds are not dissimilar to the income flow to developing countries from mining agreements, which usually include taxes additional to those mentioned above. In Indonesia, for example, the tax rate on mining companies is 35% of taxable income for the first 10 years of operation and 45% thereafter. In addition to this corporation tax, the mining company must bear property tax, dead rent and stamp duties.[249]

---

[242] See U.S. Congressional Research Services, OCEAN MANGANESE NODULES (1975), at 216–217. See also Leigh S. Ratiner and Rebecca L. Wright, *The Billion Dollar Decision: Is Deepsea Mining a Prudent Investment?* (1978) 10 LAWYER AM. 713.

[243] Ibid. See also Ronald S. Katz, *A Method for Evaluating the Deep Seabed Mining Provisions of the Law of the Sea Treaty* (1981) 7 YALE J. WORLD PUB. ORD. 114 at 122 and Bradley Larschan, *The International Legal Status of the Contractual Rights of Contractors under the Deep Seabed Mining Provisions (Part XI) of the Third United Nations Convention on the Law of the Sea* (1986) 14 DENVER J. INT'L. L. & POL'Y. 207.

[244] Convention on the Law of the Sea, Annex III, Article 13.

[245] Ibid. See also Ronald S. Katz, *Financial Arrangements for Seabed Mining Companies: An NIEO Case Study* (1979) 13 J. WORLD TRADE L. 209 at 218.

[246] David N. Smith and Louis T. Wells, Jr., NEGOTIATING THIRD WORLD MINERAL AGREEMENTS (1975), at 234. See also Wolfgang Hauser, *An International Fiscal Regime for Deep Seabed Mining: Comparisons to Land-Based Mining* (1978) 19 HARV. INT'L. L. J. 759.

[247] Convention on the Law of the Sea, Annex III, Article 13(6).

[248] Ibid.

[249] Katz, supra note 243, at 124.

5.6.7 The Ultimate Compromise

5.6.7.1 Adoption of the 1994 Agreement

The political and economic environment changed after 1990, with the collapse of the Soviet Union and the end of the Cold War, enthusiasm for the NIEO philosophy waned and economic projections for deep seabed mining subsided.[250] Further, as Baslar noted, "the decline of socialism and centrally-planned economies [was increasingly] replaced by free market economies and liberalism", making a free-market approach to deep seabed mining more acceptable on a worldwide basis.[251] In response to these developments, Javier Pérez de Cúellar, then Secretary-General of the United Nations, began informal consultations in 1990 with a view of achieving a compromise over the Convention on the Law of the Sea.[252] The imminent entry into force of the Convention on the Law of the Sea only added urgency to the resolution of the common heritage issues relating to deep seabed mining.[253]

In 1994, after extensive re-negotiations on controversial issues in Part XI of the Convention on the Law of the Sea, as mediated by Boutros Boutros-Ghali, then Secretary-General of the United Nations, the General Assembly adopted the Agreement Relating to the Implementation of Part XI of the United Nations Convention on the Law of the Sea of 10 December 1982, opened for signature on 28 July 1994. While reaffirming that the deep seabed and its mineral resources are the common heritage of mankind, the 1994 Agreement revises the rules and procedures governing the exploitation of seabed resources and rectifies some of the criticisms levelled at the initial version of Part XI of the Convention on the Law of the Sea.[254] While there was the potential that a significant number of States that have ratified the Convention on the Law of the Sea would refuse to ratify the 1994 Agreement, it is more likely that all of the Member States would agree to implement Part XI of the Convention on the Law of the Sea in accordance with the 1994 Agreement instead of the original terms of Part XI, regardless of the status of ratifications to the 1994 Agreement.

---

[250] Baslar, supra note 1, at 216–217, 336; Cook, supra note 1, at 682–685; and Elisabeth Mann Borgese, *The Role of the International Seabed Authority in the 1980s* (1981) 18 SAN DIEGO L. REV. 395.

[251] Baslar, supra note 1, at 216–217.

[252] Brian M. Hoffstadt, *Moving the Heavens: Lunar Mining and the "Common Heritage of Mankind" in the Moon Treaty* (1994) 42 U.C.L.A. L. REV. 575, at 598–600.

[253] Ibid.

[254] 1994 Agreement, Preamble. See Renate Platzöder, *Substantive Changes in a Multilateral Treaty Before Its Entry into Force: The Case of the 1982 United Nations Convention on the Law of the Sea* (1993) 4 EUR. J. INT'L. L. 390; Bernard H. Oxman, *Law of the Sea Forum: The 1994 Agreement and the Convention* (1994) 88 AM. J. INT'L. L. 687; A. Rohan Perera, *A Dawn of a New Era of the Oceans or a Return to the Grotian Ocean? Some Reflections as the Law of the Sea Convention Enters into Force* (1994) 6 SRI LANKA J. INT'L. L. 157; Louis B. Sohn, *International Law Implications of the 1994 Agreement* (1994) 88 AM. INT'L. L. 696; Jonathan I. Charney, *U.S. Provisional Application of the 1994 Deep Seabed Agreement* (1994) 88 AM. J. INT'L. L. 705; and Kenneth R. Simmonds, *The Law of the Sea* (1990) 24 INT'L. LAWYER 931.

5.6.7.2 Addressing the Objections of the United States and Other Industrialised States

The U.S. Government, under the administration of President Reagan, had raised six objectives that formed the policy basis upon which the Convention on the Law of the Sea must be reformed in order for it to be acceptable to the United States and other industrialised States:

(1) not deter development of deep seabed mineral resources to meet national and international demand;

(2) guarantee domestic access to these resources to enhance U.S. security of supply, avoid monopolisation of the resources by the Enterprise and to promote the economic development of the deep seabed;

(3) provide a decision-making role for the Council that reflects and protects the political and economic interests as well as financial contributions of the participating States;

(4) not allow amendments to the Convention on the Law of the Sea to come into force without consensus;

(5) not to set an undesirable precedent for other international regimes; and

(6) not contain provisions for the compulsory transfer of technology for funding for national liberation movements in order to secure ratification by the Senate of the United States.[255]

In an effort to address these objectives raised by the United States, the 1994 Agreement contained provisions that eventually received acceptance by the United States and other industrialised States:

(1) decision-making power in the ISA are concentrated in the Council so that the Assembly was to ratify or reject the Council's recommendations;[256]

(2) the State that had the largest economy in terms of gross domestic product in the world on the date of entry into force of the Convention on the Law of the Sea was guaranteed a seat on the Council;[257]

---

[255] The White House, *Statement by the President: U.S. Policy and the Law of the Sea*, 29 January 1982, DEPT. STATE BULL., at 54. See Roger Tansey Hertz, *The Reagan Administration and the Law of the Sea: Objections to the 1980 Draft Convention* (1982) 3 B. C. THIRD WORLD L. J. 70; Elliot L. Richardson, *Superpowers Need Law: A Response to the United States Rejection of the Law of the Sea Treaty* (1983) 17 GEO. WASH. J. INT'L. L. & EC. 1; Susan M. Banks, *Protection of Investment in Deep Seabed Mining: Does the United States Have a Viable Alternative to Participation in UNCLOS?* (1984) 2 B. U. INT'L. L. J. 267; George V. Galdorisi and James G. Stavridis, *United Nations Convention on the Law of the Sea: Time of a U.S. Reevaluation?* (1992) 40 NAVAL L. REV. 229; and D. Brian Hufford, *Ideological Rigidity vs. Political Reality: A Critique of Reagan's Policy on the Law of the Sea* (1984) 2 YALE L. & POL'Y. REV. 127.

[256] 1994 Agreement, Annex, Section 3, Paragraphs 1 and 4.

[257] Ibid., Section 3, Paragraph 15(a). This State, of course, is the United States.

(3) Member States on the Council are divided into "chambers" of States with particular interests, with two of the four-member chambers likely to be controlled by major industrialised States, and decision-making by the Council was either by consensus or by two-thirds majority, provided that such majority decisions are not opposed by a majority in any one of the chambers, thus giving the industrialised States an effective veto;[258]

(4) financial and budgetary matters must be decided based on recommendations of the Finance Committee, which operated by consensus and on which the United States and other States that are major contributors of the budget are guaranteed membership;[259]

(5) the production ceiling, production limitations, commodity agreements and production authorisation and selection provisions in the Convention on the Law of the Sea would no longer apply;[260]

(6) the provisions dealing with compulsory technology transfer would no longer apply and, instead, the industrialised States are to cooperate with the Enterprise and developing States that are unable to obtain the relevant technology on the open market or through joint ventures;[261]

(7) access to different areas of the deep seabed by applicants is to be on a first-come, first-served basis;[262]

(8) the Enterprise is subject to the same rules and regulations as commercial ventures and is to begin operations in joint ventures until it is able to function independently and commercially;[263] and

(9) the application fee is reduced and much of the financial obligations are eliminated, including the annual production charges.[264]

Most importantly in relation to the financial effects of the common heritage of mankind concept, the Convention on the Law of the Sea provided that the equitable sharing of surplus revenues was to take into particular consideration "the interests and needs of the developing States and peoples who have not attained full independence or other self-governing status".[265] The United States had objected to this formulation as it would, *inter alia*, allow for funding for national liberation groups such as the Palestine Liberation Organisation and the South West Africa People's Organisation.[266] In practice, however, political developments in the Middle East and

---

[258] Ibid., Section 3, Paragraph 5.

[259] Ibid., Section 3, Paragraphs 5, 10 and 15.

[260] Ibid., Section 6, Paragraph 7.

[261] Ibid., Section 5, Paragraphs 1 and 2.

[262] Ibid., Section 6, Paragraph 7.

[263] Ibid., Section 2, Paragraph 4.

[264] Ibid., Section 8, Paragraphs 2 and 3.

[265] Convention on the Law of the Sea, Articles 160 and 172.

[266] U.S. Congress, *Statement by Ambassador James L. Malone, Special Representative of the President, Before the House Foreign Affairs Committee*, 12 August 1982, at 48–50.

Namibia has made the issue moot and, in any event, any such distribution would happen only if the proceeds exceeded the administrative expenses of the ISA, its assistance to adversely affected land-based mineral producers and such distributions are agreed to by consensus of the Finance Committee and the Council.[267]

5.6.7.3 Application and Implementation of the 1994 Agreement

The compromise reached over the 1994 Agreement has achieved remarkable success in securing the acceptance of the industrialised States to the Convention on the Law of the Sea. As at 21 July 2009, there were 159 Member States to the ISA and the Convention on the Law of the Sea, though the United States has not ratified it.[268] Since the 1994 Agreement, eight 15-year exploration contracts had been signed between the ISA and a number of Member States or contractors that provide them with the exclusive right to explore an initial area of 150,000 km^2, with half of each of the areas to be relinquished after the first 8 years of the contract.[269] Seven of these exploration areas are located in the Pacific Ocean south and southeast of Hawaii and the remaining one is in the Indian Ocean.[270] The exploitation of mineral resources by these ventures will be subject to the regulations recently adopted by the ISA.[271]

---

[267] Convention on the Law of the Sea, Articles 161 and 173; and 1994 Agreement, Annex, Section 3, Paragraph 7 and Section 9, Paragraphs 7 and 8.

[268] Since the 1994 Agreement, there has been a significant number of industrialised States that had become Member States to the ISA, including Australia, Germany, Singapore, Italy, Slovenia, Austria, Greece, Republic of Korea, Monaco, France, Slovakia, Japan, Czech Republic, Finland, Ireland, Norway, Sweden, the Netherlands, New Zealand, Spain, Russia, United Kingdom, Portugal, South Africa, Belgium, Poland, Luxembourg, Hungary, Canada, Denmark, Estonia, Latvia, Lithuania, Croatia and Switzerland: International Seabed Authority, *Member States*, 21 July 2009, at <http://www.isa.org.jm/en/about/members/states>, last accessed on 15 November 2009.

[269] The eight contractors were the Government of India, Institute français de recherche pour l'exploitation de la mer (IFREMER) of France, the Deep Ocean Resources Development Company (DORD) of Japan, the State Enterprise Yuzhmorgeologiya of Russia, the China Ocean Mineral Resources Research and Development Association (COMRA) of China, the Interoceanmetal Joint Organisation (IOM) of Bulgaria, Cuba, Czech Republic, Poland, Russia and Slovakia, the Government of the Republic of Korea and the Federal Institute for Geosciences and Natural Resources of Germany: International Seabed Authority, INTERNATIONAL SEABED AUTHORITY HANDBOOK 2009 (2009), at 29–34.

[270] Ibid.

[271] International Seabed Authority, *Regulations on Prospecting and Exploration for Polymetallic Nodules in the Area*, at <http://www.isa.org.jm/files/documents/EN/Regs/MiningCode.pdf>, 13 July 2000, last accessed on 16 November 2009. See Jason C. Nelson, *The Contemporary Seabed Mining Regime: A Critical Analysis of the Mining Regulations Promulgated by the International Seabed Authority* (2005) 16 COLO. J. INT'L. ENVT'L. L. & POL'Y. 27; Michael W. Lodge, *International Seabed Authority's Regulations on Prospecting and Exploration for Polymetallic Nodules in the Area* (2002) 20 J. ENERGY NAT. RES. L. 270; and Dionysia-Theodora Avgerinopoulou, *The Lawmaking Process at the International Seabed Authority as a Limitation on Effective Environmental Management* (2005) 30 COLUM. J. ENVT'L. L. 565.

It must be said that the compromise represented by the 1994 Agreement was heavily skewed in favour of the industrialised States in an effort to obtain their acceptance of the Convention on the Law of the Sea.[272] To date, however, despite these efforts, the U.S. Senate has not ratified it and it is unlikely to do so before the end of the first decade of the present century.[273] It is highly unlikely that the developing States would be willing to make concessions to a similar extent in relation to any other area that is subject to the common heritage of mankind concept, such as the celestial bodies of the Solar System and their mineral resources.[274]

5.7 Evolution of Article 11 of the Moon Agreement

5.7.1 Early Controversies

Since the launch of *Sputnik-I* in 1957, considerable success in the formulation and codification of the principles of outer space has been achieved within the framework of the United Nations. In 1961 the General Assembly recognised that international law, including the United Nations Charter, applied to outer space and that the Moon and other celestial bodies were not subject to national appropriation.[275] In 1963 the General Assembly adopted the Principles Declaration, as proposed by COPUOS.[276]

---

[272] See Philip A. Burr, *The International Seabed Authority* (2006) 29 SUFFOLK TRANSNAT'L. L. REV. 271; Tullio Scovazzi, *Mining, Protection of the Environment, Scientific Research and Bioprospecting: Some Considerations on the Role of the International Seabed Authority* (2004) 19 INT'L. J. MARINE & COASTAL L. 383; and S. P. Jagota, *Developments in the Law of the Sea Between 1970 and 1998: A Historical Perspective* (2000) 2 J. HIST. INT'L. L. 91.

[273] See John Alton Duff, *UNCLOS and the New Deep Seabed Mining Regime: The Risks of Refuting the Treaty* (1996) 19 SUFFOLK TRANSNAT'L. L. REV. 1; Candace L. Bates, *U.S. Ratification of the U.N. Convention on the Law of the Sea: Passive Acceptance Is Not Enough to Protect U.S. Property Interests* (2006) 31 N. C. J. INT'L. L. & COM. REG. 745; and John Alton Duff, *The United States and the Law of the Sea Convention: Sliding Back from Accession and Ratification* (2006) 11 OCEAN & COASTAL L. J. 1.

[274] See Pierre-François Mercure, *L'échec des Modèles de Gestion des Ressources Naturelles Selon les Charactéristiques du Concept de Patrimoine Commun de l'Humanité* (1997) 28 OTTAWA L. REV. 45; Chatura Randeniya, *Sharing the World's Resources: Equitable Distribution and the North-South Dialogue in the New Law of the Sea* (2003) 15 SRI LANKA J. INT'L. L. 149; William C. Aceves, *Critical Jurisprudence and International Legal Scholarship: A Study of Equitable Distribution* (2001) 39 COLUM. J. TRANSNAT'L. L. 299; Pierre-François Mercure, *La Proposition d'un Modèle de Gestion Intégrée des Ressources Naturelles Communes de l'Humanité* (1998) 36 CAN. Y. B. INT'L. L. 41; and Alan Beesley, *The Negotiating Strategy of UNCLOS III: Developing and Developed Countries as Partners – A Pattern for Future Multilateral International Conferences?* (1983) 46 L. & CONTEMP. PROBS. 183.

[275] General Assembly Resolution 1721 (XIV). See Jonathan C. Thomas, *Spatialis Liberum* (2006) 7 FL. COASTAL L. REV. 579.

[276] General Assembly Resolution 1962 (XVIII).

5.7 Evolution of Article 11 of the Moon Agreement

Most of the principles embodied in the Declaration have become part of the Outer Space Treaty.[277]

From the time of the adoption of the Outer Space Treaty, there has been recognition among the early commentators that celestial bodies ought to be subject to international jurisdiction and control instead of being subject to unilateral control. As Smirnoff had summarised that:[278]

> The diversity of those arguments is very interesting and we should remind the reader that many of the lawyers quoted, like Faria and Goedhuis, use the example of the Antarctic Treaty 1959 to show the identity of reasons which do not admit the claims of the sovereignty of Earth countries to celestial bodies.[279] Others like Meyer think that "der Himmelaraum und die in ihm befindlichen Himmelskorper sich al seine Sache darstellen die der Gesamtheit aller Mitglieder der menschlichen Gesellschaft gehorten".[280] Faria goes a step further and adds "and also to all rational creatures of other civilised planets".[281] Jessup and Taubenfeld and also Smirnoff advocate an international solution without any right of sovereignty of individual States but with the competence of the specialised agency of the United Nations.[282] Buckling talks about the "Interplanetarisches Kooperationsrecht",[283] and Valladao is proclaiming the creation of the "Jus Inter Gentes Planetarum".[284] Menter is for the international regime of the celestial bodies with the jurisdiction of the United Nations over those bodies.[285] Quigg seeks international control for the Moon but does not think that it could be treated like Antarctica.[286] Weinmann makes a very substantial argument against the possibility of the notion of discovery as applied to the Moon for the simple reason that everybody can see the Moon every night.[287]

When Apollo 11 landed on the Moon in 1969, there was a realisation by the international community that the general principles in Outer Space Treaty were

---

[277] See, for example, Herbert Reis, *Some Reflections on the Liability Convention for Outer Space* (1978) 6 J. SP. L. 161.

[278] Michael Smirnoff, *The Legal Status of Celestial Bodies* (1962) 28 J. AIR L. & COM. 385 at 390–391.

[279] J. Escobar Faria, *Draft to an International Covenant for Outer Space* (1960) 3 PROC. COLL. L. OUTER SP. 122 and Daniel Goedhuis, *Air Sovereignty and the Legal Status of Outer Space*, report presented at the Conference of the International Law Association, August 1960, in Hamburg, Germany.

[280] Alex Meyer, *Volkerrechtliche Probleme des Weltraumgebiets* (1960) INT'L. R. & S. AB. 326.

[281] Faria, supra note 279, at 122.

[282] Philip C. Jessup and Howard J. Taubenfeld, *Controls for Outer Space*, in U.S. Senate (ed.), LEGAL PROBLEMS OF SPACE EXPLORATION (1961), at 553–570 and Michael Smirnoff, *The Role of the IAF in the Elaboration of the Norms of Future Space Law* (1959) 2 PROC. COLL. L. OUTER SP. 147 at 151.

[283] Adrian Bückling, *Interplanetarisches Kooperationsrecht* (1960) 55 DIE FRIEDENSWARTE 305.

[284] Haraldo Valladao, *The Law of Interplanetary Space* (1959) 2 PROC. COLL. L. OUTER SP. 156.

[285] Martin Menter, *Astronautical Law* (student thesis, Industrial College of the Armed Forces, 1959).

[286] Philip W. Quigg, *Open Skies and Open Space* (1958) 37 FOREIGN AFF. 95.

[287] Eric Weinmann, *The Law of Space* (1958) 35 FOREIGN SERVICE J. 2 at 22–25.

258   5 Exploitation Rights: Evolving from the "Province of Mankind" to the "Common ...

insufficient to regulate future exploitative activities on the Moon.[288] Consequently it was generally accepted that a new treaty was needed. Argentina and the Soviet Union proposed draft treaties for the Legal Sub-Committee of COPUOS in 1970 and 1971 respectively.[289] Although agreement was reached on some provisions by 1972, there remained many issues that were not resolved until later in the decade.

The Soviet Union, along with Bulgaria, Egypt, France, Japan and Poland, were of the view that the Moon Agreement should deal with the Moon only since it held a special place in the catalogue of objects in the solar system.[290] The United States, along with Australia, Belgium, Canada, Iran, Romania and United Kingdom supported the view that the agreement should apply to the Moon and all celestial bodies.[291] In the end, agreement was reached that the Moon Agreement would apply to the Moon and other celestial bodies until other treaties established regulations that were more specific in nature.[292] This agreement became embodied in Article 1 of the Moon Agreement. This article provides that the provisions of the Agreement:

> shall also apply to other celestial bodies within the solar system, other than the Earth, except insofar as specific legal norms enter into force with respect to any of these celestial bodies.

There was disagreement also within the COPUOS regarding the scope of the Moon Agreement or, specifically, if the Moon Agreement ought to apply only to the Moon and circumlunar space, to the celestial bodies in the Solar System but also to the remainder of the Milky Way galaxy or even to distant galaxies. Despite the strong arguments of the United States otherwise, it was agreed that the Agreement would be limited in scope in relation to the Solar System.[293] As to circumlunar space, the problem was not with the concept but with the precise definition of what circumlunar space encompassed. In the end, a formula was adopted, with Article I referring to orbits around and other trajectories to or around the Moon.[294]

After Malta suggested that the common heritage of mankind doctrine ought to apply to the use and exploitation of the deep seabed and its mineral resources in 1967, Ambassador Aldo Armando Cocca of Argentina proposed in the same year to COPUOS that the same doctrine ought also to apply to celestial bodies in outer

---

[288] For example, this view was espoused in E. R. C. van Bogaert, ASPECTS OF SPACE LAW (1986) at 76.

[289] See U.N.Doc. A/AC.105/C2/L.71 and U.N.Doc. A/8391.

[290] See Mark Robson, *Soviet Legal Approach to Space Law Issues at the United Nations* (1980) 3 LOYOLA L.A. INT'L. & COMP. L. ANN. 99.

[291] See Manfred A. Dauses, *Zur Rechtslage des Mondes und anderen Himmelkörper* (1975) 24 ZEIT. LUFT. WELT. 223; and Nicholas M. Matte, *Legal Principles Relating to the Moon*, in Nandasiri Jasentuliyana and Roy S. Lee (eds.), MANUAL ON SPACE LAW (1979) at 254–255.

[292] See U.N.Doc. A/AC.105/S.R.187–S.R.188 and J. Alan Beesley, *Canadian Practice in International Law During 1972 as Reflected Mainly in Public Correspondence and Statements of the Department of External Affairs* [1973] CAN. Y. B. INT'L. L. 294–295.

[293] See U.N.Doc. A/AC.105/196.

[294] This was partly the result of the failure to make a spatial delimitation of circumlunar space: see Matte, supra note 291, at 258.

space, which he considered to be *res communis humanitatus* in its draft Agreement on the Principles Governing Activities on the Use of Natural Resources of the Moon and Other Celestial Bodies.[295] Soon after, the General Assembly of the United Nations requested that the COPUOS prepare a draft treaty on lunar activities based on the existing principles of the Outer Space Treaty.[296]

After a number of years of negotiations, a joint working paper was submitted to COPUOS jointly by Argentina, Brazil, Chile, Indonesia, Mexico, Nigeria, Romania, Sierra Leone and Venezuela, with the support of Egypt, India, Italy, the Soviet Union and the United States, that included a provision that celestial bodies ought to be subject to the common heritage of mankind doctrine.[297] Further difficult negotiations between States followed, which were conducted in parallel with those in UNCLOS III, on the content and effect of the common heritage of mankind doctrine, whether a present moratorium was desirable and the interests of the developing States.[298]

In 1979, a difficult compromise was achieved in COPUOS when the developed States put forward the proposal from Brazil that was eventually contained in Article 11(1) of the Moon Agreement. Minola commented on the compromise reached as:

> In effect the developed States have agreed that the common heritage principle means that an international regime should control resource exploitation. In exchange for this concession, the developing countries agreed not to insist on a provision imposing a moratorium on exploitation pending the establishment of the international regime. Thus the Moon Treaty expresses no moratorium, and none is implied by its legislative history.[299]

The failure of the international community to reach a true agreement over the international legal regime to be created to regulate the use and mineral exploitation of celestial bodies under the Moon Agreement has resulted in the absence of any substantial acceptance of it among the international community. In order to consider what is needed to create a new international framework for the use and mineral exploitation of celestial bodies, with or without the Moon Agreement, it is prudent

---

[295] COPUOS, *Draft Agreement on the Principles Governing Activities in the Use of Natural Resources of the Moon and Other Celestial Bodies* (1970) U.N. Doc. A/AC.105/C.2/L.71. See Aldo Armando Cocca, *The Principles of the "Common Heritage of Mankind" as Applied to Natural Resources from Outer Space and Celestial Bodies* (1974) 16 PROC. COLL. L. OUTER SP. 174; Carl Q. Christol, *The Moon Treaty and the Allocation of Resources* (1997) 22 ANN. AIR & SP. L. 31 at 33; and Virgiliu Pop, WHO OWNS THE MOON? EXTRATERRESTRIAL ASPECTS OF LAND AND MINERAL RESOURCES OWNERSHIP (2008), at 121.

[296] General Assembly Resolution 2779 (XXVI).

[297] COPUOS, *Report of the Legal Sub-committee of the Work of its Fifteenth Session* (1976), U.N. Doc. A/AC.105/171, Annex 1, at 3 and COPUOS, *Report of the Legal Sub-committee of the Work of its Sixteenth Session* (1977), U.N. Doc. A/AC.105/196, Annex 1, at 45.

[298] See Adrian Bückling, *The Strategy of Semantics and the "Mankind Provisions" of the Space Treaty* (1979) 7 J. SP. L. 15; Fasan, supra note 67; Daniel Goedhuis, *The Changing Legal Regime of Air and Outer Space* (1978) 27 INT'L. & COMP. L. Q. 576; and Carl Q. Christol, *Space Joint Ventures: The United States and Developing Nations* (1975) 8 AKRON L. REV. 398.

[299] Minola, supra note 66, at 468 and S. Neil Hosenball, *The United Nations Committee on the Peaceful Uses of Outer Space: Past Accomplishments and Future Challenges* (1979) 7 J. SP. L. 95 at 103–104.

260   5   Exploitation Rights: Evolving from the "Province of Mankind" to the "Common ...

to consider what the content and effect of the Moon Agreement are at present and how the common heritage of mankind doctrine ought to be applied in practice.

5.7.2 Provisions of the Moon Agreement

Setting aside for a moment the provisions of Article 11 of the Moon Agreement that embodies the common heritage of mankind doctrine, there are a number of other provisions that are relevant to the use and exploitation of the mineral resources on the Moon and other celestial bodies.[300] First, the Moon Agreement specifically extends the freedom to explore and use the Moon and other celestial bodies to private entities, though it is arguable that this was already anticipated under Article VI of the Outer Space Treaty.[301] Second, the Moon Agreement grants title and ownership to any mineral resources on the surface of the Moon or that of a celestial body that are no longer *in situ*.[302] Third, the Moon Agreement reaffirmed the principle contained in Article VIII of the Outer Space Treaty that States and their private entities to retain ownership of equipment and installations that they bring to the Moon and other celestial bodies.[303] Fourth, it has been suggested that the Moon Agreement does not prohibit a State or a private entity from making a profit from its activities on the Moon or other celestial bodies, though it ought to be noted that the Moon Agreement also does not explicitly permit this, which means that the controversy remains over the potentially opposite effect of Article I of the Outer Space Treaty.[304] However, a proposal from Italy during the negotiations over the Moon Agreement that no State be entitled to "exclusive economic profit" from their activities on the Moon and other celestial bodies was not accepted by the Legal Subcommittee of COPUOS.[305]

---

[300] See James R. Wilson, *Regulation of the Outer Space Environment Through International Accord: The 1979 Moon Treaty* (1991) 2 FORDHAM ENVT'L. L. REP. 173.

[301] Moon Agreement, Article 14. See Art Dula, *Private Sector Activities in Outer Space* (1985) 19 INT'L. LAWYER 159; Hoffstadt, supra note 252, at 586; and James J. Trimble, *The International Law of Outer Space and Its Effect on Commercial Space Activity* (1983) 11 PEPP. L. REV. 521 at 560.

[302] Moon Agreement, Article 11(3). See Hoffstadt, supra note 252, at 586.

[303] Moon Agreement, Article 12.

[304] See Hoffstadt, supra note 252, at 586; Armel Kerrest, *Commercial Use of Space, Including Launching* (2004), in China Institute of Space Law, 2004 SPACE LAW CONFERENCE: PAPER ASSEMBLE 199; Ricky J. Lee, *Commentary Paper on Discussion Paper Titled "Commercial Use of Space, Including Launching" by Prof. Dr. Armel Kerrest* (2004), in China Institute of Space Law, 2004 SPACE LAW CONFERENCE: PAPER ASSEMBLE 220; Carl Q. Christol, *The American Bar Association and the 1979 Moon Treaty: The Search for a Position* (1981) 9 J. SP. L. 77; and Martin Menter, *Commercial Space Activities Under the Moon Treaty* (1979) 7 SYRACUSE J. INT'L. L. & COM. 213 at 220.

[305] U.N. Doc. A/AC.105/171 (28 May 1976), Annex 1, at 2 and U.N. Doc. A/AC.105/196 (11 April 1977), Annex 1, at 4. See Carl Q. Christol, THE MODERN INTERNATIONAL LAW OF OUTER SPACE (1982), at 296.

5.7 Evolution of Article 11 of the Moon Agreement

The remaining provisions of the Moon Agreement can be seen to be either the reaffirmation or elaboration of the principles of the Outer Space Treaty. In particular:

- activities on the Moon and other celestial bodies are to be conducted in compliance with public international law, including the Charter of the United Nations and the Declaration on Principles of International Law concerning Friendly Relations and Cooperation among States in accordance with the Charter of the United Nations;[306]
- the Moon and other celestial bodies are to be used exclusively for peaceful purposes and the use of force, weaponisation of orbits and trajectories around celestial bodies and the establishment of military installations on celestial bodies are prohibited;[307]
- States are to operate on the bases of cooperation and mutual assistance in their activities on celestial bodies, including disclosure to the international community of the details relating to the activities;[308] and
- States are to bear international responsibility for their activities on celestial bodies, including those carried out by non-governmental entities, which they are required to authorise and continually supervise.[309]

It is apparent that the provisions of the Moon Agreement, other than the common heritage of mankind provisions, would not be a matter of controversy either for the developing States or for the industrialised States. This is because they either reaffirm or elaborate on the existing provisions of the Outer Space Treaty, which has formed the basis for the principles of international space law. As Christol pointed out in 1980 when commenting on the Moon Agreement:

> Its terms, properly understood, will provide a regime supportive of the [Outer Space Treaty]. The [Outer Space Treaty] focuses on the exploration, use and exploitation of the space environment consisting of outer space, per se, the Moon and celestial bodies. The Moon Treaty, unlike the [Outer Space Treaty], makes specific provision for the exploitation of the natural resources of the Moon and celestial bodies. The Moon Treaty, while preserving the provision contained in Article 2 of the [Outer Space Treaty] that there may not be a sovereign appropriation of the Moon and celestial bodies, does enable defined juridical and natural persons to obtain proprietary rights in certain natural resources on and of the Moon and celestial bodies.[310]

---

[306] Moon Agreement, Article 2; Outer Space Treaty, Article III; and General Assembly Resolution 2625 (XXV), Annex.

[307] Moon Agreement, Article 3 and Outer Space Treaty, Article IV. See Stephen E. Doyle, *Confidence Building Measures Using Space Resources* (1998) 41 PROC. COLL. L. OUTER SP. 108.

[308] Moon Agreement, Articles 4 and 5 and Outer Space Treaty, Articles X and XI.

[309] Moon Agreement, Article 14 and Outer Space Treaty, Article VI.

[310] Carl Q. Christol, *The Common Heritage of Mankind Provision in the 1979 Agreement Governing the Activities of States on the Moon and Other Celestial Bodies* (1980) 14 INT'L. LAWYER 429 at 429–430 and Heidi Keefe, *Making the Final Frontier Feasible: A Critical Look at the Current Body of Outer Space Law* (1995) 11 SANTA CLARA COMPUTER & HIGH TECH. L. J. 345.

It is clear that the common heritage of mankind provisions of the Moon Agreement are the foci of the conflict between the industrialised States and the developing States. In the traditional view of most scholars, the "common heritage of mankind" principle in relation to outer space does not apply the *res communis* principle. On the contrary, it transforms into something that creates specific obligations on states utilising this area. Under the doctrine, areas designated as the common heritage of mankind, or *terra communis humanitatis*, would be owned by no one and yet theoretically managed by everyone. Sovereignty does not exist here and legally the international community as a whole would manage the area.[311] States would have no role in the management of these areas except as representatives of all mankind.

The common heritage of mankind doctrine also requires any use to be limited to peaceful purposes.[312] For the purpose of scientific research, however, free access to any *res communis humanitatis* would be permissible provided the benefits of such research are available to anyone expressing a genuine interest in them.[313] In other words, even if the research were financed by a state or a group of states, the fruits of the research would be available freely to the international community, as has been the case in Antarctica.[314] It is crucial to recognise that the doctrine requires any benefits derived from the exploitation of natural resources to be shared internationally. As a result, exploitation by commercial entities would be deemed inappropriate unless their efforts contributed to the common benefit of all mankind. The extent of this sharing of benefits was never specifically defined and as such uncertainty remains on the extent of this obligation.

The earlier proposals that were under consideration by the Legal Sub-Committee applied the common heritage of mankind concept to the mineral resources on the celestial bodies and not to the celestial bodies themselves.[315] However, on 30 March

---

[311] See Christopher C. Joyner, *Legal Implications of the Concept of the Common Heritage of Mankind* (1986) 35 INT'L. & COMP. L. Q. 190 at 191; Kelly M. Zullo, *The Need to Clarify the Status of Property Rights in International Space Law* (2002) 90 GEORGETOWN L. J. 2413; Kurt Anderson Baca, *Property Rights in Outer Space* (1993) 58 J. AIR L. & COM. 1041; Carol R. Buxton, *Property in Outer Space: The Common Heritage of Mankind Principle vs. The "First in Time, First in Right" Rule of Property Law* (2004) 69 J. AIR L. & COM. 689; and Brandon C. Gruner, *A New Hope for International Space Law: Incorporating Nineteenth Century First Possession Principles into the 1967 Space Treaty for the Colonisation of Outer Space in the Twenty-First Century* (2005) 35 SETON HALL L. REV. 299.

[312] See Timothy M. Zadalis, *"Peaceful Purposes" and Other Relevant Provisions of the Revised Composite Negotiating Text: A Comparative Analysis of the Existing and Proposed Military Regime for the High Seas* (1979) 7 SYRACUSE J. INT'L. L. & COM. 1 and Emilio Jaksetic, *The Peaceful Uses of Outer Space: Soviet Views* (1979) 28 AM. U. L. REV. 483.

[313] Moon Agreement, Article 5.

[314] Ibid., Article 3.

[315] See the proposal by Argentina for a "Draft Agreement on the Principles Governing Activities in the Use of Natural Resources of the Moon and other Celestial Bodies" in U.N. Doc. A/AC.105/C.2/L.71 at 1; and U.N. Doc. A/AC.105/85, Annex 2, at 1. See also Aldo Armando Cocca, *Legal Status of the Natural Resources of the Moon and Other Celestial Bodies* (1971) 13 PROC. COLL. L. OUTER SP. 146.

5.7 Evolution of Article 11 of the Moon Agreement

1973, Argentina changed its own position and proposed that both the celestial bodies and their mineral resources are to be the common heritage of mankind.[316] At the same time, an alternative proposal put forward by the Soviet Union on 21 May 1971 did not refer to the common heritage of mankind concept, yet it intrinsically embodied the concept by recognising that the surface and subsoil of celestial bodies, including any mineral resources contained therein, were *res communis*.[317] Meanwhile, the United States submitted a working paper to COPUOS on 13 April 1972 accepting the formulation in the original proposal of Argentina, even though this formulation was by then inconsistent with General Assembly Resolution 2749 (XXV) of 17 December 1970 that applied the common heritage of mankind concept to the deep seabed and celestial bodies.[318] After negotiations over a number of years concerning the application of the common heritage of mankind concept to celestial bodies, a joint working paper was submitted in 1976 by Argentina, Brazil, Chile, Indonesia, Mexico, Nigeria, Romania, Sierra Leone and Venezuela that urged the States Parties to create an international legal regime on the basis that the celestial bodies and their mineral resources are the common heritage of mankind.[319]

It soon became apparent that three conflicting interests in the international community had emerged during the negotiations. As Chen observed:

> Certain members considered "common heritage of mankind" as a philosophical concept lacking legal content which had no place in a legal instrument, while others maintained that it was a legal concept and a prerequisite for the elaboration of a treaty relating to the Moon. Some members held the view that activities should be permitted only for scientific purposes and that no commercial exploitation of the natural resources should take place before the establishment of the international regime; others thought that utilisation of the Moon and its natural resources should also be allowed for other experimental purposes; still others were of the opinion that utilisation should be allowed for any peaceful purpose pending the establishment of the international regime.[320]

---

[316] U.N. Doc. A/AC.105/101, 11 May 1972, at 6 and U.N. Doc. A/AC.105/196, 11 April 1977, Annex I at 13. For remarks and comments on, inter alia, this change in the negotiating position on the proposed Moon Agreement, see Vladímir Kopal, *Legal Questions Relating to the Draft Treaty Concerning the Moon* (1973) 16 PROC. COLL. L. OUTER SP. 180; Stephen Gorove, *Property Rights in Outer Space: Focus on the Proposed Moon Treaty* (1973) 16 PROC. COLL. L. OUTER SP. 177; Laszlo Szaloky, *The Way of the Further Perfection on the Legal Regulation Concerning the Moon and Other Celestial Bodies, Especially Regarding the Exploitation of Natural Resources of the Moon and Other Celestial Bodies* (1973) 16 PROC. COLL. L. OUTER SP. and Francesco Rusconi and C. Paz-Perina, *Proyecto de Tratado Relativo a la Luna Usos Pacíficos y Desarme: Dos Aspectos de una Misma Realidad* (1973) 16 PROC. COLL. L. OUTER SP. 190.

[317] U.N. Doc. A/8391, 4 June 1971. See discussion in Gennady P. Zhukov, *The Legal Regime for the Moon (Problems and Prospects)* (1972) 14 PROC. COLL. L. OUTER SP. 50.

[318] U.N. Doc. A/AC.105/C.2 and U.N. Doc. A/AC.105/196, 11 April 1977, Annex 1, at 23. See Christol, supra note 310, at 457.

[319] U.N. Doc. A/AC.171, 28 May 1976, Annex 1, at 3 and U.N. Doc. A/AC.105/196, 11 April 1977, Annex 1, at 4–5.

[320] Kwen Chen, *Pending Issues Before the Legal Sub-committee of the United Nations Committee on the Peaceful Uses of Outer Space* (1977) 5 J. SP. L. 30.

264   5 Exploitation Rights: Evolving from the "Province of Mankind" to the "Common ...

These conflicting interests eventually became embodied in the provisions contained Articles 6, 11 and 18 of the Moon Agreement, as discussed below.

5.7.3 Article 6 of the Moon Agreement

The Moon Agreement expressly provides for the collection of "moon rocks", or geological or mineralogical samples, when engaging in scientific investigations of the Moon and other celestial bodies, without granting them title or ownership to the samples so collected. Specifically, the Moon Agreement provides that:

> In carrying out scientific investigations and in furtherance of the provisions of this Agreement, the States Parties shall have the right to collect on and remove from the Moon samples of its mineral and other substances. Such samples shall remain at the disposal of those States Parties which caused them to be collected and may be used by them for scientific purposes. States Parties shall have regard to the desirability of making a portion of such samples available to other interested States Parties and the international scientific community for scientific investigation. States Parties may in the course of scientific investigations also use mineral and other substances of the Moon in quantities appropriate for the support of their missions.[321]

It is prudent to note that there are no restrictions on the exclusive collection and use of mineral samples and, more importantly, there are no restrictions on the exclusive use of resources, including in particular mineral resources, for the conduct of scientific investigations. As Christol noted, the nature and extent of such scientific use can be somewhat wide-ranging and, as such, substantial exclusive use of mineral resources from celestial bodies are allowed for scientific investigations.[322] The rights granted under Article 6 of the Moon Agreement extend to both the scientific investigations of States as well as those of international organisations and private entities, though the "appropriate" State would have the general duties to authorise and continually supervise the activities of the relevant private entities.[323]

5.7.4 Article 11 of the Moon Agreement

5.7.4.1 The Premise

As has been noted and commented upon by many since its adoption, there is no other aspect of the Moon Agreement that is more controversial than the provision declaring the Moon and other celestial bodies to be the "common heritage of mankind".[324] Such a requirement, alongside the express and implied obligations of the common heritage of mankind doctrine as imposed on states, continues to

---

[321] Moon Agreement, Article 6(2).

[322] Christol, supra note 310, at 465–466.

[323] Moon Agreement, Article 14 and Outer Space Treaty, Article VI.

[324] Moon Agreement, Article 11(1). See Francis Lyall, *On the Moon* (1998) 26 J. Sp. L. 129.

serve as a major inhibiting factor in relation to the general acceptance of the Moon Agreement. Paragraph 1 of Article 11 of the Moon Agreement provides that:

> The Moon and its natural resources are the common heritage of mankind, which finds its expression in the provisions of this Agreement and in particular in paragraph 5 of this article.

Paragraph 5 of Article 11 of the Moon Agreement provides that:

> States Parties to this Agreement hereby undertake to establish an international regime, including appropriate procedures, to govern the exploitation of the natural resources of the Moon as such exploitation is about to become feasible. This provision shall be implemented in accordance with article 18 of this Agreement.

The idea that outer space and celestial bodies cannot be subject to the sovereign ownership of any state is not a new one. Article II of the Outer Space Treaty prohibits "national appropriation", even "by means of occupation". This in effect outlaws *imperium*, the form of public ownership that establishes sovereign rights in relation to certain areas by virtue of the fact that celestial bodies are *res communis*.[325] However, *dominium* and ownership by a state of materials exploited would continue to be possible for private entities. This is particularly relevant in the creation and acquisition of intellectual property rights, as they may be regarded either as the common heritage of mankind or as a benefit derived from the celestial bodies, and therefore subject to sharing among all States.

5.7.4.2 Content and Effect of Article 11 of the Moon Agreement

Article 11 of the Moon Agreement is, at its very highest, an interim arrangement with a declaration of principle in relation to the common heritage of mankind doctrine. Except for the specific provisions discussed above, the Moon Agreement does not provide for the practical implementation of the common heritage of mankind doctrine but merely foreshadows the establishment of a new regime when the commercial exploitation of mineral resources from celestial bodies are about to become feasible.[326] The Moon Agreement does not specify the terms of that regime and also does not indicate upon what basis any determination by the State Parties to the Moon Agreement as to the feasibility of the commercial exploitation of mineral resources from celestial bodies is to be made.

In effect, Article 11 of the Moon Agreement does no more than designate celestial bodies and their mineral resources as the "common heritage of mankind" but does not impose any practical obligations on the part of States and private entities in conducting exploitation activities arising from such a doctrine. Article 11(5) of the Moon Agreement defers the creation of a regulatory framework that would impose such practical obligations until a later time when such exploitation become feasible and imminent. In effect, the designation of the celestial bodies and their mineral

---

[325] See Ian Brownlie, Principles of Public International Law (5th ed., 1998) at 105.

[326] Moon Agreement, Article 11(5).

resources as the common heritage of mankind in the Moon Agreement in practice required no more than requiring such use be exclusively for peaceful purposes and prohibiting the assertion or maintenance of property rights or territorial sovereignty over celestial bodies, which are legal provisions that already found expression in the terms of the Outer Space Treaty.[327] Until the foreshadowed international regulatory regime is implemented, the Moon Agreement and its common heritage of mankind doctrine would not impose any additional practical obligations on the part of participating States or private entities under their supervision.

5.7.5 Article 18 of the Moon Agreement

The Moon Agreement provides that, 10 years after the Moon Agreement enters into force, the General Assembly is to consider a review of the provisions of the Moon Agreement.[328] In addition to this prescribed review, one-third of the Member States can request the convening of a general conference of the States Parties to the Moon Agreement, particularly in relation to the consideration of the appropriate regime for the implementation of Article 11 of the Moon Agreement.[329] In theory, this would provide an opportunity for the States to consider the appropriate timing for the creation and implementation of a regime in accordance with the Moon Agreement. As Reynolds had commented in 1992:

> Because this review will allow all nations – not just those few that actually ratified the Moon Treaty – to discuss proposed remedies for the Treaty's flaws, it will provide an opportunity to consider revisions. In particular, the United States should play a major part by proposing amendments to the Treaty that recognise the important role played by private property rights in promoting development of outer space. Land-grant type mechanisms, administered by the United Nations or by individual nations in accordance with agreed international principles, might reward private development efforts with long-term leases or permanent property rights in space resources they develop.[330]

The Moon Agreement entered into force on 11 July 1984, being 30 days after the fifth instrument of ratification was deposited.[331] In 1994, the tenth anniversary of the entry into force of the Moon Agreement was reached and passed without much impulse or motivation on the part of the States Parties to the Moon Agreement to undertake a review of its provisions. Given the entrenched views of the industrialised and developing States on the content and effect of the common heritage of mankind

---

[327] Outer Space Treaty, Articles I and IV. See, for example, Detlev Wolter, *The Peaceful Purpose Standard of the Common Heritage of Mankind Principle in Outer Space Law* (1985) 9 A.S.I.L.S. INT'L. L. J. 117.

[328] Moon Agreement, Article 18.

[329] Ibid.

[330] Glenn Harlan Reynolds, *International Space Law: Into the Twenty-First Century* (1993) 25 VAND. J. TRANSNAT'L. L. 225 at 233 and Stephen Gorove, *Exploitation of Space Resources and the Law* (1984) 3 PUB. L. FORUM 29.

[331] Moon Agreement, Article 19(3).

5.7 Evolution of Article 11 of the Moon Agreement

concept to celestial bodies and their mineral resources, it is unlikely that any legal regime can be created in the near future. This can only be understood with reference to their contradictory positions.

5.7.6 *Attempts at Resolving the Political Impasse*

On 21 March 1994, with the tenth anniversary of the entry into force of the Moon Agreement approaching, the Legal Sub-Committee of COPUOS noted that a review of the status of the Moon Agreement was to occur in that year.[332] Coincidentally, this was the same period of time as when the negotiations over the revised terms for the implementation of the Convention on the Law of the Sea was taking place in the General Assembly, a fact that was noted by a number of States participating in the debates concerning the Moon Agreement, most notably the comments made by the representations of Portugal, Russia and Spain.[333] The General Assembly in Resolution 49/34 adopted the recommendation of COPUOS and the Legal Sub-Committee that, in considering whether to revise the terms of the Moon Agreement, the General Assembly "should take no action at the present time".[334] As Christol noted:

> While the confrontational behaviour that had been present in the negotiations of the 1960s and 1970s was not longer evident, this factor did not appreciably improve the working processes of the [Legal Sub-Committee] or the [COPUOS]. An outlook of "let well enough alone" pervaded the thinking of some countries. On the whole these were the space-resource States. But, as has been noted, the developing countries did not put forward a plan that might have provoked a meaningful analysis.[335]

Although Resolution 49/34 represented the end of any formal process to review and perhaps revise the terms of the Moon Agreement, there remain persistent efforts on the part of some States for the issue to be further addressed by the international community.[336] The fact that there are inconsistencies between the Moon Agreement and the Outer Space Treaty means that there are now in effect

---

[332] U.N. Doc. A/AC.105/C.2/SR.572, 28 March 1994, at 3.

[333] See U.N. Doc. A/AC.105/C.2/SR.577, 7 April 1994, at 4 (Portugal); U.N. Doc. A/AC.105/C.2/SR.576, 7 April 1994, at 9 (Russia); and U.N. Doc. A/AC.105/C.2/SR.578, 7 April 1994, at 4 (Spain).

[334] General Assembly Resolution 49/34 and U.N. Doc. A/A.4/49/L.11, 11 November 1994, at 9.

[335] Carl Q. Christol, *The 1979 Moon Agreement: Where Is It Today?* (1999) 27 J. Sp. L. 1 at 25.

[336] After the adoption of Resolution 49/34, there were notable comments from Spain: U.N. Doc. A/AC.3105/C.2/SR.580, 31 March 1995, at 7; Romania: U.N. Doc. A/AC.105/PV.413, 19 June 1995, at 5; Mexico: U.N. Doc. A/AC.105/639, 11 April 1996, at 40; U.N. Doc. A/AC.105/C.2/L.206 Rev. 1, 4 April 1997; and U.N. Doc. A/AC.105/674, 14 April 1997, at 22; France and Germany: U.N. Doc. A/AC.105/639, 3 June 1996, at 32; and Nigeria: U.N. Doc. A/AC.105/C.2/SR.592, 27 March 1996, at 2. See René Oosterlink, *Tangible and Intangible Property in Outer Space* (1997) 39 Proc. Coll. L. Outer Sp. 277; and ibid. at 26–29; and Nandasiri Jasentuliyana, *International Space Law and Cooperation and the Mining of Asteroids* (1990) 15 Ann. Air & Sp. L. 343.

two competing regimes under international law for the exploitation of mineral resources from celestial bodies, specifically one for those 13 States who are parties to both the Outer Space Treaty and the Moon Agreement and another for those who are party to the Outer Space Treaty but not the Moon Agreement.[337] With the prospect of the exploitation of mineral resources from celestial bodies potentially becoming an imminent development, there is much incentive for the international community to compromise and reach agreement on the terms of an international regulatory framework to forestall possible unilateral regulation on the part of individual States.[338]

The principal objections that have been raised by the industrialised States to the Moon Agreement, in particular those of the United States, are similar to those that were raised in opposition to the terms of the Convention on the Law of the Sea as it was originally adopted.[339] After all, in the context of the Cold War, there was the perception that the Moon Agreement was implementing some form of international socialism with the common heritage of mankind.[340] With the risk of generalisation, the practical objections of the industrialised States include:

(1) the absence of property rights that are essential for commercial development of mineral resources on celestial bodies;[341]
(2) the potential need to have "a hefty share of the proceeds going to less-developed countries regardless of whether they have any investment in the activity or not";[342]

---

[337] The 13 States that have ratified the Moon Agreement are Australia, Austria, Belgium, Chile, Kazakhstan, Lebanon, Mexico, Morocco, the Netherlands, Pakistan, Peru, Philippines and Uruguay. See Christol, supra note 335, at 32.

[338] See, for example, Anthony R. Filiato, *The Commercial Space Launch Act: America's Response to the Moon Treaty?* (1987) 10 FORDHAM INT'L. L. J. 763; Francis H. Esposito, *The Commercial Exploitation of Space* (1985) 25 A. F. L. REV. 159; and Julie A. Jiru, *Star Wars and Space Malls: When the Paint Chips Off a Treaty's Golden Handcuffs* (2001) 42 S. TEX. L. REV. 155.

[339] See, for example, Carl Q. Christol, *The American Bar Association and the 1979 Moon Treaty: The Search for a Position* (1981) 9 J. SP. L. 77 and Michael J. Listner, *The Ownership and Exploitation of Outer Space: A Look at Foundational Law and Future Legal Challenges to Current Claims* (2003) 1 REGENT J. INT'L. L. 75.

[340] Stephen D. Mau, *Equity, the Third World and the Moon Treaty* (1984) 8 SUFFOLK TRANSNAT'L. L. J. 221 at 258. This is in addition to the existing difficulties in attracting sufficient investment in commercial space endeavours: see Stacey A. Davis, *Unifying the Final Frontier: Space Industry Financing Reform* (2001) 106 COM. L. J. 455 and William Lee Andrews III, *A Mighty Stone for David's Sling: The International Space Company* (2003) 1 REGENT J. INT'L. L. 5.

[341] See Michael E. Davis, *Reaching for the Moon* (1997) NEW SCIENTIST, 23 August 1997, at 46 and Ryan Hugh O'Donnell, *Staking a Claim in the Twenty-First Century: Real Property Rights on Extra-Terrestrial Bodies* (2007) 32 U. DAYTON L. REV. 461.

[342] See Glenn Harlan Reynolds, *Key Objections to the Moon Treaty* (2003), National Space Society Chapters Network, <http://www.nsschapters.org/hub/pdf/MoonTreatyObjections.pdf>, 28 April 2003, last accessed on 28 November 2009 and Peter D. Nesgos, *Rights and Obligations of Participants in Space Materials Processing Activities*, paper presented at the ICC/IBA Symposium

5.7 Evolution of Article 11 of the Moon Agreement

(3) the establishment of an international organisation that would regulate mineral exploitation activities on celestial bodies with the potential for licensing processes to be "slow, cumbersome and prone to blackmail";[343]

(4) the potential for the compulsory transfer of relevant technology from the industrialised States to developing States, as was the case under Part XI of the Convention on the Law of the Sea;[344]

(5) the potential for the creation of an entity similar to the Enterprise created under Part XI of the Convention on the Law of the Sea that would actively compete with commercial interests in the exploitation of mineral resources from celestial bodies; and

(6) the implied moratorium on the commercial exploitation of mineral resources from celestial bodies until the creation and implementation of the international regulatory regime foreshadowed under Article 11 of the Moon Agreement.[345]

For the developing States, private capitalisation of commercial exploitation of mineral resources on celestial bodies is incompatible with the values of the common heritage of mankind principle.[346] For these developing States, as in the negotiations in UNCLOS III, the nature of the common heritage of mankind mandates some form of international regulation of management in their use to ensure that celestial bodies and their mineral resources are used for the benefit of all humankind.

Only in compromise between these two opposing philosophical positions can any agreement be reached over the creation of a new international regulatory regime. As discussed above, both the industrialised States including, in particular the United States, and the developing States have reached a compromise over the Convention on the Law of the Sea and their interpretation and application of the common heritage of mankind doctrine in that context. Consequently, it is possible that a similar compromise may eventually be acceptable to international community.

---

on Research and Invention in Outer Space and their Commercial Exploitation: Liability and Intellectual Property Rights, 6–7 December 1990, in Paris, France.

[343] See Stephen E. Doyle, *Using Extraterrestrial Resources Under the Moon Agreement of 1979* (1998) 26 J. SP. L. 111 and Reynolds, supra note 342.

[344] See Colin B. Picker, *A View from 40,000 Feet: International Law and the Invisible Hand of Technology* (2002) 23 CARDOZO L. REV. 149.

[345] See Stanley B. Rosenfield, *The Moon Treaty: The United States Should Not Become a Party* (1980) 74 AM. SOC'Y. INT'L. L. PROC. 162 at 165; Ty S. Twibell, *Space Law: Legal Restraints on Commercialisation and Development of Outer Space* (1997) 65 U.M.K.C. L. REV. 589; Nancy L. Griffin, *Americans and the Moon Treaty* (1981) 46 J. AIR L. & COM. 729; Robert P. Merges and Glenn H. Reynolds, *Space Resources, Common Property and the Collective Action Problem* (1998) 6 N. Y. U. ENVT'L. L. J. 107; and Thomas D. Halket, Valnora Leister, Eric A. Savage, Jerry V. Lephart and Arthur Miller, *Report on the Proposed Agreement Governing the Activities of States on the Moon and Other Celestial Bodies* (1983) 23 JURIMETRICS J. 259.

[346] See Mark Orlove, *Spaced Out: The Third World Looks For a Way in to Outer Space* (1989) 4 CONN. J. INT'L. L. 597; Art Dula, *Free Enterprise and the Proposed Moon Treaty* (1980) 2 HOUS. J. INT'L. L. 3; and Heim, supra note 9, at 834–835.

5.8 Conclusions

The evolution of some philosophical construct of common property over non-sovereign areas that would otherwise be *res extra commercium* or *res nullius* has taken some twists and turns from its application to Antarctica, outer space, the deep seabed and celestial bodies. It is apparent, however, that the common heritage of mankind concept is not more than such a philosophical doctrine, for the practical content and effects of the concept are no more than a malleable construct that had been adapted by the States negotiating over the legal regimes in practice.[347] In other words, the concepts of "common property", "province of mankind" and "common heritage of mankind" are in reality no more than doctrinal labels and it is in the practical expression of these constructs into specific rights, duties and obligations that make them controversial among members of the international community.[348]

In the earliest case of Antarctica, these doctrinal concepts found expression in the notions of common management, exclusivity of their use for peaceful purposes and the deferral of sovereign territorial claims over parts of the Antarctic continent. There is no universal participation in the regulatory regime under the ATS and, further, there is no declaration to the effect that Antarctica was not to be subject to territorial sovereignty or private ownership. With outer space and celestial bodies under the Outer Space Treaty, the "province of all mankind" does not, without more, impose practical obligations or restrictions. It is the other provisions of the Outer Space Treaty that rights and duties, such as the freedom of exploration, use and scientific investigation under Article I and the prohibitions of ownership and territorial sovereignty under Article II, are imposed.[349]

---

[347] See, for example, Manfred Lachs, *Legal Framework of an International Community* (1992) 6 EMORY INT'L. L. REV. 329; Armel Kerrest, *Exploitation of the Resources of the High Sea and Antarctica: Lessons for the Moon?*, paper presented at the IISL/ECSL Space Law Symposium on New Developments and the Legal Framework Covering the Exploitation of the Resources of the Moon, 29 March 2004, in Vienna, Austria; Jennifer Frakes, *The Common Heritage of Mankind Principle and the Deep Seabed, Outer Space and Antarctica: Will Developed and Developing Nations Reach a Compromise?* (2003) 21 WIS. INT'L. L. J. 409; Graham Nicholson, *The Common Heritage of Mankind and Mining: An Analysis of the Law as to the High Seas, Outer Space, the Antarctic and World Heritage* (2002) 6 N. Z. J. ENVT'L. L. 177; and J. Henry Glazer, *Astrolaw Jurisprudence in Space as a Place: Right Reason for the Right Stuff* (1985) 11 BROOKLYN J. INT'L. L. 1.

[348] See Stephen Gorove, *The Concept of "Common Heritage of Mankind": A Political, Moral or Legal Innovation?* (1972) 9 SAN DIEGO L. REV. 390; Richard B. Bilder, *International Law and Natural Resources Policies* (1980) 22 NAT. RES. J. 451; Scott Ervin, *Law in a Vacuum: The Common Heritage Doctrine in Outer Space Law* (1984) 7 B. C. INT'L. & COMP. L. REV. 403; James Fawcett, *The Concepts of Outer Space and the Deep Seabed in International Law: Some Comparisons* (1984) 2 NOTRE DAME INT'L. & COMP. L. J. 71; and Christie Condara, *Outer Space, Like the Sea and the Air, Whose Frontier? Incredible Potential with Inscrutable Obstacles* (1984) 6 HOUS. J. INT'L. L. 175.

[349] See Jefferson H. Weaver, *Illusion or Reality? State Sovereignty in Outer Space* (1992) 10 B. U. INT'L. L. J. 203; Linda R. Sittenfeld, *The Evolution of a New and Viable Concept of Sovereignty for Outer Space* (1981) 4 FORDHAM INT'L. L. J. 199; and Gbenga Oduntan, *Imagine There are*

5.8 Conclusions

The deep seabed has seen two different sets of obligations under the common heritage of mankind concept. The provisions in the Convention on the Law of the Sea prohibiting ownership and territorial sovereignty and requiring international regulation and management, the "equitable" sharing of benefits, mandatory transfer of technology and the creation of the Enterprise as a competing industry participant are all said to be practical expressions of the common heritage of mankind doctrine. However, the subsequent1994 Agreement had converted the regulatory regime form a comprehensive to a minimalist approach, providing only for the prohibition of territorial sovereignty and ownership and the international regulation and management of activities relating to commercial exploitation of the deep seabed.[350] This minimalist approach to the regulation of exploitation activities is nevertheless said to be an expression of the common heritage of mankind doctrine, clearly illustrating that the doctrine itself does not carry with it practical implications.

The Moon Agreement defers the formulation of these practical provisions until a later time when the exploitation of mineral resources from celestial bodies becomes feasible and imminent. On the one hand, this gave an opportunity for the international community to formulate such provisions while divorcing itself from the need to debate over the philosophical nature of the common heritage of mankind concept. On the other hand, the deferral has illustrated the deep divisions that existed between different stakeholders as to the appropriate rights and duties that ought to be imposed in the commercial exploitation activities on celestial bodies. It is clear that the creation of any international regulatory regime, whether as part of the Moon Agreement framework or otherwise, will require the balancing of the competing interests of these stakeholders in the unwritten principles of the common heritage of mankind doctrine.

No Possessions: Legal and Moral Basis of the Common Heritage Principle in Space Law (2005) 2 MAN. J. INT'L. ECON. L. 30.

[350] See Lotta E. Viikari, *The Legal Regime for Moon Resource Utilisation and Comparable Solutions Adopted for Deep Seabed Activities* (2003) 31 ADV. SP. RES. 2427.

Chapter 6
Meeting the Challenges and Balancing the Competing Interests in Creating a Legal and Regulatory Framework

6.1 Introduction

Space law, especially the provisions of the Agreement Governing the Activities of States on the Moon and Other Celestial Bodies (the "Moon Agreement"), has been influenced by developments adopted in the law of the sea and the treaty framework relating to Antarctica and the deep seabed.[1] While each has ultimately been developed separately from the others, outer space, the deep seabed, and Antarctica share several key features. For example, each constitutes an international spatial area, which has potentially valuable and exploitable natural resources and an inhospitable environment, which necessitates technological sophistication and exorbitant financial investment to effectively exploit these resources.[2] Further, the international community has regarded the preservation of both environments for future generations as an important priority.[3] More importantly, all three areas are deemed to be in the common interest of humankind and face dilemmas with respect to the application of common heritage of mankind and related principles.

It has often been suggested that the principal factor restraining industrialised States from signing and ratifying the Moon Agreement is the result of their fear for the practical and financial implications of the common heritage of mankind

---

[1] Agreement Governing the Activities of States on the Moon and Other Celestial Bodies (the "Moon Agreement"), opened for signature on 18 December 1979, 1363 U.N.T.S. 3; 18 I.L.M. 1434 (entered into force on 11 July 1984). See Eric Husby, *Sovereignty and Property Rights in Outer Space* (1994) 3 DETROIT COLL. L. J. INT'L. L. & PRAC. 359 at 362 and Kevin V. Cook, *The Discovery of Lunar Water: An Opportunity to Develop a Workable Moon Treaty* (1999) 11 GEORGETOWN INT'L. ENVT'L. L. REV. 647 at 677.

[2] Cook, supra note 1, at 677 and Grier C. Raclin, *From Ice to Ether: The Adoption of a Regime to Govern Resource Exploitation in Outer Space* (1986) 7 J. INT'L. L. & BUS. 727, at 728–730.

[3] Thomas M. Franck and Dennis M. Sughrue, *Symposium: The International Role of Equity-as-Fairness* (1993) 81 GEORGETOWN L. J. 563 at 590–594 and Kemal Baslar, THE CONCEPT OF THE COMMON HERITAGE OF MANKIND IN INTERNATIONAL LAW (1998) at 264–265.

R.J. Lee, *Law and Regulation of Commercial Mining of Minerals in Outer Space*, Space Regulations Library 7, DOI 10.1007/978-94-007-2039-8_6, © Springer Science+Business Media B.V. 2012

doctrine.[4] It is respectfully submitted that it is the moratorium imposed on commercial use and exploitation of mineral resources until the regulatory regime foreshadowed under Article 11 of the Moon Agreement has been implemented. Such a moratorium and other restrictive provisions of the Moon Agreement do not exist under the Treaty on Principles Governing the Activities of States in the Exploration and Use of Outer Space, including the Moon and other Celestial Bodies (the "Outer Space Treaty").[5] Accordingly, a State that ratifies the Moon Agreement would have the effect of voluntarily subjecting itself to a moratorium on commercial mineral exploitation activities on celestial bodies.

Given the inconsistency between the terms of the Outer Space Treaty and the Moon Agreement, the low level of ratifications for the Moon Agreement has, in effect, created two parallel bodies of international law concerning activities of States on the Moon and other celestial bodies. The States that have ratified the Moon Agreement are clearly bound by its terms, which would take precedence over the provisions of the Outer Space Treaty to the extent that they are inconsistent.[6] The States that have not ratified the Moon Agreement would be bound only by the terms of the Outer Space Treaty and not by the provisions of the Moon Agreement and thus the low level of ratifications for the Moon Agreement would have the effect of imposing binding legal restrictions on its States Parties without international recognition of the rights granted under the Moon Agreement.[7]

Clearly, as with the Convention on the Law of the Sea,[8] any new international framework for the exploitation of the celestial bodies will have to achieve a requisite degree of universality in order for there to be sufficient legal and regulatory certainty for participating States and private entities to attract the requisite level of investment required for such a large-scale endeavour. This can only be achieved if the divergent interests and concerns of the international community can be balanced in an effort

---

[4] See, for example, Michael J. Listner, *The Ownership and Exploitation of Outer Space: A Look at Foundational Law and Future Legal Challenges to Current Claims* (2003) 1 REGENT J. INT'L. L. 75; Stephen D. Mau, *Equity, the Third World and the Moon Treaty* (1984) 8 SUFFOLK TRANSNAT'L. L. J. 221; and Glenn Harlan Reynolds, *Key Objections to the Moon Treaty* (2003), National Space Society Chapters Network, <http://www.nsschapters.org/hub/pdf/MoonTreatyObjections.pdf>, 28 April 2003, last accessed on 28 November 2009.

[5] Treaty on Principles Governing the Activities of States in the Exploration and Use of Outer Space, including the Moon and other Celestial Bodies (the *"Outer Space Treaty"*), opened for signature on 27 January 1967, 610 U.N.T.S. 205; 18 U.S.T. 2410; T.I.A.S. 6347; 6 I.L.M. 386 (entered into force on 10 October 1967).

[6] Vienna Convention on the Law of Treaties, opened for signature on 23 May 1969, 1155 U.N.T.S. 331; 1980 U.K.T.S. 58 (entered into force on 27 January 1980), Article 30(2). For the States that have signed but not ratified the Moon Agreement, they are nevertheless required to refrain from acts that would defeat its object and purpose: Vienna Convention on the Law of Treaties, Article 18.

[7] This was part of the motivation for some States Parties to the Moon Agreement, most notably Australia and Chile, to seek an examination of it by the Legal Sub-committee of the Committee on the Peaceful Uses of Outer Space: U.N. Doc. COPUOS/LEGAL/T.632, 3 April 2000, at 3.

[8] United Nations Convention on the Law of the Sea, opened for signature on 10 December 1982, 1833 U.N.T.S. 3; 21 I.L.M. 1261 (entered into force on 16 November 1994).

6.2 Need for Balancing Competing Interests 275

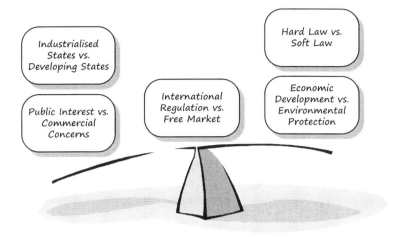

Fig. 6.1 Competing interests to be balanced in the creation of the new international regulatory framework

to find an appropriate compromise that can be acceptable to most, if not all, States in the international community (Fig. 6.1).

It is important to recognise that the present controversy over the terms of the Moon Agreement is somewhat unimportant given that it specifically states that a new regulatory framework would need to be negotiated and implemented if and when mining on celestial bodies is about to become feasible.[9] When that happens, the international community will need to balance and compromise the competing interests and concerns of various stakeholders to consider the terms upon which an international regulatory framework for the exploitation of mineral resources on celestial bodies.

6.2 Need for Balancing Competing Interests

6.2.1 Overview

In relation to the issues arising from the possible exploitation of mineral resources on celestial bodies, a number of competing interests can be identified and they can be categorised into five dichotomies, as illustrated in Table 6.1.

---

[9] Moon Agreement, Article 11.

276   6 Meeting the Challenges and Balancing the Competing Interests in Creating a . . .

Table 6.1 Specific competing interests and concerns

Dichotomy	Interest/issue/concern
Industrialised states vs. developing States	Sharing of benefits derived from exploitation Transferring technology to developing States Creation of a competing enterprise Provision of property and licensing rights
Economic development vs. environmental protection	Preservation of the environment of the Moon and other celestial bodies Restrictions on exhaustive exploitation Use of nuclear and radioisotopic power sources and propulsion systems
International regulation vs. free market	Exclusivity of licences Minimum work requirements Essential protection of industrial property rights Controlling the economic effects of mineral exploitation on commodity markets
Public interest vs. commercial concerns	Satisfying the baseline resource needs of the least developed States Protection of developing States with terrestrial mineral resource production
Hard law vs. soft law	Need for binding legal principles Need for enforcement mechanisms Avoidance of unilateral regulation by States Need for flexible and adaptive rule-making

6.2.2 Industrialised States vs. Developing States

6.2.2.1 Property Rights and the Non-appropriation Principle

Commentators from some industrialised States, particularly those of the United States, have suggested that the Moon Agreement is unacceptable because it does not provide for any property rights for use or exploitation by States and private entities.[10] In fact, the Moon Agreement prohibits any form of private ownership or territorial sovereignty over celestial bodies.[11] It has been suggested that, unless such private property rights are made available, either unilaterally or through an international organisation, there would be insufficient legal certainty to attract private investment in the commercial space sector.[12] It is prudent to note that the prohibitions contained in the Moon Agreement in relation to property rights are already

---

[10] See, for example, Reynolds, supra note 4.

[11] Moon Agreement, Article 11.

[12] See Stacey A. Davis, *Unifying the Final Frontier: Space Industry Financing Reform* (2001) 106 COM. L. J. 455 and William Lee Andrews III, *A Mighty Stone for David's Sling: The International Space Company* (2003) 1 REGENT J. INT'L. L. 5.

6.2 Need for Balancing Competing Interests

provided for, either expressly or impliedly, under the Outer Space Treaty.[13] This was a position supported by the United States at the time, albeit that this was during the Cold War in the 1960s when it was perceived to be likely that the Soviet Union would beat the United States in their race to the Moon.[14]

Despite the end of the Cold War, it is unlikely that the United States and the other industrialised States would seek a reverse course and abandon the position adopted in the Outer Space Treaty. Indeed, it has been suggested that the prohibitions of private ownership and territorial sovereignty continue to attain widespread support among both industrialised and developing States in the international community.[15] It is likely that, with the adoption of an international regulatory framework that provides for the grant of exclusive prospecting, exploration and mining licences by the new international regulator, which in the interest of convenience will be referred to as the International Space Development Authority (the "Authority"), substantial opposition on the part of the industrialised States on this issue will whittle away into irrelevance, as the legal certainty needed to attract sufficient commercial investment would have been attained without the need to disturb the existing prohibitions on sovereignty and title in the Outer Space Treaty and the Moon Agreement.

6.2.2.2 Obligation to Share the Benefits Derived

Determining the Rate of Contributions

The most controversial of all obligations customarily associated with the common heritage of mankind doctrine is the requirement that there be an equitable sharing of the benefits derived from the exploitation and use of the common heritage of mankind.[16] This principle found expression in the Convention on the Law of the

---

[13] Outer Space Treaty, Article II. See Stephen Gorove, *Sovereignty and the Law of Outer Space Re-examined* (1977) 2 ANN. AIR & SP. L. 311 at 316; Ezra J. Reinstein, *Owning Outer Space* (1999) 20 NW. J. INT'L. L. & BUS. 59; and Rosanna Sattler, *Transporting a Legal System for Property Rights: From the Earth to the Stars* (2005) 6 CHI. J. INT'L. L. 23; Leslie I. Tennen, *Second Commentary on Emerging System of Property Rights in Outer Space* (2003) United Nations, PROCEEDINGS OF THE UNITED NATIONS / REPUBLIC OF KOREA WORKSHOP ON SPACE LAW 342 at 343; and Patricia M. Sterns and Leslie I. Tennen, *Privateering and Profiteering on the Moon and Other Celestial Bodies: Debunking the Myth of Property Rights in Space* (2003) 31 ADV. SPACE RES. 2433.

[14] See generally Nandasiri Jasentuliyana and Roy S. Lee (eds.), MANUAL ON SPACE LAW (1979), vol. 1.

[15] See Tennen, supra note 13 and Reinstein, supra note 13.

[16] See Frederick Arnold, *Toward a Principled Approach to the Distribution of Global Wealth: An Impartial Solution to the Dispute over Seabed Manganese Nodules* (1980) 17 SAN DIEGO L. REV. 557; Elisabeth Mann Borgese, *A Constitution for the Oceans: Comments and Suggestions Regarding Part XI of the Informal Composite Negotiating Text* (1978) 15 SAN DIEGO L. REV. 371; James R. Silkenat, *Solving the Problem of the Deep Seabed: The Informal Composite Negotiating Text for the First Committee of UNCLOS III* (1977) 9 N. Y. U. J. INT'L. L. & POL'Y. 177; Robert B. Krueger, *Policy Options in the Law of the Sea Negotiations* (1978) 6 INT'L. BUS. LAWYER 89; and Robert F. Pietrowski, Jr., *Hard Minerals on the Deep Ocean Floor: Implications for American Law and Policy* (1978) 19 WM. & MARY L. REV. 43.

Sea in relation to the exploitation of mineral resources in the deep seabed, which provides for a share for the International Seabed Authority ("ISA") of 35–50% before the mining concern has recovered all its development costs (including interest) and of 40–70% after all development costs have been recovered.[17] These shares of proceeds have proven to be unacceptable to the industrialised States.

Under the 1994 Agreement relating to the Implementation of Part XI of the United Nations Convention on the Law of the Sea of 10 December 1982 (the "1994 Agreement"), the scale of contributions was replaced with the following formulation:

> The rates of payments under the system shall be within the range of those prevailing in respect of land-based mining of the same or similar minerals in order to avoid giving deep seabed miners an artificial competitive advantage or imposing on them a competitive disadvantage. The system should not be complicated and should not impose major administrative costs on the [ISA] or on a contractor. Consideration should be given to the adoption of a royalty system or a combination of a royalty and profit-sharing system.[18]

Conceptually, it is relatively easy to see the appeal of this approach to free market economists of industrialised States and also to treasury officials of developing States that derive much of their revenue from the mining sector. In practice, however, it is worth noting that the total effective tax rate that is payable by terrestrial mining concerns, which form the basis of the comparison in the formulation above, do not differ substantially from the fixed contribution shares in the prescribed original Convention on the Law of the Sea. This is particularly so keeping in mind that almost all States in the world have zero-rated or exempted all minerals from export duties.[19]

Given the divergent range of royalty and taxation rates applicable to different mineral resources in different States throughout the world, the conceptual formulation found in the 1994 Agreement may have the potential to allow for too much discretion in the hands of a regulating authority, while fixed rates may have the potential to be too rigid in adapting to changes in global economic conditions (Table 6.2).

It may be preferable to fix the contribution rate with reference to a basket of domestic royalty rates in relation to selected mineral resources, in much the same way as some currencies were floated with reference to a "basket" of other currencies in the 1970s and 1980s.[20] The royalty rates included in the "basket" may be

---

[17] Convention on the Law of the Sea, Annex III, Article 13.

[18] Agreement Relating to the Implementation of Part XI of the United Nations Convention on the Law of the Sea of 10 December 1982 (the "*1994 Agreement*"), opened for signature on 28 July 1994, 1836 U.N.T.S. 3; 33 I.L.M. 1309 (entered into force on 28 July 1996), Annex, Section 8.

[19] James Otto, Craig Andrews, Fred Cawood, Michael Doggett, Pietro Guj, Frank Stermole, John Stermole and John Tilton, MINING ROYALTIES: A GLOBAL STUDY OF THEIR IMPACT ON INVESTORS, GOVERNMENT AND CIVIL SOCIETY (2006), at 35–37.

[20] See Eiji Ogawa and Takatoshi Ito, *On the Desirability of a Regional Basket Currency Arrangement* (2002) 16 J. JAPAN. & INT'L. ECON. 317; Hali J. Edison and Erling Vårdal, *Optimal Currency Baskets for Small, Developed Economies* (1990) 92 SCAND. J. ECON. 559; and Lars Hörngren and Anders Vredin, *Exchange Risk Premia in a Currency Basket System* (1989) 125 REV. WORLD ECON. 311.

Table 6.2 Comparative taxation on a model copper mine in selected states

State	Total effective tax rate (%)
Sweden	28.6
Western Australia, Australia	36.4
Chile	36.6
Zimbabwe	39.8
Argentina	40.0
China	41.7
Papua New Guinea	42.7
Bolivia	43.1
South Africa	45.0
Philippines	45.3
Kazakhstan	46.1
Peru	46.5
Tanzania	47.8
Poland	49.6
Arizona, United States	49.9
Mexico	49.9
Greenland, Denmark	50.2
Indonesia	52.2
Ghana	54.4
Mongolia	55.0
Uzbekistan	62.9
Côte d'Ivoire	62.4
Ontario, Canada	63.8

Source: James Otto et al. (2006, 36)

adjusted every few years by the Authority with reference to the competitive position of extraterrestrial mining *vis-à-vis* terrestrial mining on a similar basis as provided under the 1994 Agreement. This would provide some fixed reference for the purpose of providing certainty to the commercial venture while being flexible enough to adapt to medium and long term trends in the market.

Collection and Management of the Contributions

The collection of the mandatory contributions is to be done by the Authority or by a Member State on behalf of the Authority in the case of a private entity being responsible for the mining operation. The amount of the contributions to be paid to the Authority must be capable of being enforced in a practical and meaningful manner and, further, must be able to be verified by the conduct of an independent audit.

The financial contributions collected by the Authority ought to be paid into two separate funds, the "Administrative Fund" and the "Common Heritage Fund". The Administrative Fund is to provide the funding necessary for all of the functions of the Authority, while the Common Heritage Fund is to be used to finance projects for the benefit of the whole of humankind or for large-scale development and infrastructure projects for developing States. The issues relating to the management

6.2.2.3 Mandatory Transfer of Technology

In the Convention of the Law of the Sea, there is a provision for the mandatory transfer of the technology used by a licensee to the Enterprise for its use in the exploitation of mineral resources from the deep seabed.[21] Under the 1994 Agreement, this mandatory obligation was transformed into a general duty on the part of the industrialised States to facilitate the acquisition of technology by the Enterprise.[22] It is reasonable for the developing States to assume that the mandatory transfer of technology would form part of the obligations imposed on the industrialised States under any new international regime concerning celestial bodies arising from both the NIEO principles and the common heritage of mankind doctrine.[23]

To some extent, such a transfer is not difficult to achieve, considering much of the technology required is likely to be the subject of patents and other forms of registered industrial property and, as such, would be published at the registration authority and available for inspection by the public at large. Given the long time-frames involved in any mining venture on celestial bodies, all that would be required to achieve this result would be for the Implementation Agreement to provide for the publication and dissemination of patented or other registered technologies upon the expiry of any exclusivity or protection period as prescribed under international and domestic law.[24]

Unlike mining activities in the deep seabed, however, the technological capabilities needed for mining activities on celestial bodies are both advanced and sensitive. Based on the phases of a commercial mining venture on a Near Earth Asteroid, one would expect the technological requirements to include:

- transportation technologies, both in the outbound journey for the equipment and the inbound journey for the processed ores;
- advanced automated rocketry capabilities;
- landing capabilities;
- automated or robotic space mining equipment;
- space materiai processing capabilities;
- advanced computerised targeting capabilities; and
- advanced space power generation and propulsion systems.

---

[21] Convention on the Law of the Sea, Article 144.

[22] 1994 Agreement, Annex, Section 5.

[23] See, for example, Mark Orlove, *Spaced Out: The Third World Looks for a Way in to Outer Space* (1989) 4 CONN. J. INT'L. L. 597.

[24] Since 1995, most States have enacted patent protection laws that provide protection from the date of grant of the patent to 20 years from the filing date of the application: Agreement on Trade-Related Aspects of Intellectual Property Rights, opened for signature on 15 April 1994, 1869 U.N.T.S. 299 (entered into force on 1 January 1995), Article 33.

6.2 Need for Balancing Competing Interests

It is apparent that the industrialised States would not conceivably allow these technological capabilities to be transferred to an international commercial entity or to developing States, as much of the technological capabilities referred to above clearly fall within the scope of existing military and dual-use technology control regimes, such as the Missile Technology Control Regime (the "MTCR") and the Wassenaar Arrangement on Export Controls for Conventional Arms and Dual-Use Goods and Technologies (the "Wassenaar Arrangement").[25] It is noteworthy that many developing States have agreed to the principle that such technological capabilities ought not be proliferated internationally by their ratification and participation in the MTCR and the Wassenaar Arrangement.[26]

Accordingly, it is both inconceivable and unrealistic for developing States to expect any form of technology transfer for mining of celestial bodies would be acceptable to the industrialised States, regardless of whether such transfer is to be on a mandatory or a voluntary basis. However, it is improbable that such a requirement would be pressed by the developing States, knowing that there would not be any forthcoming compromise by the industrialised States on this issue.[27] It is more likely that the developing States would be content with a significant share of the benefits derived from such exploitation activities without experiencing the desire to undertake such exploitation activities for themselves.

6.2.2.4 Competing and Participating International Enterprise

The concept of creating an international commercial entity, controlled and operated by the relevant regulatory authority, to actively compete with commercial ventures regulated by that authority has proven to be completely unacceptable to the industrialised States. Accordingly, the creation of the Enterprise to participate in deep seabed mining had proven to be a significant obstacle to the acceptance of the

---

[25] Wassenaar Arrangement, BASIC DOCUMENTS (2009), at <http://www.wassenaar.org/publicdocuments/2009/Basic%20Documents%20-%20Jan%202009.pdf>, 20 January 2009, last accessed on 30 November 2009.

[26] The Member States of the Wassenaar Arrangement are Argentina, Australia, Austria, Belgium, Bulgaria, Canada, Croatia, Czech Republic, Denmark, Estonia, Finland, France, Germany, Greece, Hungary, Ireland, Italy, Japan, Latvia, Lithuania, Luxembourg, Malta, the Netherlands, New Zealand, Norway, Portugal, Republic of Korea, Romania, Russia, Slovakia, Slovenia, South Africa, Spain, Sweden, Switzerland, Turkey, Ukraine, the United Kingdom and the United States: Wassenaar Arrangement, *Participating States*, at <http://www.wassenaar.org/participants/index.html>, last accessed on 28 November 2009; and the MTCR Partners are Argentina, Australia, Austria, Belgium, Brazil, Canada, Czech Republic, Denmark, Finland, France, Germany, Greece, Hungary, Iceland, Ireland, Italy, Japan, Luxembourg, the Netherlands, New Zealand, Norway, Poland, Portugal, Republic of Korea, Russia, South Africa, Spain, Sweden, Switzerland, Turkey, Ukraine, the United Kingdom and the United States: Missile Technology Control Regime, *MTCR Partners*, at <http://www.mtcr.info/english/partners.html>, last accessed on 28 November 2009.

[27] See Colin B. Picker, *A View from 40,000 Feet: International Law and the Invisible Hand of Technology* (2002) 23 CARDOZO L. REV. 149.

Convention on the Law of the Sea by the industrialised States.[28] It would be reasonable to assume that the industrialised States have not had a change of heart on the issue when considering the regulatory framework for celestial bodies.

In fact, the difficulties and risks posed by the international proliferation of advanced space technology would only be more problematic in relation to the Enterprise. If the industrialised States would hesitate before transferring such technologies to other States, this hesitation would only multiply in relation to transferring such technologies to an intergovernmental organisation to be dominated by developing States. It would not be surprising that the industrialised States would prefer to surrender a larger share of the revenue derived from the exploitation of mineral resources than to be placed under an obligation to allow other States to acquire such advanced technology.

6.2.3 Economic Development vs. Environmental Safeguards

6.2.3.1 Preservation of the Environment of Celestial Bodies

The Outer Space Treaty requires States to avoid harmful contamination of outer space, the Moon and other celestial bodies in their exploration and use.[29] However, the term "harmful contamination" is not defined in the Outer Space Treaty. This obligation is expanded in the Moon Agreement, which requires States to:

> ... take measures to prevent the disruption of the existing balance of its environment, whether by introducing adverse changes in that environment, by its harmful contamination through the introduction of extra-environmental matter or otherwise.[30]

If the obligations of the Outer Space Treaty and the Moon Agreement required strict compliance, then much would turn on the definitions of "harmful contamination" and "adverse changes". There is as much potential to give narrow definitions to these terms, so as to make it virtually impossible to have practical use of celestial bodies, as it would be to define them broadly so that only activities that have the potential to threaten the wholesale destruction of the celestial body would need to be prevented. This absence of meaningful definitions has been the subject of much criticism in terms of the need to protect the environment of celestial bodies.[31] However, this *lacuna* does give the Authority with the opportunity to develop and

---

[28] Convention on the Law of the Sea, Article 170. See Reynolds, supra note 4 and Stephen E. Doyle, *Using Extraterrestrial Resources Under the Moon Agreement of 1979* (1998) 26 J. SP. L. 111.

[29] Outer Space Treaty, Article IX.

[30] Moon Agreement, Article 7(1).

[31] See, for example, Paul G. Dembling and Swadesh S. Kalsi, *Pollution of Man's Last Frontier: Adequacy of Present Space Environmental Law in Preserving the Resource of Outer Space* (1973) 20 NETH. INT'L. L. REV. 125; Marta Miklody, *Some Remarks to the Legal Status of Celestial Bodies and Protection of Environment* (1982) 25 PROC. COLL. L. OUTER SP. 117; and Raymond T. Swenson, *Pollution of the Extraterrestrial Environment* (1985) 25 A. F. L. REV. 70.

adopt detailed rules and guidelines on the protection of the environment of celestial bodies that would be consistent with the requirements of the Outer Space Treaty and the Moon Agreement.

In particular, the Authority ought to adopt appropriate rules that require Member States and their private entities that undertake prospecting, exploration and extraction activities on celestial bodies to:

- remove all equipment, fixtures and installations at the end of their licensed period and either return them to the Earth, transport them to another site on the same celestial body, transport them to another celestial body or to dispose of them by burning them up through a planetary atmosphere or in the Sun;
- remove all pollutants and waste materials at the end of their licensed period and either return them to the Earth or to dispose of them through the atmosphere of another planet or the Sun;
- minimise the generation of space debris in orbits around the Earth and the celestial body; and
- to the extent feasible, make good any environmental damage caused to the surface and subsoil of the celestial body during the licensed period.

While these requirements may appear to be somewhat stringent in nature, there are a number of factors that would mean that the burden imposed by these rules would be minimal in practice. First, prospecting, exploration and extraction activities on celestial bodies are likely to be robotic rather than manual in nature, as this would dispense with the need to send life support systems, living quarters, food and water supplies, a tavern and satellite-relayed screenings of Monday Night Football from the surface of the Earth, significantly reducing waste production. Second, as the energy needed to send materials back to the Earth is negligible compared to the energy cost of sending materials from the surface of the Earth, it is likely that ore processing would take place on the Earth, thus minimising the generation of pollutants. Third, the significant equipment manufacturing and transportation costs incurred make it highly likely that most of the equipment would be transferred or sold to another mining site, either on the same celestial body or otherwise. These factors would combine to reduce the actual cost of complying with these environmental preservation rules and, in any event, are comparable to the cost of complying with the environmental law applicable to mining activities in many terrestrial States.

6.2.3.2 Prevention of Harmful Contamination of the Earth

The Outer Space Treaty requires States to avoid harmful contamination and adverse changes to the environment of the Earth as a result of the introduction of extra-terrestrial matter.[32] This requirement is reaffirmed in the Moon Agreement.[33]

---

[32] Outer Space Treaty, Article IX.

[33] Moon Agreement, Article 7(1).

Christol has noted the limited effectiveness of these requirements as it leaves "ample room for States to obstruct international cooperation in space and to take arbitrary decisions".[34] Further, Dembling and Kalsi have observed that Article IX of the Outer Space Treaty "does not prevent harmful space conduct especially that result in Earth pollution" as it is "self-judging, self-imposed and self-policed".[35]

However, the dawn of the age of space mining will bring with it the significant risk that much harm may be done to the Earth, its biological diversity or its environment through the introduction of extra-terrestrial matter. This concern may be compounded by recent scientific studies that suggested that there was life on Mars.[36] While the risk of a biological contamination of the Earth resulting from mining of celestial bodies may be somewhat remote, the risk of chemical contamination of the Earth cannot be underestimated. Accordingly, if deemed appropriate, the Authority ought to adopt rules concerning the compulsory quarantine and sample testing and analysis of materials intended to be returned to the Earth.[37]

6.2.3.3 Restrictions on Exhaustive Exploitation of Celestial Bodies

As a definitional issue, the Moon Agreement and the other relevant treaties do not stipulate what type of celestial bodies it applies to or, more pertinently, what would constitute "celestial bodies" for the purposes of international space law. After all, bodies that exist in the Solar System range in size from to microparticles to Jupiter, in composition from dense metallic solids such as a number of asteroids to frozen lumps of ice and rock such as most comets, and in distance to the Earth as close as the Moon and as far as distant Kuiper Belt Objects.[38] The absence of such definitional limits raises a particular problem in the circumstance where technology enables a mining operation to move or even completely consume a celestial body in the Solar System, such as a comet or a small Near Earth Asteroid.

Considering the divergent views on what may constitute a "celestial body", it may be more prudent for any new international legal regime to ensure that all bodies in the Solar System be considered "celestial bodies" for the purposes of the international treaties and simply prescribe restrictions on the "exhaustive" or "complete" exploitation of smaller bodies, which involves the total destruction of the

---

[34] Carl Q. Christol, THE MODERN INTERNATIONAL LAW OF OUTER SPACE (1982), at 140.

[35] Dembling and Kalsi, supra note 31, at 141.

[36] See Hannah Devlin, *Evidence of Life on Mars Lurks Beneath Surface of Meteorite, NASA Experts Claim*, THE TIMES (London), 27 November 2009, at <http://www.timesonline.co.uk/tol/news/science/space/article6934078.ece>, last accessed on 30 November 2009.

[37] See George S. Robinson, *Earth Exposure to Martian Matter: Back Contamination Procedures and International Quarantine Regulations* (1976) 15 COLUM. J. TRANSNAT'L. L. 17.

[38] See, for example, R. Lynne Allen, Brett Gladman, J. J. Kavelaars, Jean-Marc Petit, Joel W. Parker and Philip D. Nicholson, *Discovery of a Low-Eccentricity, High Inclination Kuiper Belt Object at 58 AU* (2006) 640 ASTROPHYSICS J. 83 and Jane X. Luu and David C. Jewitt, *Kuiper Belt Objects: Relics from the Accretion Disk of the Sun* (2002) 30 ANN. REV. ASTRON. & ASTROP. 63.

small object. In other words, it may be more economically and physically convenient and feasible to move an entire small asteroid from its natural orbit to Earth orbit, where its resources can then be processed and dispatched to the surface of the Earth. For example, these restrictions may include prescribing the maximum diameter and mass of objects that may be removed from their natural orbit for any purpose, including mineral resource exploitation. Such limitations would also have the benefit of making it unlawful to move an object of significant size being moved to Earth orbit, which may pose significant risk to life and property on Earth by its accidental or even deliberate impact with the Earth.[39]

6.2.3.4 Nuclear Power Sources and Propulsion Systems

Nuclear and Radioisotopic Power Sources

Although solar power remains the most efficient and effective means of generating electrical power in the inner Solar System, such power sources may need to be supplemented by other means of electricity generation considering the likely distance from the Sun and also the tremendous needs for electrical power for mining operations. Nuclear reactors are thus very strong candidates for deployment on such missions, as they can provide a large amount of electrical power without the need to require much mass to be launched from the surface of the Earth.

In 1992, the General Assembly of the United Nations adopted the Principles Relevant to the Use of Nuclear Power Sources in Outer Space (the "NPS Principles").[40] Elaborating on the requirements of the Outer Space Treaty, the NPS Principles impose conditions that have to be met when designing nuclear and radioisotopic power sources onboard a spacecraft. In particular, the NPS Principles requires that:

(1) the probability of accidents onboard with serious radiological consequences must be kept extremely low;
(2) any foreseeable safety-related failures or malfunctions onboard the spacecraft must be capable of being corrected or counteracted by procedural or automatic means;
(3) the design of the spacecraft must be done in a manner that ensures, with a high degree of confidence, that the hazards in foreseeable operational or accidental circumstances are kept below acceptable levels, with reference to appropriate standards imposed by the International Commission on Radiological Protection

---

[39] See, for example, Leonard David, *Space Weapons for Earth Wars*, at <http://www.space.com/businesstechnology/technology/space_war_020515-1.html>, 15 May 2002, last accessed on 30 November 2009 and Robert Preston, Dana J. Johnson, Sean J. A. Edwards, Michael D. Miller and Calvin Shipbaugh, SPACE WEAPONS, EARTH WARS (2002), at Appendix C.

[40] General Assembly Resolution 47/68. See also Jason Reiskind, *Toward a Responsible Use of Nuclear Power in Outer Space – The Canadian Initiative in the United Nations* (1981) 4 ANN. AIR & SP. L. 461.

and relevant international radiological protection guidelines to limit exposure in accidents; and

(4) spacecraft design must restrict radiation exposure geographically and to individuals to the limit of 1 millisievert per year.[41]

One of the issues of particular concern in the use of nuclear and radioisotopic fuel cells has a different condition, namely that of ultimate disposal. Such spacecraft must be designed with a system of containment that will withstand the heat and other conditions of re-entry and impact on the surface of the Earth or water to ensure that no radioactive material may be scattered into the atmosphere, the ocean or the soil. These concerns stem from the long half-life and the potency of plutonium-238 and other likely radioisotopic materials to be used. Further, new guidelines ought to be adopted by the international community to ensure that such materials and equipment are not left on the surface of celestial bodies but are instead sent into a trajectory towards the Sun for ultimate disposal, preferably without being in close proximity to the Earth when crossing Earth orbit.

Nuclear and Radioisotopic Propulsion Systems

Given the distances that need to be travelled in any commercial space mining venture, nuclear and radioisotopic propulsion systems would be strong candidates for many such ventures. However, the preamble to the NPS Principles specifies that the sphere of their application includes only utilisation of nuclear power sources in space to generate "electric power on board space objects for non-propulsive purposes". At the same time, it leaves the option to revise the Principles, as new nuclear power applications emerge and international recommendations on radiological protection evolve. In practice, this means that for the time being only the existing general principles as contained in the relevant treaties would apply to a spacecraft using nuclear and radioisotopic propulsion systems and the safety requirements of the NPS Principles would have no application.

As nuclear propulsion systems are not limited by the quantity of chemical fuels that may be carried into space, such systems can generate continuous thrust to achieve much faster speeds than those based on the existing chemical rocket technologies. In 1998, the United States deployed an ion propulsion engine using xenon gas onboard the probe *Deep Space 1*.[42] This example clearly shows that there is the

---

[41] NPS Principles, Principle 3.

[42] See Michael J. Patterson, John E. Foster, Thomas Haag, Vincent K. Rawlin, George C. Soulas and Robert F. Roman, *NEXT: NASA's Evolutionary Xenon Thruster* (2002), paper presented at the 38th Joint Propulsion Conference and Exhibition, 7–10 July 2002, in Indianapolis, IN, USA and Ivana Hrbud, Melissa van Dyke, Mike Houts and Keith Goodfellow, *End-to-End Demonstrator of the Safe Affordable Fission Engine (SAFE) 30: Power Conversion and Ion Engine Operation*, paper presented at the Space Technologies Applications International Forum Conference, 3–7 February 2002, Albuquerque, NM, USA.

potential to develop nuclear electric propulsion systems that will make space travel and transportation faster, cheaper and more energy efficient.

As nuclear propulsion systems are much more efficient and effective when deployed on deep space and interplanetary missions, the lack of specific legal regulation of such propulsion systems would, in practice, pose minimum risk to the Earth and its environment as well as to human health, both in terms of their operation, waste residues and disposal.[43] The new international framework for celestial bodies ought to prescribe appropriate safety rules and guidelines, particularly in relation to:

(1) the need for immediate, public and full disclosure of any relevant information by the responsible country about a spacecraft with a nuclear or radioisotopic propulsion system onboard in cases of its malfunction or possible re-entry into the Earth atmosphere;
(2) imposing the same design safeguards on nuclear propulsion systems as those relating to nuclear power sources under the NPS Principles;
(3) providing for the full and absolute entitlement to reimbursement of all reasonable costs incurred in the recovery, cleanup and return of the spacecraft and of any environmental damage caused or sustained; and
(4) require for their planned ultimate disposal by a trajectory that would ensure their eventual capture by the gravity of the Sun.[44]

6.2.4 Regulation vs. Free Market

6.2.4.1 Creation of an International Regulatory Authority

One of the fundamental aspects of the common heritage of mankind doctrine is the international regulation and control of their use and exploitation. Consequently, the unilateral regulatory approaches taken by various industrialised States, particularly the United States, in opposition to the Convention on the Law of the Sea was a

---

[43] See, for example, Roger M. Myers, Eric J. Pencil, Vincent K. Rawlin, Michael Kussmaul and Katessha Oden, *NSTAR Ion Thruster Plume Impact Assessments*, paper presented at the 31st Joint Propulsion Conference and Exhibition, 10–12 July 1995, in San Diego, CA, USA; John S. Synder, John R. Anderson, Jonathan L. van Noord and George C. Soulas, *Environmental Testing of the NEXT PM1 Ion Engine*, paper presented at the 43rd Joint Propulsion Conference and Exhibition, 8–11 July 2007, in Cincinnati, OH, USA; and James S. Sovey, Joyce A. Dever and John L. Power, *Retention of Sputtered Molybdenum on Ion Engine Discharge Chamber Surfaces*, paper presented at the 27th International Electric Propulsion Conference, 14–19 October 2001, in Pasadena, CA, USA.

[44] See Ricky J. Lee and Catherine Doldirina, *Legal and Policy Issues Arising from the Use of Nuclear and Radioisotopic Power Sources and Propulsion Systems in Outer Space*, paper presented at the 60th International Astronautical Congress, 12–16 October 2009, in Daejeon, Republic of Korea.

serious blow to the advocates of international regulation.[45] This was particularly so as the unilateral regulatory frameworks, which operated between States on the basis of mutual and reciprocal recognition and comity, presented an alternative regulatory approach for mining activities in the deep seabed that was practical and feasible.

It is inconceivable for the developing States to agree to abandon their demand for an international regulatory framework administered by an international authority and, instead, to accept unilateral regulation of mining activities on celestial bodies on the basis of reciprocal recognition. This is particularly so in light of the experience in relation to the Convention on the Law of the Sea. Accordingly, it is evident that a regulatory regime can only be implemented through the creation of the Authority that will have the right to control and regulate activities in relation to exploitation of mineral resources from celestial bodies.

6.2.4.2 Exclusivity in Licensing

As part and parcel of the effort to persuade industrialised States to concede the role of regulating mining activities on celestial bodies to the Authority instead of doing so by means of unilateral approaches, the Authority must be able to grant exclusive licensing rights to States and their private entities. Exclusivity would form the fundamental basis on which sufficient legal certainty can be found to attract investment.

Although such a system may be subject to abuse and certain measures would need to be adopted in order to eliminate or minimise such abuses, such exclusivity would not be a negotiable element for the industrialised States. In any event, the absence of exclusivity would only lead to potential disputes arising between Member States or their private entities as a result of overlapping rights. This only greatly increases the risk of harmful interference in their respective space activities, about which the Outer Space Treaty has required States to prevent and consult with each other.[46]

6.2.4.3 Minimum Work Obligations

One of the significant lessons learned from the regulation of the use of the geostationary orbit by the International Telecommunication Union is the need to avoid the "paper satellite" problem. The problem became particularly acute when Tonga sought to abuse the first-come first-served process of allocating orbital slots on the geostationary orbit by filing for 16 orbital slots between Asia and the Americas.[47]

---

[45] See Robert Everett Bostrom, *The United States' Legislative Response to the Third United Nations Conference on the Law of the Sea Deadlock* (1979) 2 B. C. INT'L. & COMP. L. J. 409 and Charles Douglas Oliver, *Interim Deep Seabed Mining Legislation: An International Environmental Perspective* (1981) 8 J. LEGIS. 73.

[46] Outer Space Treaty, Article IX.

[47] See Edmund L. Andrews, *Tiny Tonga Seeks Satellite Empire in Space*, THE NEW YORK TIMES, 28 August 1990, at A1; Albert N. Delzeit and Robert F. Beal, *The Vulnerability of the Pacific*

Although Tonga was eventually compelled to limit its claim to six orbital slots, it is clear that conferring orbital slots to most developing States would only allow these States to "lease" their orbital slots to commercial operators for profit.[48]

Such problems can be overcome by the Authority being proactive in assessing the feasibility of a proposed mining operation and granting exclusive licences for prospecting, exploration and/or extraction activities only on condition that the licensee would be required to satisfy minimum work requirements that prescribe the minimum amount of ore that must be extracted within a certain timeframe or the licensee may face monetary penalties and the forfeiture of the licence. These arrangements are quite common under the domestic laws and regulations of most States. However, a balance must be struck between the need to ensure licences are not granted for frivolous licence applications and allowance for the enormous amount of financial investment and timeframe involved in extra-terrestrial mining operations.

6.2.4.4 Protection of the Global Commodity Markets

One of the unique features of extra-terrestrial mining ventures is that, unlike terrestrial mining or even deep seabed mining, there is unlikely to be a regular or gradual production of mineral resources from such mining activities. Instead, it is more likely that the entire production of the mining operation would be transported in a single "shipment" from the celestial body to the Earth, or for the extracted ores to be transported to the Earth in large tranches. Unless the mining venture sought to store the ores on the Earth and gradually release them onto the commodity markets, which it is unlikely to do given the anticipated need to recover the capital investment as quickly as possible, such a large influx of supply of a particular commodity would have a significant impact on the price of that commodity in the global market.[49] However, any intervention on the part of the Authority in restricting the amount of mineral resources that may be released onto the commodity markets may

---

Rim Orbital Spectrum Under International Space Law (1996) 9 N. Y. INT'L. L. Rev. 69; and Jonathan Ira Ezor, *Costs Overhead: Tonga's Claiming of Sixteen Geostationary Orbital Sites and the Implications for U.S. Space Policy* (1993) 24 L. & POL'Y. INT'L. BUS. 915.

[48] See, for example, Jannat C. Thompson, *Space for Rent: The International Telecommunication Union, Space Law and Orbit/Spectrum Licensing* (1996) 62 J. AIR L. & COM. 279 and Henry Wong, *The Paper "Satellite" Chase: The ITU Prepares for Its Final Exam in Resolution 18* (1998) 63 J. AIR L. & COM. 849.

[49] See, for example, Milton Friedman, *The Reduction of Fluctuations in the Incomes of Primary Producers: A Critical Comment* (1954) 64 ECON. J. 698; David Bevan, Paul Collier and Jan Willem Gunning, TEMPORARY TRADE SHOCKS IN DEVELOPING COUNTRIES: CONSEQUENCES AND POLICY RESPONSES (1991); Jeffrey M. Davis, *The Economic Effects of Windfall Gains in Export Earnings 1975–1978* (1983) 11 WORLD DEV. 119; and Angus Deaton and Ron Miller, *International Commodity Prices, Macroeconomic Performance and Politics in Sub-Saharan Africa*, at Princeton University, <http://www.princeton.edu/~deaton/downloads/International_Commodity_Prices.pdf>, October 1995, last accessed on 29 November 2009.

be seen by the industrialised States to be an unwelcome intrusion in the workings of a functioning global free market.

One solution may be for the Authority to buy the mineral resources produced from celestial bodies at the market price of those mineral resources, after deducting a small discount. The Authority would then slowly release the resources onto the global commodity market in accordance with an established plan set out by the Authority. This would have the benefit of providing the extra-terrestrial mining venture with the best and quickest means of realising full value for their mineral resources. The Authority, meanwhile, would have effectively shielded both the relevant venture and the States with significant terrestrial mining ventures from a sudden depreciation of the commodity prices resulting from a sudden large increase in supply. However, to implement such a strategy, the Authority would need significant cash reserves that it may not have available to it without resort to the Common Heritage Fund.

6.2.5 Public Interest vs. Commercial Concerns

6.2.5.1 Satisfaction of Baseline Demand of the Least Developed States for Essential and Scarce Mineral Resources

The principal motivation for mining of mineral resources from celestial bodies will be economic needs driven by the physical and economic scarcity of such resources in the Earth's crust. Consequently, it is foreseeable that eventually the entire global demand for most mineral resources will be met by the supply from celestial bodies. This is particularly so with the increasing consciousness among the international community as to the environmental and climatic impact of terrestrial mining activities that may lead to the socio-economic exhaustion of mineral resources on Earth before the physical exhaustion of such resources.

If the mineral resources of the Earth will eventually be depleted by mining activities, then there may be philosophical as well as economic objections on the part of the developing States, particularly the poorest and least developed among them. Principally, it would be seen as an attempt by the industrialised States to deplete the mineral resources of celestial bodies to fuel their industrial needs, just as they have done with those in the Earth's crust since the Industrial Revolution. Since these resources are to be extracted from the common heritage of mankind, a strong argument may be made that such mining activities ought to ensure that the poorest and least developed States of the world are ensured a baseline supply of essential mineral resources to ensure their survival and basic economic development. Such baseline supply obligations can be compared with the universal service obligations that existed in International Telecommunication Satellite Organisation ("INTELSAT") and International Maritime Satellite Organisation ("INMARSAT") prior to their corporatisation near the end of the previous century.[50]

---

[50] See John C. Panzar, *A Methodology for Measuring the Costs of Universal Service Obligations* (2000) 12 INFO. ECON. & POL'Y. 211 and Kenneth Katkin, *Communication Breakdown?*

6.2 Need for Balancing Competing Interests 291

The United Nations defines a Least Developed Country ("LDC") based on:

(1) has gross national income per capita of less than US $745.00;
(2) human factors, such as high percentage of undernourished population, high infant mortality rate, low secondary school enrolment ratio and low literacy rate; and
(3) economic factors, such as population, geographical remoteness, merchandise export concentration, share of primary production in gross domestic product, homelessness and economic and export instability.[51]

As in November 2009, there are currently 49 LDCs as defined by the United Nations.[52] It is suggested that, in relation to certain mineral resources identified by the Authority as being essential for survival and basic economic development, that the Authority would ensure a baseline supply of such mineral resources to the LDCs. If the Authority is to purchase the entire production output of mining activities from celestial bodies that has been returned to the Earth, then such a supply would not be difficult to be procured. Further, the financing for such supplies would be provided by the discounting applied by the Authority to the commodity price when purchasing the mineral resources from the extra-terrestrial mining operation.

6.2.5.2 Economic Protection of Terrestrial Mining Activities

In the Convention on the Law of the Sea, the economic interests of developing States with significant terrestrial mining activities are protected by certain economic measures. These measures include limitations placed on production of mineral resources from the deep seabed and payments by deep seabed mining operations to a compensation fund to assist such developing States in adjusting to any resulting adverse economic conditions.

However, deep seabed mining and the framework for mining of celestial bodies as proposed differ in a number of significant aspects and, accordingly, it may not

---

The Future of Global Connectivity After the Privatisation of INTELSAT (2005) 38 VAND. J. TRANSNAT'L. L. 1323.

[51] United Nations Office of the High Representative for the Least Developed Countries, Landlocked Developing Countries and the Small Island Developing States, *Criteria for Identification of LDCs*, at <http://www.unohrlls.org/en/ldc/related/59/>, last accessed on 27 November 2009.

[52] The 49 LDCs are Afghanistan, Angola, Bangladesh, Benin, Bhutan, Burkina Faso, Burundi, Cambodia, Central African Republic, Chad, Comoros, Democratic Republic of the Congo, Djibouti, Equatorial Guinea, Eritrea, Ethiopia, Gambia, Guinea, Guinea-Bissau, Kiribati, Laos, Lesotho, Liberia, Madagascar, Malawi, Maldives, Mali, Mauritania, Mozambique, Myanmar, Nepal, Niger, Rwanda, Samoa, São Tomé and Príncipe, Senegal, Sierra Leone, Solomon Islands, Somalia, Sudan, Tanzania, Timor Leste, Togo, Tuvalu, Uganda, Vanuatu, Yemen and Zambia: United Nations Office of the High Representative for the Least Developed Countries, Landlocked Developing Countries and the Small Island Developing States, *Least Developed Countries: Country Profiles*, at <http://www.unohrlls.org/en/ldc/related/62/>, last accessed on 27 November 2009.

be necessary or prudent for the similar measures to be implemented in relation to mining of celestial bodies. These major differences include:

(1) the mining of celestial bodies is likely to be prompted by a decline in the supply of mineral resources from the Earth's crust, which is unlikely to be a strong economic factor in the pursuit of deep seabed mining;
(2) the Contribution to be paid to the Administrative Fund and the Common Heritage Fund would be comparable to royalties payable to any of the Member States, had the mining operation taken place within its territory;
(3) the Authority will purchase the entire production of mining operations on celestial bodies that are returned to the Earth and gradually release the mineral resources onto the commodity markets to avoid price shocks and other sudden changes to the global markets; and
(4) the costs incurred in undertaking a mining operation on celestial bodies is significantly and substantially higher than equivalent costs of deep seabed mining and terrestrial mining, particularly taking into account the costs to be incurred in transportation and robotic mining equipment.

6.2.5.3 Management and Control of the Common Heritage Fund by the International Monetary Fund and the World Bank

One of the potentially controversial aspects of the implementation of the new regulatory framework will be the control and management of the Common Heritage Fund. Instead of having such management functions carried out by the Authority or a part of the Authority, it may be more appropriate to take advantage of existing international institutions and have the Common Heritage Fund managed and controlled jointly by the International Monetary Fund (the "IMF") and the International Bank for Reconstruction and Development (the "World Bank").

The World Bank is tasked with providing financial and technical assistance to developing States, usually by means of low-interest loans, interest-free credits and grants to developing States for infrastructure and social projects.[53] Decision-making power is held by the Board of Directors, which comprises one Executive Director from each of France, Germany, Japan, the United Kingdom and the United States and another 19 Executive Directors elected by the other Member States, with voting rights weighted by the shareholding of the Member States that appointed them.[54]

The IMF was established for the purpose of, inter alia, facilitating the balanced growth of international trade, provide financial resources to assist States in balance of payments difficulties and to assist in the reduction of poverty through funding to States in financial difficulties and provide technical assistance to assist in improving economic management.[55] Similar to the World Bank, the IMF is managed by

---

[53] Articles of Agreement of the International Bank for Reconstruction and Development, opened for signature on 27 December 1945, 2 U.N.T.S. 134 (entered into force on 27 December 1945), Article I.

[54] Ibid., Article V(4).

[55] Articles of Agreement of the International Monetary Fund, opened for signature on 27 December 1945, 2 U.N.T.S. 39 (entered into force on 27 December 1945), Article I.

its Executive Board, on which voting rights are determined based on the special drawing rights held by the Member State appointing that Executive Director.[56]

The proposal to have the Contributions paid into a fund such as the Common Heritage Fund, which is to be administered by the World Bank, is not a new one.[57] However, though some of the funds in the Common Heritage Fund may be distributed to developing States for their own use or to assist them in the exploration and use of outer space, it may be prudent to limit such payments to a minor portion of the use of the Common Heritage Fund.[58] Instead, the majority of the Common Heritage Fund ought to be administered jointly by the World Bank and the IMF to assist developing States in specific funding for retiring short-term and high-interest sovereign debt, providing funding for infrastructure, health and education projects and to correct balance of payment problems.

6.2.6 Hard Law vs. Soft Law

Since the adoption of the Moon Agreement, the principal law-making activities of the United Nations in relation to outer space and celestial bodies have shifted focus from the drafting and adoption of multilateral treaties to General Assembly declarations. While treaties are binding on States that have ratified them, it remains a matter of academic and intergovernmental controversy if resolutions of the General Assembly can be considered binding instruments of international law without the intervention of customary principles. Even with the five multilateral treaties, the international community appears to have deliberately avoided the creation of binding dispute settlement and enforcement mechanisms, most notably in the nature of the Claims Commission provided for under the Convention on the International Liability for Damage Caused by Space Objects (the "Liability Convention").[59]

For the Authority and the provisions of the Outer Space Treaty and the Moon Agreement to have any international legal credibility in regulating the prospecting, exploration, extraction and ancillary activities on celestial bodies, enforcement mechanisms and dispute settlement mechanisms must be created as an integral part of the Authority. To that end, it would be necessary for its Member States to adopt binding and compulsory dispute settlement mechanisms and for enforcement mechanisms involving imposition of penalties, forfeiture of the Permits or even seizure and confiscation of mineral resources from celestial bodies (Fig. 6.2).

---

[56] Ibid., Article XII.

[57] See Brian M. Hoffstadt, *Moving the Heavens: Lunar Mining and the "Common Heritage of Mankind" in the Moon Treaty* (1994) 42 U.C.L.A. L. REV. 575 at 615.

[58] Ibid.; and Elena P. Kamenetskaya, *On the Establishment of a World Space Organisation: Some Considerations and Remarks* (1989) 32 PROC. COLL. L. OUTER SP. 358.

[59] Convention on International Liability for Damage Caused by Space Objects (the "*Liability Convention*"), opened for signature on 29 March 1972, 961 U.N.T.S. 187, 24 U.S.T. 2389, T.I.A.S. 7762; 1975 A.T.S. 5 (entered into force on 1 September 1972), Article XIX.

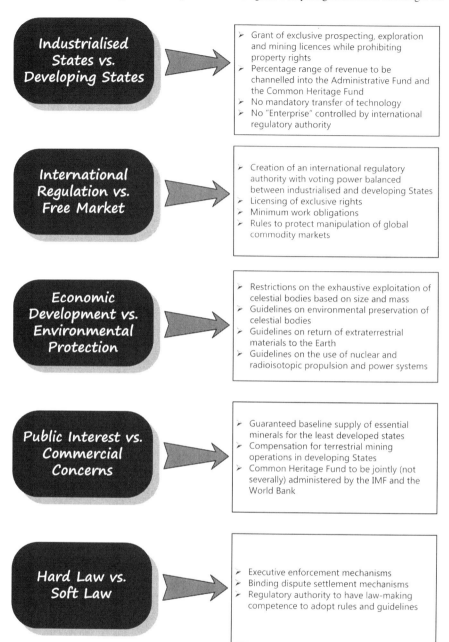

Fig. 6.2 Summary of proposed compromises in the new legal framework

Only through such mechanisms can the developing States have confidence in the ability of the Authority to ensure compliance with legal principles and specific rules and guidelines of the Authority by the industrialised States and their nationals and private entities.

In addition to the multilateral treaties, it will be necessary for the Authority to have quasi-legislative powers in drafting and enacting specific regulations, rules and guidelines dealing with various aspects of the regulation of mining activities on celestial bodies. These issues include preservation of the environment of celestial bodies, protection of the environment of the Earth and the use of nuclear and radioisotopic power sources and propulsion systems. It is preferable for the Authority to regulate these and similar issues by way of regulations, rules and guidelines rather than by treaty provisions as the former provides for much-needed flexibility and adaptability to technological advancement and innovation and shifting international socio-economic and environmental concerns. Consequently, there must be some means by which the Authority can have such powers to make regulations and for their approval and implementation by the Member States of the Authority.

6.3 Practical Implementation of the New Legal Framework

6.3.1 Overview

With the legal and policy issues resolved, the creation of an international legal framework for commercial mining activities in outer space would not be complete without resolving also the practical and administrative issues in relation to the implementation and operation of such a new framework. While the new international legal framework can be created through the adoption of the Implementation Agreement, the framework would have no practical effect until it is implemented and the mechanisms for doing so are created. It would become apparent that the practical implementation of this new framework cannot be achieved without the establishment of a new organisation that would administer the new regulatory framework.

The essential elements of the structure, composition, functions and powers of the organs of this new organisation will need to be considered in turn to enable its establishment and the implementation of the revised legal principles to be adopted in the new framework. This is in addition to considerations that must be made in relation to the financing and budgetary requirements of creating and operating such an organisation, potentially prior to any revenues are generated from the exploitation of resources from celestial bodies.

It is envisaged that the resolution of the practical issues arising from the implementation of a new international legal framework would improve the acceptability of the new proposed framework for the international community. Further, this would also reduce the legal uncertainty relating to commercial space mining ventures and provide incidentally an administrative framework for the regulation of other future activities in outer space.

6.3.2 Implementation Agreement

Just as the 1994 Agreement had provided for the necessary compromises concerning the position of various States on the Convention on the Law of the Sea, it is appropriate to implement the new international legal framework by means of a multilateral treaty (the "Implementation Agreement") embodying the legal principles to be applied in creating the framework under the Outer Space Treaty and the Moon Agreement. As such, it is clear that any State that desired to participate in the Authority must also be party to the Outer Space Treaty and the Moon Agreement.

There are four principal reasons why adoption by means of the Implementation Agreement is preferable to drafting and adopting a treaty de novo to replace the Moon Agreement. First, most of the principles in the Moon Agreement would be acceptable to both industrialised and developing States if agreement can be reached as to the practical and financial content and effect of the common heritage of mankind obligations under Article 11. Second, the majority of the principles contained in the Moon Agreement merely reaffirm or elaborate the existing legal principles as contained in the Outer Space Treaty and may have even crystallised into principles of customary international law. Third, the existence of the Moon Agreement is one of the major driving forces behind any impetus to create a new regulatory framework and, in the absence of the Moon Agreement, it is possible that such efforts would lose some sense of motivation and urgency in adopting a new framework. Fourth, it is perhaps a matter of fairness to the 13 States Parties to the Moon Agreement that they ought to retain the rights and benefits of being party to the Moon Agreement as the new framework is no more than an implementation of the regime foreshadowed in Article 11 of the Moon Agreement.

6.3.3 Organisational Structure

The Authority will have to be established by multilateral treaty, presumably the same treaty that would adopt the new legal and policy principles to be implemented by the Authority. The proposed objects of the Authority would include:

(1) the participation of all of its Member States in the formulation and revision of legal and regulatory provisions of the relevant treaties dealing with space activities, particularly concerning space mining ventures;
(2) the regulation of space mining ventures on celestial bodies through the licensing of various relevant activities and continuing surveillance and monitoring of activities conducted by licensees, as well as the collection and distribution of a share of the financial benefits as derived; and

6.3 Practical Implementation of the New Legal Framework

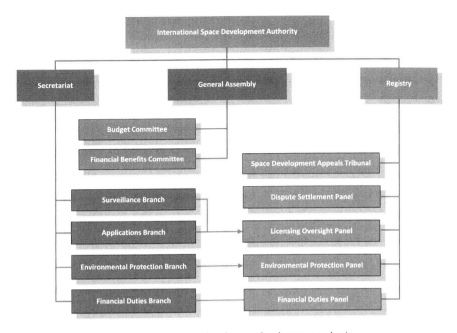

Fig. 6.3 Structure of the proposed international space development authority

(3) the peaceful and fair settlement of disputes between the Authority and the States and between States themselves in relation to mining activities in outer space and the provision of judicial oversight of the licensing activities of the Authority.

Accordingly, the Authority is proposed to have three major administrative organs. The General Assembly of the Authority is to act as the quasi-legislative body and would be the principal deliberative body of the Authority. The operational and administrative work of the Authority is to be conducted by the Secretariat, which also facilitates the work of the General Assembly, the Branches of which are responsible for different aspects of the practical implementation of the new legal and regulatory framework. The dispute settlement role of the Authority is to be facilitated by the Registry, which provides administrative support and coordination for a number of arbitral panels and the appellate body, which for convenience may be called the Space Development Appeals Tribunal (the "Tribunal") (Fig. 6.3).

6.3.4 Membership of the Authority

6.3.4.1 Different Approaches to Membership

On the subject of qualification for membership of the Authority, there are clearly at least four possible approaches to determining membership of the Authority:

(1) membership is limited to space-faring States only with equal voting rights;
(2) membership is limited to States that have made a financial investment into the commercial activities of the Authority only;
(3) membership is open to all Member States of the United Nations; or
(4) membership is open to all States regardless of their status of membership of the United Nations.

There are some precedents for intergovernmental organisations that limit access to States that meet a certain qualifying criteria. The 1959 Antarctic Treaty, for example, limits participation to States that have demonstrated their interest in Antarctica by "conducting substantial research activity there, such as the establishment of a scientific station or the despatch of a scientific expedition".[60] The 1994 Final Act and Agreement Establishing the World Trade Organisation provides that accession to the World Trade Organisation ("*WTO*") would be on terms to be agreed between the prospective Member State and each interested Member State by consensus.[61] Although there is the possibility for limiting membership to the Authority only to space-faring States, the existing space treaties would undermine the legal effectiveness of any of the legal principles contained in the new legal framework. This is particularly the case with the Outer Space Treaty and the Moon Agreement. The existing principles of non-appropriation and common heritage of mankind, among other legal principles, would cast doubt and uncertainty on the effect of legal provisions limited in their application to an exclusive club of Member States. With a legal framework created with the intention of providing legal certainty to the international community at large, such limited membership to the new legal framework is clearly unsuitable.

Similarly, it is probably not feasible or practicable for membership of the Authority to be limited to States that make a substantial financial investment in the Authority. Certain international intergovernmental institutions, such as the IMF, INTELSAT or INMARSAT, all require some substantial contribution to the financial capital of the organisation. In the case of the IMF, for example, membership and voting rights are tied to the financial contribution of its Member States.[62] Similar provisions can be found in the respective constituent documents of INTELSAT and INMARSAT, before their subsequent corporatisation to become multinational business concerns.[63]

---

[60] Antarctic Treaty, opened for signature on 1 December 1959, 402 U.N.T.S. 71; 12 U.S.T. 794; 19 I.L.M. 860 (entered into force on 23 June 1961), Article IX.

[61] Final Act and Agreement Establishing the World Trade Organisation, opened for signature on 15 April 1994, 1867 U.N.T.S. 154; 33 I.L.M. 1144 (entered into force on 1 January 1995), Article XII.

[62] Articles of Agreement of the International Monetary Fund, Articles III and XII(5).

[63] Agreement on the International Telecommunications Satellite Organisation, opened for signature on 20 August 1971, 23 U.S.T. 3810 (entered into force on 12 February 1973), Articles V and Convention on the International Maritime Satellite Organisation, opened for signature on 3 September 1976, 31 U.S.T. 1; T.I.A.S. 905 (entered into force on 16 July 1979), Articles 5.

6.3 Practical Implementation of the New Legal Framework

Such a model for the membership of the Authority is not feasible because it is not carrying on an enterprise of some kind. For example, INTELSAT and INMARSAT were intergovernmental organisations created to enable a common business enterprise with very high infrastructure costs and potential global benefits to be established and financed.[64] The IMF was established to act as a lender to poorer States in monetary crises and endemic liquidity issues and, as such had to be capitalised by contributions from its Member States.[65] The Authority can conceivably carry on a common global enterprise in the pursuit of mineral wealth in outer space, in a similar fashion to the earlier creation and development of INTELSAT and INMARSAT. However, such a development for the deep seabed had proved to be widely unpopular among members of the international community.[66] It is thus likely that taking a similar approach to the mining of celestial bodies may prove to be equally unpopular, if not more so.

There is a significant number of multilateral treaties and conventions that are only open to members of the United Nations, such as the Convention on the Prevention and Punishment of the Crime of Genocide or the International Covenant of Civil and Political Rights.[67] There are certainly some positive reasons why such a restriction may be prudent, especially within the context of an international treaty dealing with the peaceful exploitation of mineral resources in outer space. First, the Charter of the United Nations requires its Member States to refrain from the use of force in resolving international disputes.[68] Second, the Member States of the United Nations are expressly prohibited from intervening in the domestic affairs of other

---

[64] Agreement on the International Telecommunications Satellite Organisation, Article III and Convention on the International Maritime Satellite Organisation, Article 3. See Alan Beesley, Edward McWhinney Q.C., Dallas W. Smythe, Barry Mawhinney and A. E. Gotlieb, *The Legal Problems of International Telecommunications with Special Reference to INTELSAT* (1970) 20 UNI. TORONTO L. J. 287; and Steven A. Levy, *INTELSAT: Technology, Politics and the Transformation of a Regime* (1975) 29 INT'L. ORG. 655.

[65] Articles of Agreement of the International Monetary Fund, Article III.

[66] James B. Morell, THE LAW OF THE SEA: AN HISTORICAL ANALYSIS OF THE 1982 TREATY AND ITS REJECTION BY THE UNITED STATES (1992); David Silverstein, *Proprietary Protection for Deepsea Mining Technology in Return for Technology Transfer: New Approach to the Seabed Controversy* (1978) 60 J. PAT. OFF. SOC'Y. 135; and Yuwen Li, TRANSFER OF TECHNOLOGY FOR DEEP SEABED MINING: THE 1982 LAW OF THE SEA CONVENTION AND BEYOND (1994).

[67] Convention on the Prevention and Punishment of the Crime of Genocide, opened for signature on 9 December 1948, 78 U.N.T.S. 277 (entered into force on 12 January 1951), Article 11 and International Covenant on Civil and Political Rights, opened for signature on 16 December 1966, 999 U.N.T.S. 171 (entered into force on 23 March 1976), Article 48.

[68] Charter of the United Nations, opened for signature on 26 June 1945, 1 U.N.T.S. xvi; 1946 U.K.T.S. 67 (entered into force on 24 October 1945), Article 2(4). Article 2(4) of the Charter of the United Nations states that "All Members shall refrain in their international relations from the threat or use of force against the territorial integrity or political independence of any state, or in any other manner inconsistent with the Purposes of the United Nations."

States.[69] Third, States are bound to comply with decisions of the Security Council.[70] These decisions include decisions imposing trade sanctions and embargoes and the use of military force against States that are considered by the Security Council to be a threat to international peace and security.[71] Limiting the membership of the Authority to that of the United Nations would mean that the Member States of the Authority would have to comply with the international legal norms relating to peaceful settlement of disputes and other relevant principles in their activities in outer space. However, this benefit may be redundant given that those norms have probably crystallised into customary international law and, in any event, any State that is party to the Outer Space Treaty would already be required to comply with the Charter of the United Nations.[72] On the other hand, this requirement can potentially create much legal controversy, such as that over the status of the membership of Serbia and Montenegro to the Genocide Convention when it was deemed not to have succeeded to Yugoslavia's membership of the United Nations at the relevant time.[73]

If membership of the Authority is open to all States, some of the problems arising from the limitations suggested above, whether based on substantial activity, financial investment or membership of the United Nations, would be overcome or avoided. Considering the biggest obstacle to the adoption of the Moon Agreement and the Convention on the Law of the Sea have been the lack of widespread acceptance by the international community, opening up the Authority to universal membership can only help to persuade States to become members of the Authority and participate in its mechanisms.[74] Although some industrialised States may find discouragement in

---

[69] Ibid., Article 2(7). Article 2(7) of the Charter of the United Nations provides that "Nothing contained in the present Charter shall authorize the United Nations to intervene in matters which are essentially within the domestic jurisdiction of any state or shall require the Members to submit such matters to settlement under the present Charter; but this principle shall not prejudice the application of enforcement measures under Chapter VII."

[70] Ibid., Article 25.

[71] Ibid., Articles 41–42.

[72] Outer Space Treaty, Article III.

[73] *Application for Revision of the Judgment of 11 July 1996 in the Case concerning Application of the Convention on the Prevention and Punishment of the Crime of Genocide (Bosnia and Herzegovina v. Yugoslavia), Preliminary Objections (Yugoslavia v. Bosnia and Herzegovina)* [2003] I.C.J. Rep. 7. See also Ricky J. Lee, *Application for Revision of the Judgment of 11 July 1996 in the Case concerning Application of the Convention on the Prevention and Punishment of the Crime of Genocide (Yugoslavia v. Bosnia and Herzegovina)* [2003] AUS. INT'L. L. J. 205 and John R. Crook, *The 2003 Judicial Activity of the International Court of Justice* (2004) 98 AM. J. INT'L. L. 309.

[74] See Morell, supra note 66; Stanley B. Rosenfield, *The Moon Treaty: The United States Should Not Become A Party* (1980) 74 AM. SOC'Y. INT'L. L. PROC. 162; Carl Q. Christol, *The 1979 Moon Agreement: Where Is It Today?* (1999) 27 J. SP. L. 1; Doug Bandow, *UNCLOS III: A Flawed Treaty* (1981) 19 SAN DIEGO L. REV. 475; and Steven J. Molitor, *The Provisional Understanding Regarding Deep Seabed Matters: An Ill-Conceived Regime for U.S. Deep Seabed Mining* (1987) 20 CORNELL INT'L. L. J. 223.

6.3 Practical Implementation of the New Legal Framework

the potential voting power held by developing States in the Authority, the structure, composition and processes outlined below may alleviate some of the concerns of the industrialised States.

6.3.4.2 Proposed Process for Accession to Membership

In the absence of a comprehensive multilateral convention dealing with the law of outer space, membership of the Authority would thus involve the prerequisites of being party to the Outer Space Treaty, the Moon Agreement and the Implementation Agreement. Accordingly, the process for acceding to membership of the Authority would reasonably involve four essential criteria or steps:

(1) the prospective Member State being a party to the Outer Space Treaty, if it is not already an existing party;
(2) the prospective Member State being a party to the Moon Agreement, if it is not already an existing party;
(3) the prospective Member State being a party to the Implementation Agreement; and
(4) the prospective Member State being accepted by the General Assembly of the Authority as a Member State.

Accession to the Outer Space Treaty is done simply by the prospective Member State signing and ratifying it and then depositing the instrument of ratification with the designated depositories, namely the Governments of the Russian Federation, the United Kingdom and the United States.[75] Similarly, the Moon Agreement can be acceded to in the same way, with a prospective Member State signing and ratifying it and then depositing its instrument of ratification with the Secretary-General of the United Nations.[76] One would suggest that a prospective Member State may accede to the Implementation Agreement in the same manner, unless the Implementation Agreement prescribes a different specific procedure.

The next issue that arises is whether the grant of new membership to a State would require approval by the General Assembly of the Authority and, if so, by how large a majority of Member States of the Authority before membership can be approved. In the case of the United Nations, for example, acceptance of a new Member State requires a two-thirds majority of its General Assembly and a majority of nine out of the fifteen members of the Security Council, including the concurrence of its five permanent members.[77] In the case of most intergovernmental organisations, such as the Organisation for the Prohibition of Chemical Weapons ("OPCW") or the United Nations Educational, Scientific and Cultural Organisation

---

[75] Outer Space Treaty, Article XIV.

[76] Moon Agreement, Article 19.

[77] Charter of the United Nations, Article 4.

302   6 Meeting the Challenges and Balancing the Competing Interests in Creating a …

Table 6.3 Summary of admission procedures for major organisations

Automatic admission on accession	Two-thirds majority vote
International Civil Aviation Organisation[a]	Food and Agriculture Organisation
International Labour Organisation[b]	International Monetary Fund[c]
International Maritime Organisation[d]	World Tourism Organisation[e]
International Telecommunication Union[f]	
OPCW	
UNESCO	
United Nations Industrial Development Organisation[g]	*Special rules*
Universal Postal Union[h]	International Atomic Energy Agency
World Health Organisation[i]	IFAD
World Intellectual Property Organisation[j]	World Bank[k]
World Meteorological Organisation[l]	World Trade Organisation[m]

[a] Chicago Convention on International Civil Aviation, opened for signature on 7 December 1944, 15 U.N.T.S. 295 (entered into force on 4 April 1947), Article 92.

[b] Constitution of the International Labour Organisation, opened for signature on 28 June 1919, 15 U.N.T.S. 35; 1948 U.K.T.S. 47 (entered into force on 10 January 1920), Article 1.

[c] Terms of membership is to be determined by the Board of Governors: Articles of Agreement of the International Monetary Fund, Article II.

[d] Convention on the International Maritime Organisation, opened for signature on 6 March 1948, 289 U.N.T.S. 48; 9 U.S.T. 621 (entered into force on 17 March 1958), Article 5.

[e] Statutes of the World Tourism Organisation, Article 5.

[f] Constitution and Convention of the International Telecommunication Union, opened for signature on 22 December 1992, 1825 U.N.T.S. 3; 28 U.S.T. 7645 (entered into force on 1 July 1994), Article 2.

[g] Constitution of the United Nations Industrial Development Organisation, opened for signature on 8 April 1979, 1401 U.N.T.S. 3; 1985 A.T.S. 18 (entered into force on 21 June 1985), Article 3.

[h] Constitution of the Universal Postal Union, opened for signature on 10 July 1964, 999 U.N.T.S. 171 (entered into force on 1 January 1966), Article 11.

[i] Constitution of the World Health Organisation, opened for signature on 22 July 1946, 14 U.N.T.S. 185 (entered into force on 7 April 1948), Article 4.

[j] Convention Establishing the World Intellectual Property Organisation, opened for signature on 14 July 1967, 1160 U.N.T.S. 231; 28 U.S.T. 7645 (entered into force on 26 April 1970), Article 5.

[k] Membership is open to the Member States of the International Monetary Fund: Articles of Agreement of the International Bank for Reconstruction and Development, Article II.

[l] Convention of the World Meteorological Organisation, opened for signature on 11 October 1947, 77 U.N.T.S. 143; 1950 A.T.S. 5 (entered into force on 23 March 1950), Article 3.

[m] Membership is to be on specific terms approved by a two-thirds majority of the Ministerial Conference: Agreement Establishing the World Trade Organisation, opened for signature on 15 April 1994, 1867 U.N.T.S. 154; 33 I.L.M. 1144 (entered into force on 1 January 1995), Article XII.

("UNESCO"), admission to membership is automatic for Member States of the United Nations upon ratification of the constituent treaty governing the organisation.[78] In the case of some intergovernmental organisations, such as the Food and Agriculture Organisation and the World Tourism Organisation, a two-thirds majority

---

[78] Convention on the Prohibition of the Development, Production, Stockpiling and Use of Chemical Weapons and on their Destruction, opened for signature on 13 January 1993, 1015 U.N.T.S. 163; 32 I.L.M. 800 (entered into force on 29 April 1997), Article XX and Constitution of

of its existing Member States as present and voting would be necessary before a new Member State can be admitted to membership of the organisation.[79] There are also international organisations with special membership rules, such as the International Fund for Agricultural Development ("IFAD") and the International Atomic Energy Agency, that are peculiar to the organisations themselves.[80]

The requirements for membership to some international organisations are summarised in Table 6.3.

It is suggested that the Authority would be better classified with the organisations that provide for automatic admission upon accession to the Moon Agreement and the Implementation Agreement, given the desire to attain universal membership among the international community. This is similar to the requirement for membership to the ISA, which is automatic upon accession to the Convention on the Law of the Sea and the 1994 Agreement.[81]

6.4 Administrative and Dispute Settlement Mechanisms

6.4.1 Enforcement and Dispute Settlement Mechanisms

After the adoption and implementation of administrative processes for the operation of the Authority, it is prudent to then turn to the need for dispute settlement mechanisms within the Authority. This is because disputes will inevitably arise between applicants and the Authority, the permit-holders and the Authority as well as between permit-holders themselves. There is also the additional need for judicial accountability in the operation and administration of the Authority. Accordingly, it

---

the United Nations Educational, Scientific and Cultural Organisation, opened for signature on 16 November 1945, 4 U.N.T.S. 275 (entered into force on 4 November 1946), Article II(1).

[79] Constitution of the Food and Agriculture Organisation, opened for signature on 16 October 1945, 12 U.S.T. 980 (entered into force on 1 November 1945), Article II(2) and Statutes of the World Tourism Organisation, opened for signature on 27 September 1970, 27 U.S.T. 2211 (entered into force on 2 January 1975), Article 5.

[80] Membership to the International Fund for Agricultural Development requires approval by the Governing Council, in which voting rights are based on membership and financial contribution: Agreement Establishing the International Fund for Agricultural Development, opened for signature on 20 December 1976, 28 U.S.T. 8435; 15 I.L.M. 922 (entered into force on 30 November 1977), Articles 3 and 6. Membership subject to approval by a simple majority of the General Conference upon recommendation by the Board of Governors: Statute of the International Atomic Energy Agency, opened for signature on 26 October 1956, 276 U.N.T.S. 3; 8 U.S.T. 1093 (entered into force on 29 July 1957), Articles IV and V.

[81] Convention on the Law of the Sea, opened for signature on 10 December 1982, 1833 U.N.T.S. 3; 21 I.L.M. 1261 (entered into force on 16 November 1994) and Agreement Relating to the Implementation of Part XI of the United Nations Convention on the Law of the Sea of 10 December 1982, opened for signature on 28 July 1994, 1836 U.N.T.S. 3; 33 I.L.M. 1309 (entered into force on 28 July 1996).

is clear that judicial mechanisms must be created to enable the peaceful, effective and judicious settlement of such disputes.

To that end, it is probably preferable to create separate arbitral panels (the "Panels") for the resolution of different types of disputes, allowing for different specialist expertise to be employed in the determination of different types of disputes. It is envisaged that the following bodies would be created within the Authority, each to be vested with judicial and quasi-executive functions relevant to their area of specialist expertise or competence:

(1) the *Licensing Oversight Panel* to hear and resolve disputes between the Authority and applicants or permit-holders in decisions relating to the grant or refusal of applications, the enforcement of conditions of permits and considerations of contraventions of legal or regulatory provisions of the relevant international law;
(2) the *Environmental Protection Panel* to provide for surveillance, monitoring and enforcement of environmental protection safeguards in the protection of the environment of the Earth, the mitigation of space debris, contamination and pollution as well as the remediation works on celestial bodies;
(3) the *Financial Duties Panel* for the assessment and settlement of disputes over the quantum of contributions payable into the Administrative Fund and the Common Heritage Fund;
(4) the *Dispute Settlement Panel* for the settlement of commercial disputes between applicants or permit-holders in relation to activities subject to the premises of the Authority; and
(5) the *Space Development Appeals Tribunal* that hears appeals from the Panels, including but not limited to the Dispute Settlement Panel.

The jurisdiction, powers, functions, composition and procedures of each of these Panels and the Space Development Appeals Tribunal are outlined below.

6.4.2 Licensing Oversight Panel

The Licensing Oversight Panel has the primary responsibility within the Authority for dealing with Exploration Permits, Mining Permits or Occupation Permits (together the "Permits").[82] In fulfilling this responsibility, the functions of the Licensing Oversight Panel would include:

(1) considering and evaluating applications for the Permits or their renewals;
(2) exercising the discretion to grant or refuse applications for the Permits or their renewals;

---

[82] In this context, Occupation Permits are granted for areas that are not being explored or mined but are occupied for ancillary purposes, such as an electric power plant.

(3) prescribing general or specific limitations or conditions on the Permits granted; and
(4) if merited, revoking any of the Permits granted for non-compliance with any limitation or condition imposed by the Permits.

As the Licensing Oversight Panel would be tasked with the most essential and controversial aspect of the operations of the Authority, its composition and decision-making process will undoubtedly be the subject of much debate in the negotiations. This is because both the industrialised and developing States will seek to have majority control. Given the history of the negotiations over the composition and decision-making power of the Council of the International Seabed Authority, it is unlikely that the industrialised States would allow the developing States to attain control over the decisions concerning the Permits. Given that the industrialised States would be surrendering their ability to regulate domestically and unilaterally the mining activities on celestial bodies as well as their agreement to make substantial contributions to the Administrative Fund and the Common Heritage Fund, this may be considered to be an appropriate compromise for the developing States (Fig. 6.4).

6.4.3 Financial Duties Panel

The purpose of the Financial Duties Panel is to assess and levy the financial contributions payable by space mining ventures (the "Contribution") to the Administrative Fund and the Common Heritage Fund, in implementation of the common heritage of mankind doctrine under Article 11 of the Moon Agreement. This is perhaps the

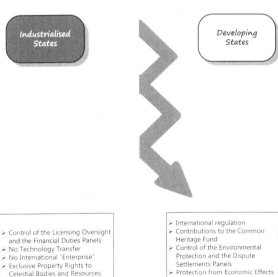

Fig. 6.4 Balancing some of the opposing interests between the industrialised states and the developing states

most controversial of all issues arising from the Moon Agreement and, as a matter of practical reality, also the most controversial of all aspects of the Authority and its operations. It is expected that the Financial Duties Panel would have to strive towards the following policy objectives and principles, even though such objectives and principles may be conflicting or even contradictory in nature:

(1) there should be no Contribution payable on a space mining venture to the extent that the mineral resources are to be used for global public interest or scientific purposes;
(2) the Contribution must not significantly discourage the financing and undertaking of commercial space mining ventures for mineral resources that are particularly rare in the Earth's crust and have significant economic or physical demand on Earth;
(3) the physical and environmental impact of any space mining venture must be recognised and duly compensated for through the Contribution;
(4) the economic impact of mineral resources from outer space on mining activities in developing States must be adjusted for and minimised; and
(5) the Contribution must be substantial and not token compensation for the extraction and exploitation of mineral resources from the celestial bodies in pursuance to the common heritage of mankind doctrine.

In pursuit of these objectives, it is apparent that the Financial Duties Panel would have to take into account the following relevant considerations in determining the financial duty payable for any given space mining venture:

(1) the extent to which the mineral resources are to be used for global public interest or scientific purposes instead of commercial profit-making;
(2) the reasonable costs incurred in the venture, including the reasonable costs associated in obtaining finance for the venture;
(3) the stability and the extent of price fluctuations of the world commodity markets for the relevant mineral resources;
(4) the profits that may be reasonably derived from the venture, taking into account good, prudent and reasonable industry and business practices;
(5) the abundance, or lack thereof, of the relevant mineral resources in the Earth's crust and the reasonable accessibility of known reserves;
(6) the economic and physical necessity of such mineral resources;
(7) the physical impact of the mining activity on the celestial body in question and its availability and suitability for future exploration, scientific investigation and use by third parties; and
(8) the economic impact of the release of the mineral resources from the venture on the world commodity markets, especially the impact on the economies of developing States that engage in mining activities for the same or similar mineral resources on Earth.

As with the Licensing Oversight Panel, it is probable that the industrialised States would not concede control of the Financial Duties Panel as they would not want the risk of the Financial Duties Panel acting arbitrarily in prescribing an unreasonably high rate of the Contributions to be made. This would be the case despite the fact that the control of the Common Heritage Fund itself would be devolved to the World Bank and the IMF under the present proposal.

6.4.4 Environmental Protection Panel

The Environmental Protection Panel would be responsible for ensuring that holders of various Permits comply with the conditions and obligations imposed under the Outer Space Treaty, the Moon Agreement and the Implementation Agreement as well as any specific rules or guidelines adopted by the Authority. To that end, the Environmental Protection Panel would be responsible for enforcing such rules and issuing penalty assessments in the event that the holder of a Permit has failed to comply with any such obligation. The Environmental Protection Panel would also have the responsibility of directing the Secretariat of the Authority to report to the Secretary-General of the United Nations in the event of any risk of the harmful contamination of the environment of the Earth or a celestial body as required under the Moon Agreement.[83]

6.4.5 Dispute Settlement Panel and the Space Development Appeals Tribunal

The Dispute Settlement Panel is intended to be the arbitral authority of first instance in resolving disputes between States in relation with their activities that are connected with the Moon Agreement or the Implementation Agreement. As such, it is modelled after the Dispute Settlement Body within the framework of the World Trade Organisation (the "WTO") and the Seabed Disputes Chamber of the International Tribunal on the Law of the Sea (the "ITLOS").[84] The decisions of the Dispute Settlement Panel are to be binding on the parties but are subject to appeal to the Tribunal or the International Court of Justice.

The Tribunal is to act as the highest arbitral authority in the enforcement of the international law relating to the exploitation and use of celestial bodies. The Tribunal may be modelled after the Appellate Body within the WTO, the ITLOS

---

[83] Moon Agreement, Article 7.

[84] Final Act and Agreement Establishing the World Trade Organisation, WTO Understanding on Rules and Procedures Governing the Settlement of Disputes, opened for signature on 15 April 1994, 1867 U.N.T.S. 154; 33 I.L.M. 1144 (entered into force on 1 January 1995) and the Convention on the Law of the Sea, Annex VI – Statute of the International Tribunal on the Law of the Sea, opened for signature on 10 December 1982, 1833 U.N.T.S. 3; 21 I.L.M. 1261 (entered into force on 28 July 1994). See Shabtai Rosenne, *Establishing the International Tribunal for the Law of the Sea* (1995) 89 AM. J. INT'L. L. 806.

and, of course, the International Court of Justice.[85] The jurisdiction of the Tribunal comprises all appeals from decisions of the Panels. It is proposed that the Tribunal is not to have original jurisdiction, given the suggested jurisdictional competence of the Panels and the potential concurrent jurisdiction of the International Court of Justice.[86]

The capacity of State Parties that are parties of a decision in any of the Panels to appeal to the Tribunal may be an automatic right per se or, alternatively, may require the prior leave of the Tribunal. In the case of the Appellate Body of the WTO, the right to appeal is a right that is automatic and exists per se.[87] It is suggested that, given the nature of the functions of the Panels and the importance of their decisions, the right to appeal to the Tribunal should be automatic. This is particularly the case given the substantial costs and time that are often required to conduct a case before international courts and tribunals.

The Convention on the Law of the Sea offers a choice to its Member States in the compulsory settlement of their disputes between the ITLOS, the International Court of Justice or ad hoc arbitration.[88] Each State is free to decide by written declaration for itself the means by which the interpretation or application of the Convention on the Law of the Sea is to be decided.[89] This so-called Montreux Formula during the negotiations on the Convention on the Law of the Sea provides that a plaintiff would be required to go to the forum chosen by the defendant to be bound.[90] In the case of disputes arising under Part XI of the Convention on the Law of the Sea in relation to mining activities in the deep seabed, the Seabed Disputes Chamber of the ITLOS is vested with specific jurisdiction.[91]

Although the Montreux Formula may be appropriate in the case of the Convention on the Law of the Sea, it is suggested that it may be more appropriate for:

- the Dispute Settlement Panel to have exclusive and compulsory jurisdiction in determining any dispute between the Authority and a Member State or a particular applicant or holder of a Permit;

---

[85] The Statute of the International Tribunal for the Law of the Sea; the WTO Understanding on Rules and Procedures Governing the Settlement of Disputes; and the Statute of the International Court of Justice, opened for signature on 26 June 1945, 1945 U.K.T.S. 67 (entered into force on 24 October 1945).

[86] The Appellate Body of the WTO similarly only has appellate jurisdiction and no original jurisdiction: WTO Understanding on Rules and Procedures Governing the Settlement of Disputes, Article 17(1). The International Court of Justice has original jurisdiction and no appellate jurisdiction, though it does have competence to hear applications for revisions to judgments where a decisive fact is newly discovered: Statute of the International Court of Justice, Articles 36 and 61.

[87] WTO Understanding on Rules and Procedures Governing the Settlement of Disputes, Article 17(4).

[88] Convention on the Law of the Sea, Article 287.

[89] Ibid.

[90] Andronico O. Adede, THE SYSTEM FOR SETTLEMENT OF DISPUTES UNDER THE UNITED NATIONS CONVENTION ON THE LAW OF THE SEA (1987), at 53–54.

[91] Convention on the Law of the Sea, Article 189.

6.4 Administrative and Dispute Settlement Mechanisms

- the Dispute Settlement Panel to have non-exclusive but compulsory jurisdiction in determining any dispute between Member States and their nationals in relation to any activity to which the Implementation Agreement relates;
- the Tribunal to have exclusive and compulsory jurisdiction in determining any appeals from the Dispute Settlement Panel in relation to any dispute between the Authority and a Member State or a particular applicant or holder of a Permit; and
- the Tribunal to have non-exclusive and non-compulsory jurisdiction in determining disputes between Member States and their nationals in relation to any activity to which the Implementation Agreement relates.

Despite the availability of both the Dispute Settlement Panel and the Tribunal, it is anticipated that Member States may nevertheless elect to resolve relevant disputes between them through binding arbitration or determination by the International Court of Justice. To this end, it would be appropriate to implement the Montreux Formula to the extent that it would require Member States to make a written declaration to submit to the compulsory jurisdiction of the Tribunal, the International Court of Justice or ad hoc arbitration to determine disputes with other Member States.

Both the Dispute Settlement Panel and the Tribunal would be composed of a number of independent individual members, with the total numbers of both the Tribunal and the Dispute Settlement Panel being sufficient to take into account the needs of impartiality, workload and being representative of the principal legal systems and geopolitical distribution of the world. The International Tribunal for the Law of the Sea, for example, is composed of 21 individual members, while the Appellate Body of the WTO comprises seven individual members and the International Court of Justice comprises 15 members.[92] Specifically, it is suggested on the basis of those two model bodies that the members of the Tribunal:

(1) are nominated by State Parties to the Implementation Agreement as being recognised experts in the field of international law and space law and recognised for having a reputation of fairness and integrity;[93]
(2) are elected by a two-thirds majority of the present and voting members of the General Assembly of the Authority;[94]

---

[92] WTO Understanding on Rules and Procedures Governing the Settlement of Disputes, Article 17(1); Statute of the International Tribunal for the Law of the Sea, Article 2; and Statute of the International Court of Justice, Article 3.

[93] See, for example, Statute of the International Tribunal for the Law of the Sea, Articles 2(1) and 4(1); WTO Understanding on Rules and Procedures Governing the Settlement of Disputes, Article 17(3); and Statute of the International Court of Justice, Article 2.

[94] Members are of the ITLOS are elected by a two-thirds majority of a meeting of the State Parties to the Convention on the Law of the Sea, provided that a quorum of two-thirds of the State Parties is constituted and the vote represents a simple majority of State Parties: Statute of the International Tribunal for the Law of the Sea, Article 4(4). Members of the Appellate Body

(3) are elected to terms of 9 years, except that one-third of the original members of the Tribunal would have terms of 3 years and another one-third of their number would have terms of 6 years, as determined by lot or specific election;[95]

(4) may be re-elected to further terms;[96]

(5) have no political or administrative function or any active association or financial interest in any activity that are connected with the Moon Agreement or the Implementation Agreement;[97]

(6) enjoy diplomatic immunities and privileges when engaged on the business of the Tribunal;[98]

(7) may not hear a case in which they were previously agent, counsel or advocate for one of the parties or in which they had previously heard as a member of any domestic or international court, tribunal or any other capacity;[99] and

---

of the WTO are appointed by the Dispute Settlement Body: WTO Understanding on Rules and Procedures Governing the Settlement of Disputes, Article 17(2). Members of the International Court of Justice are elected by an absolute majority of the General Assembly and the Security Council of the United Nations, with the two votes conducted independently of each other: Statute of the International Court of Justice, Articles 8 and 10.

[95] The same structure exists for the ITLOS: Statute of the International Tribunal for the Law of the Sea, Article 5(1) and (2). The members of the Appellate Body of the WTO are elected for terms of 4 years, of which three of the seven original members are to be limited to a 2-year term as determined by lot: WTO Understanding on Rules and Procedures Governing the Settlement of Disputes, Article 17(2). Similarly, the International Court of Justice adopts a similar structure, with judges elected to terms of 9 years, with five of the original judges elected for 3 year terms and another five of that number elected for 6 year terms as determined by lot: Statute of the International Court of Justice, Article 13(1) and (2).

[96] This is also the case for the ITLOS, the Appellate Body of the WTO and the International Court of Justice: Statute of the International Tribunal for the Law of the Sea, Article 5(1); WTO Understanding on Rules and Procedures Governing the Settlement of Disputes, Article 17(2); and Statute of the International Court of Justice, Article 13(1).

[97] Similar restrictions apply to members of the ITLOS, the Appellate Body of the WTO and the International Court of Justice: see Statute of the International Tribunal for the Law of the Sea, Article 7; WTO Understanding on Rules and Procedures Governing the Settlement of Disputes, Article 17(3); and Statute of the International Court of Justice, Article 16.

[98] The same applies to members of the ITLOS and the International Court of Justice: Statute of the International Tribunal for the Law of the Sea, Article 10 and Statute of the International Court of Justice, Article 19. No equivalent provision exists in the WTO Understanding on Rules and Procedures Governing the Settlement of Disputes. Examples of the diplomatic immunities and privileges enjoyed can be found in the Vienna Convention on Diplomatic Relations, opened for signature on 18 April 1961, 500 U.N.T.S. 95 (entered into force on 24 April 1964); and the Vienna Convention on Consular Relations, opened for signature on 24 April 1963, 596 U.N.T.S. 261 (entered into force on 19 March 1967).

[99] Similar restrictions apply to members of the ITLOS, the Appellate Body of the WTO and the International Court of Justice: see Statute of the International Tribunal for the Law of the Sea, Article 8; WTO Understanding on Rules and Procedures Governing the Settlement of Disputes, Article 17(3); and Statute of the International Court of Justice, Article 17.

(8) may be removed from office if, in the unanimous opinion of the other members of the Tribunal, that particular member of the Tribunal does not comply with the conditions or the qualifications of the office.[100]

6.5 Financing the Proposed Authority

The Authority, with its proposed structure and composition, is expected to incur substantial expenditure in its establishment and administration, which would have to be met either from direct contributions from its Member States or from the regular budget of the United Nations. For example, the budget for the ISA until 1997 was derived from the regular budget of the United Nations, which provided US $776,000.00 in 1994–1995 and US $2,627,100.00 for 1995–1996.[101] By comparison, the World Trade Organisation, which has a much higher degree of complexity in structure and administrative burden than the Authority as proposed, cost its Member States US $75,500,000.00 in 1995 when it was established.[102] In the case of the ISA and the ITLOS, the cost incurred was directly levied on their Member States on an assessed basis until they can be financed by contributions from seabed mining activities.[103]

There are at least four potential approaches to the way in which the costs of establishing the Authority can be levied on its Member States. First, the costs can be apportioned equally among the Member States of the Authority. The benefit of this approach is that the cost burden incurred by each State is the same and reflects the voting power that each State has in the Authority. This is the approach taken in some regional organisations, such as the Association of Southeast Asian Nations ("ASEAN"),[104] and the Southern Africa Development Community (the "SADC").[105]

Second, the costs can be apportioned on the basis of the likelihood that the Member States would participate in activities that may involve the mineral exploitation of the Moon and other celestial bodies. Some may perceive this to be the implementation of a "user pays" system for the establishment of the Authority, in

---

[100] This is the provision for the removal of the members of the ITLOS: Statute of the International Tribunal for the Law of the Sea, Article 9. Judges of the International Court of Justice may be removed from office by the unanimous decision of the other judges if a particular judge is considered not to comply with the conditions or qualifications of that office: Statute of the International Court of Justice, Article 18.

[101] Michael C. Wood, *International Seabed Authority: The First Four Years* (1999) 3 MAX PLANCK U.N.Y.B. 173 at 214.

[102] The Economist, *Crunch! Budget Problems at the World Trade Organisation*, THE ECONOMIST, 4 April 1998.

[103] Convention on the Law of the Sea, Article 171.

[104] Rodolfo C. Severino, SOUTHEAST ASIA IN SEARCH OF AN ASEAN COMMUNITY: INSIGHTS FROM THE FORMER ASEAN SECRETARY-GENERAL (2006), at 33.

[105] Katharina Pichler Coleman, INTERNATIONAL ORGANISATIONS AND PEACE ENFORCEMENT: THE POLITICS OF INTERNATIONAL LEGITIMACY (2007), at 125–126.

that those who benefit from the establishment and the use of the Authority are to bear their costs. The adoption of this justification, however, fails to recognise that the purpose of the common heritage of mankind doctrine is, among others, the sharing of the benefits derived from the exploitation of mineral resources from the Moon and other celestial bodies. Accordingly, the practical effect of the doctrine is that all Member States of the Authority, in particular the developing States, are the main beneficiaries from the establishment and use of the Authority. It stands to reason that, if the intention is to adopt a "user pays" approach to financing the cost of establishing the Authority, then all Member States of the Authority and not just the industrialised States are to bear some of the burden of this cost.

Third, the costs can be apportioned on the basis of the size of the economies of the Member States, in a similar manner to which the expenditure budget of the United Nations is apportioned on the basis of the capacity of the Member States to pay.[106] The principal benefit of this approach is that a poorer State is not precluded from participating in the United Nations or other intergovernmental organisations simply because it cannot afford to pay an equal contribution to the budget of the organisation.[107] The principal deficiency with this approach lies in the fact that, although the developing States are likely to be significant financial beneficiaries of the operations of the Authority, they would be bearing a disproportionately light financial burden in relation to the establishment of the Authority.

Fourth, the costs can be apportioned on the basis of the size of the economies of the Member States but a prescribed minimum contribution is required of all Member States regardless of their economic power.[108] Similarly, a prescribed maximum

---

[106] Article 17 of the Charter of the United Nations provides that "The expenses of the Organisation shall be borne by the Members as apportioned by the General Assembly". Rule 160 of the General Assembly Rules of Procedure provides that the Committee on Contributions is to advise the General Assembly on the scale of assessment upon which the apportionment of the financial contributions are based, which is to be reviewed regularly: General Assembly Resolution 58/1(B). On 21 December 2007, the General Assembly of the United Nations adopted a biennial budget of US$4.17 billion for 2008–2009: United Nations, "Concluding Main Part of Session, General Assembly Adopts $4.17 Billion Budget" (press release, 21 December 2007), at <http://www.un.org/News/Press/docs/2007/ga10684.doc.htm>, last accessed on 19 May 2008. See also Warren Hoge, *Despite U.S. Opposition, United Nations Budget is Approved*, THE NEW YORK TIMES, 23 December 2007, at <http://www.nytimes.com/2007/12/23/world/23nations.html>, last accessed on 4 May 2008.

[107] See Irving B. Kravisand and Michael W. S. Davenport, *The Political Arithmetic of International Burden Sharing* (1963) 71 J. POL. ECON. 309; Gi-Heon Kwon, *The Declining Role of Western Powers in International Organisations: Exploring a New Model of U.N. Burden Sharing* (1995) 15 J. PUB. POL'Y. 65; Michele Fratianni and John Pattison, *The Economics of International Organisations* (1982) 35 KYKLOS 244; and J. Diamond and J. R. Dodsworth, *Normative and Positive Theories of International Cost Sharing: The Case of the Netherlands* (1977) 124 DE ECONOMIST 403.

[108] The United Nations has adopted a similar approach, prescribing the minimum contribution of a Member State to be one-thousandth percent (0.001%) of its operating budget, though in order to avoid imposing an overly high financial burden on some countries, a prescribed ceiling of one hundredth precent (0.01%) is imposed for least developed States: United Nations, *Briefing on Methodology of the Scale of Assessment* (2006), at <http://www.un.org/ga/61/fifth/scale-method. pps>, last accessed on 19 May 2008.

contribution is also determined to avoid the handful of the largest economies in the world shouldering an overly large financial burden vis-à-vis all other States.[109] In this manner, a balance is struck between the desire to provide some equitable sharing of the burden of establishing the Authority while taking into account the economic and financial weaknesses of developing States in contrast to the industrialised States.

6.6 Conclusions

The present uncertainty in the legal environment in which commercial mining activities on celestial bodies may take place cannot attain any more degree of entrenchment or permanence than it already has. With the Moon Agreement adopted by the General Assembly more than 30 years ago, the time has come for the international community to negotiate in good faith, along with an appreciation of political realities and commercial practicalities, with a view to reaching agreement on the terms of the Implementation Agreement that would create the Authority.

For such an effort to succeed, there must be a strong sense of universality achieved in spirit and in practice with the creation of this new regulatory framework for the Implementation Agreement to be able to provide international legal certainty in theory as well as practice and ideology. Though balancing the diverse interests and concerns of the various stakeholders in the international community, this proposal may then be successful in reaching an appropriate compromise that can be acceptable to most, if not all, States in the international community.

---

[109] The United Nations has adopted a similar approach, to cap the contribution of the United States, the largest economy in the world as measured by gross domestic product, at twenty-two percent (22%) of its operating budget: ibid.

Chapter 7
Concluding Observations

7.1 Quod Erat Demonstrandum

7.1.1 Elements of Proving the Hypothesis

The central hypothesis to be demonstrated in this work is that one of the major inhibiting factors to the commercial exploitation of mineral resources from celestial bodies is the absence of an appropriate legal framework to govern such activities. In order to demonstrate that the hypothesis is satisfied, it is contended that the following assumptions must be proven:

(1) there is an economic scarcity of mineral resources in the Earth's crust;
(2) an analysis of the physical and technical requirements to conduct activities in the exploration, prospecting and exploitation of mineral resources from celestial bodies, particularly the Near Earth Asteroids, shows that such activities are feasible within the capabilities of existing technology;
(3) an analysis of the socio-economic factors relating to the supply of various mineral resources on Earth and the financial requirements for the commercial exploitation of mineral resources from celestial bodies, particularly the Near Earth Asteroids, shows that such exploitation is economically and financially feasible;
(4) generally and specifically, the absence of a legal framework giving certain rights and duties to private actors in the space sector inhibits the financing and development of such private or commercial ventures; and
(5) at present, an appropriate legal framework governing relevant activities on celestial bodies does not exist.

Each of these elements are discussed in greater detail below.

R.J. Lee, *Law and Regulation of Commercial Mining of Minerals in Outer Space*, Space Regulations Library 7, DOI 10.1007/978-94-007-2039-8_7, © Springer Science+Business Media B.V. 2012

7.1.2 Economic Scarcity of Resources in the Earth's Crust

It is undisputed that the mineral resources that exist in the Earth's crust are finite and, consequently, it is only a matter of time that such mineral resources would be physically exhausted by continued human consumption and use. While the physical scarcity of mineral resources would certainly, without more, create an economic scarcity of such resources, it is more likely that economic exhaustion of such mineral resources would precede their physical exhaustion. This is because there would eventually be a point on the supply and demand curve where consumers would refuse to pay the high prices resulting from diminishing reserves, increasing costs of extraction and need for mitigation of environmental costs, or when producers refuse to sell at government-imposed low prices.

Accordingly, given the physical scarcity of mineral resources in the Earth's crust, it is a necessary consequence that there is an economic scarcity of mineral resources on Earth if such resources are not supplemented by further physical reserves to be found. It is clear that such mineral resources may be found and, indeed, can only be found on the celestial bodies of the Solar System, particularly asteroids and other smaller celestial bodies such as comets that pass near the orbit of the Earth.

7.1.3 Physical and Technical Feasibility of Asteroid Mining

Within the orbit of Jupiter, there is a large number of asteroids orbiting around the Sun in orbits of various shapes, eccentricities and distances. Most asteroids are concentrated in the main Asteroid Belt between Mars and Jupiter, which are referred to as the Main Belt asteroids. In addition to the Main Belt asteroids, there are a few classes of asteroids that have orbits that are near, or even crossing, the heliocentric orbit of the Earth. These asteroids are collectively referred to as the Near Earth Asteroids and they present themselves as strong candidates for future mining activities when the mineral resources on Earth are being depleted. This is because the gravity, travel time and launch costs to these asteroids would be comparatively low for economically viable exploitation to take place.

The two most important considerations in determining the viability of mining a particular asteroid is the time factors as well as the energy costs required for the mission, assuming that the existing human technology for propulsion, spacecraft design and construction as well as mining infrastructure can be adapted for asteroid mining purposes. The ideal asteroid for mining purposes would be one that is close to the Earth and allows for a long mining season without requiring large amounts of energy to reach it or to return to Earth from it. In this context, the mining of some Near Earth Asteroids would be preferable over others as some energy efficient trajectories, such as Hohmann transfer trajectories, allow for only very short mining seasons on the asteroid. These factors mean that, from physical and technical perspectives, the exploitation of mineral resources from Near Earth Asteroids is certainly feasible within the confines of present technology.

7.1.4 Economic and Financial Feasibility of Commercial Mining on Near Earth Asteroids

The important factors in determining the economic and financial feasibility of the mining of mineral resources from Near Earth Asteroids include:

(1) the cost of the operation;
(2) the time between the launch of the operation and the eventual sale or use of the mineral resources;
(3) various risks of failure; and
(4) the expected return on the investment.

After a brief analysis, it is apparent that the predominant factor that is the most variable in the evaluation of the financial viability of a commercial venture to extract and exploit mineral resources from celestial bodies is the evaluation of the risks. This is especially the case in relation to risks that cannot be resolved by technical or financial means. For such a venture to be profitable or at least financially feasible, there must be legal and regulatory certainty for the conduct of such activities.

7.1.5 Absence of an Appropriate Legal Framework

It is already commonly acknowledged that the existing body of space law is very ill adapted to the commercial realities in the space industry today. When the Outer Space Treaty was drafted and adopted in the early 1960s, the present commercialisation and development in space was inconceivable to all except the most devoted science fiction writers and filmmakers. It was not until the Moon Agreement that the possible mineral exploitation of other celestial bodies was considered. However, even in the Moon Agreement it is quite apparent that its drafters did not anticipate the development of space mining ventures until the distant future, as Article 11 of the Moon Agreement deferred the creation of a specific intergovernmental organisation until such time as deemed necessary by the parties to the Moon Agreement when space mining activities become imminent.

It is in the exploration and extraction segments of the mining operation that would encounter most of the legal obstacles. In the exploration segment, ore samples are either robotically analysed in situ or are returned to Earth for further and more detailed analysis. While this exercise of gathering samples may arguably be no different to the collection of lunar rocks during the U.S. Apollo missions, the crucial distinction is that in the case of mineral prospecting, the samples are collected for ultimate private commercial rather than public scientific gain. Accordingly, this raises issues as to the lawfulness of such activities.

In the extraction segment, the legal obstacles involved are more complex and difficult to resolve. The freedom of access and the principle of non-appropriation found in the Outer Space Treaty, in particular, makes mineral extraction activities on celestial bodies difficult, if not impossible, to justify in law. This is because the conduct of mineral extraction activities must have, as a necessary requirement, some degree of exclusionary right in the area of asteroid being mined, a right that would be contrary to those legal principles. Further, the act of extraction itself may contravene, by its very nature, the principle of non-appropriation.

It is thus apparent that, even if the Moon Agreement is to be considered an integral and applicable part of space law, the existing international legal framework cannot provide an adequate regime for the regulation of mineral exploration and extraction activities on celestial bodies. The fact that only 13 States have signed and ratified the Moon Agreement have led commentators to suggest that the impasse in the international community over the effect and application of its terms is unlikely to be broken in the near future.[1] However, unless such regulation is provided by means of new legal instruments, the conclusion can be easily reached that there is indeed a *lacuna* of law dealing with such activities on celestial bodies.

7.2 Quod Erat Meliorandum

7.2.1 Conceptual Evolution of the Common Heritage of Mankind

In the 1970s, the increasing international clout of the developing world has led to an attempt of a revolution in international law and international economic relations. The New International Economic Order that was introduced at that time also included the invention of the concept of the common heritage of mankind and the increasing conflict between the industrialised States and the developing States over the content and effect of the application of this concept, particularly to the deep seabed and celestial bodies.

The doctrine of the common heritage of mankind not only prohibits sovereignty and ownership over the celestial bodies, it also creates positive rights on the part of the States in the international community to insist that they have a role in the management of the areas that are the common heritage of mankind as representatives of all humankind. As such, the doctrine requires any benefits so derived from the exploitation of mineral resources from the common heritage of mankind to be shared internationally or is for the common benefit of all humankind. Given that the terms and the extent of the sharing of such benefits are not specified, much uncertainty and debate remains on the practical effects of the doctrine.

---

[1] United Nations Office of Outer Space Affairs, *Status of International Agreements Relating to Activities in Outer Space as at 1 January 2010*, 1 January 2010, at <http://www.oosa.unvienna.org/pdf/publications/ST_SPACE_11_Rev2_Add3E.pdf>, last accessed on 22 April 2010.

7.2.2 Past Failures over the Deep Seabed and Antarctica

Because of this differing opinion, both the Moon Agreement and the original 1982 United Nations Convention on the Law of the Sea received very little support among States and, consequently, the world reached an impasse over the law relating to deep seabed and celestial bodies. The industrialised States refused to be part of any intergovernmental system that set a precedent for international taxation, while the developing States did not want any agreement that did not clarify the extent of their rights and benefits derived from the common heritage of mankind. However, when the Convention on the Law of the Sea was about to enter into force in the early 1990s, strenuous diplomatic efforts were made to reach a compromise. In 1994, this resulted in the adoption of the Agreement relating to the Implementation of Part XI of the United Nations Convention on the Law of the Sea, which was acceptable to both industrialised and developing States by significantly reducing the obligations of the deep seabed mining States.

The idea that States may not own a particular spatial area to the exclusion of other States is not a new one, for as far back as the Seventeenth Century it was already recognised that the sovereignty of States does not extend to the high seas. However, the idea that a spatial area may be subject to the universal ownership of humankind is certainly a new concept. For example, it was proposed in the early Twentieth Century that the Antarctic continent should be a sanctuary for all humankind. Although the Antarctic Treaty does not stipulate as such, it is certainly the intention of the Antarctic Treaty System for some form of common management to take place over the exploration and scientific work conducted in Antarctica by deferring the territorial claims of the interested States.

7.2.3 Attempt to Create a Legal Framework for the Exploitation of Mineral Resources on Celestial Bodies

The industrialised States saw the concept of celestial bodies being declared the "common heritage of mankind" as providing that all States shall have access to the benefits derived from the resources contained in those spatial areas and nothing more. The developing States, on the other hand, believed that the industrialised States would be required to share their profits derived from the exploitation of these spatial areas with the developing States on an equitable basis. This is because, by exploiting resources in the common property of humankind, the industrialised States are depriving the developing States of the resources of which they are proud part owners.

Because of this differing opinion, the Moon Agreement received very little support among States and, consequently, the world reached an impasse over the law relating to deep seabed and celestial bodies. The industrialised States refused to be part of any intergovernmental system that set a precedent for international taxation, while the developing States did not want any agreement that did not clarify the extent of their rights and benefits derived from the common heritage of mankind. Resulting from the uncertainty over the implications of the Moon Agreement, it is clear that the Moon Agreement or some form of it must be fully implemented and made acceptable to most States for commercial space mining ventures to take place once the economic and technical conditions are present. Without legal certainty, the entrepreneurs or their investors would not be satisfied that title could be safely asserted over the mining site and for the ore to be extracted and sold on Earth.

No similar efforts have been made with the implementation of Article 11 of the Moon Agreement as with the treatment of the deep seabed under the 1994 Implementation Agreement. Meanwhile, views differ over the legal ramifications of the present Article 11 in the interim period before the Article is implemented. Without agreement on the application of Article 11 of the Moon Agreement, there is no legal framework that is applicable to the exploitation of mineral resources from celestial bodies, posing as a major obstacle to the future resource exploitation in outer space.

7.2.4 Balancing the Interests of Competing Stakeholders

If the international impasse over the practical effects of the common heritage of mankind concept as applied to celestial bodies is to be broken, there must be some attempt at reaching a compromise by balancing the competing interests of the various stakeholders on the issue. In fact, the need the balance competing interests have become the most important step in removing the legal shackles to the future commercial mining of asteroids.

Considering the enormous financial commitments needed to invest in the technological advances necessary to enable such mining developments, the absence of any legal certainty will only inhibit such commitments, with the incidental detriment of continuing the degradation of the global environment through the exploration and extraction of mineral resources from environmentally sensitive areas of the Earth's crust. Accordingly, with the mining of asteroids the next necessary step in the economic evolution of the human civilisation, the adoption of a practical legal framework for the commercial exploitation of mineral resources in outer space will transform this present science fiction into a future reality.

7.3 Quod Erat Faciendum

7.3.1 Creating a New Legal Framework

Resulting from the uncertainty over the implications of the Moon Agreement, it is a fundamental cornerstone of this work that the Moon Agreement or some form of it must be fully implemented and made acceptable to most States for commercial space mining ventures to take place once the economic and technical conditions are present. This is because without some form of international property rights available, it is unlikely that the entrepreneurs or their investors would be satisfied that title could be safely asserted over the mining site and for the ore to be extracted and sold on Earth.

There have been different suggestions for possible frameworks in implementing Article 11 of the Moon Agreement, ranging from the minimalist approach of a subsidiary treaty clarifying the contents of Article 11 to the replacement of the entire treaty with the creation of an intergovernmental agency or organisation to govern all activities of States involving the use of outer space. However, controversy and disagreements remain over the implementation of the common heritage of mankind provision of the Moon Agreement, for which there are several options available to the international community.

However, the international community may opt to create a unique regime for the development of resources in outer space. There are some significant differences between the terms of Article 11 of the Moon Agreement and Part XI of the Convention on the Law of the Sea, for example, which may imply that a different form of sharing of benefits may be required.

7.3.2 Practical Aspects of the New Legal Framework

Perhaps more importantly, it is important for any new intergovernmental organisation dealing with space activities to clarify the powers and functions of the organisation in relation to four broad areas that are relevant to commercial space mining activities. These areas are the implementation of the "common heritage of mankind" principle, the provision of property "licences", and the resolution of some jurisdictional and liability issues applicable to a space mining venture. It is therefore fitting that this work concludes by providing for ideas and proposals for the creation of such an organisation to allow space mining to take place, in order to give a practical demonstration of what needs to be done.

7.3.3 How the World May Change

There is no need to list the many challenges facing humanity as it enters the next century. Be they environmental, political, economic or social, the problems are both obvious and immediate. ... For the environmentalists, the space option is the ultimate environmental solution. For the Cornucopians, it is the technological fix that they are relying on. For the hard core space community, the obvious by-product would be the eventual exploration and settlement of the Solar System. For most of humanity, however, the ultimate benefit is having a realistic hope in a future with possibilities. Indeed, the space option is humanity's most optimistic approach to its future.

Our civilisation is at its peak – we have the means today to implement the space option but not yet the commitment. However, if our species does not soon embrace this unique opportunity with sufficient commitment, it may miss its one and only chance to do so. Humanity could soon be overwhelmed by one or more of the many challenges it now faces. The window of opportunity is closing as fast as the population is increasing. As the 21st Century draws near, the main challenge to the space community will be informing and then convincing the public of the viability of the space option as the only optimistic alternative to the other current approaches to human destiny because our future will be either a Space Age or a Stone Age.

Arthur R. Woods and Marco C. Bernasconi[2]

---

[2] Arthur R. Woods and Marco C. Bernasconi, *Choosing a Space Age or a Stone Age* (1995), SPACE NEWS, 2–8 October 1995.

References

Treaties

ACP-EEC Convention of Lomé, opened for signature on 28 February 1975 (1976) O.J. L25/2 (entered into force on 1 April 1976).

Agreed Measures for the Conservation of Antarctic Flora and Fauna, opened for signature on 13 June 1964, 17 U.S.T. 991; 1998 A.T.S. 6 (entered into force on 1 November 1982).

Agreement Among the Government of Canada, Governments of the Member States of the European Space Agency, the Government of Japan, the Government of the Russian Federation and the Government of the United States of America Concerning Cooperation on the Civil International Space Station, opened for signature on 29 January 1998, Temp. State Dep't No. 01-52, CTIA No. 10073.000 (entered into force on 27 March 2001).

Agreement Between the Government of Australia and the Government of the Russian Federation on Cooperation in the Field of Exploration and Use of Outer Space for Peaceful Purposes, opened for signature on 23 May 2001 (2004) A.T.S. 17 (entered into force on 12 July 2004).

Agreement Establishing the International Fund for Agricultural Development, opened for signature on 20 December 1976, 28 U.S.T. 8435; 15 I.L.M. 922 (entered into force on 30 November 1977).

Agreement Governing the Activities of States on the Moon and Other Celestial Bodies, opened for signature on 18 December 1979, 1363 U.N.T.S. 3; 18 I.L.M. 1434 (entered into force on 11 July 1984).

Agreement on the Rescue of Astronauts, the Return of Astronauts and the Return of Objects Launched into Outer Space, opened for signature on 22 April 1968, 672 U.N.T.S. 119, T.I.A.S. 6599, 19 U.S.T. 7570; 1986 A.T.S. 8 (entered into force on 3 December 1968).

Agreement on Trade-Related Aspects of Intellectual Property Rights, opened for signature on 15 April 1994, 1869 U.N.T.S. 299 (entered into force on 1 January 1995), Article 33.

Agreement Relating to the Implementation of Part XI of the United Nations Convention on the Law of the Sea of 10 December 1982, opened for signature on 28 July 1994, 1836 U.N.T.S. 3; 33 I.L.M. 1309 (entered into force on 28 July 1996).

Amended Convention on the International Mobile Satellite Organisation, opened for signature on 24 April 1998 (2001) A.T.S. 11 (entered into force on 31 July 2001).

Antarctic Treaty, opened for signature on 1 December 1959, 402 U.N.T.S. 71; 12 U.S.T. 794; 19 I.L.M. 860 (entered into force on 23 June 1961).

Articles of Agreement of the International Bank for Reconstruction and Development, opened for signature on 27 December 1945, 2 U.N.T.S. 134 (entered into force on 27 December 1945).

Articles of Agreement of the International Monetary Fund, opened for signature on 22 July 1944, 1 U.N.T.S. 39; 1946 U.K.T.S. 21; 29 U.S.T. 2203; T.I.A.S. 8937 (entered into force on 27 December 1945).

324 References

Brussels Convention Relating to the Distribution of Programme-Carrying Signals Transmitted by Satellite, opened for signature on 21 May 1974, 1144 U.N.T.S. 3; 13 I.L.M. 1444 (entered into force on 25 August 1979).

Charter of the United Nations, opened for signature on 26 June 1945, 1 U.N.T.S. xvi; 1945 U.K.T.S. 67 (entered into force on 24 October 1945).

Chicago Convention on International Civil Aviation, opened for signature on 7 December 1944, 15 U.N.T.S. 295 (entered into force on 4 April 1947).

Constitution and Convention of the International Telecommunication Union, opened for signature on 22 December 1992, 1825 U.N.T.S. 3; 28 U.S.T. 7645 (entered into force on 1 July 1994).

Constitution of the Food and Agriculture Organisation, opened for signature on 16 October 1945, 12 U.S.T. 980 (entered into force on 1 November 1945).

Constitution of the International Labour Organisation, opened for signature on 28 June 1919, 15 U.N.T.S. 35; 1948 U.K.T.S. 47 (entered into force on 10 January 1920).

Constitution of the United Nations Educational, Scientific and Cultural Organisation, opened for signature on 16 November 1945, 4 U.N.T.S. 275 (entered into force on 4 November 1946).

Constitution of the United Nations Industrial Development Organisation, opened for signature on 8 April 1979, 1401 U.N.T.S. 3; 1985 A.T.S. 18 (entered into force on 21 June 1985).

Constitution of the Universal Postal Union, opened for signature on 10 July 1964, 999 U.N.T.S. 171 (entered into force on 1 January 1966).

Constitution of the World Health Organisation, opened for signature on 22 July 1946, 14 U.N.T.S. 185 (entered into force on 7 April 1948).

Convention Establishing the World Intellectual Property Organisation, opened for signature on 14 July 1967, 1160 U.N.T.S. 231; 28 U.S.T. 7645 (entered into force on 26 April 1970).

Convention for the Conservation of Antarctic Seals, opened for signature on 1 June 1972, 29 U.S.T. 441; 11 I.L.M. 251 (entered into force on 11 March 1978).

Convention for the Establishment of a European Space Agency, opened for signature on 30 May 1975, 14 I.L.M. 864 (entered into force on 30 October 1980).

Convention of the World Meteorological Organisation, opened for signature on 11 October 1947, 77 U.N.T.S. 143; 1950 A.T.S. 5 (entered into force on 23 March 1950).

Convention on International Liability for Damage Caused by Space Objects, opened for signature on 29 March 1972, 961 U.N.T.S. 187; 24 U.S.T. 2389; T.I.A.S. 7762; 1975 A.T.S. 5 (entered into force on 1 September 1972).

Convention on Registration of Objects Launched into Outer Space, opened for signature on 14 January 1975, 1023 U.N.T.S. 15; T.I.A.S. 8480; 28 U.S.T. 695; 1986 A.T.S. 5 (entered into force on 15 September 1976).

Convention on the Conservation of Antarctic Marine Living Resources, opened for signature on 20 May 1980, 1329 U.N.T.S. 47; 33 U.S.T. 3476 (entered into force on 7 April 1982).

Convention on the Conservation of Antarctic Marine Living Resources, opened for signature on 20 May 1980, 33 U.S.T. 3476 (entered into force on 7 April 1982).

Convention on the High Seas, opened for signature on 29 April 1958, 450 U.N.T.S. 11 (entered into force on 30 September 1962).

Convention on the International Maritime Organisation, opened for signature on 6 March 1948, 289 U.N.T.S. 48; 9 U.S.T. 621 (entered into force on 17 March 1958).

Convention on the International Maritime Satellite Organisation (INMARSAT), opened for signature on 3 September 1976; 31 U.S.T. 1; T.I.A.S. 9605 (entered into force on 16 July 1979).

Convention on the Limitation of Liability for Maritime Claims, opened for signature on 19 November 1976, 1456 U.N.T.S. 221 (entered into force on 1 December 1986).

Convention on the Prevention and Punishment of the Crime of Genocide, opened for signature on 9 December 1948, 78 U.N.T.S. 277 (entered into force on 12 January 1951).

Convention on the Prohibition of the Development, Production, Stockpiling and Use of Chemical Weapons and on their Destruction, opened for signature on 13 January 1993, 1015 U.N.T.S. 163; 32 I.L.M. 800 (entered into force on 29 April 1997).

Final Act and Agreement Establishing the World Trade Organisation, opened for signature on 15 April 1994, 1867 U.N.T.S. 154; 33 I.L.M. 1144 (entered into force on 1 January 1995).

Fourth ACP-EEC Convention of Lomé, opened for signature on 15 December 1989 (1991) O.J. L229/3 (entered into force on 1 March 1990).

General Agreement on Tariffs and Trade, opened for signature on 30 October 1947, 55 U.N.T.S. 187 (entered into force on 1 January 1948).

International Covenant on Civil and Political Rights, opened for signature on 16 December 1966, 999 U.N.T.S. 171 (entered into force on 23 March 1976).

International Telecommunications Satellite Organization (INTELSAT) Agreement Between the United States of American and Other Governments and Operating Agreement, opened for signature on 20 August 1971, 23 U.S.T. 3813; T.I.A.S. 7532 (entered into force on 12 February 1973).

Madrid Protocol on Environmental Protection to the Antarctic Treaty, opened for signature on 4 October 1991, 30 I.L.M. 1455 (entered into force on 14 January 1998).

Partnership Agreement Between the Members of the African, Caribbean and Pacific Group of States, of the One Part, and the European Community and its Member States, of the Other Part, opened for signature on 15 December 2000 (2000) O.J. L317/3 (entered into force on 1 April 2003).

Protocol Amending the General Agreement on Tariffs and Trade, opened for signature on 8 February 1965, 572 U.N.T.S. 320 (entered into force on 27 June 1966).

Protocol on Environmental Protection to the Antarctic Treaty, opened for signature on 4 October 1991, 30 I.L.M. 1461 (entered into force on 14 January 1998).

Second ACP-EEC Convention of Lomé, opened for signature on 31 October 1979 (1980) O.J. L347/1 (entered into force on 1 January 1981).

Statute of the International Atomic Energy Agency, opened for signature on 26 October 1956, 276 U.N.T.S. 3; 8 U.S.T. 1093 (entered into force on 29 July 1957).

Statute of the International Court of Justice, opened for signature on 26 June 1945, 1945 U.K.T.S. 67 (entered into force on 24 October 1945).

Statutes of the World Tourism Organisation, opened for signature on 27 September 1970, 27 U.S.T. 2211 (entered into force on 2 January 1975).

Third ACP-EEC Convention of Lomé, opened for signature on 31 March 1986 (1986) O.J. L86/3 (entered into force on 1 May 1986).

Treaty Banning Nuclear Weapon Tests in the Atmosphere, in Outer Space and Under Water, opened for signature on 5 August 1963, 480 U.N.T.S. 43; 14 U.S.T. 1313 (entered into force on 10 October 1963).

Treaty on Principles Governing the Activities of States in the Exploration and Use of Outer Space, including the Moon and other Celestial Bodies, opened for signature on 27 January 1967, 610 U.N.T.S. 205; 18 U.S.T. 2410; T.I.A.S. 6347; 6 I.L.M. 386 (entered into force on 10 October 1967).

United Nations Convention on the Law of the Sea, opened for signature on 10 December 1982, 1833 U.N.T.S. 3; 21 I.L.M. 1261 (entered into force on 16 November 1994).

Vienna Convention on Consular Relations, opened for signature on 24 April 1963, 596 U.N.T.S. 261 (entered into force on 19 March 1967).

Vienna Convention on Diplomatic Relations, opened for signature on 18 April 1961, 500 U.N.T.S. 95 (entered into force on 24 April 1964).

Vienna Convention on the Law of Treaties, opened for signature on 23 May 1969, 1155 U.N.T.S. 331; 1980 U.K.T.S. 58 (entered into force on 27 January 1980).

Warsaw Convention for the Unification of Certain Rules Relating to International Carriage by Air, opened for signature on 12 October 1929, 137 L.N.T.S. 11; 1933 U.K.T.S. 11 (entered into force on 13 February 1933).

Wassenaar Arrangement, BASIC DOCUMENTS (2009), at <http://www.wassenaar.org/publicdocuments/2009/Basic%20Documents%20-%20Jan%202009.pdf>, 20 January 2009, last accessed on 30 November 2009.

Wellington Convention on the Regulation of Antarctic Mineral Resource Activities, opened for signature on 2 June 1988, 21 I.L.M. 859 (not in force).

United Nations Documents

Circular Letter C.N. 418.1993.TREATIES-7 (14 January 1994).

COPUOS, Draft Agreement on the Principles Governing Activities in the Use of Natural Resources of the Moon and Other Celestial Bodies (1970) U.N. Doc. A/AC.105/C.2/L.71.

COPUOS, Report of the Legal Sub-Committee of the Work of its Fifteenth Session (1976), U.N. Doc. A/AC.105/171.

COPUOS, Report of the Legal Sub-Committee of the Work of its Sixteenth Session (1977), U.N. Doc. A/AC.105/196.

General Assembly Resolution 1514 (1960), U.N. G.A.O.R. Supp. (No. 16) at 66, U.N. Doc. A/4684.

General Assembly Resolution 1710 (1961), U.N. G.A.O.R. Supp. (No. 17) at 17, U.N. Doc. A/5100.

General Assembly Resolution 1721 (XIV).

General Assembly Resolution 1803 (1962), U.N. G.A.O.R. Supp. (No. 17) at 14, U.N. Doc. A/5217.

General Assembly Resolution 1962 (XVIII).

General Assembly Resolution 2222 (XXI).

General Assembly Resolution 2345 (XXII).

General Assembly Resolution 2574D (XXIV).

General Assembly Resolution 2625 (XXV).

General Assembly Resolution 2749 (XXV).

General Assembly Resolution 2777 (XXVI).

General Assembly Resolution 2779 (XXVI).

General Assembly Resolution 3201 (S-VI), U.N. G.A.O.R. Supp. (No. 1) at 3, U.N. Doc. A/9559 (1974).

General Assembly Resolution 3202 (S-VI), U.N. G.A.O.R. Supp. (No. 1) at 5, U.N. Doc. A/9559 (1974).

General Assembly Resolution 3235 (XXIX).

General Assembly Resolution 3281 (XXIX), U.N. G.A.O.R. Supp. (No. 30) at 50, U.N. Doc. A/9631 (1974).

General Assembly Resolution 49/34.

General Assembly Resolution A/16/1721B (1961).

General Assembly Resolution A/34/68 (1979).

U.N. Doc. A/6695 (1967).

U.N. Doc. A/8391 (4 June 1971).

U.N. Doc. A/A.4/49/L.11 (11 November 1994).

U.N. Doc. A/AC.105/101 (11 May 1972).

U.N. Doc. A/AC.105/171 (28 May 1976).

U.N. Doc. A/AC.105/196 (11 April 1977).

U.N. Doc. A/AC.105/271.

U.N. Doc. A/AC.105/58.

U.N. Doc. A/AC.105/6.

U.N. Doc. A/AC.105/639 (11 April 1996).

U.N. Doc. A/AC.105/674 (14 April 1997).

U.N. Doc. A/AC.105/735.

U.N. Doc. A/AC.105/737.

U.N. Doc. A/AC.105/740.

U.N. Doc. A/AC.105/762.

U.N. Doc. A/AC.105/806 (22 August 2003).

U.N. Doc. A/AC.105/85, Annex 2.

U.N. Doc. A/AC.105/C.2/13.

U.N. Doc. A/AC.105/C.2/85.

U.N. Doc. A/AC.105/C.2/L.10.

References

U.N. Doc. A/AC.105/C.2/L.10/Rev.1.
U.N. Doc. A/AC.105/C.2/L.12.
U.N. Doc. A/AC.105/C.2/L.13.
U.N. Doc. A/AC.105/C.2/L.2.
U.N. Doc. A/AC.105/C.2/L.206 Rev. 1 (4 April 1997).
U.N. Doc. A/AC.105/C.2/L.21.
U.N. Doc. A/AC.105/C.2/L.23.
U.N. Doc. A/AC.105/C.2/L.28.
U.N. Doc. A/AC.105/C.2/L.4.
U.N. Doc. A/AC.105/C.2/L.45.
U.N. Doc. A/AC.105/C.2/L.53.
U.N. Doc. A/AC.105/C.2/L.54.
U.N. Doc. A/AC.105/C.2/L.7.
U.N. Doc. A/AC.105/C.2/L.71.
U.N. Doc. A/AC.105/C.2/L.8.
U.N. Doc. A/AC.105/C.2/L.82.
U.N. Doc. A/AC.105/C.2/L.83.
U.N. Doc. A/AC.105/C.2/L.9.
U.N. Doc. A/AC.105/C.2/SR.576 (7 April 1994).
U.N. Doc. A/AC.105/C.2/SR.577 (7 April 1994).
U.N. Doc. A/AC.105/C.2/SR.578 (7 April 1994).
U.N. Doc. A/AC.105/C.2/SR.592 (27 March 1996).
U.N. Doc. A/AC.105/C.2/SR.71 (1966).
U.N. Doc. A/AC.105/C2/L.71.
U.N. Doc. A/AC.105/L.2.
U.N. Doc. A/AC.105/PV.173 (1977).
U.N. Doc. A/AC.105/PV.413 (19 June 1995).
U.N. Doc. A/AC.171 (28 May 1976), Annex 1.
U.N. Doc. A/AC.3105/C.2/SR.580 (31 March 1995).
U.N. Doc. A/C.1/879.
U.N. Doc. A/C.1/881.
U.N. Doc. A/C.1/L.301.
U.N. Doc. A/C.1/PV.1492 (1966).
U.N. Doc. A/C.1/PV.1524.
U.N. Doc. A/C.1/PV.1525.
U.N. Doc. A/C.1/PV.1526.
U.N. Doc. A/C.1/PV.1527.
U.N. Doc. A/C.1/PV.1530.
U.N. Doc. A/C.1/SR.4393 (1966).
U.N. Doc. A/C.1/SR.1210.
U.N. Doc. A/C1/881 (14 October 1962).
U.N. Doc. A/CONF.62/30/Rev. 3.
U.N. Doc. A/CONF.62/65 (1978).
U.N. Doc. A/CONF.62/BUR/SR.41 (1978).
U.N. Doc. A/CONF.62/SR.102 (1978).
U.N. Doc. A/PV.2229 (1 May 1974).
U.N. Doc. COPUOS/LEGAL/T.632 (3 April 2000).
U.N. Doc. ST/SG/Ser.E/417 (25 September 2002).
U.N. G.A.O.R. 37th Sess. (10th Meeting), U.N. Doc. A/37/PV.10 (1982).
UNCLOS III, *Official Records*, vol. XVI.
UNCLOS III, *Official Records*, vol. XVII (U.N. Doc. A/CONF.62/122).

International Cases

Application for Revision of the Judgment of 11 July 1996 in the Case Concerning Application of the Convention on the Prevention and Punishment of the Crime of Genocide (Bosnia and Herzegovina v. Yugoslavia), Preliminary Objections (Yugoslavia v. Bosnia and Herzegovina) (2003) I.C.J. Rep. 7.
BP Exploration Company (Libya) Ltd. v. Libya (1978) 53 I.L.R. 297.
Case Concerning Military and Paramilitary Activities in and against Nicaragua (Merits) (Nicaragua v. United States) (1986) I.C.J. Rep. 14.
Chorzów Factory (Indemnity) (Merits) (1928) P.C.I.J. Rep., Ser. A, No. 17.
Flexi-Van Leasing Inc. v. Iran (1986) 12 Iran-U.S.C.T.R. 335.
Foremost Tehran Inc. v. Iran (1986) 10 Iran-U.S.C.T.R. 228.
Hyatt International Corporation v. Government of the Islamic Republic of Iran (1985) 9 Iran-U.S.C.T.R. 72.
Kuwait v. The American Independent Oil Co. (1982) 66 I.L.R. 518.
Legality of the Threat or Use of Nuclear Weapons (1996) I.C.J. Rep. 226.
Libyan American Oil Co. v. Libya (1982) 62 I.L.R. 140.
Lighthouses (1956) 12 R.I.A.A. 155.
Massey (United States v. Mexico) (1927) 4 R.I.A.A. 155.
Mobil Oil Iran, Inc. v. Iran (1987) 16 Iran-U.S.C.T.R. 3.
North Sea Continental Shelf Cases (Germany v. Denmark; Germany v. The Netherlands) (1969) I.C.J. Rep. 3.
Prosecutor v. Tadic (1999) 38 I.L.M. 1518.
S.S. Lotus (France v. Turkey) (1927) P.C.I.J. Rep, Ser. A, No. 10.
South West Africa Case (Ethiopia and Liberia v. South Africa) (1966) I.C.J. Rep. at 292.
Texaco Overseas Petroleum Co. v. Libya (1978) 53 I.L.R. 389.
U.S. Diplomatic and Consular Staff in Tehran (1980) I.C.J. Rep. 3.

Domestic Legislation and Regulations

Australia	*Space Activities Act 1998* (Cth.)
	Space Activities Regulations 2001 (Cth.)
Belgium	*Law on the Activities of Launching, Flight Operations and Guidance of Space Objects 2005*
Brazil	*Resolution on Commercial Launching Activities from Brazilian Territory* (Res. No. 51 of 26 January 2001)
	Regulation on Procedures and on Definition of Necessary Requirements for the Request, Evaluation, Issuance, Follow-up and Supervision of Licences for Carrying out Launching Space Activities on Brazilian Territory (No. 27)
France	*Space Operations Act* (No. 2008-518 of 3 June 2008)
	Law on the Exploration and Exploitation of Mineral Resources of the Deep Seabed (81-1135)
	Decree on the Authorisations Issued in Accordance with French Act No. 2008-518 of 3 June 2008 Relating to Space Operations (No. 2009-643)
Germany	*Act of Interim Regulation of Deep Seabed Mining 1981*
Hong Kong	*Outer Space Ordinance 1997* (No. 65 of 1997)
Japan	*Basic Space Law* (No. 43 of 28 May 2008)

Netherlands	*Space Activities Act* (13 June 2006)
Norway	*Act on Launching Objects from Norwegian Territory into Outer Space* (No. 38 of 13 June 1969)
Republic of Korea	*Space Liability Act* (No. 8852 of 21 December 2007)
Russia	*Law on Space Activities 1993* (Decree 5663-1)
	Statute on Licensing Space Operations (No. 104)
South Africa	*Space Affairs Act 1993*
Sweden	*Act on Space Activities* (1982:963)
	Decree on Space Activities (1982:1069)
Ukraine	*Ordinance of the Supreme Soviet of Ukraine on Space Activity* (15 November 1996)
United Kingdom	*Cornwall Submarine Mines Act 1858*
	Deep Sea Mining Act 1981
	Outer Space Act 1986
United States	*Commercial Space Launch Act 1994* (49 U.S.C. 701)
	Commercial Space Transportation Regulations (14 C.F.R. chap. III)
	Deep Seabed Hard Mineral Resources Act 1980
	House Resolutions 816–824, 828–830, 834–835, 837, 839–840, 843–844, 854–857, 865, 876, 881 and 916, 90th Cong., 1st Sess. (1967).

Domestic Cases

Gann v. Free Fishers of Whitstable (1865) 11 E.R. 1305.
Martin v. Lessee of Waddell (1842) 41 U.S. 367.
Ward v. Creswell (1741) 125 E.R. 1165.

Secondary Sources

Abe, Mukai, Hirata, Barnouin-Jha, Cheng, Demura, Gaskell, Hashimoto, Hiraoka, Honda, Kubota, Matsuoka, Mizuno, Nakamura, Scheeres and Yoshikawa, *Mass and Local Topography Measurements of Itokawa by Hayabusa* (2006) 312 SCIENCE 1344.

Aceves, *Critical Jurisprudence and International Legal Scholarship: A Study of Equitable Distribution* (2001) 39 COLUM. J. TRANSNAT'L. L. 299.

Adams, *Mission Design Study for the 2003/2005 Mars Sample Return Mission* (2002), Colorado Centre for Astrodynamics Research, University of Colorado at Boulder, at <http://ccar. colorado.edu/asen5050/projects/projects_2002/adams/>, 12 December 2002, last accessed on 23 December 2004.

Adams, *The Outer Space Treaty: An Interpretation in Light of the No-Sovereignty Provision* (1968) 9 HARV. INT'L. L. J. 140.

Adede, *Law of the Sea – Developing Countries' Contribution to the Development of the Institutional Arrangements for the International Seabed Authority* (1978) 4 BROOKLYN J. INT'L. L. 1.

Adede, THE SYSTEM FOR SETTLEMENT OF DISPUTES UNDER THE UNITED NATIONS CONVENTION ON THE LAW OF THE SEA (1987).

Adede, *The System for the Exploitation of the "Common Heritage of Mankind" at the Caracas Conference* (1975) 69 AM. J. INT'l. L. 31.

AEA Technologies, *Platinum and Hydrogen for Fuel Cell Vehicles* (2002), U.K. Department for Transportation, at <http://www.dft.gov.uk/stellent/groups/dft_roads/documents/pdf/dft_roads_ pdf_024056.pdf>, last accessed on 3 January 2004.

Agosto, *Beneficiation and Powder Metallurgical Processing of Lunar Soil Metal*, paper presented at the 5th Princeton/AIAA Conference on Space Manufacturing, May 1981, in Princeton, NJ, USA.

Ahrens and Harris, *Deflection and Fragmentation of Near-Earth Asteroids*, in Gehrels (ed.), HAZARDS DUE TO COMETS AND ASTEROIDS (1994), at 897–927.

Akehurst, *Custom as a Source of International Law* (1974–1975) 47 BRIT. Y. B. INT'L. L. 1.

Akinsanya and Davies, *Third World Quest for a New International Economic Order: An Overview* (1984) 33 INT'L. COMP. L. Q. 208.

Alberts, *Technology Transfer and its Role in International Environmental Law: A Structural Dilemma* (1992) 6 HARV. J. L. & TECH. 63.

Alberts and Papp (eds.), THE INFORMATION AGE: AN ANTHOLOGY ON ITS IMPACTS AND CONSEQUENCES (1998).

Alexander, *Legal Aspects: Exploitation of Antarctic Resources: A Recommended Approach to the Antarctic Resource Problem* (1978) 33 U. MIAMI L. REV. 371.

Alexander, *Measuring Damages Under the Convention on International Liability for Damage Caused by Space Objects* (1978) 6 J. SP. L. 151.

Alexander and Hodgson, *The Impact of the 200-Mile Economic Zone on the Law of the Sea* (1975) 12 SAN DIEGO L. REV. 570.

Alexander, Cameron and Nixon, *The Costs of Failure at the Third Law of the Sea Conference* (1978) 9 J. MARIT. L. & COM. 1.

Allen, *Protecting the Oceanic Gardens of Eden: International Law Issues in Deep-Sea Vent Resource Conservation and Management* (2001) 13 GEO. INT'L. ENVT'L. L. REV. 563.

Allen, Gladman, Kavelaars, Petit, Parker and Nicholson, *Discovery of a Low-Eccentricity, High Inclination Kuiper Belt Object at 58 AU* (2006) 640 ASTROPHYSICS J. 83.

Alley, *New Zealand and Antarctica* (1984) 39 INT'l. J. 911.

Allott, *Power Sharing in the Law of the Sea* (1983) 77 AM. J. INT'L. L. 1.

Amarasuriya, *The Third United Nations Conference on the Law of the Sea: An Introductory Insight* (2001) 13 SRI LANKA J. INT'L. L. 137.

American Meteor Society, *Definition of Terms by the IAU Commission 22, 1961,* at <http://www.amsmeteors.org/define.html>, last accessed on 27 December 2004.

Amos, *Messenger spies iron on Mercury*, British Broadcasting Corporation (4 November 2009), at <http://news.bbc.co.uk/2/hi/science/nature/8342000.stm>, last accessed on 5 November 2009.

Anand, NEW STATES AND INTERNATIONAL LAW (1972).

Anand, ORIGIN AND DEVELOPMENT OF THE LAW OF THE SEA (1983).

Andem, *The 1968 Rescue Agreement and the Commercialisation of Outer Space Activities During the 21st Century – Some Reflections* (1998) 41 PROC. COLL. L. OUTER SP. 75.

Andrews, *A Mighty Stone for David's Sling: The International Space Company* (2003) 1 REGENT J. INT'L. L. 5.

Andrews, *Tiny Tonga Seeks Satellite Empire in Space*, THE NEW YORK TIMES, 28 August 1990, at A1.

Armitage, *Making the Empire British: Scotland in the Atlantic World 1542–1707* (1997) 155 PAST & PRESENT 34.

Arnold, *Toward a Principled Approach to the Distribution of Global Wealth: An Impartial Solution to the Dispute over Seabed Manganese Nodules* (1980) 17 SAN DIEGO L. REV. 557.

Arrow, *Seabeds, Sovereignty and Objective Regimes* (1984) 7 FORDHAM INT'L. L. J. 169.

Arrow, *The "Alternative" Seabed Mining Regime: 1981* (1982) 5 FORDHAM INT'L. L. J. 1.

Arrow, *The Proposed Regime for the Unilateral Exploitation of Deep Seabed Mineral Resources by the United States* (1980) 21 HARV. INT'L. L. J. 337.

Asiz, *The New International Order: Search for Common Ground* in Ghosh (ed.), NEW INTERNATIONAL ECONOMIC ORDER: A THIRD WORLD PERSPECTIVE (1984).

Asphaug and Benz, *The Tidal Evolution of Strengthless Planetesimals* (1995), paper presented at the 26th Lunar and Planetary Sciences Conference, March 1995, in Houston, TX, USA.

Auburn, *The International Seabed Area* (1971) 20 INT'L. & COMP. L. Q. 173.

References

Australian Antarctic Division, *Antarctic Law & Treaty: Treaty Parties*, Australian Department of the Environment, Water, Heritage and the Arts (19 August 2009), at <http://www.aad.gov.au/default.asp?casid=80>, last accessed on 14 November 2009.

Australian Department of Foreign Affairs and Trade, *Final Protocol and Partial Revision of the 1998 Radio Regulations, as incorporated in the Final Acts of the World Radiocommunication Conference (WRC-2000), done at Istanbul on 2 June 2000* (2001) A.T.N.I.A. 32, <http://www.austlii.edu.au/au/other/dfat/nia/2001/32.html>, last accessed on 18 April 2004.

Avgerinopoulou, *The Lawmaking Process at the International Seabed Authority as a Limitation on Effective Environmental Management* (2005) 30 COLUM. J. ENVT'l. L. 565.

Awford, *Commercial Space Activities: Legal Liability Issues*, in Mani, Bhatt and Reddy (eds.), RECENT TRENDS IN INTERNATIONAL SPACE LAW AND POLICY (1997).

Awford, *Legal Liability Arising from Commercial Activities in Outer Space*, paper presented at the Annual Conference of the International Bar Association, December 1990, in Paris, France.

Baca, *Property Rights in Outer Space* (1993) 58 J. AIR L. & COM. 1041.

Back-Impallomeni, *The Article VI of the Outer Space Treaty*, in United Nations, PROCEEDINGS OF THE UNITED NATIONS/REPUBLIC OF KOREA WORKSHOP ON SPACE LAW (2003), at 348–351.

Bailey, *Australia and the Law of the Sea* (1962) 1 ADEL. L. REV. 1.

Balassa, *The Tokyo Round and the Developing Countries* (1980) 14 J. WORLD TRADE L. 93.

Balch, *The Arctic and Antarctic Regions and the Law of* Nations (1910) 4 AM. J. INT'L. L. 265.

Ball, *Many Moons*, NATURE SCIENCE UPDATE (1 October 1999), at <http://www.nature.com/nsu/991007/991007-2.html#1>, last accessed on 5 October 2001.

Balsano, *Industrial Property Rights in Outer Space in the International Governmental Agreement (IGA) on the Space Station and the European Partner* (1992) 35 PROC. COLL. L. OUTER SP. 216.

Bandow, *UNCLOS III: A Flawed Treaty* (1981) 19 SAN DIEGO L. REV. 475.

Bankes, *Environmental Protection in Antarctica: A Comment on the Convention on the Conservation of Antarctic Marine Living Resources* (1981) 19 CAN. Y. B. INT'L. L. 303.

Banks, *Protection of Investment in Deep Seabed Mining: Does the United States Have a Viable Alternative to Participation in UNCLOS?* (1984) 2 B. U. INT'L. L. J. 267.

Barbier, ECONOMICS, NATURAL RESOURCE SCARCITY AND DEVELOPMENT (1989).

Barkenbus, *Seabed Negotiations: The Failure of United States Policy* (1977) 14 SAN DIEGO L. REV. 623.

Barnet, THE LEAN YEARS: POLITICS IN THE AGE OF SCARCITY (1980).

Barnett and Morse, SCARCITY AND GROWTH: THE ECONOMICS OF NATURAL RESOURCE AVAILABILITY (1963).

Barnett, van Muiswinkel, Schechter and Myers, *Global Trends in Non-Fuel Minerals*, in Simon and Kahn (eds.), THE RESOURCEFUL EARTH: A RESPONSE TO GLOBAL 2000 (1984), at 316–338.

Barrie, *The Antarctic Treaty Forty Years On* (1999) 116 S. AFR. L. J. 173.

Bartlett and van Rensburg, *Technical, Economic and Institutional Constraints on the Production of Minerals from the Deep Seabed* (1986) ACTA JURIDICA 69.

Barucci, Capria, Coradini and Fulchignoni, *Classification of Asteroids using G-Mode Analysis* (1987) 72 ICARUS 304.

Baslar, THE CONCEPT OF THE COMMON HERITAGE OF MANKIND IN INTERNATIONAL LAW (1998).

Bates, *U.S. Ratification of the U.N. Convention on the Law of the Sea: Passive Acceptance is Not Enough to Protect U.S. Property Interests* (2006) 31 N. C. J. INT'L. L. & COM. REG. 745.

Beck, *Regulating One of the Last Tourism Frontiers: Antarctica* (1990) 10 APP. GEOG. 343.

Beckman, *1968 Rescue Agreement – An Overview*, in United Nations, PROCEEDINGS OF THE UNITED NATIONS/REPUBLIC OF KOREA WORKSHOP ON SPACE LAW (2003), at 370–378.

Beeby, *An Overview of the Problems Which Should be Addressed in the Preparation of a Regime Governing the Mineral Resources of Antarctica*, in Orrego-Vicuna (ed.), ANTARCTIC RESOURCES POLICY: SCIENTIFIC, LEGAL AND POLITICAL ISSUES (1983).

Beesley, *Canadian Practice in International Law During 1972 as Reflected Mainly in Public Correspondence and Statements of the Department of External Affairs* (1973) CAN. Y. B. INT'L. L. 294–295.

Beesley, McWhinney, Smythe, Mawhinney and Gotlieb, *The Legal Problems of International Telecommunications with Special Reference to INTELSAT* (1970) 20 UNI. TORONTO L. J. 287.

Beesley, *The Negotiating Strategy of UNCLOS III: Developing and Developed Countries as Partners – A Pattern for Future Multilateral International Conferences?* (1983) 46 L. & CONTEMP. PROBS. 183.

Beeton, Buckley, Jones, Morgan, Reichelt and Trewin, *Australia State of the Environment 2006: Independent Report of the Australian Government Minister for the Environment and Heritage* (6 December 2006), Australian Government Department of the Environment and Heritage, at <http://www.deh.gov.au/soe/2006/publications/report/index.html>, last accessed on 27 January 2007.

Beletskii, Ivanov and Otstavnov, *Model Problem of a Space Elevator* (2005) 43 COSMIC RESEARCH 152–156.

Bell, *Mineralogical Evolution of Meteorite Parent Bodies* (1986) 17 LUNAR PLANET. SCI. 985.

Bell, *Q-Class Asteroids and Ordinary Chondrites*, paper presented at the 26th Lunar and Planetary Sciences Conference, March 1995, in Houston, TX, USA.

Bell, Davis, Hartmann and Gaffey, *Asteroids: The Big Picture*, in Binzel, Gehrels and Matthews (eds.), ASTEROIDS II (1987), at 921–945.

Benedek, *The Lomé Convention and the International Law of Development: A Concretisation of the New International Economic Order?* (1982) 26 J. AFR. L. 74.

Benkö and Schrogl, *The 1998 European Initiative in the UNCOPUOS Legal Sub-Committee to Improve the Registration Convention* (1998) 41 PROC. COLL. L. OUTER SP. 58.

Benkö, de Graaff and Reijnen, SPACE LAW IN THE UNITED NATIONS (1985).

Bentham, *Antarctica: A Minerals Regime* (1990) 8 J. ENERGY NAT. RES. L. 120.

Bentham and Quine, *The Status of the Law of the Sea Negotiations* (1978) 6 INT'L. BUS. LAWYER 76.

Bentley, Booth, Burton, Coleman, Sellwood and Whitfield, *Perspectives on the Future of Oil* (2000) 18 ENERGY EXPLOR. & EXPLOIT. 147–206.

Bergin, *Antarctica, the Antarctic Treaty Regime and Legal and Geopolitical Implications of Natural Resource Exploration and Exploitation* (1988) 4 FL. INT'L. L. J. 1.

Bergman, *The Regulation of Seabed Mining Under the Reciprocating States Regime* (1981) 30 AM. U. L. REV. 477.

Berkowitz, JOHN SELDEN'S FORMATIVE YEARS: POLITICS AND SOCIETY IN EARLY SEVENTEENTH CENTURY (1988).

Bernhardt, *Sovereignty in Antarctica* (1975) 5 CAL. W. INT'L. L. J. 297.

Bernstein, Dworkin, Sandford and Allamandola, *Ultraviolet Irradiation of Naphthalene in H_2O Ice: Implications for Meteorites and Biogenesis* (2001) 36 METEOR. & PLANET. SCI. 351.

Bertrand and Foliard, *Low-Thrust Optimal Trajectories for Rendezvous with Near Earth Asteroids*, paper presented at the 18th International Symposium on Space Flight Dynamics, 11–15 October 2004, in Munich, Germany.

Bethill, *People's China and the Law of the Sea* (1974) 8 INT'L. LAWYER 724.

Beuttler, *The Composite Text and Nodule Mining – Over-Regulation as a Threat to the "Common Heritage of Mankind"* (1978) 1 HASTINGS INT'L. & COMP. L. REV. 167.

Bevan, Collier and Gunning, TEMPORARY TRADE SHOCKS IN DEVELOPING COUNTRIES: CONSEQUENCES AND POLICY RESPONSES (1991).

B.H.P. Billiton Ltd., *BHP Billiton Diamonds: History*, at <http://ekati.bhpbilliton.com/repository/aboutMine/history.asp>, last accessed on 28 April 2007.

Bhagwati (ed.), THE NEW INTERNATIONAL ECONOMIC ORDER: THE NORTH-SOUTH DEBATE (1977).

Bhatt, *Legal Control of the Exploration and Use of the Moon and Celestial Bodies* (1968) 8 INDIAN J. INT'L. L. 38.

Biblowit, *Deep Seabed Mining: The United States and the United Nations Convention on the Law of the Sea* (1984) 58 ST. JOHN'S L. REV. 267.

References

Biggs, *Deepsea's Adventures: Grotius Revisited* (1975) 9 INT'L. LAWYER 271.

Bilder, *International Law and Natural Resources Policies* (1980) 22 NAT. RES. J. 451.

Binzel and Xu, *Chips Off of Asteroid 4 Vesta: Evidence for the Parent Body of Basaltic Achondrite Meteorites* (1993) 260 SCIENCE 186.

Binzel, Xu, Bus, Skrutskie, Meyer, Knezek and Barker, *Discovery of a Main Belt Asteroid Resembling Ordinary Chondrite Meteorites* (1993) 262 SCIENCE 1541.

Blair, *The Role of Near-Earth Asteroids in Long-Term Platinum Supply*, paper presented at the Second Space Resources Roundtable, Colorado School of Mines, 8–10 November 2000, in Boulder, CO, USA, as at <http://www.mines.edu/research/srr/Presentations/blair-platinum. PDF>, last accessed on 8 December 2004.

Bland, Berry, Jull, Smith, Bevan, Cadogan, Sexton, Franchi and Pillinger, ^{57}Fe *Mössbauer Spectroscopy Studies of Meteorites: Implications for Weathering Rates, Meteorite Flux and Early Solar System Processes* (2003) 142 HYPERFINE INTERACTIONS 481.

Bloomfield, *The Arctic: Last Unmanaged Frontier* (1982) 60 FOREIGN AFF. 87.

Blum, *The Deep Freeze: Torts, Choice of Law, and the Antarctic Treaty Regime* (1994) 8 EMORY INT'L. L. REV. 667.

Böckstiegel, *Reconsideration of the Legal Framework for Commercial Space Activities* (1990) 33 PROC. COLL. L. OUTER SP. 3.

Böckstiegel, *Settlement of Disputes Regarding Space Activities* (1993) 21 J. SP. L. 1.

Böckstiegel, *The Term "Launching State" in International Space Law* (1994) 37 PROC. COLL. L. OUTER SP. 80.

Böckstiegel, *The Terms "Appropriate State" and "Launching State" in the Space Treaties: Indications of State Responsibility and Liability for State and Private Space Activities* (1992) 35 PROC. COLL. L. OUTER SP. 15.

Boczek, *Ideology and the Law of the Sea: The Challenge of the New International Economic Order* (1984) 7 B. C. INT'L. & COMP. L. REV. 1.

Boczek, *The Soviet Union and the Antarctic Regime* (1984) 78 AM. J. INT'L. L. 834.

Borgerson, *Arctic Meltdown: The Economic and Security Implications of Global Warming* (2008) 87:2 FOREIGN AFF. 63.

Borgese, *A Constitution for the Oceans: Comments and Suggestions Regarding Part XI of the Informal Composite Negotiating Text* (1978) 15 SAN DIEGO L. REV. 371.

Borgese, *Boom, Doom and Gloom over the Oceans: The Economic Zone, the Developing Nations and the Conference on the Law of the Sea* (1974) 11 SAN DIEGO L. REV. 541.

Borgese, *The New International Economic Order and the Law of the Sea* (1977) 14 SAN DIEGO L. REV. 584.

Borgese, *The Role of the International Seabed Authority in the 1980s* (1981) 18 SAN DIEGO L. REV. 395.

Borgsmidt, *The Generalised System of Preferences in Favour of Developing Countries against the Historical Background in the Light of Public International Trade Law and the New International Economic Order* (1985) 54 NORDISK TIDS. INT'L. RET. 33.

Bostrom, *The United States' Legislative Response to the Third United Nations Conference on the Law of the Sea Deadlock* (1979) 2 B. C. INT'L. & COMP. L. J. 409.

Bostwick, *Going Private with the Judicial System: Making Creative Use of ADR Procedures to Resolve Commercial Space Disputes* (1995) 23 J. SP. L. 1.

Bourély, *Rules of International Law Governing the Commercialisation of Space Activities* (1986) 29 PROC. COLL. L. OUTER SP. 157.

Breaux, *The Diminishing Prospects for an Acceptable Law of the Sea Treaty* (1979) 19 VA. J. INT'L. L. 257.

Brierly, *The Lotus Case* (1928) 44 L. Q. REV. 154.

Brilmayer and Klein, *Land and Sea: Two Sovereignty Regimes in Search of a Common Denominator* (2001) 33 N. Y. U. J. INT'L. L. & POL'Y. 703.

Briscoe and Lambert, *Seabed Mineral Discoveries Within National Jurisdiction and the Future of the Law of the Sea* (1984) 18 U. S. F. L. REV. 433.

Britt, Kring and Bell, *The Density/Porosity of Asteroids*, paper presented at the 26th Lunar and Planetary Sciences Conference, March 1995, in Houston, TX, USA.

Brody, *Thinking Differently with Space Elevators* (2006) AD ASTRA, Summer 2006, at 34.

Brooke, *The Current Status of Deep Seabed Mining* (1984) 24 VA. J. INT'L. L. 361.

Brooks, *Control and Use of Planetary Resources* (1969) 11 PROC. COLL. L. OUTER SP. 342.

Brooks, *National Control of Natural Planetary Bodies: Preliminary Considerations* (1966) 32 J. AIR L. & COM. 315.

Brooks and Hampshire, *Multiple Asteroid Flyby Missions*, in Gehrels (ed.), PHYSICAL STUDIES OF MINOR PLANETS (1981) at 527–537.

Brower, *The Charter of Economic Rights and Duties of Rights and the American Constitutional Tradition: A Bicentennial Perspective on the "New International Economic Order"* (1976) 10 INT'L. LAWYER 701.

Brower and Tepe, *The Charter of Economic Rights and Duties of States: A Reflection or Rejection of International Law?* (1975) 9 INT'L. LAWYER 295.

Brown, *The UN Convention on the Law of the Sea 1982* (1984) 2 J. ENERGY NAT. RES. L. 259.

Brown and Wolk, *Natural Resource Scarcity and Technological Change* (2000:1) EC. & FIN. REV. 2.

Brownlie, PRINCIPLES OF PUBLIC INTERNATIONAL LAW (6th ed., 2003).

Brunet, *Musing on the Bottom: Economic and Legal Implications of the United States' Proposed Draft United Nations Convention on the International Seabed* (1974) U. ILL. L. F. 251.

Bückling, *Interplanetarisches Kooperationsrecht* (1960) 55 DIE FRIEDENSWARTE 305.

Bückling, *The Strategy of Semantics and the "Mankind Provisions" of the Space Treaty* (1979) 7 J. SP. L. 15.

Burbine and Bell, *Asteroid Taxonomy: Problems and Proposed Solutions*, in Milani, Di Martino and Cellino (eds.) INTERNATIONAL ASTRONOMICAL UNION SYMPOSIUM 160: ASTEROIDS, COMETS AND METEORS (1993), at 49.

Burbine and Bell, *How Diverse is the Asteroid Belt?* (1993) 24 LUNAR PLANET. SCI. 223.

Burbine and Binzel, *Asteroid Spectroscopy and Mineralogy* in Milani, Di Martino and Cellino (eds.), INTERNATIONAL ASTRONOMICAL UNION SYMPOSIUM 160: ASTEROIDS, COMETS AND METEORS (1993) at 255.

Burbine, Gaffey and Bell, *S-Class Asteroids 387 Aquitania and 980 Anacostia: Possible Fragments of the Breakup of a Spiral-Bearing Parent Body with CO_3/CV_3 Affiliates* (1992) 27 METEORITICS 424.

Burr, *The International Seabed Authority* (2006) 29 SUFFOLK TRANSNAT'L. L. REV. 271.

Burton, *Freedom of the Seas: International Law Applicable to Deep Seabed Mining Claims* (1977) 29 STANFORD L. REV. 1135.

Burton, *New Stresses on the Antarctic Treaty: Toward International Legal Institutions Governing Antarctic Resources* (1979) 65 VA. L. REV. 421.

Butler, *Grotius and the Law of the Sea*, in Bull, Kingsbury and Roberts (eds.), HUGO GROTIUS AND INTERNATIONAL RELATIONS (1992), at 229–220.

Butler, *Socialist International Institutions and the New International Economic Order* (1984) 3 PUB. L. FORUM 152.

Butte, *International Norms in the Antarctic Treaty* (1992) 3 INT'L. LEG. PERSP. 1.

Butte, *The Law of the Sea – Breakers Ahead* (1972) 6 INT'L. LAWYER 237.

Buxton, *Property in Outer Space: The Common Heritage of Mankind Principle vs. The "First in Time, First in Right" Rule of Property Law* (2004) 69 J. AIR L. & COM. 689.

Buzan, *Seabed Issues at the Law of the Sea Conference: The Caracas Session* (1974) 12 CAN. Y. B. INT'L. L. 222.

Byrum, *An International Seabed Authority: The Impossible Dream?* (1978) 10 CASE W. RES. J. INT'L. L. 621.

Cable News Network, *SpaceShipOne Captures X Prize* (4 October 2004), at <http://edition.cnn.com/2004/TECH/space/10/04/spaceshipone.attempt.cnn/>, last accessed on 21 December 2004.

Caicedo Perdomo, *La Teoría del Jus Cogens en Derecho Internacional a la Luz de la Convención de Viena Sobre el Derecho de los Tratados* (1975) 206–207 REV. ACA. COLOM. JURIS. 259.

Cameron, *Origin of the Solar System* (1988) 26 ANN. REV. ASTRON. & ASTROP. 441.

References

Campbell, THE COMING OIL CRISIS (1997).

Canmann, *Antarctic Oil Spills of 1989: A Review of the Application of the Antarctic Treaty and the New Law of the Sea to the Antarctic Environment* (1990) 1 COLO. J. INT'L. ENVT'L. L. & POL'Y. 211.

Carasco, *A Nationalisation Compensation Framework in the New International Economic Order* (1983) THIRD WORLD LEG. STUD. 49.

Carlson, *Soviet Policy on the Seabed and the Ocean Floor* (1973) 1 SYR. J. INT'L. L. & COM. 104.

Caron, *Municipal Legislation for Exploitation of the Deep Seabed* (1980) 8 OCEAN DEV. & INT'L. L. J. 259.

Carpenter, *Warm is the New Cold: Global Warming, Oil, UNCLOS Article 76 and How an Arctic Treaty Might Stop a New Cold War* (2009) 39 ENVT'L. L. 215.

Carrington, *Spacecraft Snatches First Samples from Asteroid* (26 November 2005), NEW SCIENTIST, at <http://space.newscientist.com/article.ns?id=dn8380>, last accessed on 28 January 2007.

Carroll, *Of Icebergs, Oil Wells and Treaties: Hydrocarbon Exploitation Offshore Antarctica* (1983) 19 STAN. J. INT'L. L. 207.

Carson, Mitchell, Hanemann, Kopp, Presser and Ruud, *Contingent Valuation and Lost Passive Use: Damages from the Exxon Valdez Oil Spill* (2003) 25 ENVT'L. & RES. EC. 257.

Carusi, Valsecchi, D'Abramo and Boattini, *Deflecting NEOs in Route of Collision with the Earth* (2002) 159 ICARUS 417.

Castillo Argañarás, *Benefits Arising From Space Activities and the Needs of Developing Countries* (2000) 43 PROC. COLL. L. OUTER SP. 50.

Chapman, *S-Type Asteroids, Ordinary Chondrites and Space Weathering: The Evidence from Galileo's Fly-bys of Gaspra and Ida* (1996) 31 METEOR. & PLANET. SCI. 699.

Charlesworth, *Customary International Law and the Nicaragua Case* (1988) 11 AUST. Y. B. INT'L. L. 1.

Charney, *Customary International Law in the Nicaragua Case Judgment on the Merits* (1988) 1 HAGUE Y. B. INT'L. L. 16.

Charney, *Entry into Force of the 1982 Convention on the Law of the Sea* (1995) 35 VA. J. INT'L. L. 381.

Charney, *Law of the Sea: Breaking the Deadlock* (1977) 55 FOREIGN AFF. 598.

Charney, *The Equitable Sharing of Revenues from Seabed Mining* (1975) 8 STUD. TRANSNAT'L. LEG. POL'Y. 53.

Charney, *The Law of the Deep Seabed Post UNCLOS III* (1984) 63 OR. L. REV. 19.

Charney, *U.S. Provisional Application of the 1994 Deep Seabed Agreement* (1994) 88 AM. J. INT'L. L. 705.

Charney, *United States Interests in a Convention on the Law of the Sea: The Case for Continued Efforts* (1978) 11 VAND. J. TRANSNAT'L. L. 39.

Charney, *Universal International Law* (1993) 87 AM. J. INT'L. L. 529.

Chen, *Pending Issues Before the Legal Sub-Committee of the United Nations Committee on the Peaceful Uses of Outer Space* (1977) 5 J. SP. L. 30.

Cheng, *"Space Objects", "Astronauts" and Related Expressions* (1991) 34 PROC. COLL. L. OUTER SP. 17.

Cheng, *Article VI of the Outer Space Treaty Revisited: "International Responsibility", "National Activities" and "The Appropriate State"* (1998) 26 J. SP. L. 10.

Cheng, STUDIES IN INTERNATIONAL SPACE LAW (1998).

Cheng, *The 1967 Outer Space Treaty: Thirtieth Anniversary* (1998) 23 AIR & SP. L. 156.

Cheng, *The Commercial Development of Space: The Need for New Treaties* (1991) 19 J. SP. L. 17.

Cheng, *United Nations Resolutions on Outer Space: "Instant" International Customary Law?* (1965) 5 INDIAN J. INT'L. L. 23.

Chivers, *Japan Plans Giant Solar Power Station in Space* (10 November 2009), The Telegraph (United Kingdom), at <http://www.telegraph.co.uk/earth/energy/solarpower/6536752/Japan-plans-solar-power-station-in-space.html>, last accessed on 13 November 2009.

Chopra, *Antarctica as a Commons Regime: A Conceptual Framework for Cooperation and Coexistence*, in Joyner and Chopra (eds.), THE ANTARCTIC LEGAL REGIME (1988), at 163–186.

Chopra, *Antarctica in the United Nations: Rethinking the Problems and Prospects* (1986) 80 AM. SOC. INT'L. L. PROC. 269.

Christol, *Article 2 of the 1967 Principles Treaty Revisited* (1984) 9 ANN. AIR & SP. L. 217.

Christol, *International Liability for Damage Caused by Space Objects* (1980) 74 AM. J. INT'L. L. 346.

Christol, *Protection of Space from Environmental Harms* (1979) 4 ANN. AIR & SP. L. 433.

Christol, *Space Joint Ventures: The United States and Developing Nations* (1975) 8 AKRON L. REV. 398.

Christol, SPACE LAW: PAST, PRESENT AND FUTURE (1991).

Christol, *The 1979 Moon Agreement: Where is it Today?* (1999) 27 J. SP. L. 1.

Christol, *The American Bar Association and the 1979 Moon Treaty: The Search for a Position* (1981) 9 J. SP. L. 77.

Christol, *The Common Heritage of Mankind Provision in the 1979 Agreement Governing the Activities of States on the Moon and other Celestial Bodies* (1980) 14 INT'L. LAWYER 429.

Christol, THE MODERN INTERNATIONAL LAW OF OUTER SPACE (1982), at 140.

Christol, *The Moon Treaty and the Allocation of Resources* (1997) 22 ANN. AIR & SP. L. 31.

Christou, *Coorbital Objects in the Main Asteroid Belt* (2000) 356 ASTRON. & ASTROPHYS. 71.

Christy, *Property Rights in the World Ocean* (1975) 15 NAT. RES. J. 695.

Chu, *The United States and UNCLOS III in the New Decade: Is it Time for a Compromise?* (1992) 4 J. CONTEMP. LEG. ISSUES 253.

Clancy, *The Tragedy of the Global Commons* (1998) 5 IND. J. GLOBAL LEGAL STUD. 601.

Cloutis and Gaffey, *The Constituent Minerals in Calcium-Aluminium Inclusions: Spectral Reflectance Properties and Implications for CO Carbonaceous Chondrites and Asteroids* (1993) 105 ICARUS 568.

Cocca, *Legal Status of the Natural Resources of the Moon and Other Celestial Bodies* (1971) 13 PROC. COLL. L. OUTER SP. 146.

Cocca, *The Principles of the "Common Heritage of Mankind" as Applied to Natural Resources from Outer Space and Celestial Bodies* (1974) 16 PROC. COLL. L. OUTER SP. 174.

Cocca, *The Advances in International Law through the Law of Outer Space* (1981) 9 J. SP. L. 13.

Collins, AFTER SPUTNIK: FIFTY YEARS OF THE SPACE AGE (2007).

Collins, *Deep Seabed Hard Mineral Resources Act – Matrix for United States Deep Seabed Mining* (1981) 13 NAT. RES. LAWYER 571.

Collins, *Implications of Reduced Launch Costs for Commercial Space Law*, in Tatsuzawa (ed.), LEGAL ASPECTS OF SPACE COMMERCIALISATION (1992).

Collins, *Mineral Exploitation of the Seabed: Problems, Progress and Alternatives* (1979) 12 NAT. RES. LAWYER 599.

Collins, *One World . . . One Telephone: Iridium, One Look at the Making of a Global Age* (2005) 21 HIST. & TECH. 301.

Collins and Isozaki, JRS RESEARCH ACTIVITIES FOR SPACE TOURISM (1995), paper presented at the 6th International Conference of Pacific Basin Space Societies, December 1995, in Marina del Rey, CA, USA.

Collins, Iwasaki, Kanayama and Ohmuki, *Commercial Implications of Market Research on Space Tourism* (1994), paper presented at the 19th International Symposium on Space Technology and Science, May 1994, in Yokohama, Japan.

Colorado School of Mines, *The Henderson Mine* (2004), at <http://cause.mines.edu/media/UNO_ Henderson.pdf>, last accessed on 24 July 2007.

Colson, *The United States Position on Antarctica* (1986) 19 CORNELL INT'L. L. J. 291.

Commoner, THE CLOSING CIRCLE: MAN, NATURE AND TECHNOLOGY (1972).

Conant and Conant, *Resource Development and the Seabed Regime of UNCLOS III: A Suggestion for Compromise* (1978) 18 VA. J. INT'L. L. 61.

References 337

Condara, *Outer Space, Like the Sea and the Air, Whose Frontier? Incredible Potential with Inscrutable Obstacles* (1984) 6 HOUS. J. INT'L. L. 175.

Conforti, *Territorial Claims in Antarctica: A Modern Way to Deal with an Old Problem* (1986) 19 CORNELL INT'L. L. J. 249.

Cook, *The Discovery of Lunar Water: An Opportunity to Develop a Workable Moon Treaty* (1999) 11 GEORGETOWN INT'L. ENVT'L. L. REV. 647.

Costa, *Pacific Resources and Ocean Law: A Latin American Perspective* (1989) 16 ECOLOGY L. Q. 245.

Cox and Brogan, *Law of the Sea – Proposed Deep Seabed Hard Mineral Resources Act* (1979) 9 GA. J. INT'L. & COMP. L. 641.

Craven, *Technology and the Law of the Sea: The Effect of Prediction and Misprediction* (1985) 45 LA. L. REV. 1143.

Crawford and Rothwell, *Legal Issues Confronting Australia's Antarctica* (1991) 13 AUST. Y. B. INT'L. L. 53.

Criswell, *Lunar Solar Power System: Scale and Cost*, paper presented at the 45th International Astronautical Congress, 9–14 October 1995, in Jerusalem, Israel.

Cronin, Pizzarello and Cruikshank, *Organic Matter in Carbonaceous Chondrites, Planetary Satellites, Asteroids and Comets*, in Kerridge and Matthews (eds.), METEORITES IN THE EARLY SOLAR SYSTEM (1988).

Crook, *The 2003 Judicial Activity of the International Court of Justice* (2004) 98 AM. J. INT'L. L. 309.

Crovisier, *Molecular Abundances in Comets*, in Milani, Di Martino and Cellino (eds.), INTERNATIONAL ASTRONOMICAL UNION SYMPOSIUM 160: ASTEROIDS, COMETS AND METEORS (1993) at 313–326.

Cruikshank, Allamandola, Hartmann, Tholen, Brown, Matthews and Bell, *Solid C≡N Bearing Material on Outer Solar System Bodies* (1991) 94 ICARUS 345.

D'Amato, *An Alternative to the Law of the Sea Convention* (1983) 77 AM. J. INT'L. L. 281.

D'Sa, *The "Right to Development" and the New International Economic Order with Special Reference to Africa* (1984) THIRD WORLD LEG. STUD. 140

Dalfen, Bissonnette, Juneau and Vlasic, *International Legal Problems of Direct Satellite Broadcasting* (1970) 20 U. TORONTO L. J. 314.

Dalton, *The Chilean Mar Presencial: A Harmless Concept or a Dangerous Precedent?* (1993) 8 INT'L. J. MARINE & COASTAL L. 397.

Daly, STEADY-STATE ECONOMICS (1977).

Daly and Townsend, VALUING THE EARTH: ECONOMICS, ECOLOGY AND ETHICS (1993).

Danilenko, *International Jus Cogens: Issues of Law-Making* (1991) 2 EUR. J. INT'L. L. 42.

Danilenko, *Outer Space and the Multilateral Treaty-Making Process* (1989) 4 BERKELEY TECH. L. J. 217.

Danilenko, *The Concept of the "Common Heritage of Mankind" in International Law* (1988) 13 ANN. AIR & SP. L. 247.

Danzig, *A Funny Thing Happened to the Common Heritage on the Way to the Sea* (1975) 12 SAN DIEGO L. REV. 655.

Darman, *The Law of the Sea: Rethinking U.S. Interests* (1978) 56 FOREIGN AFF. 373.

Darwin, *The Outer Space Treaty* (1967) 42 BRIT. Y. B. INT'L. L. 282.

Dauses, *Zur Rechtslage des Mondes und anderen Himmelkörper* (1975) 24 ZEIT. LUFT. WELT. 223.

Davenport, EUROPEAN TREATIES BEARING ON THE HISTORY OF THE UNITED STATES AND ITS DEPENDENCIES TO 1648 (1917).

David, *Space Cooperation: The China Factor* (5 January 2003), at <http://www.space.com/news/china_cooperation_030121.html>, last accessed on 11 July 2006.

David, *Space Weapons for Earth Wars* (15 May 2002), at <http://www.space.com/businesstechnology/technology/space_war_020515-1.html>, last accessed on 30 November 2009.

David, *The Space Elevator Comes Closer to Reality* (2002), at <http://www.space.com/businesstechnology/technology/space_elevator_020327-1.html>, last accessed on 15 July 2006.

Davidow and Chiles, *The United States and the Issue of the Binding or Voluntary Nature of International Codes of Conduct Regarding Restrictive Business Practices* (1978) 72 AM. J. INT'L. L. 247.

Davis, *Protecting Antarctica: Will a Minerals Agreement Guard the Door or Open the Door to Commercial Exploitation?* (1990) 23 GEO. WASH. J. INT'L. L. & EC. 733.

Davis, *Reaching for the Moon* (1997) NEW SCIENTIST, 23 August 1997, at 46.

Davis, *The Economic Effects of Windfall Gains in Export Earnings 1975–78* (1983) 11 WORLD DEV. 119.

Davis, *Unifying the Final Frontier: Space Industry Financing Reform* (2001) 106 COM. L. J. 455.

Davis and Lee, *Twenty Years After: The Moon Agreement and its Legal Controversies* (1999) AUST. INT'L. L. J. 9.

Davis and Turekian, METEORITES, COMETS AND PLANETS (2005)

de Jager and Reijnen, *Mesospace: The Region between Airspace and Outer Space* (1975) 18 PROC. COLL. L. OUTER SP. 107.

de Selding, *Virgin Galactic Customers Parting with their Cash* (3 April 2006), *Space News*, at <http://www.space.com/spacenews/businessmonday_060403.html>, last accessed on 27 January 2007.

de Vattel, LE DROIT DES GENS OU PRINCIPES DE LA LOI NATURELLE, APPLIQUES A LA CONDUITE ET AUX AFFAIRES DES NATIONS ET DES SOUVERAINS (1758).

Deaton and Miller, *International Commodity Prices, Macroeconomic Performance and Politics in Sub-Saharan Africa*, at Princeton University (October 1995), at <http://www.princeton.edu/~deaton/downloads/International_Commodity_Prices.pdf>, last accessed on 29 November 2009.

Debehogne and d Freitas Mourão, *Positions of Comet P/Ashbrook-Jackson in 1978 – A Note on Perturbations by Jupiter* (1979) 29 ACTA ASTRON. 301.

Deffeyes, BEYOND OIL: THE VIEW FROM HUBBERT'S PEAK (2005).

Deihl, *Antarctica: An International Laboratory* (1991) 18 B. C. ENVT'L. AFF. L. REV. 423.

Dellapenna, *Treaties as Instruments for Managing Internationally-Shared Water Resources: Restricted Sovereignty vs. Community of Property* (1994) 26 CASE W. RES. J. INT'L. L. 27.

Delzeit and Beal, *The Vulnerability of the Pacific Rim Orbital Spectrum Under International Space Law* (1996) 9 N. Y. INT'L. L. REV. 69.

Dembling, *Cosmos 954 and the Space Treaties* (1978) 6 J. SP. L. 129.

Dembling and Kalsi, *Pollution of Man's Last Frontier: Adequacy of Present Space Environmental Law in Preserving the Resource of Outer Space* (1973) 20 NETH. INT'L. L. REV. 125.

DeSaussure and Haanappel, *A Unified Multinational Approach to the Application of Tort and Contract Principles to Outer Space* (1978) 6 SYRACUSE J. INT'L. L. 1.

Devine, *Southern Africa and the Law of the Sea: Problems Common, Uncommon and Unique* (1986) ACTA JURIDICA 29.

Diamond and Dodsworth, *Normative and Positive Theories of International Cost Sharing: The Case of the Netherlands* (1977) 124 DE ECONOMIST 403.

Dias, *The NIEO Revisited* (1996) 8 SRI LANKA J. INT'L. L. 27.

Dickey, *Should the Law of the Sea Conference be Saved?* (1978) 12 INT'L. LAWYER 1.

Dickinson, *Introductory Comment to the Harvard Research Draft Convention on Jurisdiction with Respect to Crime* (1935) 29 AM. J. INT'L. L. SUPP. 443.

Diederiks-Verschoor, AN INTRODUCTION TO SPACE LAW (2nd ed., 1999).

Doyle, *Confidence Building Measures Using Space Resources* (1998) 41 PROC. COLL. L. OUTER SP. 108.

Doyle, *Using Extraterrestrial Resources under the Moon Agreement of 1979* (1998) 26 J. SP. L. 111.

Dubner, *A Proposal for Accommodating the Interests of Archipelagic and Maritime States* (1976) 8 N. Y. U. J. INT'L. L. & POL'Y. 39.

Dubner, *On the Basis for Creation of a New Method of Defining International Jurisdiction in the Arctic Ocean* (2005) 13 MO. ENVT'L. L. & POL'Y. REV. 1.

References   339

Dubner, *The Caspian: Is it a Lake, a Sea or an Ocean and Does it Matter? The Danger of Utilising Unilateral Approaches to Resolving Regional/International Issues* (2000) 18 DICK. J. INT'L. L. 253.

Dubow, *The Third United Nations Conference on the Law of the Sea: Questions of Equity for American Business* (1982) 4 NW. J. INT'L. L. & BUS. 172.

Duff, *The United States and the Law of the Sea Convention: Sliding Back from Accession and Ratification* (2006) 11 OCEAN & COASTAL L. J. 1.

Duff, *UNCLOS and the New Deep Seabed Mining Regime: The Risks of Refuting the Treaty* (1996) 19 SUFFOLK TRANSNAT'L. L. REV. 1.

Dula, *Free Enterprise and the Proposed Moon Treaty* (1980) 2 HOUS. J. INT'L. L. 3.

Dula, *Private Sector Activities in Outer Space* (1985) 19 INT'L. LAWYER 159.

Dupuy, *The Notion of the Common Heritage of Mankind Applied to the Seabed* (1983) 18 ANN. AIR & SP. L. 347.

Dupuy and Vignes, A HANDBOOK ON THE NEW LAW OF THE SEA (Volume 2, 1991).

Duruigbo, *Permanent Sovereignty and Peoples' Ownership of Natural Resources in International Law* (2006) 38 GEO. WASH. INT'L. L. REV. 33.

Edison and Vårdal, *Optimal Currency Baskets for Small, Developed Economies* (1990) 92 SCAND. J. ECON. 559.

Edwards, *Responses of the International Monetary Fund and the World Bank to the Call for a "New International Economic Order": Separating Substance from Rhetoric* (1984) 3 PUB. L. FORUM 89.

Edwards and Westling, THE SPACE ELEVATOR: A REVOLUTIONARY EARTH-TO-SPACE TRANSPORTATION SYSTEM (2003).

Ehrlich, THE POPULATION BOMB (1970).

Eichelberger, *The Seabed Question in Context: One of the Many Issues Massing for the 1973 Conference* (1971) 8 SAN DIEGO L. REV. 653.

Eichelberger, *The United Nations and the Bed of the Sea* (1969) 6 SAN DIEGO L. REV. 339.

Eichelberger and Christy, *Comments on International Control of the Sea's Resources*, in Alexander (ed.), THE LAW OF THE SEA (1967) at 299–309.

Eilingsfeld and Schaetzler, *The Cost of Capital for Space Tourism Ventures*, paper presented at the 51st International Astronautical Congress, 2–6 October 2000, in Rio de Janeiro, Brazil.

Ellis, *The New International Economic Order and General Assembly Resolutions: The Debate over the Legal Effects of General Assembly Resolutions Revisited* (1985) 15 CAL. W. INT'L. L. J. 647.

Eltman, *A Peace Zone on the High Seas: Managing the Commons for Equitable Use* (1993) 5 INT'L. LEGAL PERSP. 47.

Ely, *Potential Regimes for Deep Seabed Mining* (1978) 6 INT'L. BUS. LAWYER 93.

Ely, *The Draft United Nations Convention on the International Seabed Area – American Bar Association Position* (1971) 4 NAT. RES. LAWYER 60.

Ely, *United States Seabed Minerals Policy* (1971) 4 NAT. RES. LAWYER 597.

Encrenaz, *The Chemical Atmospheric Composition of the Giant Planets* (1994) 67 EARTH, MOON & PLANETS 77.

Encrenaz, Drossart, Feuchtgruber, Lellouch, Bézard, Fouchet and Atreya, *The Atmospheric Composition and Structure of Jupiter and Saturn from ISO Observations: A Preliminary Review* (1999) 47 PLANET. & SPACE SCI. 1225.

Encrenaz, Kallenbach, Owen and Sotin (eds.), THE OUTER PLANETS AND THEIR MOONS: COMPARATIVE STUDIES OF THE OUTER PLANETS (2005).

Engle, *The Failure of the Nation State and the New International Economic Order: Multiple Converging Crises Present Opportunity to Elaborate a New Jus Gentium* (2004) 16 ST. THOM. L. REV. 187.

Enzanbacher, *The Regulation of Antarctic Tourism*, in Hall and Johnson (eds.), POLAR TOURISM: TOURISM IN THE ARCTIC AND ANTARCTIC REGIONS (1995).

Enzenbacher, *Antarctic Tourism: An Overview of 1992/93 Season Activity, Recent Developments and Emerging Issues* (1994) 30 POLAR REC. 105.

Ervin, *Law in a Vacuum: The Common Heritage Doctrine in Outer Space Law* (1984) 7 B. C. INT'L. & COMP. L. REV. 403.

Esposito, *The Commercial Exploitation of Space* (1985) 25 A. F. L. REV. 159.

European Space Agency, *All About E.S.A.* (12 September 2008), at <http://www.esa.int/SPECIALS/About_ESA/SEMW16ARR1F_0.html>, last accessed on 1 November 2008.

European Space Agency, *Not-So-Soft Landings on Other Worlds* (2001), at <http://www.esa.int/export/esaCP/ESAZXCZ84UC_FeatureWeek_0.html>, last accessed on 23 December 2004.

Exxon Mobil Corporation, *Exploration in Developing Countries* (1978), paper presented at the Energy Committee Seminar, Aspen Institute of Humanistic Studies, 16–20 July 1978, in Boulder, CO, USA.

Ezor, *Costs Overhead: Tonga's Claiming of Sixteen Geostationary Orbital Sites and the Implications for U.S. Space Policy* (1993) 24 L. & POL'Y. Int'l. BUS. 915.

Fabricio dos Santos, *Developing Countries and the UNIDROIT Protocol on Space Property* (2002) 45 PROC. COLL. L. OUTER SP. 23.

Fanale and Salvail, *Evolution of the Water Regime of Phobos* (1990) 88 ICARUS 380.

Faria, *Draft to an International Covenant for Outer Space* (1960) 3 PROC. COLL. L. OUTER SP. 122.

Farinella, Froeschlé, Froeschlé, Gonczi, Hahn, Morbidelli and Valsecchi, *Asteroids Falling into the Sun* (1994) 371 NATURE 314.

Farrar, *How to Harvest Solar Power? Beam it Down from Space!* (1 June 2008), CABLE NEWS NETWORK, at <http://edition.cnn.com/2008/TECH/science/05/30/space.solar/index.html>, last accessed on 13 November 2009.

Fasan, *Asteroids and Other Celestial Bodies – Some Legal Differences* (1998) 26 J. SP. L. 33.

Fasan, *Large Space Structures and Celestial Bodies* (1984) 27 PROC. COLL. L. OUTER SP. 243.

Fasan, *The Meaning of the Term "Mankind" in Space Legal Language* (1974) 2 J. SP. L. 125.

Fawcett, INTERNATIONAL LAW AND THE USES OF OUTER SPACE (1968).

Fawcett, *The Concepts of Outer Space and the Deep Seabed in International Law: Some Comparisons* (1984) 2 NOTRE DAME INT'L. & COMP. L. J. 71.

Feder, *A Legal Regime for the Arctic* (1978) 6 ECOLOGY L. Q. 785.

Feierberg, Larsen, Fink and Smith, *Spectrascopic Evidence for Two Achondrite Parent Bodies: Asteroids, 349 Dembowska and 4 Vesta* (1980) 45 GEOCHIM. & COSMOCHIM. ACTA 971.

Ferguson, *The New International Economic Order* (1980) U. ILL. L. F. 693.

Feuer, *La Charte des Droits et Devoirs Economiques des Etats* (1975) 79 REVUE GEN. DR. INT'L. PUB. 2272.

Fidler, *Revolt Against or From Within the West? TWAIL, the Developing World and the Future Direction of International Law* (2003) CHI. J. INT'L. L. 29.

Filiato, *The Commercial Space Launch Act: America's Response to the Moon Treaty?* (1987) 10 FORDHAM INT'L. L. J. 763.

Finkelstein and Sanford, *Learning from Corporate Mistakes: The Rise and Fall of Iridium* (2000) 29 ORGANISATIONAL DYNAMICS 138.

Finlay, *United States Policy with Respect to High Seas Fisheries and Deep Seabed Minerals – A Study in Contrasts* (1976) 9 NAT. RES. LAWYER 629.

Finlay and McKnight, *Law of the Sea: Its Impact on the International Energy Crisis* (1974) 6 L. & POL'Y. Int'l. BUS. 639.

Firestone, *Problems in the Resolution of Disputes Concerning Damage Caused in Outer Space* (1985) 59 TUL. L. REV. 747.

Floren, *Antarctic Mining Regimes: An Appreciation of the Attainable* (2001) 16 J. ENVT'L. L. & LIT. 467.

Forrester, WORLD DYNAMICS (1970).

Foster, *The Convention on International Liability for Damage Caused by Space Objects* (1972) 10 CAN. Y. B. INT'L. L. 137.

Foust, *Virgin Galactic and the Future of Commercial Spaceflight* (23 May 2005), *Ad Astra*, at <http://www.space.com/adastra/050523_virgin_nss.html>, last accessed on 27 January 2007.

References 341

Frakes, *The Common Heritage of Mankind Principle and the Deep Seabed, Outer Space and Antarctica: Will Developed and Developing Nations Reach a Compromise?* (2003) 21 WIS. INT'L. L. J. 409.

Francioni, *Legal Aspects of Mineral Exploitation in Antarctica* (1986) 19 CORNELL INT'L. L. J. 163.

Francioni, *Resource Sharing in Antarctica: For Whose Benefit?* (1990) 1 EUR. J. INT'L. L. 258.

Franck and Chesler, *An International Regime for the Seabed Beyond National Jurisdiction* (1975) 13 OSGOODE HALL L. J. 579.

Franck and Sughrue, *Symposium: The International Role of Equity-as-Fairness* (1993) 81 GEORGETOWN L. J. 563.

Franck, El Baradei and Aron, *The New Poor: Land-locked, Shelf-locked and Other Geographically Disadvantaged States* (1974) 7 N. Y. U. J. INT'L. L. & POL. 33.

Frank and Jenett, *Murky Waters: Private Claims to Deep Ocean Seabed Minerals* (1975) 7 L. & POL'Y. INT'L. BUS. 1237.

Frankle, *International Regulation of Orbital Debris* (2000) 43 PROC. COLL. L. OUTER SP. 369

Fratianni and Pattison, *The Economics of International Organisations* (1982) 35 KYKLOS 244.

Friedheim, *UNCLOS and the New International Economic Order* (1987) 4 J. L. & ENV'T. 17.

Friedheim and Akaha, *Antarctic Resources and International Law: Japan, the United States and the Future of Antarctica* (1989) 16 ECOLOGY L. Q. 119.

Friedman, *The Reduction of Fluctuations in the Incomes of Primary Producers: A Critical Comment* (1954) 64 ECON. J. 698.

Friedman and Williams, *The Group of 77 at the United Nations: An Emergent Force in the Law of the Sea* (1979) 16 SAN DIEGO L. REV. 555.

Friedmann, *Selden Redivivus – Towards a Partition of the Seas?* (1971) 65 AM. J. INT'L. L. 757.

Froeschlé and Rickman, *A Monte Carlo Investigation of Jovian Perturbations on Short-Period Comet Orbits* (1981) 46 ICARUS 400.

Fujiwara, Kawaguchi and Sasaki, *Hayabusa Mission to Asteroid Itokawa: In-Situ Observation and Sample Return*, paper presented at the Workshop on Dust in Planetary Systems, 26–30 September 2005, in Hawaii, USA, at <http://www.lpi.usra.edu/meetings/dust2005/pdf/4024. pdf>, last accessed on 28 January 2007.

Fujiwara, Kawaguchi, Uesugi, Yeomans, Saito, Abe, Mukai, Kato, Okada, Yoshikawa, Yano, Demura, Scheeres, Gaskell, Barnouin-Jha, Cheng, Miyamoto, Hirata, Nakamura, Sasaki and Nakamura, *Global Properties of 25143 Itokawa Observed by Hayabusa*, paper presented at the 37th Annual Lunar and Planetary Science Conference, 13–17 March 2006, in League City, TX, USA.

Fujiwara, Kawaguchi, Yeomans, Abe, Mukai, Okada, Saito, Yano, Yoshikawa, Scheeres, Barnouin-Jha, Cheng, Demura, Gaskell, Hirata, Ikeda, Kominato, Miyamoto, Nakamura, Nakamura, Sasaki and Uesugi, *The Rubble-Pile Asteroid Itokawa as Observed by Hayabusa* (2006) 312 SCIENCE 1330.

Fujiwara, Mukai, Kawaguchi and Uesugi, *Sample Return Mission to NEA: MUSES-C* (2000) 25 ADV. SPACE RES. 231.

Fulton, THE SOVEREIGNTY OF THE SEA (1911).

Futron Corporation, *Space Transportation Costs: Trends in Price Per Pound to Orbit 1999–2000* (2002), at <http://www.futron.com/pdf/resource_center/white_papers/FutronLaunchCostWP. pdf>, 6 September 2002, last accessed on 24 April 2007.

Gaffey and McCord, *Mineralogical and Petrological Characterisations of Asteroid Surface Materials*, in Gehrels (ed.), ASTEROIDS (1979), at 688–723.

Gaffey, Bell and Cruikshank, *Reflectance Spectroscopy and Asteroid Surface Mineralogy* in Binzel, Gehrels and Matthews (eds.), ASTEROIDS II (1987), at 98–127.

Gaffey, Bell, Brown, Burbine, Piatek, Reed and Chaky, *Mineralogic Variations within the S-Type Asteroid Class* (1993) 106 ICARUS 83.

Gaffey, Reed and Kelley, *Relationship of E-Type Asteroid 3103 (1982BB) to the Enstatite Achondrite Meteorites and the Hungaria Asteroids* (1992) 100 ICARUS 95.

Gál, SPACE LAW (1969).

Gál, *Space Treaties and Space Technology: Questions of Interpretation* (1972) 15 PROC. COLL. L. OUTER SP. 105.

Galdorisi and Stavridis, *United Nations Convention on the Law of the Sea: Time of a U.S. Reevaluation?* (1992) 40 NAVAL L. REV. 229.

Galloway, *Agreement Governing the Activities of States on the Moon and Other Celestial Bodies* (1980) 5 ANN. AIR & SP. L. 481.

Gamble, *Assessing the Reality of the Deep Seabed Regime* (1985) 22 SAN DIEGO L. REV. 779.

Gamble, *The Third United Nations Conference on the Law of the Sea and the New International Economic Order* (1983) 6 LOYOLA L. A. INT'L. & COMP. L. J. 65.

Gamble and Frankowska, *International Law's Response to the New International Economic Order: An Overview* (1986) 9 B. C. INT'L. & COMP. L. REV. 257.

Gamble and Frankowska, *The 1982 Convention and Customary Law of the Sea: Observations, a Framework and a Warning* (1984) 21 SAN DIEGO L. REV. 491.

García-Amador, *The Proposed New International Economic Order: A New Approach to the Law Governing Nationalisation and Compensation* (1980) 12 LAWYER AM. 2.

Gardner, *Telecom Cable Repairs Under Way in Wake of Asian Earthquake* (3 January 2007), Information Week, at <http://www.informationweek.com/management/showArticle. jhtml?articleID=196800845>, last accessed on 27 January 2007.

General Agreement on Tariffs and Trade, BASIC INSTRUMENTS AND SELECTED DOCUMENTS, 18th Supp. (1971).

General Agreement on Tariffs and Trade, BASIC INSTRUMENTS AND SELECTED DOCUMENTS, 26th Supp. (1979).

Geon, *A Right to Ice? The Application of International and National Water Laws to the Acquisition of Iceberg Rights* (1998) 19 MICH. J. INT'L. L. 277.

Gerlach, *Profitably Exploiting Near-Earth Object Resources* (2005), paper presented at the 2005 International Space Development Conference, 19–22 May 2005, in Washington, DC, USA, at Gerlach Space Systems L.L.C., <http://gerlachspace.com/resources/NEO%20Resources.pdf>, last accessed on 25 April 2007.

Gertsch and Gertsch, *Economic Analysis Tools for Mineral Projects in Space*, Colorado School of Mines, Golden, Colorado, the United States, at <http://www.mines.edu/research/srr/rgertsch. pdf>, last accessed on 31 October 2001.

Gibson, *Five Days in October: "Tracking" Sputnik I at Redstone Arsenal* (2001), Cold War Museum, at <http://www.coldwar.org/text_files/gibson.pdf>, last accessed on 7 December 2004.

Gibson, SPAIN IN AMERICA (1966).

Gidel, LE DROIT INTERNATIONAL PUBLIC DE LA MER (1932).

Gilpin and Gilpin, GLOBAL POLITICAL ECONOMY: UNDERSTANDING THE INTERNATIONAL ECONOMIC ORDER (2001).

Gladman, Michel and Froeschlé, *The Near-Earth Object Population* (2000) 146 ICARUS 176.

Glaser, SPACE INDUSTRIALISATION (1982).

Glassner, *The Status of Developing Landlocked States Since 1965* (1973) 5 LAWYER AM. 480.

Glazer, *Astrolaw Jurisprudence in Space as a Place: Right Reason for the Right Stuff* (1985) 11 BROOKLYN J. INT'L. L. 1.

Glazer, *The Maltese Initiatives Within the United Nations – A Blue Planet Blueprint for Transnational Space* (1975) 4 ECOLOGY L. Q. 279.

Goedhuis, *Air Sovereignty and the Legal Status of Outer Space*, report presented at the Conference of the International Law Association, August 1960, in Hamburg, Germany.

Goedhuis, *Influence of the Conquest of Outer Space on National Sovereignty: Some Observations* (1978) 6 J. SP. L. 36.

Goedhuis, *Reflections on the Evolution of Space Law* (1966) 13 NED. TIJD. INT'L. RECHT 109.

Goedhuis, *The Changing Legal Regime of Air and Outer Space* (1978) 27 INT'L. & COMP. L. Q. 576.

Goh, *Coping with the Lack of a Mechanism for the Settlement of Disputes Arising in Relation to Space Commercialisation* (2001) 5 SINGAPORE J. INT'L. & COMP. L. 180.

Goldberg, *The State of the Negotiations on the Law of the Sea* (1980) 31 HASTINGS L. J. 1091.

Goldie, *A General International Law Doctrine for Seabed Régimes* (1973) 7 INT'L. LAWYER 796.

Goldie, *A Note on Some Diverse Meanings of "The Common Heritage of Mankind"* (1983) 10 SYRACUSE J. INT'L. L. & COM. 69.

Goldie, *Mining Rights and the General International Law Regime of the Deep Ocean Floor* (1975) 2 BROOKLYN J. INT'L. L. 1.

Goldsmith, Allen, Allaby, Davoll and Lawrence, BLUEPRINT FOR SURVIVAL (1972).

Górbiel, *Remarques sur la définition de l'espace extra-atmosphérique* (1978) 21 PROC. COLL. L. OUTER SP. 89.

Gore, EARTH IN THE BALANCE (1992).

Gormley, *The Development and Subsequent Influence of the Roman Legal Norm of "Freedom of the Seas"* (1963) 40 U. DET. L. J. 561.

Gorove, *Definitional Issues Pertaining to "Space Object"* (1994) 37 PROC. COLL. L. OUTER SP. 87.

Gorove, *Exploitation of Space Resources and the Law* (1984) 3 PUB. L. FORUM 29.

Gorove, *Freedom of Exploration and Use in the Outer Space Treaty* (1971) 1 DENVER J. INT'L. L. & POL'Y. 93.

Gorove, *Implications of International Space Law for Private Enterprise* (1982) 7 ANN. AIR & SP. L. 319.

Gorove, *Interpreting Article II of the Outer Space Treaty* (1969) 37 FORDHAM L. REV. 349.

Gorove, *Interpreting Salient Provisions of the Agreement on the Rescue of Astronauts, the Return of Astronauts and Return of Objects Launched into Outer Space* (1968) 11 PROC. COLL. L. OUTER SP. 93.

Gorove, *Liability in Space Law: An Overview* (1983) 8 ANN. AIR & SP. L. 373.

Gorove, *Property Rights in Outer Space: Focus on the Proposed Moon Treaty* (1973) 16 PROC. COLL. L. OUTER SP. 177.

Gorove, *Sovereignty and the Law of Outer Space Re-Examined* (1977) 2 ANN. AIR & SP. L. 311.

Gorove, *Space Transportation Systems: Some International Legal Considerations* (1981) 24 PROC. COLL. L. OUTER SP. 117.

Gorove, *The Concept of "Common Heritage of Mankind": A Political, Moral or Legal Innovation?* (1972) 9 SAN DIEGO L. REV. 390.

Gorove, *Toward a Clarification of the Term "Space Object": An International Legal and Policy Imperative?* (1993) 21 J. SP. L. 11.

Grabar (trans. Butler), THE HISTORY OF INTERNATIONAL LAW IN RUSSIA 1647–1917 (1991).

Gray, Arabshahi, Lamassoure, Okino and Andringa, *A Real Options Framework for Space Mission Design*, paper presented at the I.E.E.E. Aerospace Conference, 5–12 March 2005, in Big Sky, MT, USA.

Greeley, Collins, Spaun, Sullivan, Moore, Senske, Tufts, Johnson, Belton and Tanaka, *Geologic Mapping of Europa* (2000) 104 J. GEOPHYS. RES. 22559.

Greenwald, *Problems of Legal Security of the World Hard Minerals Industry in the International Ocean* (1971) 4 NAT. RES. LAWYER 639.

Griffin, *Americans and the Moon Treaty* (1981) 46 J. AIR L. & COM. 729.

Grotius, DE JURE BELLI AC PACIS. LIBRI TRES (1646).

Gruner, *A New Hope for International Space Law: Incorporating Nineteenth Century First Possession Principles into the 1967 Space Treaty for the Colonisation of Outer Space in the Twenty-First Century* (2005) 35 SETON HALL L. REV. 299.

Guntrip, *The Common Heritage of Mankind: An Adequate Regime for Managing the Deep Seabed?* (2003) 4 MELB. J. INT'L. L. 376.

Haanappel, *Airspace, Outer Space and Mesospace* (1976) 19 PROC. COLL. L. OUTER SP. 160.

Haanappel, *Some Observations on the Crash of the Cosmos 954* (1978) 6 J. SP. L. 147.

Haight, *The New International Economic Order and the Charter of Economic Rights and Duties of States* (1975) 9 INT'L. LAWYER 591.

Haimbaugh, *Technological Disparity and the United Nations Seabed Debates* (1973) 6 IND. L. REV. 690.

Haji, *Finance, Money, Developing Countries and UNCTAD*, in Michael Zammit Cutajar and Waldek R. Malinowski, UNCTAD AND THE SOUTH-NORTH DIALOGUE: THE FIRST TWENTY YEARS (1985).

Hale, DE JURE MARIS ET BRACHIORUM EJUSDEM (1667).

Halket, Leister, Savage, Lephart and Miller, *Report on the Proposed Agreement Governing the Activities of States on the Moon and Other Celestial Bodies* (1983) 23 JURIMETRICS J. 259.

Hall, ESSAY ON THE RIGHTS OF THE CROWN AND THE PRIVILEGES OF THE SUBJECT IN THE SEA SHORES OF THE REALM (2nd ed., 1875).

Hall and McArthur, *Ecotourism in Antarctica and Adjacent Sub-Antarctic Islands: Development, Impacts, Management and Prospects for the Future* (1993) 14:2 TOURISM MAN. 117.

Hamilton, *Deimos* (2000), at <http://www.solarviews.com/french/deimos.htm>, last accessed on 15 July 2006.

Hamilton, *Phobos* (2000), at <http://www.solarviews.com/french/phobos.htm>, last accessed on 15 July 2006.

Hanessian, *General Principles of Law in the Iran-U.S. Claims Tribunal* (1989) 27 COLUM. J. TRANSNAT'L. L. 309.

Hardin, *The Tragedy of the Commons* (1968) 162 SCIENCE 1243.

Harry, *The Deep Seabed: The Common Heritage of Mankind or Arena for Unilateral Exploitation?* (1992) 40 NAVAL L. REV. 207.

Hart, THE NEW INTERNATIONAL ECONOMIC ORDER: CONFLICT AND COOPERATION IN NORTH-SOUTH ECONOMIC RELATIONS 1974–1977 (1983).

Hartley and Medlock, *Political and Economic Influences on the Future World Market for Natural Gas* (2005), at James A. Baker III Institute for Public Policy, Geopolitics of Gas Working Paper Series, at <http://www.rice.edu/energy/publications/docs/GAS_PoliticalEconomicInfluences.pdf>, last accessed on 28 April 2007.

Hassan, *Third Law of the Sea Conference Fishing Rights of Landlocked States* (1976) 8 LAWYER AM. 686.

Hauser, *An International Fiscal Regime for Deep Seabed Mining: Comparisons to Land-Based Mining* (1978) 19 HARV. INT'L. L. J. 759.

Hayashi, *The Antarctica Question in the United Nations* (1986) 19 CORNELL INT'L. L. J. 275.

Hayatsu, Matsuoka, Scott, Studier and Anders, *Origin of Organic Matter in the Early Solar System: The Organic Polymer in Carbonaceous Chondrites* (1977) 41 GEOCHIM. & COSMOCHIM. ACTA 1325.

He, *Review of Definitional Issues in Space Law in the Light of Development of Space Activities* (1991) 34 PROC. COLL. L. OUTER SP. 32.

He, *The Outer Space Treaty in Perspective* (1997) 25 J. SP. L. 93.

Head, Chapman, Domingue, Hawkins, McClintock, Murchie, Prockter, Robinson, Strom and Watters, *The Geology of Mercury: The View Prior to the MESSENGER Mission* (2007) 131 SPACE SCI. REV. 41.

Heim, *Exploring the Last Frontier for Mineral Resources: A Comparison of International Law Regarding the Deep Seabed, Outer Space and Antarctica* (1990) 23 VAND. J. TRANSNAT'L. L. 819.

Helin and Shoemaker, *Earth Approaching Asteroids as Targets for Exploration*, in Morrison and Wells (eds.), ASTEROIDS: AN EXPLORATION ASSESSMENT (1978).

Henkin, Schachter, Pugh and Smit, INTERNATIONAL LAW: CASES AND MATERIALS (3rd ed., 1993).

Henri, *Orbit/Spectrum Allocation Procedures Registration Mechanism*, paper presented at the International Telecommunication Union Biennial Seminar of the Radiocommunication Bureau, 15–19 November 2004, in Geneva, Switzerland.

References 345

Herczeg, *Interpretation of the Space Treaty of 1967 (Introductory Report)* (1967) 10 PROC. COLL. L. OUTER SP. 105.

Herrick, *Exploration and 1994 Exploitation of Geographos* in Gehrels (ed.), ASTEROIDS (1979) at 212–221.

Hertz, *The Reagan Administration and the Law of the Sea: Objections to the 1980 Draft Convention* (1982) 3 B. C. THIRD WORLD L. J. 70.

Hickman and Adams, *Future Launch Systems* (2003) 5:1 CROSSLINK 42, as at Aerospace Corporation, <http://www.aero.org/publications/crosslink/pdfs/V5N1.pdf>, last accessed on 23 April 2007.

Hinkley, *Protecting American Interests in Antarctica: The Territorial Claims Dilemma* (1990) 39 NAVAL L. REV. 43.

Hoagland, *The Conservation and Disposal of Ocean Hard Minerals: A Comparison of Ocean Mining Codes in the United States* (1988) 28 NAT. RES. J. 451.

Hodgkins, *International Cooperation in the Peaceful Uses of Outer Space*, Remarks on Agenda Item 75 in the Fourth Committee of the United Nations General Assembly, 9 October 2002, at <http://www.state.gov/g/oes/rls/rm/2002/14362.htm>, last accessed on 9 April 2004.

Hodgkins, *Procedures for Return of Space Objects Under the Agreement on the Rescue of Astronauts, the Return of Astronauts and the Return of Objects Launched into Outer Space*, in United Nations, PROCEEDINGS OF THE UNITED NATIONS/INTERNATIONAL INSTITUTE OF AIR AND SPACE LAW WORKSHOP ON CAPACITY BUILDING IN SPACE LAW (2002).

Hoffstadt, *Moving the Heavens: Lunar Mining and the "Common Heritage of Mankind" in the Moon Treaty* (1994) 42 U.C.L.A. L. REV. 575.

Hoge, *Despite U.S. Opposition, United Nations Budget is Approved*, THE NEW YORK TIMES, 23 December 2007, at <http://www.nytimes.com/2007/12/23/world/23nations.html>, last accessed on 4 May 2008.

Holliday, Marcou and Vickerman, THE CHANNEL TUNNEL: PUBLIC POLICY, REGIONAL DEVELOPMENT AND EUROPEAN INTEGRATION (1991).

Hoover, *Law and Security in Outer Space from the Viewpoint of Private Industry* (1983) 11 J. SP. L. 115.

Hope-Thompson, *The Third World and the Law of the Sea: The Attitude of the Group of 77 Toward the Continental Shelf* (1980) 1 B. C. THIRD WORLD L. J. 37.

Horan, Walker and Morgan, *High Precision Measurements of Pt and Os in Chondrites*, paper presented at the 30th Annual Lunar and Planetary Science Conference, 15–29 March 1999, in Houston, TX, USA.

Hörl and Hermida, *Change of Ownership, Change of Registry? Which Objects to Register, What Data to be Furnished, When, and Until When?*, in United Nations, PROCEEDINGS OF THE I.I.S.L./E.C.S.L. SYMPOSIUM: REINFORCING THE REGISTRATION CONVENTION (2003), at 15.

Horn, *Normative Problems of a New International Economic Order* (1982) 16 J. WORLD TRADE L. 338.

Hörngren and Vredin, *Exchange Risk Premia in a Currency Basket System* (1989) 125 REV. WORLD ECON. 311.

Hornsey, *International Law – All at Sea* (1978) 29 N. IR. LEG. Q. 250.

Hornyak, *Farming Solar Energy in Space* (July 2008), SCIENTIFIC AMERICAN, <http://www. scientificamerican.com/article.cfm?id=farming-solar-energy-in-space>, last accessed on 13 November 2009.

Hosenball, *Current Issues of Space Law Before the United Nations* (1974) 2 J. SP. L. 8.

Hosenball, *Space Law, Liability and Insurable Risks* (1976) 12 THE FORUM 141.

Hosenball, *The United Nations Committee on the Peaceful Uses of Outer Space: Past Accomplishments and Future Challenges* (1979) 7 J. SP. L. 95.

Howell, Merényi and Lebofsky, *Classification of Asteroid Spectra Sing a Neural Network* (1994) 99 J. GEOPHYS. RES. 10.

Hrbud, van Dyke, Houts and Goodfellow, *End-to-End Demonstrator of the Safe Affordable Fission Engine (SAFE) 30: Power Conversion and Ion Engine Operation*, paper presented at the Space Technologies Applications International Forum Conference, 3–7 February 2002, in Albuquerque, NM, USA.

Hubbert, *Energy from Fossil Fuels* (1956) 109 SCIENCE 103.

Hubbert, *Energy Resources*, in National Academy of Sciences, RESOURCES AND MAN (1969), at 157–242.

Hudes, *Towards a New International Economic Order* (1976) 2 YALE STUD. WORLD PUB. ORD. 88.

Hudson, THREE SCENARIOS: THE LAW OF THE SEA, OCEAN MINING AND THE NEW INTERNATIONAL ECONOMIC ORDER (1977).

Huebert, *Canada and the Law of the Sea Convention* (1997) 52 INT'L. J. 69.

Hufford, *Ideological Rigidity vs. Political Reality: A Critique of Reagan's Policy on the Law of the Sea* (1984) 2 YALE L. & POL'Y. REV. 127.

Humphrey and Nase, *The Cape Town Convention 2001: An Australian Perspective* (2006) 31 AIR & SP. L. 5.

Husby, *Sovereignty and Property Rights in Outer Space* (1994) 3 DETROIT COLL. L. J. INT'L. L. & PRAC. 359.

Hutsemékers, Manfroid, Jehin, Arpigny, Cochran, Schulz, Stüwe and Zucconi, *Isotopic Abundance of Nitrogen and Carbon in Distant Comets* (2005) 432 ASTRON. & ASTROPHYS. 5.

IAU Minor Planet Centre, *List of Amor Minor Planets* (19 November 2009), at <http://www.cfa.harvard.edu/iau/lists/Amors.html>, last accessed on 19 November 2009.

IAU Minor Planet Centre, *List of Apollo Minor Planets* (19 November 2009), at <http://www.cfa.harvard.edu/iau/lists/Apollos.html>, last accessed on 19 November 2009.

IAU Minor Planet Centre, *List of Aten Minor Planets* (19 November 2009), at <http://www.cfa.harvard.edu/iau/lists/Atens.html>, last accessed on 19 November 2009.

Ingram, INTERNATIONAL ECONOMIC PROBLEMS (2nd ed., 1970).

Institute of Space and Astronautical Science, *Hayabusa* (2004) at Japan Aerospace Exploration Agency, <http://www.isas.jaxa.jp/e/enterp/missions/hayabusa/index.shtml>, last accessed on 10 January 2005.

International Academy of Astronautics, *The International Exploration of Mars* (1996), at <http://www.iaanet.org/p_papers/mars.html>, last accessed on 23 December 2004.

International Seabed Authority, *Regulations on Prospecting and Exploration for Polymetallic Nodules in the Area* (13 July 2000), at <http://www.isa.org.jm/files/documents/EN/Regs/MiningCode.pdf>, last accessed on 16 November 2009.

Irwin, GIANT PLANETS OF OUR SOLAR SYSTEM: AN INTRODUCTION (2006).

Island One Society, *LMF Mining Robots* (25 July 1999), at <http://www.islandone.org/MMSG/aasm/AASM5D.html>, last accessed on 7 December 2004.

Jackson, *Deepsea Ventures: Exclusive Mining Rights to the Deep Seabed as a Freedom of the Sea* (1976) 28 BAYLOR L. REV. 170.

Jacobs, Atherton and Wallenstein, THE DYNAMICS OF INTERNATIONAL ORGANISATIONS (1972), at 415.

Jacobsen and Delucchi, *A Path to Sustainable Energy by 2030* (2009) 301:5 SCIENTIFIC AMERICAN 58.

Jacobson and Hanlon, *Regulation of Hard Mineral Mining on the Continental Shelf* (1971) 50 OR. L. REV. 425.

Jaffé, Ross, Lo, Marsden, Farrelly and Uzer, *Statistical Theory of Asteroid Escape Rates* (2002) 89:1 PHYS. REV. LETTERS 11001.

Jagota, *Developments in the Law of the Sea between 1970 and 1998: A Historical Perspective* (2000) 2 J. HIST. INT'L. L. 91.

Jaksetic, *The Peaceful Uses of Outer Space: Soviet Views* (1979) 28 AM. U. L. REV. 483.

Jansen, MONETARISM, ECONOMIC CRISIS AND THE THIRD WORLD (1983).

Jasentuliyana (ed.), INTERNATIONAL SPACE LAW AND THE UNITED NATIONS (1999).

Jasentuliyana, *International Space Law and Cooperation and the Mining of Asteroids* (1990) 15 ANN. AIR & SP. L. 343.

References 347

Jasentuliyana, *The Role of Developing Countries in the Formulation of Space Law* (1995) 20:2 ANN. AIR & SP. L. 95.

Jasentuliyana and Lee (eds.), MANUAL ON SPACE LAW (1979).

Jedicke, Nesvorny, Whiteley, Ivezic and Juric, *An Age-Colour Relationship for Main Belt S-Complex Asteroids* (2004) 429 NATURE 275.

Jenks, SPACE LAW (1965).

Jenks, A NEW WORLD OF LAW? A STUDY OF THE CREATIVE IMAGINATION IN INTERNATIONAL LAW (1969).

Jennings, THE ACQUISITION OF TERRITORY IN INTERNATIONAL LAW (1963).

Jennings, *The United States Draft Treaty on the International Seabed Area – Basic Principles* (1971) 20 INT'L. & COMP. L. Q. 433.

Jericho and McCracken, *Space Law: Is it the Last Legal Frontier?* (1985) 51 J. AIR L. & COM. 791.

Jessup and Taubenfeld, *Controls for Outer Space*, in U.S. Senate (ed.), LEGAL PROBLEMS OF SPACE EXPLORATION (1961), at 553–570.

Jiru, *Star Wars and Space Malls: When the Paint Chips Off a Treaty's Golden Handcuffs* (2001) 42 S. TEX. L. REV. 155.

Johnson, *A Comparison of Energy Requirements Between Terrestrial Metal Extraction and Recovery of Asteroid Metal Resources*, at <http://www.erie.net/~fjohnson/AsteroidPaper.html>, last accessed on 3 October 2001.

Johnson, *The New International Economic Order* (1983) Y. B. WORLD AFF. 204.

Johnson, *The Public Trust Doctrine*, in Canning and Scott (eds.), THE PUBLIC TRUST DOCTRINE IN WASHINGTON STATE: PROCEEDINGS OF THE SYMPOSIUM, NOVEMBER 18, 1992 (1993).

Johnson Matthey (2004) at <http://www.platinum.matthey.com/production/africa.html>, last accessed on 21 December 2004.

Johnson Matthey, *Platinum 2004 Interim Report* (2004) at <http://www.platinum.matthey.com/publications/1100682070.html>, last accessed on 21 December 2004.

Johnston, *Deep Seabed Mineral Resources Act* (1980) 20 NAT. RES. J. 163.

Jones, Eppler, Davis, Friedlander, McAdams and Krikalev, *Human Exploration of Near Earth Asteroids*, in Gehrels, Matthews and Schumann (eds.), HAZARDS DUE TO COMETS AND ASTEROIDS (1994).

Jones, Lebofsky, Lewis and Marley, *The Composition and Origin of the C, P and D Asteroids: Water as a Tracer of Thermal Evolution in the Outer Belt* (1990) 88 ICARUS 172.

Jørgensen-Dahl and Østreng, THE ANTARCTIC TREATY SYSTEM IN WORLD POLITICS (1991).

Joyner, *Antarctica and the Law of the Sea: An Introductory Overview* (1983) 13 OCEAN DEV. & INT'L. L. 277.

Joyner, *Antarctica and the Law of the Sea: Rethinking the Current Legal Dilemmas* (1981) 18 SAN DIEGO L. REV. 415.

Joyner, *Ice-Covered Regions in International Law* (1991) 31 NAT. RES. J. 213.

Joyner, *Japan and the Antarctic Treaty System* (1989) 16 ECOLOGY L. Q. 155.

Joyner, *Legal Implications of the Concept of the Common Heritage of Mankind* (1986) 35 INT'L. & COMP. L. Q. 190.

Joyner, *Oceanic Pollution and the Southern Ocean: Rethinking the International Legal Implications for Antarctica* (1984) 24 NAT. RES. J. 1.

Joyner, *The Antarctic Minerals Negotiating Process* (1987) 81 AM. J. INT'L. L. 888.

Joyner, *The Antarctic Treaty System and the Law of the Sea – Competing Regimes in the Southern Ocean?* (1995) 10 INT'L. J. MARINE & COASTAL L. 301.

Joyner, *The Southern Ocean and Marine Pollution: Problems and Prospects* (1985) 17 CASE W. RES. J. INT'L. L. 165.

Joyner, *Towards a Legal Regime for the International Seabed: The Soviet Union's Evolving Perspective* (1975) 15 VA. J. INT'L. L. 871.

Joyner, *U.N. General Assembly Resolutions and International Law: Rethinking the Contemporary Dynamics of Norm Creation* (1981) 11 CAL. W. INT'L. L. J. 445.

Joyner and Ewing, *Antarctica and the Latin American States: The Interplay of Law, Geopolitics and Environmental Priorities* (1992) 4 GEORGETOWN INT'L. ENVT'L. L. REV. 1.

Joyner and Lipperman, *Conflicting Jurisdictions in the Southern Ocean: The Case of an Antarctic Minerals Regime* (1987) 27 VA. J. INT'L. L. 1.

Joyner and Theis, EAGLE OVER THE ICE: THE U.S. IN THE ANTARCTIC (1997).

Joyner and Theis, *The United States and Antarctica: Rethinking the Interplay of Law and Interests* (1987) 20 CORNELL INT'L. L. J. 65.

Kamenetskaya, *On the Establishment of a World Space Organisation: Some Considerations and Remarks* (1989) 32 PROC. COLL. L. OUTER SP. 358.

Kamienski, *Researches on the Motion of Comet P/Wolf I: Perturbations Due to Venue, Earth, Mars, Jupiter, Saturn and Uranus During the Period 1950 October – 1959 March* (1957) 7 ACTA ASTRON. 5.

Kamrad and Ernst, *An Economic Model for Evaluating Mining and Manufacturing Ventures with Output Yield Uncertainty* (2001) 49 OPERATIONS RESEARCH 690.

Karev, *The Russian Federation and the UN Conference on the Law of the Sea* (1995) 89 AM. SOC'Y. INT'L. L. PROC. 455.

Kargel, *Market Value of Asteroidal Precious Metals in an Age of Diminishing Terrestrial Resources*, in Johnson (ed.), ENGINEERING, CONSTRUCTION AND OPERATIONS IN SPACE V: PROCEEDINGS OF THE FIFTH INTERNATIONAL CONFERENCE ON SPACE (1996).

Kargel, *Metalliferous Asteroids as Potential Sources of Precious Metals* (1994) 99 J. GEOPHYS. RES. 21129.

Katkin, *Communication Breakdown? The Future of Global Connectivity After the Privatisation of INTELSAT* (2005) 38 VAND. J. TRANSNAT'L. L. 1323.

Katz, *A Method for Evaluating the Deep Seabed Mining Provisions of the Law of the Sea Treaty* (1981) 7 YALE J. WORLD PUB. ORD. 114.

Katz, *Financial Arrangements for Seabed Mining Companies: An NIEO Case Study* (1979) 13 J. WORLD TRADE L. 209.

Kay and Mirrlees, *The Desirability of Natural Resource Depletion*, in David W. Pearce (ed.), THE ECONOMICS OF NATURAL RESOURCE DEPLETION (1975).

Keefe, *Making the Final Frontier Feasible: A Critical Look at the Current Body of Outer Space Law* (1995) 11 SANTA CLARA COMPUTER & HIGH TECH. L. J. 345.

Kerrest, *Commercial Use of Space, including Launching* (2004), in China Institute of Space Law, 2004 SPACE LAW CONFERENCE: PAPER ASSEMBLE 199.

Kerrest, *Exploitation of the Resources of the High Sea and Antarctica: Lessons for the Moon?*, paper presented at the IISL/ECSL Space Law Symposium on New Developments and the Legal Framework Covering the Exploitation of the Resources of the Moon, 29 March 2004, in Vienna, Austria.

Kerrest, *Launching Spacecraft from the Sea and the Outer Space Treaty: The Sea Launch Project* (1997) 40 PROC. COLL. L. OUTER SP. 264.

Kerrest, *Remarks on the Responsibility and Liability for Damage Other Than Those Caused by the Fall of a Space Object* (1997) 40 PROC. COLL. L. OUTER SP. 134.

Kerridge, *Carbon, Hydrogen and Nitrogen in Carbonaceous Chondrites: Abundances and Isotopic Composition in Bulk Samples* (1985) 49 GEOCHIM. & COSMOCHIM. ACTA 1707.

Kim and Yang, *Carbon Isotope Analyses of Individual Hydrocarbon Molecules in Bituminous Coal, Oil Shale and Murchison Meteorite* (1998) 15 J. ASTRON. & SPACE SCI. 163.

Kindt, *Ice-Covered Areas and the Law of the Sea: Issues Involving Resource Exploitation and the Antarctic Environment* (1988) 14 BROOKLYN J. INT'L. L. 27.

Kindt and Parriott, *Ice-Covered Areas: The Competing Interests of Conservation and Resource Exploitation* (1984) 21 SAN DIEGO L. REV. 941.

Kiss, *Conserving the Common Heritage of Mankind* (1990) 59 REV. JUR. U. P. R. 773.

Klass, Burrows and Beggs, INTERNATIONAL MINERAL CARTELS AND EMBARGOES: POLICY AND IMPLICATIONS FOR THE U.S. (1980).

Klingelhöfer and Fegley, *Iron Mineralogy of Venus's Surface Investigated by Mössbauer Spectroscopy* (2000) 147 ICARUS 1.

References

Klingelhöfer, Morris, Yen, Ming, Schröder and Rodionov, *Mineralogy on Mars at Gusev Crater and Meridiani Planum as seen by Iron Mössbauer Spectroscopy* (2006) 70 Supp 1 GEOCHEM. COSM. ACTA A325.

Knight, *The Deep Seabed Hard Mineral Resources Act – A Negative View* (1973) 10 SAN DIEGO L. REV. 446.

Knight, *The Draft United Nations Conventions on the International Seabed Area: Background, Description and Some Preliminary Thoughts* (1971) 8 SAN DIEGO L. REV. 459.

Knight and McKee, *Hayabusa Touched Asteroid Itokawa After All* (23 November 2005), NEW SCIENTIST, at <http://www.newscientist.com/article.ns?id=dn8362>, last accessed on 27 January 2007.

Koch, *The Antarctic Challenge: Conflicting Interests, Cooperation, Environmental Protection and Economic Development* (1984) 15 J. MARIT. L. & COM. 117.

Koh, *Negotiating a New World Order for the Sea* (1984) 24 VA. J. INT'L. L. 761.

Koh, *The Origins of the 1982 Convention on the Law of the Sea* (1987) 29 MALAYA L. REV. 1.

Kopal, *Legal Questions Relating to the Draft Treaty Concerning the Moon* (1973) 16 PROC. COLL. L. OUTER SP. 180.

Kopal, *Problems Arising from the Interpretation of Agreement on the Rescue of Astronauts, Return of Astronauts and Return of Objects Launched into Outer Space* (1968) 11 PROC. COLL. L. OUTER SP. 98.

Kopal, *Some Remarks on Issues Relating to Legal Definitions of "Space Object", "Space Debris" and "Astronaut"* (1994) 37 PROC. COLL. L. OUTER SP. 99.

Kopal, *The Role of United Nations Declarations of Principles in the Progressive Development of Space Law* (1988) 16 J. SP. L. 5.

Kopal, *Vladimír Mandl – Founder of Space Law* (1968) 11 PROC. COLL. L. OUTER SP. 357.

Koskenniemi and Lehto, *The Privilege of Universality: International Law, Economic Ideology and Seabed Resources* (1996) 65 NORDIC J. INT'L. L. 533.

Kosmo, *The Commercialisation of Space: A Regulatory Scheme that Promotes Commercial Ventures and International Responsibility* (1988) 61 S. CAL. L. REV. 1055.

Kotz, *The Common Heritage of Mankind: Resource Management of the International Seabed* (1978) 6 ECOLOGY L. Q. 65.

Kowal, ASTEROIDS: THEIR NATURE AND UTILISATION (1988).

Krasner, *Think Again: Sovereignty* (2001) 122 FOREIGN POLICY 20.

Kravisand and Davenport, *The Political Arithmetic of International Burden Sharing* (1963) 71 J. POL. ECON. 309.

Kronmiller, *Reconciling Public and Private Interests in Multilateral Negotiations: The Law of the Sea Conference* (1981) 1 PUB. L. F. 159.

Kronmiller, THE LAWFULNESS OF DEEP SEABED MINING (1981).

Krstic, *Customary Law Rules in Regulating Outer Space Activities* (1977) 20 PROC. COLL. L. OUTER SP. 320.

Krueger, *Policy Options in the Law of the Sea Negotiations* (1978) 6 INT'L. BUS. LAWYER 89.

Kuck, *Exploitation of Space Oases*, in Faughnan (ed.), SPACE MANUFACTURING PATHWAYS TO THE HIGH FRONTIER: PROCEEDINGS OF THE TWELFTH S.S.I.-PRINCETON CONFERENCE (1995), at 136–156.

Kuck, *Near-Earth Extraterrestrial Resources* (1979), paper presented at the 4th Princeton/AIAA Conference on Space Manufacturing, May 1979, in Princeton, NJ, USA.

Kuck, *The Exploitation of Space Oases*, paper presented at the Princeton-AIAA Conference on Space Manufacturing, May 1995, in Princeton, NJ, USA.

Kuck and Gillett, *Extraterrestrial Resources: Implications from Terrestrial Experience* (1991) in University of Arizona, RESOURCES OF NEAR-EARTH SPACE: ABSTRACTS, at 11.

Kumar, *A World Stage: Global Factors are Pushing Platinum Group Metals to Historically High Price Levels* (1 September 2006), Recycling Today, at <http://www.thefreelibrary.com/A+world+stage%3a+global+factors+are+pushing+platinum+group+metals+to...-a0152513170>, last accessed on 27 January 2007.

Kwon, *The Declining Role of Western Powers in International Organisations: Exploring a New Model of U.N. Burden Sharing* (1995) 15 J. PUB. POL'Y. 65.

La Que, *Different Approaches to International Regulation of Exploitation of Deep-Ocean Ferromanganese Nodules* (1978) 15 SAN DIEGO L. REV. 477.

Lachs, *Legal Framework of an International Community* (1992) 6 EMORY INT'L. L. REV. 329.

Lachs, THE LAW OF OUTER SPACE: AN EXPERIENCE IN CONTEMPORARY LAW MAKING (1972).

Lafferranderie, *L'application par l'Agence Spatial Européenne de la Convention sur l'immatriculation des objets lancés dans l'espace extra-atmospherique* (1986) 11 ANN. AIR & SP. L. 229.

Lála, *The United Nations Register of Objects Launched into Outer Space*, in United Nations, PROCEEDINGS OF THE UNITED NATIONS/INTERNATIONAL INSTITUTE OF AIR AND SPACE LAW WORKSHOP ON CAPACITY BUILDING IN SPACE LAW (2002).

Landis, *Colonisation of Venus*, paper presented at the Conference on Human Space Exploration, Space Technology and Applications International Forum, 2–6 February 2003, in Albuquerque, NM, USA.

Larschan, *The International Legal Status of the Contractual Rights of Contractors under the Deep Seabed Mining Provisions (Part XI) of the Third United Nations Convention on the Law of the Sea* (1986) 14 DENVER J. INT'L. L. & POL'Y. 207.

Larschan and Brennan, *The Common Heritage of Mankind Principle in International Law* (1983) 21 COLUM. J. TRANSNAT'L. L. 305.

Larsen, *Critical Issues in the UNIDROIT Draft Space Protocol* (2002) 45 PROC. COLL. L. OUTER SP. 2.

Larson, *Determination of Meteor Showers on Other Planets Using Comet Ephemerides* (2001) 121 THE ASTRO. J. 1722.

Larson, *The United States Position on the Deep Seabed* (1979) 3 SUFFOLK TRANSNAT'L. L. J. 1.

Lau and Hulkower, *Accessibility of Near-Earth Asteroids* (1987) 10 J. GUID. CON. & DYN. 225.

Lau and Hulkower, *On the Accessibility of Near Earth Asteroids*, paper presented at the AAS-AIAA Astrodynamics Conference, 12–15 August 1985, in Vail, CO, USA.

Lawless, *Implementation of the Deep Seabed Hard Mineral Resources Act*, paper presented at the Offshore Technology Conference, 3–6 May 1982, in Houston, TX, USA.

Ledec, *Minimising Environmental Problems from Petroleum Exploration and Development in Tropical Forest Areas*, paper presented at the Proceedings of the First International Symposium on Oil and Gas Exploration and Production Waste Management Practices, 10–13 September 1990, in New Orleans, LA, USA.

Lee, *Application for Revision of the Judgment of 11 July 1996 in the Case concerning Application of the Convention on the Prevention and Punishment of the Crime of Genocide (Yugoslavia v. Bosnia and Herzegovina)* (2003) AUS. INT'L. L. J. 205.

Lee, *Commentary Paper on Discussion Paper Titled "Commercial Use of Space, Including Launching" by Prof. Dr. Armel Kerrest*, in China Institute of Space Law, PROCEEDINGS OF THE 2004 SPACE LAW CONFERENCE (2004), at 220–231.

Lee, *Costing and Financing a Commercial Asteroid Mining Venture* (2003), paper presented at the 54th International Astronautical Congress, 1–3 October 2003, in Bremen, Germany.

Lee, *Deep Seabed Mining and Developing Countries* (1979) 6 SYRACUSE J. INT'L. L. & COM. 213.

Lee, *Definitions of "Exploration" and "Scientific Investigation" with Focus on Mineralogical Prospecting and Exploration Activities* (2005), paper presented at the 56th International Astronautical Congress, 17–21 October 2005, in Fukuoka, Japan.

Lee, *Liability Arising from Article VI of the Outer Space Treaty: States, Domestic Law and Private Operators* (2005) 48 PROC. COLL. L. OUTER SP. 216.

Lee, *Property and Mining Rights for Lunar Mining Operations in the Absence of International Consensus on the Moon Agreement* (2003), paper presented at the 54th International Astronautical Congress, 29 September 2003 to 3 October 2003, in Bremen, Germany.

Lee, *Reconciling Space Law and the Commercial Realities of the Twenty-First Century* (2000) 4 SING. J. INT'L. & COMP. L. 194.

References   351

Lee, *The Law of the Sea Convention and Third States* (1983) 77 AM. J. INT'L. L. 541.

Lee, *The Liability Convention and Private Space Launch Services – Domestic Regulatory Responses* (2006) 31 ANN. AIR & SP. L. 351.

Lee, *Transferring Registration of Space Objects: The Interpretative Solution*, paper presented at the 47th Colloquium on the Law of Outer Space, 4–8 October 2004, in Vancouver, Canada.

Lee and Doldirina, *Legal and Policy Issues Arising from the Use of Nuclear and Radioisotopic Power Sources and Propulsion Systems in Outer Space*, paper presented at the 60th International Astronautical Congress, 12–16 October 2009, in Daejeon, South Korea.

Lee and Freeland, *The Crystallisation of General Assembly Space Declarations into Customary International Law* (2003) 46 PROC. COLL. L. OUTER SP. 122.

Lee and Freeland, *The Impact of Arms Limitation Agreements and Export Control Regulations on International Commercial Launch Activities* (2002) 45 PROC. COLL. L. OUTER SP. 321.

Lee and Roach, *The Importance of Private Property Rights for Selected Commercial Applications in Lunar and Martian Settlements*, paper presented at the 58th International Astronautical Congress, 2–6 October 2006, in Valencia, Spain.

Lee and Yao, *Abundance of Chemical Elements in the Earth's Crust and its Major Tectonic Units* (1970) INT'L. GEOL. REV. 778.

Lee, Braham, Gladman, Mungas, Silver, Thomas and West, *Mars Indirect: Phobos as a Critical Step in Human Mars Exploration*, paper presented at the International Space Development Conference, May 2005, in Washington, DC, USA.

Lee, Braham, Mungas, Silver, Thomas and West, *Phobos: A Critical Link between Moon and Mars Exploration*, paper presented at the Space Resources Roundtable VII: Lunar Exploration Analysis Group Conference on Lunar Exploration, 25–28 October 2005, in League City, TX, USA, also at <http://www.lpi.usra.edu/meetings/leag2005/pdf/2049.pdf>, last accessed on 20 April 2007.

Leeb, THE OIL FACTOR: HOW OIL CONTROLS THE ECONOMY AND YOUR FINANCIAL FUTURE (2004).

Leich, DIGEST OF UNITED STATES PRACTICE IN INTERNATIONAL LAW (1980).

Lemon, *Earthquake Disputes Internet Access in Asia* (27 December 2006), at <http://www.computerworld.com/action/article.do?command=viewArticleBasic%articleId=9006819>, last accessed on 27 January 2007.

Leonhard, *Ixtoc I: A Test for the Emerging Concept of the Patrimonial Sea* (1980) 17 SAN DIEGO L. REV. 617.

Lerche, GEOLOGICAL RISK AND UNCERTAINTY IN OIL EXPLORATION: UNCERTAINTY, RISK AND STRATEGY (1997).

Levinson, *Technology as the Cutting Edge of Cosmic Evolution*, in Durbin (ed.), RESEARCH IN PHILOSOPHY AND TECHNOLOGY (1985).

Levy, Grayson and Wolf, *The Organic Analysis of the Murchison Meteorite* (1973) 37 GEOCHIM. & COSMOCHIM. ACTA 475.

Levy, *INTELSAT: Technology, Politics and the Transformation of a Regime* (1975) 29 INT'L. ORG. 655.

Lewis, *Logistical Implications of Water Extraction from Near Earth Asteroids* (1993), paper presented at the AIAA/SSI Space Manufacturing Conference, May 1993, in Princeton, NJ, USA.

Lewis, MINING THE SKY: UNTOLD RICHES FROM THE ASTEROIDS, COMETS AND PLANETS (1997).

Lewis, *Resources of the Asteroids* (1997) 50 J. BRIT. INTERPLANETARY SOC. 51.

Lewis and Hutson, *Asteroidal Resource Opportunities Suggested by Meteorite Data*, in Lewis, Matthews and Guerrieri (eds.), RESOURCES OF NEAR-EARTH SPACE (1993), at 534.

Lewis and Meinel, *Asteroid Mining and Space Bunkers* (1983) DEF. SCI. 2000.

Lewis, Ramohalli and Triffet, *Extraterrestrial Resource Utilisation for Economy in Space Missions*, paper presented at the 41st International Astronautical Congress, 6–12 October 1990, in Dresden, Germany.

Li, *Sovereignty at Sea: China and the Law of the Sea Conference* (1979) 15 STANFORD J. INT'L. STUD. 225.

Li, Transfer of Technology for Deep Seabed Mining: The 1982 Law of the Sea Convention and Beyond (1994).

Lillich, *Economic Coercion and the "New International Economic Order": A Second Look at Some First Impressions* (1976) 16 Va. J. Int'l. L. 233.

Listner, *The Ownership and Exploitation of Outer Space: A Look at Foundational Law and Future Legal Challenges to Current Claims* (2003) 1 Regent J. Int'l. L. 75.

Lodge, *International Seabed Authority's Regulations on Prospecting and Exploration for Polymetallic Nodules in the Area* (2002) 20 J. Energy Nat. Res. L. 270.

Logue, *The Nepal Proposal for a Common Heritage Fund* (1979) 9 Cal. W. Int'l. L. J. 598.

Lohmeier, *Keeping Cool Amidst the Ice: Addressing the Challenge of Antarctic Mineral Resources* (1988) 2 Emory J. Int'l. Disp. Resol. 141.

Love and Ahrens, *Catastrophic Impacts on Gravity Dominated Asteroids* (1996) 124 Icarus 141.

Lowder, *A State's International Legal Role: From the Earth to the Moon* (1999) 7 Tulsa J. Comp. & Int'l. L. 253.

Lowe, *Reflections on the Waters: Changing Conceptions of Property Rights in the Law of the Sea* (1986) 1 Int'l. J. Estuarine & Coastal L. 1.

Lowe, *The International Seabed and the Single Negotiating Text* (1976) 13 San Diego L. Rev. 489.

Lundquist, *The Iceberg Cometh? International Law Relating to Antarctic Iceberg Exploitation* (1977) 17 Nat. Res. J. 1.

Luoma, *A Comparative Study of National Legislation Concerning the Deep Sea Mining of Manganese Nodules* (1983) 14 J. Marit. L. & Com. 243.

Luu and Jewitt, *Kuiper Belt Objects: Relics From the Accretion Disk of the Sun* (2002) 30 Ann. Rev. Astron. & Astrop. 63.

Lyall, *On the Moon* (1998) 26 J. Sp. L. 129.

Lynch, *The Nepal Proposal for a Common Heritage Fund: Panacea or Pipedream?* (1980) 10 Cal. W. Int'l. L. J. 25.

Lytje, *Obstacles on the Way to a General Convention* (2004) 47 Proc. Coll. L. Outer Sp. 267.

Madsen, *A Certain False Security: The Madrid Protocol to the Antarctic Treaty* (1993) 4 Colo. J. Int'l. Envt'l. L. & Pol'y. 458.

Magri, Ostro, Rosema, Thomas, Mitchell, Campbell, Chandler, Shapiro, Giorgini and Yeomans, *Mainbelt Asteroids: Results of Arecibo and Goldstone Radar Observations of 37 Objects During 1980–1995* (1999) 140 Icarus 379.

Malone, *Customary International Law and the United Nations' Law of the Sea Treaty* (1983) 13 Cal. W. Int'l. L. J. 181.

Malone, *The United States and the Law of the Sea after UNCLOS III* (1983) 46 L. & Contemp. Probs. 29.

Malthus, *An Essay on the Principle of Population* (1798), in Garrett Hardin (ed.), Population, Evolution and Birth Control (1969) at 4–17.

Malthus, Principles of Political Economy: Considered with a View to Their Practical Application (1820).

Maluwa, *Southern African Landlocked States and Rights of Access Under the New Law of the Sea* (1995) 10 Int'l. J. Marine & Coastal L. 529.

Malysheva, *General Convention on Space Law: Some Arguments for Elaboration* (2004) 47 Proc. Coll. L. Outer Sp. 254.

Mandl, Das Weltraumrecht: Ein Problem der Raumfahrt (1932).

Mani, *Development of Effective Mechanism(s) for Settlement of Disputes Arising in Relation to Space Commercialisation* (2001) 5 Sing. J. Int'l. & Comp. L. 191.

Manners, *Three Issues of Mineral Policy* (1977) 125 J. Royal Soc. Arts 386.

Marceau, *Some Evidence of a New International Economic Order in Place* (1991) 22 Revue Gén. Dr. 397.

Marcus (ed.), Mining Environmental Handbook: Effects of Mining on the Environment and American Environmental Controls on Mining (1997).

References 353

Marko, *A Kinder, Gentler Moon Treaty: A Critical Review of the Current Moon Treaty and a Proposed Alternative* (1992) 8 J. NAT. RES. & ENVT'L. L. 293.

Markoff, *La Lune et le Droit International* (1964) 68 REV. GEN. DR. INT'L. PUB. 248.

Marsden and Williams (eds.), CATALOGUE OF COMETARY ORBITS (1999).

Marshall, PRINCIPLES OF ECONOMICS: AN INTRODUCTORY VOLUME (8th ed., 1949).

Martin, *Legal Aspects of Seabed Mineral Exploitation* (1975) 3 INT'L. BUS. LAWYER 148.

Martinez, *The Third United Nations Conference on the Law of the Sea: Prospects, Expectations and Realities* (1976) 7 J. MARIT. L. & COM. 253.

Mascy and Niehoff, *Sample Return Missions to the Asteroid Eros*, in Gehrels (ed.), PHYSICAL STUDIES OF MINOR PLANETS: PROCEEDINGS OF INTERNATIONAL ASTRONOMICAL UNION COLLOQUIUM 12, March 1971 in Tucson, AZ, USA.

Masters, Root and Attanasi, *World Oil and Gas Resources: Future Production Realities* (1990) 15 ANN. REV. ENERGY 23.

Matsumoto and Kawaguchi, *Optimum Trajectory for Low-Thrust Multiple Trojan Asteroid Flybys* (2004), paper presented at the 18th International Symposium on Space Flight Dynamics, 11–15 October 2004, in Munich, Germany.

Matte, AEROSPACE LAW: TELECOMMUNICATIONS SATELLITES (1982).

Matte, *Legal Principles Relating to the Moon*, in Jasentuliyana and Lee (eds.), MANUAL ON SPACE LAW (1979), at 317.

Matte, *The Draft Treaty on the Moon, Eight Years Later* (1978) 3 ANN. AIR & SP. L. 511.

Mau, *Equity, the Third World and the Moon Treaty* (1984) 8 SUFFOLK TRANSNAT'L. L. J. 221.

McCabe, Ruvolo and Wayne, *Recent Developments in the Law of the Seas II: A Synopsis* (1971) 8 SAN DIEGO L. REV. 658.

McCall, *A New Combination to Davy Jones' Locker: Melee over Marine Minerals* (1978) 9 LOYOLA U. CHI. L. J. 935.

McCloskey, *Domestic Legislation and the Law of the Sea Conference* (1979) 6 SYRACUSE J. INT'L. L. & COM. 225.

McCloskey and Losch, *The U.N. Law of the Sea Conference and the U.S. Congress: Will Pending U.S. Unilateral Action on Deep Seabed Mining Destroy Hope for a Treaty?* (1979) 1 NW. J. INT'L. L. & BUS. 240.

McClure, *Escape Velocity on the Moon*, at <http://www.idialstars.com/evme.htm>, last accessed on 15 July 2006.

McColloch, *Protocol on Environmental Protection to the Antarctic Treaty –The Antarctic Treaty – Antarctic Minerals Convention – Wellington Convention – Convention on the Regulation of Antarctic Mineral Resource Activities* (1992) 22 GA. J. INT'L. & COMP. L. 211.

McDorman, *The 1982 Law of the Sea Convention: The First Year* (1984) 15 J. MARIT. L. & COM. 211.

McDorman, *The Entry into Force of the 1982 LOS Convention and the Article 76 Outer Continental Shelf Regime* (1995) 10 INT'L. J. MARINE & COASTAL L. 165.

McDorman, *Will Canada Ratify the Law of the Sea Convention?* (1988) 25 SAN DIEGO L. REV. 535.

McDougal, Lasswell and Vlasic, LAW AND PUBLIC ORDER IN SPACE (1963).

McGarrigle, *Hazardous Biological Activities in Outer Space* (1984) 18 AKRON L. REV. 103.

McGowan, *Geographic Disadvantage as a Basis for Marine Resource Sharing between States* (1987) 13 MONASH U. L. REV. 209.

McWhinney, *The International Law-Making Process and the New International Economic Order* (1976) 14 CAN. Y. B. INT'L. L. 57

McWhinney, THE WORLD COURT AND THE CONTEMPORARY LAW-MAKING PROCESS (1979).

Meadows, Meadows and Randers, BEYOND THE LIMITS: CONFRONTING GLOBAL COLLAPSE ENVIRONING A SUSTAINABLE FUTURE (1992).

Meadows, Meadows, Randers and Behrens III, THE LIMITS TO GROWTH (1972).

Mellott, *Electronics and Sensor Cooling with a Stirling Cycle for Venus Surface Mission* (2004), paper presented at the 2nd International Energy Conversion Engineering Conference, 16–19 August 2004, in Providence, RI, USA.

Menefee, *"The Oar of Odysseus": Landlocked and "Geographically Disadvantaged" States in Historical Perspective* (1993) 23 CAL. W. INT'L. L. J. 1.

Menter, *Astronautical Law* (student thesis, Industrial College of the Armed Forces, 1959).

Menter, *Commercial Space Activities Under the Moon Treaty* (1979) 7 SYRACUSE J. INT'L. L. & COM. 213.

Mercure, *L'échec des Modèles de Gestion des Ressources Naturelles Selon les Charactéristiques du Concept de Patrimoine Commun de l'Humanité* (1997) 28 OTTAWA L. REV. 45.

Mercure, *La Proposition d'un Modèle de Gestion Intégrée des Ressources Naturelles Communes de l'Humanité* (1998) 36 CAN. Y. B. INT'L. L. 41.

Merges and Reynolds, *Space Resources, Common Property and the Collective Action Problem* (1998) 6 N. Y. U. ENVT'L. L. J. 107.

Meyer, *Volkerrechtliche Probleme des Weltraumgebiets* (1960) INT'L. R. & S. AB. 326.

Mikesell, THE WORLD COPPER INDUSTRY: STRUCTURE AND ECONOMIC ANALYSIS (1979).

Miklody, *Some Remarks to the Legal Status of Celestial Bodies and Protection of the Outer Space Environment* (1983) 25 PROC. COLL. L. OUTER SP. 13.

Milic, *Third United Nations Conference on the Law of the Sea* (1976) 8 CASE W. RES. J. INT'L. L. 168.

Mill, PRINCIPLES OF POLITICAL ECONOMY WITH SOME OF THEIR APPLICATION TO SOCIAL PHILOSOPHY (1909).

Miller and Delehant, *Deep Seabed Mining: Government Guaranteed Financing under the Maritime Aids of the Merchant Marine Act 1936 as an Alternative to Treaty-Related Loss Compensation* (1980) 11 J. MARIT. L. & COM. 453.

Minola, *The Moon Treaty and the Law of the Sea* (1981) 18 SAN DIEGO L. REV. 455.

Minor Planet Centre, Harvard University, <http://cfa-www.harvard.edu/iau/Ephemerides/Soft02. html>, last accessed on 6 July 2001.

Minta, *The Lome Convention and the New International Economic Order* (1984) 27 HOWARD L. J. 953.

Mishan, THE COSTS OF ECONOMIC GROWTH (1967).

Missile Technology Control Regime, *MTCR Partners*, at <http://www.mtcr.info/english/partners. html>, last accessed on 28 November 2009.

Molitor, *The Provisional Understanding Regarding Deep Seabed Matters: An Ill-Conceived Regime for U.S. Deep Seabed Mining* (1987) 20 CORNELL INT'L. L. J. 223.

Molitor, *The U.S. Deep Seabed Mining Regulations: The Legal Basis for an Alternative Regime* (1982) 19 SAN DIEGO L. REV. 599.

Moltz, *Moonstruck: What's Up with U.S. Space Policy?* (2 February 2004), Centre for Nonproliferation Studies, at <http://cns.miis.edu/pubs/week/040202.htm>, last accessed on 11 July 2006.

Money, *The Protocol on Environmental Protection to the Antarctic Treaty: Maintaining a Legal Regime* (1993) 7 EMORY INT'L. L. REV. 163.

Monserrat Filho, *Why and How to Define "Global Public Interest"* (2000) 43 PROC. COLL. L. OUTER SP. 22.

Moore, *The Law of the Sea and the New International Economic Order* (1984) 3 PUB. L. FORUM 13.

Moore, *The Polar Regions and the Law of the Sea* (1976) 8 CASE W. RES. J. INT'L. L. 204.

Morbidelli, Bottke, Froeschlé and Michel, *Origin and Evolution of Near-Earth Objects*, in William F. Bottke, Alberto Cellino, Paolo Paolicchi and Richard P. Binzel (eds.), ASTEROIDS III (2002) at 409–422.

Morell, THE LAW OF THE SEA: AN HISTORICAL ANALYSIS OF THE 1982 TREATY AND ITS REJECTION BY THE UNITED STATES (1992).

Morrison, *An International Program to Protect the Earth From Impact Catastrophe: Initial Steps* (1993) 30 ACTA ASTRONAUTICA 11.

References

Morrison and Niehoff, *Future Exploration of the Asteroids* in Gehrels (ed.), ASTEROIDS (1979) at 227–249.

Morse, *Low and Behold: Making the Most of Cheap Oil* (2009) 88:5 FOREIGN AFF. 36.

Moskowitz, *Space Clown Comes Back Down to Earth* (11 October 2009), at <http://www.msnbc.msn.com/id/33262374/ns/technology_and_science-space/>, last accessed on 26 October 2009.

Mosteshar, *Intellectual Property Issues in Space Activities*, in Mosteshar (ed.), RESEARCH AND INVENTION IN OUTER SPACE: LIABILITY AND INTELLECTUAL PROPERTY RIGHTS (1995), at 189–198.

Mundell, *Dollar Standards in the Dollar Era* (2007) 29 J. POLICY MODELLING 677.

Murphy, *Antarctic Treaty System – Does the Minerals Regime Signal the Beginning of the End?* (1991) 14 SUFFOLK TRANSNAT'L. L. J. 523.

Murphy, *Deep Ocean Mining: Beginning of a New Era* (1976) 8 CASE W. RES. J. INT'L. L. 46.

Murphy, *The Politics of Manganese Nodules: International Considerations and Domestic Legislation* (1979) 16 SAN DIEGO L. REV. 531.

Mwenda, *Deep Sea-Bed Mining Under Customary International Law* (2000) 7 MURDOCH U. ELEC. J. L. 2.

Myers, Pencil, Rawlin, Kussmaul and Oden, *NSTAR Ion Thruster Plume Impact Assessments*, paper presented at the 31st Joint Propulsion Conference and Exhibition, 10–12 July 1995, in San Diego, CA, USA.

Nagasawa, Ida and Tanaka, *Origin of High Orbital Eccentricity and Inclination of Asteroids* (2001) 53 EARTH PLANETS & SP. 1085.

Nagatomo, *An Approach to Develop Space Solar Power as a New Energy System for Developing Countries* (1996) 56:1 SOLAR ENERGY 111.

Namouni, Christou and Murray, *Coorbital Dynamics at Large Eccentricity and Inclination* (1999) 83 PHYS. REV. LTRS. 2506.

National Aeronautics and Space Administration, *Amalthea – Voyager 1* (2003) at <http://nssdc.gsfc.nasa.gov/imgcat/html/object_page/vg1_1638131.html>, last accessed on 6 December 2004.

National Aeronautics and Space Administration, at <http://www1.jsc.nasa.gov/er/seh/apollo_program.pdf>, last accessed on 25 July 2007.

National Aeronautics and Space Administration, *Near Earth Asteroid Rendezvous* (2001), at <http://nssdc.gsfc.nasa.gov/planetary/near.html>, last accessed on 23 December 2004.

National Aeronautics and Space Administration, *Solar System Exploration* (2006) at <http://solarsystem.nasa.gov/index.cfm>, last accessed on 15 July 2006.

Nelson, *The Contemporary Seabed Mining Regime: A Critical Analysis of the Mining Regulations Promulgated by the International Seabed Authority* (2005) 16 COLO. J. INT'L. ENVT'L. L. & POL'Y. 27.

Nesgos, *International and Domestic Law Applicable to Commercial Launch Vehicle Transportation* (1984) 27 PROC. COLL. L. OUTER SP. 98.

Nesgos, *Rights and Obligations of Participants in Space Materials Processing Activities*, paper presented at the ICC/IBA Symposium on Research and Invention in Outer Space and their Commercial Exploitation: Liability and Intellectual Property Rights, 6–7 December 1990, in Paris, France.

Newendorp and John R. Schuyler, DECISION ANALYSIS FOR PETROLEUM EXPLORATION (2nd ed., 2000).

Newlin, *An Alternative Legal Mechanism for Deep Sea Mining* (1980) 20 VA. J. INT'L. L. 257.

Newman, *The Antarctica Mineral Resources Convention: Developments From the October 1986 Tokyo Meeting of the Antarctic Treaty Consultative Parties* (1987) 15 DENVER J. INT'L. L. & POL'Y. 421.

Newmont Mining Corporation, *Batu Hijau, Indonesia*, at <http://www.newmont.com/en/operations/indonesia/batuhijau/index.asp>, last accessed on 28 April 2007.

Nicholson, *The Common Heritage of Mankind and Mining: An Analysis of the Law as to the High Seas, Outer Space, the Antarctic and World Heritage* (2002) 6 N. Z. J. ENVT'L. L. 177.

Nickel Institute, *Economics of Recycling* (2004), at <http://www.nickelinstitute.org/index.cfm/ci_id/121.htm>, last accessed on 21 December 2004.

Niehoff, *Asteroid Mission Alternatives*, in Morrison and Wells (eds.), ASTEROIDS: AN EXPLORATION ASSESSMENT (1978), at 225–244.

Niehoff, *Round-Trip Mission Requirements for Asteroids 1976AA and 1973EC* (1977) 31 ICARUS 430.

Nweihed, *Venezuela's Contribution to the Contemporary Law of the Sea* (1974) 11 SAN DIEGO L. REV. 603.

Oda, FIFTY YEARS OF THE LAW OF THE SEA (2003).

Oda, INTERNATIONAL CONTROL OF SEA RESOURCES (1989).

Oda, *Sharing of Ocean Resources – Unresolved Issues in the Law of the Sea* (1981) 3 N. Y. J. INT'L. & COMP. L. 1.

Oda, *Some Reflections on Recent Developments in the Law of the Sea* (2002) 27 YALE J. INT'L. L. 217.

O'Donnell, *Staking a Claim in the Twenty-First Century: Real Property Rights on Extra-Terrestrial Bodies* (2007) 32 U. DAYTON L. REV. 461.

O'Leary, *International Manned Missions to Mars and the Resources of Phobos and Deimos* (1992) 26:1 ACTA ASTRONAUTICA 37.

O'Leary, *Mining the Apollo and Amor Asteroids* (1977) 197 SCIENCE 363.

O'Neill, *Testimony on House Concurrent Resolution 451* (1978), U.S. House of Representatives, 25 January 1978.

Odell, *Optimal Development of the North Sea's Oil Fields – A Summary* (1977) 5:4 ENERGY POLICY 282.

Odell and Rosing, *Optimal Development of the North Sea's Oil Fields – The Reply* (1977) 5:4 ENERGY POLICY 295.

Oduntan, *Imagine There are No Possessions: Legal and Moral Basis of the Common Heritage Principle in Space Law* (2005) 2 MAN. J. INT'L. ECON. L. 30.

Oduntan, *Legality of the Common Heritage of Mankind Principle in Space Law: Reconciliation of the Views from the North and South*, paper presented at the International Institute of Space Law Symposium, 6 May 2003, in Sydney, Australia.

Oduntan, *The Never Ending Dispute: Legal Theories on the Spatial Demarcation Boundary Plane between Airspace and Outer Space* (2003) 1 HERTS. L. J. 64.

Ogawa and Ito, *On the Desirability of a Regional Basket Currency Arrangement* (2002) 16 J. JAPAN. & INT'L. ECON. 317.

Oliver, *Interim Deep Seabed Mining Legislation: An International Environmental Perspective* (1981) 8 J. LEGIS. 73.

Olsson-Steel, *Theoretical Meteor Radiants of Recently Discovered Asteroids, Comets and Twin Showers of Known Meteor Streams* (1988) AUST. J. ASTRON. 93.

Onuf and Birney, *Peremptory Norms of International Law: Their Source, Function and Future* (1974) 4 DENVER J. INT'L. L. & POL'Y. 187.

Oosterlink, *Tangible and Intangible Property in Outer Space* (1997) 39 PROC. COLL. L. OUTER SP. 277.

Oosterlink, *The Intergovernmental Space Station Agreement and Intellectual Property Rights* (1989) 17 J. SP. L. 31.

Organisation of the Petroleum Exporting Countries, WORLD OIL OUTLOOK 2009 (2009), at <http://www.opec.org/library/World%20Oil%20Outlook/pdf/WOO%202009.pdf>, last accessed on 5 November 2009.

Orlove, *Spaced Out: The Third World Looks For a Way in to Outer Space* (1989) 4 CONN. J. INT'L. L. 597.

Orrego-Vicuña, *National Laws on Seabed Exploitation: Problems of International Law* (1981) 13 LAWYER AM. 139.

Orrego-Vicuña, *The Regime for the Exploitation of the Seabed Mineral Resources and the Quest for a New International Economic Order of the Oceans: A Latin-American View* (1978) 10 LAWYER AM. 774.

References 357

Ortiz, LEGAL ASPECTS OF THE NORTH BORNEO QUESTION (1964).

Ostro, Campbell and Shapiro, *Mainbelt Asteroids: Dual Polarisation Radar Observations* (1985) 229 SCIENCE 442.

Ostro, Campbell, Chandler, Hine, Hudson, Rosema and Shapiro, *Asteroid 1986DA: Radar Evidence of a Metallic Composition* (1991) 252 SCIENCE 1399.

Ott, *An Analysis of Deep Seabed Mining Legislation* (1978) 10 NAT. RES. LAWYER 591.

Otto, Andrews, Cawood, Doggett, Guj, Stermole, Stermole and Tilton, MINING ROYALTIES: A GLOBAL STUDY OF THEIR IMPACT ON INVESTORS, GOVERNMENT AND CIVIL SOCIETY (2006).

Oxman, *Antarctica and the New Law of the Sea* (1986) 19 CORNELL INT'L. L. J. 211.

Oxman, *Law of the Sea Forum: The 1994 Agreement and the Convention* (1994) 88 AM. J. INT'L. L. 687.

Oxman, *The High Seas and the International Seabed Area* (1989) 10 MICH. J. INT'L. L. 526.

Oxman, *The Territorial Temptation: A Siren Song at Sea* (2006) 100 AM. J. INT'L. L. 830.

Oxnevad, *An Investment Analysis Model for Space Mining Ventures* (1991), paper presented at the 42nd International Astronautical Congress, 5–11 October 1991, in Montréal, Canada.

Palmer, *The United States Draft United Nations Conference on the International Seabed Area and the Accommodation of Ocean Uses* (1973) 1 SYRACUSE J. INT'L. L. & COM. 110.

Panzar, *A Methodology for Measuring the Costs of Universal Service Obligations* (2000) 12 INFO. ECON. & POL'Y. 211.

Papinian, THE DIGEST OF JUSTINIAN, vol. IV (1985).

Pardo, *An International Regime for the Deep Seabed: Developing Law or Developing Anarchy?* (1969) 5 TEX. INT'L. L. F. 204.

Pardo, *Before and After* (1983) 46 L. & CONTEMP. PROBS. 95.

Pardo, *Development of Ocean Space – An International Dilemma* (1971) 31 LA. L. REV. 45.

Pardo, *The Law of the Sea: Its Past and its Future* (1984) 63 OR. L. REV. 7.

Pardo, *Who Will Control the Seabed?* (1969) 47 FOREIGN AFF. 123.

Park, *Incremental Steps for Achieving Space Security: The Need for a New Way of Thinking to Enhance the Legal Regime for Space* (2006) 28 HOUSTON J. INT'L. L. 871.

Parker Clote, *Implications of Global Warming on State Sovereignty and Arctic Resources under the United Nations Convention on the Law of the Sea: How the Arctic is No Longer Communis Omnium Naturali Jure* (2008) 8 RICH. J. GLOBAL L. & BUS. 195.

Parriott, *Territorial Claims in Antarctica: Will the United States be Left Out in the Cold* (1986) 22 STANFORD J. INT'L. L. 67.

Patterson, Foster, Haag, Rawlin, Soulas and Roman, *NEXT: NASA's Evolutionary Xenon Thruster* (2002), paper presented at the 38th Joint Propulsion Conference and Exhibition, 7–10 July 2002, in Indianapolis, IN, USA.

Pearce and Rose (eds.), THE ECONOMICS OF NATURAL RESOURCE DEPLETION (1975).

Pearce and Walter, RESOURCE CONSERVATION: THE SOCIAL AND ECONOMIC DIMENSIONS OF RECYCLING (1977).

Peeters, *Square Peg, Round Hole: Jurisdiction Over Minerals Mining Offshore Antarctica* (2004) 1 MQ. J. INT'L. & COMP. ENVT'L. L. 217.

Pell, *A New Era in Ocean Policy* (1997) 12 INT'L. J. MARINE & COASTAL L. 1.

Perera, *A Dawn of a New Era of the Oceans or a Return to the Grotian Ocean? Some Reflections as the Law of the Sea Convention Enters into Force* (1994) 6 SRI LANKA J. INT'L. L. 157.

Perrott, *EEC Attitudes and Responses to the New International Economic Order* (1984) 3 PUB. L. FORUM 115.

Picker, *A View from 40,000 Feet: International Law and the Invisible Hand of Technology* (2002) 23 CARDOZO L. REV. 149.

Pietrowski, *Hard Minerals on the Deep Ocean Floor: Implications for American Law and Policy* (1978) 19 WM. & MARY L. REV. 43.

Pinto, *The Developing Countries and the Exploitation of the Deep Seabed* (1980) 15 COLUM. J. WORLD BUS. 30.

Platzöder, *Substantive Changes in a Multilateral Treaty Before Its Entry into Force: The Case of the 1982 United Nations Convention on the Law of the Sea* (1993) 4 EUR. J. INT'L. L. 390.

Pop, *A Celestial Body is a Celestial Body is a Celestial Body . . .* (2001) 44 PROC. COLL. L. OUTER SP. 100.

Pop, WHO OWNS THE MOON? EXTRATERRESTRIAL ASPECTS OF LAND AND MINERAL RESOURCES OWNERSHIP (2008).

Power, *The Role of Metal Mining in the Alaskan Economy* (13 February 2002), Southeast Alaska Conservation Council, at <http://www.seacc.org/Publications/MetalMiningReport.doc>, last accessed on 27 January 2007.

Predictive Mineral Discovery Cooperative Research Centre, *Utilisation and Application of the Research: Commercialisation and Links with Users* (2002), located at <http://www.pmdcrc.com.au/repspubs/annrep.html>, last accessed on 20 January 2005.

Preston, Johnson, Edwards, Miller and Shipbaugh, SPACE WEAPONS, EARTH WARS (2002).

Prialnik and Mekler, *The Formation of an Ice Crust below the Dust Mantle of a Cometary Nucleus* (1991) 366 ASTROPHYSICS J. 318.

Price, *Values and Concepts in Conservation* (1955) 45:1 ANN. ASSOC. AM. GEOG. 65.

Proudfoot, *Guarding the Treasures of the Deep: The Deep Seabed Hard Mineral Resources Act* (1973) 10 HARV. J. ON LEGIS. 596.

Pruitt, *Unilateral Deep Seabed Mining and Environmental Standards: A Risky Venture* (1982) 8 BROOKLYN J. INT'L. L. 345.

Quigg, *Open Skies and Open Space* (1958) 37 FOREIGN AFF. 95.

Raclin, *From Ice to Ether: The Adoption of a Regime to Govern Resource Exploitation in Outer Space* (1986) 7 J. INT'L. L. & BUS. 727.

Ramakrishna, *North-South Issues, Common Heritage of Mankind and Global Climate Change* (1990) 19 MILL. J. INT'L. STUD. 429.

Raman, *Transnational Corporations, International Law and the New International Economic Order* (1979) 6 SYRACUSE J. INT'L. L. & COM. 17.

Rana, *The "Common Heritage of Mankind" & the Final Frontier: A Revaluation of Values Constituting the International Legal Regime for Outer Space Activities* (1994) 26 RUTGERS L. J. 225.

Randeniya, *Sharing the World's Resources: Equitable Distribution and the North-South Dialogue in the New Law of the Sea* (2003) 15 SRI LANKA J. INT'L. L. 149.

Rankin, *Utilization of the Geostationary Orbit – A Need for Orbital Allocation* (1974) 13 COLUM. J. TRANS. L. 101.

Ratiner and Wright, *The Billion Dollar Decision: Is Deepsea Mining a Prudent Investment?* (1978) 10 LAWYER AM. 713.

Ratiner and Wright, *United States Ocean Mineral Resource Interests and the United Nations Conference on the Law of the Sea* (1973) 6 NAT. RES. LAWYER 1.

Rees, NATURAL RESOURCES: ALLOCATION, ECONOMICS AND POLICY (2nd ed., 1990).

Reinstein, *Owning Outer Space* (1999) 20 NW. J. INT'L. L. & BUS. 59.

Reintanz, *Vladimír Mandl – The Father of Space Law* (1968) 11 PROC. COLL. L. OUTER SP. 362.

Reis, *Some Reflections on the Liability Convention for Outer Space* (1978) 6 J. SP. L. 161.

Reiskind, *Toward a Responsible Use of Nuclear Power in Outer Space – The Canadian Initiative in the United Nations* (1981) 4 ANN. AIR & SP. L. 461.

Reynolds, *International Space Law: Into the Twenty-First Century* (1993) 25 VAND. J. TRANSNAT'L. L. 225.

Reynolds, *Key Objections to the Moon Treaty* (2003), National Space Society Chapters Network, <http://www.nsschapters.org/hub/pdf/MoonTreatyObjections.pdf>, 28 April 2003, last accessed on 28 November 2009.

Reynolds, *Legislative Comment: The Patents in Space Act* (1990) 3 HARV. J. LAW & TECH. 13.

Ricardo, PRINCIPLES OF POLITICAL ECONOMY AND TAXATION (1817, reprinted 1962).

Rich, *A Minerals Regime for Antarctica* (1982) 31 INT'L. & COMP. L. Q. 709.

Richardson, *Law in the Making: A Universal Regime for Deep Seabed Mining?* (1981) 8 J. LEGIS. 199.

References   359

Richardson, *Superpowers Need Law: A Response to the United States Rejection of the Law of the Sea Treaty* (1983) 17 GEO. WASH. J. INT'L. L. & EC. 1.

Richardson, *The United States Posture Toward the Law of the Sea Convention: Awkward but not Irreparable* (1983) 20 SAN DIEGO L. REV. 505.

Richardson, *United States Interests and the Law of the Sea* (1978) 10 LAWYER AM. 651.

Riddell-Dixon, *The Preparatory Commission on the International Seabed Authority: "New Realism"?* (1992) 7 INT'L. J. ESTUARINE & COASTAL L. 195.

Rieder, Wänke, Economou and Turkevich, *Determination of the Chemical Composition of Martian Soil and Rocks: The Alpha Proton X Ray Spectrometer* (1997) 102 J. GEOPHYS. RES. 4027.

Rifkin, The Hydrogen Economy (2002).

Ripley, ENVIRONMENTAL EFFECTS OF MINING (1996).

Rivkin, Britt, Howell and Lebofsky, *Hydrated E-Class and M-Class Asteroids* (1995) 117 ICARUS 90.

Rivkin, Howell and Bus, *Diversity of Types of Hydrated Minerals on C-Class Asteroids*, paper presented at the 35th Lunar and Planetary Science Conference, 15–19 March 2004, in League City, TX, USA.

Robert and Epstein, *Carbon, Hydrogen and Nitrogen Isotopic Composition of the Renazzo and Orgueil Organic Components* (1980) 15 METEORITICS 355.

Robert and Epstein, *The Concentration and Isotopic Composition of Hydrogen, Carbon and Nitrogen in Carbonaceous Meteorites* (1982) 46 GEOCHIM. & COSMOCHIM. ACTA 81.

Robertson and Vasaturo, *Recent Developments in the Law of the Sea 1981–1982* (1983) 20 SAN DIEGO L. REV. 679.

Robinson, *Earth Exposure to Martian Matter: Back Contamination Procedures and International Quarantine Regulations* (1976) 15 COLUM. J. TRANSNAT'L. L. 17.

Robinson and Taylor, *Ferrous Oxide in Mercury's Crust and Mantle* (2001) 36 METEOR. & PLANET. SCI. 841.

Robson, *Soviet Legal Approach to Space Law Issues at the United Nations* (1980) 3 LOYOLA L.A. INT'L. & COMP. L. ANN. 99.

Rode-Verschoor, *The Responsibility of States for the Damage Caused by Launched Space-Bodies* (1958) 1 PROC. COLL. L. OUTER SP. 103.

Rosenfield, *The Moon Treaty: The United States Should Not Become a Party* (1980) 74 AM. SOC'Y. INT'L. L. PROC. 162.

Rosenne, *Establishing the International Tribunal for the Law of the Sea* (1995) 89 AM. J. INT'L. L. 806.

Rosenne, *The Third United Nations Conference on the Law of the Sea* (1976) 11 ISR. L. REV. 1.

Rosenne, *The United Nations Convention on the Law of the Sea 1982: The Application of Part XI: An Element of Background*, in Rosenne, ESSAYS ON INTERNATIONAL LAW AND PRACTICE (2007).

Rosenne, *The United Nations Convention on the Law of the Sea, 1982: The Application of Part XI: An Element of Background* (1995) 29 ISR. L. REV. 491.

Ross, *Near-Earth Asteroid Mining* (2001), at Laboratory for Spacecraft and Mission Design, California Institute of Technology, <http://www.esm.vt.edu/~sdross/papers/ross-asteroid-mining-2001.pdf>, last accessed on 24 April 2007.

Rothwell, *A Maritime Analysis of Conflicting International Law Regimes in Antarctica and the Southern Ocean* (1994) 15 AUST. Y. B. INT'L. L. 155.

Rozental, *The Charter of Economic Rights and Duties of States and the New International Economic Order* (1976) 16 VA. J. INT'L. L. 309.

Rubin, *Economic and Social Human Rights and the New International Economic Order* (1986) 1 AM. U. J. INT'L. L. & POL'Y. 67.

Rusconi, *An Essay on the Lawful Concept of Heavenly Bodies* (1966) 9 PROC. COLL. L. OUTER SP. 55.

Rusconi and Paz-Perina, *Proyecto de Tratado Relativo a la Luna Usos Pacíficos y Desarme: Dos Aspectos de una Misma Realidad* (1973) 16 PROC. COLL. L. OUTER SP. 190.

Rusk, "Letter of Submittal from Secretary Rusk to President Johnson", 27 January 1967, in *Hearings on Treaty on Outer Space Before the Senate Committee on Foreign Relations* (1967), 90th Cong., 1st Sess., at 112.

Rusk and Ball, *Sea Changes and the American Republic* (1979) 9 GA. J. INT'L. & COMP. L. 1.

Ryan, *Towards a New International Economic Order* (1975) 9 U. QLD. L. J. 135.

Saffo, *The Common Heritage of Mankind: Has the General Assembly Created a Law to Govern Seabed Mining?* (1979) 53 TUL. L. REV. 492.

Saguirian, *The USSR and the New Law of the Sea Convention: In Search of Practical Solutions* (1990) 84 AM. SOC'Y. INT'L. L. PROC. 295

Sand, *Public Trusteeship for the Oceans*, in Ndiaye and Wolfrum (eds.), LAW OF THE SEA, ENVIRONMENTAL LAW AND SETTLEMENT OF DISPUTES: LIBER AMICORUM JUDGE THOMAS A. MENSAH (2007), at 521–544.

Sanders, THE INSTITUTES OF JUSTINIAN (4th ed., 1903).

Sanguin, *Geopolitical Scenarios, From the Mare Liberum to the Mare Clausum: The High Seas and the Case of the Mediterranean Basin* (1997) 2 GEOADRIA 51.

Sattler, *Transporting a Legal System for Property Rights: From the Earth to the Stars* (2005) 6 CHI. J. INT'L. L. 23.

Save Alaska, at <http://www.savealaska.com/sa_headlines.html>, last accessed on 6 December 2004.

Scheeres, Broschart, Ostro and Benner, *The Dynamic Environment about Asteroid 25143 Itokawa: Target of the Hayabusa Mission*, paper presented at the AIAA/AAS Astrodynamics Specialist Conference and Exhibit, 16–19 August 2004, in Providence, RI, USA.

Scherer and Neutsch, *Dynamical Evolution of Asteroidal Orbits* (1993) 9 ASTRON. GESELL. ABS. SER. 98.

Schneider, *Something Old, Something New: Some Thoughts on Grotius and the Marine Environment* (1978) 18 VA. J. INT'L. L. 147.

Schoonover, *The History of Negotiations Concerning the System of Exploitation of the International Seabed* (1977) 9 N. Y. U. J. INT'L. L. & POL'Y. 483.

Schrogl and Davies, *A New Look at the "Launching State": The Results of the UNCOPUOS Legal Subcommittee Working Group "Review of the Concept of the 'Launching State'" 2000–2002* (2002) 45 PROC. COLL. L. OUTER SP. 286.

Schwebel, *Letter of U.S. State Department Deputy Legal Adviser Stephen Schwebel of 25 April 1975* (1975) U.S.D.I.L. 85.

Scott, *Institutional Developments Within the Antarctic Treaty System* (2003) 52 INT'L. & COMP. L. Q. 473.

Scott, *Protecting United States Interests in Antarctica* (1989) 26 SAN DIEGO L. REV. 575.

Scott, *Universalism and Title to Territory in Antarctica* (1997) 66 NORDIC J. INT'L. L. 33.

Scovazzi, *Mining, Protection of the Environment, Scientific Research and Bioprospecting: Some Considerations on the Role of the International Seabed Authority* (2004) 19 INT'L. J. MARINE & COASTAL L. 383.

Sea Launch Company LLC, *Sea Launch User's Guide* (2003), located at <http://www.sea-launch.com/customers_webpage/sluw/>, last accessed on 19 January 2005.

Seach, *Conflicting Interests in Antarctica: People or Nature? Who Decides?* (1991) 5 TEMPLE INT'L. & COMP. L. J. 109.

Sears, *The Case for Asteroid Sample Return*, paper presented at the 61st Annual Meteoritical Society Meeting, 27–31 July 1998, in Dublin, Ireland.

Sears and Scheeres, *Asteroid Constraints on Multiple Near-Earth Asteroid Sample Return* (2001) 36 METEOR. & PLANET. SCI. 186.

Sears, Benoit, McKeever, Banerjee, Kral, Stites, Roe, Jansma and Mattioli, *Investigation of Biological, Chemical and Physical Processes on and in Planetary Surfaces by Laboratory Simulation* (2002) 50 PLANET. SP. SCI. 821.

Sears, Britt and Cheng, *Asteroid Sample Return: 433 Eros as an Example of Sample Site Selection* (2001) 36 METEOR. & PLANET. SCI. 30.

Severino, SOUTHEAST ASIA IN SEARCH OF AN ASEAN COMMUNITY: INSIGHTS FROM THE FORMER ASEAN SECRETARY-GENERAL (2006).

Sharma, TERRITORIAL ACQUISITION, DISPUTES AND INTERNATIONAL LAW (1997).

Shaw, TITLE TO TERRITORY IN AFRICA (1971).

Shihata, *Arab Oil Policies and the New International Economic Order* (1976) 16 VA. J. INT'L. L. 261.

Shoemaker and Helin, *Earth-Approaching Asteroids: Populations, Origin and Compositional Types*, in Morrison and Wells (eds.), ASTEROIDS: AN EXPLORATION ASSESSMENT (1978), at 163–175.

Shusterich, *The Antarctic Treaty System: History, Substance and Speculation* (1984) 39 INT'L. J. 800.

Silkenat, *Solving the Problem of the Deep Seabed: The Informal Composite Negotiating Text for the First Committee of UNCLOS III* (1977) 9 N. Y. U. J. INT'L. L. & POL'Y. 177.

Silverstein, *Proprietary Protection for Deepsea Mining Technology in Return for Technology Transfer: New Approach to the Seabed Controversy* (1978) 60 J. PAT. OFF. SOC'Y. 135.

Simma, *The Antarctic Treaty as a Treaty Providing for an "Objective Regime"* (1986) 19 CORNELL INT'L. L. J. 189.

Sinclair, *Environmentalism: A la Recherché du Temps Perdu-Bien Perdu?* in Cole, Freeman, Jahoda and Pavitt (eds.), MODELS OF DOOM: A CRITIQUE OF THE LIMITS TO GROWTH (1973), at 175–192.

Sinclair, THE VIENNA CONVENTION ON THE LAW OF TREATIES (2nd ed., 1984).

Sittenfeld, *The Evolution of a New and Viable Concept of Sovereignty for Outer Space* (1981) 4 FORDHAM INT'L. L. J 199.

Skelton, *UNCTAD's Draft Code of Conduct on the Transfer of Technology: A Critique* (1981) 14 VAND. J. TRANSNAT'L. L. 381.

Sloan, *General Assembly Resolutions Revisited* (1987) 58 BRIT. Y. B. INT'L. L. 39.

Smirnoff, *Report from Working Group Three on the Law of Outer Space* (1964) 7 PROC. COLL. L. OUTER SP. 352.

Smirnoff, *The Legal Status of Celestial Bodies* (1962) 28 J. AIR L. & COM. 385.

Smirnoff, *The Role of the IAF in the Elaboration of the Norms of Future Space Law* (1959) 2 PROC. COLL. L. OUTER SP. 147.

Smith, *Apostrophe to a Troubled Ocean* (1972) 5 IND. LEG. F. 267.

Smith, *Intellectual Property Rights in Outer Space Activities – Aid or Impediment?*, paper presented at the ISRO-IISL Space Law Conference 2005, 26–29 June 2005, in Bangalore, India.

Smith, *The Technical, Legal and Business Risks of Orbital Debris* (1997) 6 N. Y. U. ENVT'L. L. J. 50.

Smith and Wells, NEGOTIATING THIRD WORLD MINERAL AGREEMENTS (1975).

Smithsonian Astrophysical Observatory, <http://cfa-www.harvard.edu/cfa/ps/mpc.html>, last accessed in September 2001.

Snyder, PLEASE STOP KILLING ME (1971).

Sohn, *International Law Implications of the 1994 Agreement* (1994) 88 AM. J. INT'L. L. 696.

Sohn, *Managing the Law of the Sea: Ambassador Pardo's Forgotten Second Idea* (1998) 36 COLUM. J. TRANSNAT'L. L. 285.

Sohn, *The Development of the Charter of the United Nations: The Present State*, in Bos (ed.), THE PRESENT STATE OF INTERNATIONAL LAW AND OTHER ESSAYS (1973), at 39–60.

Solomon, *Return to the Iron Planet*, NEW SCIENTIST, 29 January 2000, at 35.

Sonter, *Near Earth Objects as Resources for Space Industrialisation* (2001) 1:1 SOLAR SYSTEM DEV. J. 1.

Sonter, *The Technical and Economic Feasibility of Mining the Near-Earth Asteroids*, paper presented at the 49th International Astronautical Congress, 5–9 October 1998, in Melbourne, Australia.

Sornarajah, *The New International Economic Order, Investment Treaties and Foreign Investment Laws in ASEAN* (1985) 27 MALAYA L. REV. 440.

Sovey, Dever and Power, *Retention of Sputtered Molybdenum on Ion Engine Discharge Chamber Surfaces,* paper presented at the 27th International Electric Propulsion Conference, 14–19 October 2001, in Pasadena, CA, USA.

Space Adventures Ltd, *Our Clients,* at <http://www.spaceadventures.com/index.cfm?fuseaction=orbital.Clients>, last accessed on 1 October 2009.

Space Daily, *Rosetta Swings by Asteroid Steins 2867 on Route to Comet Churyumov,* Space Daily, 6 September 2008, at <http://www.spacedaily.com/reports/Rosetta_Swings_By_Asteroid_Steins_2867_On_ Route_To_Comet_Churyumov_999.html>, last accessed on 7 September 2008.

Space Research Associates, *Report of Satellite Solar Power Systems* (1986) 6 SPACE POWER 1.

SpaceDev, Inc., *Near Earth Asteroid Prospector* (2004), at <http://www.spacedev.com/newsite/templates/subpage3.php?pid=191%26subNav=11%26subSel=3>, last accessed on 6 December 2004.

Spectar, *Saving the Ice Princess: NGOs, Antarctica and International Law in the New Millennium* (2000) 23 SUFFOLK TRANSNAT'L. L. REV. 57.

Stakelbeck, *Inconsistent U.S. Policy* (2 June 2006), The Washington Times, at <http://washingtontimes.com/op-ed/20060601-085037-6169r.htm>, last accessed on 11 July 2006.

Stanford, *The Cape Town Convention and the Preliminary Draft Space Protocol: An Update,* paper presented at the ISRO-IISL Space Law Conference 2005, 26–29 June 2005, in Bangalore, India.

Stang, *Political Cobwebs Beneath the Sea* (1973) 7 INT'L. LAWYER 1.

Staniland, *A Sea-Change: The United Nations Convention on the Law of the Sea* (1983) 100 S. AFR. L. J. 700

Stanley, *International Codes of Conduct for MNCs: A Skeptical View of the Process* (1981) 30 AM. U. L. REV. 973.

Staub, *The Antarctic Treaty as Precedent to the Outer Space Treaty* (1974) 17 PROC. COLL. L. OUTER SP. 282.

Steel, ROGUE ASTEROIDS AND DOOMSDAY COMETS: THE SEARCH FOR THE MILLION MEGATON MENACE THAT THREATENS LIFE ON EARTH (1995).

Stenersen, *Mining the Seabed: International v. National Control* (1993) 4 U.S.A.F.A. J. LEG. STUD. 103.

Sterns and Tennen, *Privateering and Profiteering on the Moon and Other Celestial Bodies: Debunking the Myth of Property Rights in Space* (2003) 31 ADV. SPACE RES. 2433.

Sterns and Tennen, *Resolution of Disputes in the Corpus Juris Spatialis: Domestic Law Considerations* (1993) 36 PROC. COLL. L. OUTER SP. 172.

Stevenson and Oxman, *The Third United Nations Conference on the Law of the Sea: The 1974 Caracas Session* (1975) 69 AM. J. INT'L. L. 1.

Stokke and Østreng, *The Effectiveness of ATS Regimes: Introduction,* in Stokke and Vidas (eds.), GOVERNING THE ANTARCTIC: THE EFFECTIVENESS AND LEGITIMACY OF THE ANTARCTIC TREATY SYSTEM (1996), at 113–119.

Stoller, *Protecting the White Continent: Is the Antarctic Protocol Mere Words or Real Action?* (1995) 12 AZ. J. INT'L. & COMP. L. 335.

Stone, *The United States Draft Convention on the International Seabed Area* (1971) 45 TUL. L. REV. 527.

Strati, *Deep Seabed Cultural Property and the Common Heritage of Mankind* (1991) 40 INT'L. & COMP. L. Q. 859.

Studier, Hayatsu and Anders, *Origin of Organic Matter in the Early Solar System: Further Studies of Meteoritic Hydrocarbons and a Discussion of Their Origin* (1972) 36 GEOCHIM. & COSMOCHIM. ACTA 189.

Suess, *Chemical Evidence Bearing on the Origin of the Solar System* (1965) 3 ANN. REV. ASTRON. & ASTROP. 217

Sullivan, Greeley, Pappalardo, Asphaug, Moore, Morrison, Belton, Carr, Chapman, Geissler, Greenberg, Granahan, Head, Kirk, McEwen, Lee, Thomas and Veverka, *Geology of 243 Ida* (1996) 120 ICARUS 119.

References

Surace-Smith, *United States Activity Outside of the Law of the Sea Convention: Deep Seabed Mining and Transit Passage* (1984) 84 COLUM. L. REV. 1032.

Suzuki, *Self-Determination and World Public Order: Community Responses to Territorial Separation* (1976) 16 VA. J. INT'L. L. 779.

Suzuki, THE SACRED BALANCE: REDISCOVERING OUR PLACE IN NATURE (1997).

Swenson, *Pollution of the Extraterrestrial Environment* (1985) 25 A. F. L. REV. 70.

Swing, *Who Will Own the Oceans?* (1976) 54 FOREIGN AFF. 527.

Synder, Anderson, van Noord and Soulas, *Environmental Testing of the NEXT PM1 Ion Engine*, paper presented at the 43rd Joint Propulsion Conference and Exhibition, 8–11 July 2007, in Cincinnati, OH, USA.

Szaloky, *The Way of the Further Perfection on the Legal Regulation Concerning the Moon and Other Celestial Bodies, Especially Regarding the Exploitation of Natural Resources of the Moon and Other Celestial Bodies* (1973) 16 PROC. COLL. L. OUTER SP. 196.

Sztucki, *Remarks During the Discussion on the Introductory Report* (1966) 9 PROC. COLL. L. OUTER SP. 64.

Takeda, Saika, Otsuki and Hiroi, *A New Antarctic Meteorite with Chromite, Orthopyroxene and Metal with Reference to a Formation Model of S Asteroids* (1993) 24 LUNAR PLANET. SCI. 1395.

Tan, *Towards a New Regime for the Protection of Outer Space as the "Province of All Mankind"* (2000) 25 YALE J. INT'L. L. 145.

Tannenwald, *Law Versus Power on the High Frontier: The Case for a Rule-Based Regime in Outer Space* (2004) 29 YALE J. INT'L. L. 363.

Tanzer, THE RACE FOR RESOURCES: CONTINUING STRUGGLES OVER MINERALS AND FUELS (1980).

Tate, *Near Earth Objects – A Threat and an Opportunity* (2003) 38 PHYSICS EDU. 218.

Taylor and Smith, UNITED NATIONS CONFERENCE ON TRADE AND DEVELOPMENT (UNCTAD) (2007).

Tedesco, *Evidence of Hydrated Minerals in E and M Type Asteroids*, paper presented at the 26th Lunar and Planetary Sciences Conference, March 1995, in Houston, TX, USA.

Tedesco, Matson, Veeder, Gradie and Lebofsky, *A Three-Parameter Asteroid Taxonomy* (1989) 97 ASTRON. J. 580.

Tenenbaum, *A World Park in Antarctica: The Common Heritage of Mankind* (1991) 10 VA. ENVT'L. L. J. 109.

Tennen, *Outer Space: A Preserve for All Humankind* (1979) 2 HOUS. J. INT'L. L. 145.

Tennen, *Second Commentary on Emerging System of Property Rights in Outer Space*, in United Nations, PROCEEDINGS OF THE UNITED NATIONS/REPUBLIC OF KOREA WORKSHOP ON SPACE LAW (2003), at 342.

Teson, *State Contracts and Oil Expropriations: The Aminoil-Kuwait Arbitration* (1984) 24 VA. J. INT'L. L. 323.

Tetzeli, *Allocation of Mineral Resources in Antarctica: Problems and a Possible Solution* (1987) 10 HASTINGS INT'L. & COMP. L. REV. 525.

Thamsborg and Lilje-Jensen, *Demarcation of the Area* (1998) 67 NORDIC J. INT'L. L. 215.

The Economist, *Crunch! Budget Problems at the World Trade Organisation*, THE ECONOMIST, 4 April 1998.

The White House, *Statement by the President: U.S. Policy and the Law of the Sea*, 29 January 1982, DEPT. STATE BULL., at 54.

Theutenberg, *The Arctic Law of the Sea* (1983) 52 NORDISK TIDS. INT'L. RET 3.

Thirlway, INTERNATIONAL CUSTOMARY LAW AND CODIFICATION: AN EXAMINATION OF THE CONTINUING ROLE OF CUSTOM IN THE PRESENT PERIOD OF CODIFICATION OF INTERNATIONAL LAW (1972).

Thomas, *Spatialis Liberum* (2006) 7 FL. COASTAL L. REV. 579.

Thompson, *NASA Probe See Changing Seasons on Mercury*, MSNBC, 3 November 2009, at <http://www.msnbc.msn.com/id/33609756/ns/technology_and_science-space/>, last accessed on 5 November 2009.

Thompson, *Space for Rent: The International Telecommunication Union, Space Law and Orbit/Spectrum Licensing* (1996) 62 J. AIR L. & COM. 279.

Thomson, INTRODUCTION TO SPACE DYNAMICS (1986).

Tilton, THE FUTURE OF NONFUEL MINERALS (1977).

Tomuschat, *The Charter of Economic Rights and Duties of States: Some Thoughts on the Significance of Declarations of the United Nations General Assembly* (1976) ZEIT. AUS. RECHT. VÖLK. 36444.

Torreh-Bayouth, *UNCLOS III: The Remaining Obstacles to Consensus on the Deepsea Mining Regime* (1981) 16 TX. INT'L. L. J. 79.

Travaglini, *Reconciling Natural Law and Legal Positivism in the Deep Seabed Mining Provisions of the Convention on the Law of the Sea* (2001) 15 TEMPLE INT'L. & COMP. L. J. 313.

Treat, *The United States' Claims of Customary Legal Rights under the Law of the Sea Convention* (1984) 41 WASH. & LEE L. REV. 253.

Treves, *Military Installations, Structures and Devices on the Seabed* (1980) 74 AM. J. INT'L. L. 808.

Triggs, *The Antarctic Treaty Regime: A Workable Compromise or a "Purgatory of Ambiguity"?* (1985) 17 CASE W. RES. J. INT'L. L. 195.

Trimble, *The International Law of Outer Space and its Effect on Commercial Space Activity* (1983) 11 PEPP. L. REV. 521.

Tung, *Jurisdictional Issues in International Law: Kelp Farming Beyond the Territorial Sea* (1982) 31 BUFF. L. REV. 885.

Turbeville, *American Ocean Policy Adrift: An Exclusive Economic Zone as an Alternative to the Law of the Sea Treaty* (1983) 35 U. FLA. L. REV. 492.

Tuthill, *Guidelines for the Use of Copper Alloys in Seawater* (May 1987), U.S. Copper Development Association, at <http://www.copper.org/applications/marine/seawater/seawater_corrosion.html>, last accessed on 27 January 2007.

Twibell, *Space Law: Legal Restraints on Commercialisation and Development of Outer Space* (1997) 65 U.M.K.C. L. REV. 589.

U.S. Army Corps of Engineers, *Proceedings of a Workshop on Extraterrestrial Mining and Construction* (1990) USACERL SPECIAL REPORT M-92/14.

U.S. Bureau of Mines, COMMODITY DATA SUMMARIES 1972–1976 (1977).

U.S. Bureau of Mines, MINERALS YEARBOOK (1974).

U.S. Congressional Research Services, OCEAN MANGANESE NODULES (1975).

U.S. Department of Energy, HYDROGEN POSTURE PLAN: AN INTEGRATED RESEARCH, DEVELOPMENT AND DEMONSTRATION PLAN (2004).

U.S. Department of Energy, *Isotope Uses* (2003) <http://www.ne.doe.gov/isotopes/ipuses.asp>, last accessed on 22 December 2004.

U.S. Department of Energy, *Types of Fuel Cells* (2004), at <http://www.eere.energy.gov/hydrogenandfuelcells/fuelcells/fc_types.html>, last accessed on 4 June 2004.

U.S. Department of State, *Outer Space Treaty*, <http://www.state.gov/t/ac/trt/5181.htm>, last accessed on 18 April 2004.

U.S. Senate Committee on Foreign Relations, CONVENTION ON INTERNATIONAL LIABILITY FOR DAMAGE CAUSED BY SPACE OBJECTS (1972) S. EXEC. REP. 92-38, 92nd Cong., 2nd Sess. 9 at 7.

Uchitomi, *State Responsibility/Liability for "National" Space Activities: Towards Safe and Fair Competition in Private Space Activities* (2001) 44 PROC. COLL. L. OUTER SP. 51.

Udombana, *The Third World and the Right to Development: Agenda for the Next Millennium* (2000) 22 HUMAN RIGHTS Q. 753.

United Nations Division for Ocean Affairs and the Law of the Sea, *Chronological Lists of Ratifications of, Accessions and Successions to the Convention and the Related*

Agreements (28 December 2006), at <http://www.un.org/Depts/los/reference_files/chronological_lists_of_ratifications.htm#The%20United%20Nations%20Convention%20on%20the%20Law%20of%20the%20Sea>, last accessed on 27 January 2007.

United Nations Office of Outer Space Affairs, PROCEEDINGS OF THE IISL/ECSL SYMPOSIUM: REINFORCING THE REGISTRATION CONVENTION (2003).

United Nations Office of Outer Space Affairs, *Status of International Agreements Relating to Activities in Outer Space as at 1 January 2010* (1 January 2010), at < http://www.oosa.unvienna.org/pdf/publications/ST_SPACE_11_Rev2_Add3E.pdf>, last accessed on 22 April 2010.

United Nations Office of the High Commissioner for Human Rights, *International Human Rights Instruments* (2007), at <http://www.unhchr.ch/html/intlinst.htm>, last accessed on 28 January 2007.

United Nations, "Concluding Main Part of Session, General Assembly Adopts $4.17 Billion Budget" (press release, 21 December 2007), at <http://www.un.org/News/Press/docs/2007/ga10684.doc.htm>, last accessed on 19 May 2008.

United Nations, *Briefing on Methodology of the Scale of Assessment* (2006), at <http://www.un.org/ga/61/fifth/scale-method.pps>, last accessed on 19 May 2008.

United Nations, *Report of the Ad Hoc Committee on the Peaceful Uses of Outer Space to the United Nations General Assembly* (1959) U.N.Doc. A/4141, Part III.

Valladao, *The Law of Interplanetary Space* (1959) 2 PROC. COLL. L. OUTER SP. 156.

van Bogaert, ASPECTS OF SPACE LAW (1986).

van der Essen, *The Origin of the Antarctic System* (trans. Susan Fisher), in Francioni and Scovazzi (eds.), INTERNATIONAL LAW FOR ANTARCTICA (2nd ed., 1996), at 17–30.

van Dyke and Yuen, *"Common Heritage" v. "Freedom of the High Seas": Which Governs the Seabed?* (1982) 19 SAN DIEGO L. REV. 493.

van Ettinger, King and Payoyo, *The Common Heritage of Mankind and Four Problem Areas*, the United Nations University, <http://www.unu.edu/unupress/unupbooks/uu15oe/uu15oe0q.htm>, last accessed on 12 February 2001.

van Fenema, *The Registration Convention*, in United Nations, PROCEEDINGS OF THE UNITED NATIONS WORKSHOP ON CAPACITY BUILDING IN SPACE LAW (2002), at 33.

van Overbeek, *Article 121(3) LOSC in Mexican State Practice in the Pacific* (1989) 4 INT'L. J. ESTUARINE & COASTAL L. 252

van Traa-Engelman, COMMERCIAL UTILISATION OF OUTER SPACE: LAW AND PRACTICE (1993).

van Traa-Engelman, *Settlement of Space Law Disputes* (1990) 3 LEIDEN J. INT'L. L. 139.

Vanderzwaag, Donihee and Faegteborg, *Towards Regional Ocean Management in the Arctic: From Coexistence to Cooperation* (1988) 37 U. N. B. L. J. 1.

Vasciannie, *Part XI of the Law of the Sea Convention and Third States: Some General Observations* (1989) 48 CAM. L. J. 85.

Vereshchetin and Danilenko, *Custom as a Source of International Law of Outer Space* (1985) 13 J. SP. L. 22.

Verhaag, *It is Not Too Late: The Need for a Comprehensive International Treaty to Protect the Arctic Environment* (2003) 15 GEO. INT'L. ENVT'L. L. REV. 555.

Veverka and Thomas, *Phobos and Deimos: A Preview of What Asteroids Are Like?* in Gehrels, ASTEROIDS (1979) at 628–651.

Vicuna, *From the Energy Crisis to the Concept of an Economic Heritage of Mankind: Guidelines for Reorganising the International Economic System* (1976) 1 INT'L. TRADE L. J. 87.

Vidal, *The End of Oil is Closer than You Think* (21 April 2005), The Guardian, at <http://www.guardian.co.uk/life/feature/story/0,13026,1464050,00.html>, last accessed on 27 January 2007.

Viikari, *Problems Related to Time in the Development of International Space Law* (2004) 47 PROC. COLL. L. OUTER SP. 259.

Viikari, *The Legal Regime for Moon Resource Utilisation and Comparable Solutions Adopted for Deep Seabed Activities* (2003) 31 ADV. SP. RES. 2427.

Vlasic, *The Space Treaty: A Preliminary Evaluation* (1967) 5 CAL. L. REV. 512.

Vogler, THE GLOBAL COMMONS: A REGIME ANALYSIS (1995).

Vogler, THE GLOBAL COMMONS: ENVIRONMENTAL AND TECHNOLOGICAL GOVERNANCE (2000).

Von Albertini, DEKOLONISATION (1996).

Von Mehren and Kourides, *International Arbitrations between States and Foreign Private Parties: The Libyan Nationalisation Cases* (1981) 75 AM. J. INT'L. L. 476.

Wadegaonkar, ORBIT OF SPACE LAW (1984).

Wagner, *International Regulation of the Oceans and their Resources* (1971) 37 BROOKLYN L. REV. 402.

Wälde, *A Requiem for the "New International Economic Order": The Rise and Fall in International Economic Law and a Post-Mortem with Timeless Significance*, in Hafner and Loibl (eds.), LIBER AMICORUM PROFESSOR SIEDL-HOHENVELDERN (1998) at 771–804.

Wall, Wilson and Jones, *Optimal Development of the North Sea's Oil Fields – The Criticisms* (1977) 5:4 ENERGY POLICY 284.

Walters, PLANETS (1995).

Wang, HANDBOOK ON OCEAN POLITICS & LAW (1992).

Wani, *An Evaluation of the Convention on the Law of the Sea from the Perspective of the Landlocked States* (1982) 22 VA. J. INT'L. L. 627.

Ward, *Black Gold in a White Wilderness – Antarctic Oil: The Past, Present and Potential of a Region in Need of Sovereign Environmental Stewardship* (1998) 13 J. LAND USE & ENVT'L. L. 363.

Warren, MINERAL RESOURCES (1973).

Wassenaar Arrangement, *Participating States*, at <http://www.wassenaar.org/participants/index.html>, last accessed on 28 November 2009.

Watts, INTERNATIONAL LAW AND THE ANTARCTIC TREATY SYSTEM (1992).

Weaver, *Illusion or Reality? State Sovereignty in Outer Space* (1992) 10 B. U. INT'L. L. J. 203.

Weeks, *Subsea Petroleum Resources*, U.N. Doc. A/AC.138/87 (1973).

Weinmann, *The Law of Space* (1958) 35 FOREIGN SERVICE J. 2.

Weiss, *International Environmental Law: Contemporary Issues and the Emergence of a New World Order* (1993) 81 GEORGETOWN L. J. 675.

Weiss, *The Balance of Nature and Human Needs in Antarctica: The Legality of Mining* (1995) 9 TEMPLE INT'L. & COMP. L. J. 387.

Weissberg, *International Law Meets the Short-Term National Interest: The Maltese Proposal on the Sea-Bed and Ocean Floor – Its Fate in Two Cities* (1969) 18 INT'L. & COMP. L. Q. 41.

Welch, *The Antarctic Treaty System: Is It Adequate to Regulate or Eliminate the Environmental Exploitation of the Globe's Last Wilderness?* (1992) 14 HOUSTON J. INT'L. L. 597.

Wercinski, *Mars Sample Return – A Direct and Minimum-Risk Design* (1996) 33 J. SPACECRAFT & ROCKETS 381.

Weston, *The Charter of Economic Rights and Duties of States and the Deprivation of Foreign-Owned Wealth* (1981) 75 AM. J. INT'L. L. 437.

Wetherill and Kortenkamp, *Asteroid Belt Formation with an Early Formed Jupiter and Saturn*, paper presented at the 30th Annual Lunar and Planetary Science Conference, 15–29 March 1999, in Houston, TX, USA.

Whipple and Shelus, *The Orbital Dynamics of the NEAR Mission Asteroids* (1995) 27 BULL. AM. ASTRON. SOC. 1203.

White, *A New International Economic Order* (1975) 24 INT'L. & COMP. L. Q. 542.

White, *A New International Economic Order?* (1976) 16 VA. J. INT'L. L. 323.

White, *Expropriation of the Libyan Oil Concessions: Two Conflicting International Arbitrations* (1981) 30 INT'L. & COMP. L. Q. 1.

White, *Real Property Rights in Outer Space* (1997) 40 PROC. COLL. L. OUTER SP. 370.

White House, *President Bush Announces New Vision for Space Exploration Program* (14 January 2004), at <http://www.whitehouse.gov/news/releases/2004/01/20040114-3.html>, last accessed on 11 July 2006.

White House, *President Bush Delivers State of the Union Address* (31 January 2006), at <http://www.whitehouse.gov/news/releases/2006/01/20060131-10.html>, last accessed on 27 January 2007.

References 367

Whitehouse, *Scientists Get Near the Real Eros* (21 September 2000), British Broadcasting Corporation, at <http://news.bbc.co.uk/1/hi/sci/tech/936149.stm>, last accessed on 15 July 2006.

Whiting, *Orbital Transfer Trajectory Optimisation* (2004), Massachusetts Institute of Technology, <http://ssl.mit.edu/publications/theses/SM-2004-WhitingJames.html>, last accessed on 23 December 2004.

Whitney, *Environmental Regulation of United States Deep Seabed Mining* (1978) 19 WM. & MARY L. REV. 77.

Wiegert, Innanen and Mikkola, *Cruithne* (1997) 387 NATURE 685.

Wigdor, *Canada and the New International Economic Order: Some Legal Implications* (1982) 20 CAN. Y. B. INT'L. L. 161.

Wilburn, *International Mineral Exploration Activities from 1995 through 2004* (25 August 2006), U.S. Geological Survey, at <http://pubs.usgs.gov/ds/2005/139/>, last accessed on 27 January 2007.

Wilburn and Bleiwas, *Platinum Group Metals – World Supply and Demand* (2004), U.S. Geological Survey, at <http://pubs.usgs.gov/of/2004/1224/2004-1224.pdf>, last accessed on 27 January 2007.

Williams, *Chronology of Lunar and Planetary Exploration* (2006), NASA Goddard Space Flight Centre, at <http://nssdc.gsfc.nasa.gov/planetary/chrono.html>, last accessed on 11 July 2006.

Williams, *Las Empresas Privadas en el Espacio Ultraterrestre* (1983) 8 REV. CEN. INV. DIF. AERO. ESP. 39.

Williams, *The Law of Outer Space and Natural Resources* (1987) 36 INT'L. & COMP. L. Q. 142.

Williams, *Utilisation of Meteorites and Celestial Products* (1969) 12 PROC. COLL. L. OUTER SP. 271.

Wilson, *Mining the Deep Seabed: Domestic Regulation, International Law and UNCLOS III* (1983) 18 TULSA L. J. 207.

Wilson, *Regulation of the Outer Space Environment Through International Accord: The 1979 Moon Treaty* (1991) 2 FORDHAM ENVT'L. L. REP. 173.

Wingo, MOONRUSH: IMPROVING LIFE ON EARTH WITH THE MOON'S RESOURCES (2004).

Wirin, *Practical Implications of Launching State – Appropriate State Definitions* (1994) 37 PROC. COLL. L. OUTER SP. 109.

Wirin, *Space Debris and Space Objects* (1991) 34 PROC. COLL. L. OUTER SP. 45.

Wolter, *The Peaceful Purpose Standard of the Common Heritage of Mankind Principle in Outer Space Law* (1985) 9 A.S.I.L.S. INT'L. L. J. 117.

Wong, *The Paper "Satellite" Chase: The ITU Prepares for its Final Exam in Resolution 18* (1998) 63 J. AIR L. & COM. 849.

Wood, *International Seabed Authority: The First Four Years* (1999) 3 MAX PLANCK U.N.Y.B. 173.

Woodmansee, *Opinion: Space 'Adventurers' Paving the Way for the Rest of Us* (18 September 2006), at <http://www.space.com/adastra/060918_woodsmansee_ansari.html>, last accessed on 27 January 2007.

Woods and Bernasconi, *Choosing a Space Age or a Stone Age* (1995), SPACE NEWS, 2–8 October 1995.

Woods and Klose, *Mineralogy on Venus and Areas of High Fresnel Reflection Coefficient Detected by Magellan Radar* (1991) 22 ABS. LUNAR & PLANETARY SCI. CONF. 1521.

World Bank, *Prospects for the Global Economy: Commodity Markets* (2006), at <http://web.worldbank.org/WBSITE/EXTERNAL/EXTDEC/EXTDECPROSPECTS/EXTGBLPROSPECTSAPRIL/0,contentMDK:20371216~menuPK:2300917~pagePK:2470434~piPK:2470429~theSitePK:659149,00.html>, last accessed on 16 September 2006.

Woytinsky and Woytinsky, WORLD POPULATION AND PRODUCTION: TRENDS AND OUTLOOK (1953).

Wright, *The Ownership of Antarctica, its Living and Mineral Resources* (1987) 4 J. L. & ENV'T. 49.

Yankov, *The Law of the Sea Conference at the Crossroads* (1978) 18 VA. J. INT'L. L. 31.

Yarn, *The Transfer of Technology and UNCLOS III* (1984) 14 GA. J. INT'L. & COMP. L. 121.

Yen, Gellert, Schröder, Morris, Bell, Knudson, Clark, Ming, Crisp, Arvidson, Blaney, Brückner, Christensen, DesMarais, De Souza, Economou, Ghosh, Hahn, Herkenhoff, Haskin, Hurowitz, Joliff, Johnson, Klingelhöfer, Madsen, McLennan, McSween, Richter, Rieder, Rodionov, Soderblom, Squyres, Tosca, Wang, Wyatt and Zipfel, *An Integrated View of the Chemistry and Mineralogy of Martian Soils* (2005) 436 NATURE 49.

Yoshikawa, *Mass of Asteroid (25143) Itokawa Determined by Hayabusa Spacecraft* (2006), paper presented at the 36th COSPAR Scientific Assembly, 16–23 July 2004, in Beijing, China.

Yost, *The International Seabed Authority Decision-Making Process: Does it Give a Proportionate Voice to the Participant's Interests in Deep Sea Mining?* (1983) 20 SAN DIEGO L. REV. 659.

Young, *Antarctic Resource Jurisdiction and the Law of the Sea: A Question of Compromise* (1985) 11 BROOKLYN J. INT'L. L. 45.

Yuan, *The United Nations Convention on the Law of the Sea from a Chinese Perspective* (1984) 19 TX. INT'L. L. J. 415.

Zadalis, *"Peaceful Purposes" and Other Relevant Provisions of the Revised Composite Negotiating Text: A Comparative Analysis of the Existing and Proposed Military Regime for the High Seas* (1979) 7 SYRACUSE J. INT'L. L. & COM. 1.

Zang, *Frozen in Time: The Antarctic Mineral Resource Convention* (1991) 76 CORNELL L. REV. 722.

Zellner, Tholen and Tedesco, *The Eight-Color Asteroid Survey: Results for 589 Minor Planets* (1985) 61 ICARUS 355.

Zhukov, *Delimitation of Outer Space* (1980) 23 PROC. COLL. L. OUTER SP. 221.

Zhukov, KOSMICHESKOYE PRAVO (1966).

Zhukov, *The Legal Regime for the Moon (Problems and Prospects)* (1972) 14 PROC. COLL. L. OUTER SP. 50.

Zhukov, *The Problem of the Definition of Outer Space* (1967) 10 PROC. COLL. L. OUTER SP. 271.

Zhukov, WELTRAUMRECHT (1968).

Zhukov and Kolosov, INTERNATIONAL SPACE LAW (1984).

Zou, *China's Efforts in Deep Seabed Mining: Law and Practice* (2003) 18 INT'L. J. MARINE & COASTAL L. 481.

Zubrin and Wagner, THE CASE FOR MARS: THE PLAN TO SETTLE THE RED PLANET AND WHY WE MUST (1997).

Zuleta, *The Law of the Sea after Montego Bay* (1983) 20 SAN DIEGO L. REV. 475.

Zullo, *The Need to Clarify the Status of Property Rights in International Space Law* (2002) 90 GEORGETOWN L. J. 2413.

Index

A

Antarctica, 7, 14, 16–17, 49, 52, 183–184, 204–215, 257, 262, 270, 273, 298, 319

Antarctic Treaty, 14, 183, 204–216, 257, 298, 319

Asteroids
  Amalthea, 78
  Amor, 22, 71, 75–76
  Amun, 62, 67
  Anacostia, 58
  Apollo, 62–63, 71, 74–75
  Aquitania, 58
  Arjuna, 73
  Aten, 60, 67, 74–75
  Božnêmcová, 60, 63
  Ceres, 25
  Cruithne, 67
  Eger, 59
  Eros, 62, 76, 79
  Europa, 55
  Gaspra, 7, 62
  Ida, 55
  Itokawa, 82
  Khufu, 67
  Mathilde, 55, 62
  Nysa, 59
  Pandora, 59
  Psyche, 59
  Steins, 62
  Undina, 59
  Vesta, 60, 79–80

B

Bauxite, 36, 41, 43–44

Broadcasting Principles, 116, 119–122

C

Carbon, 24, 26, 46–47, 53, 55–57, 63–64

Celestial bodies, 1–2, 5–13, 15–20, 22, 47–50, 99, 102, 106, 109–110, 128, 155–156, 159–160, 190–204

China, 5, 132, 135, 157, 160, 209–210, 238, 244, 255, 260, 279

Chondrite, 60

Chromium, 28, 42, 44, 59

Comet, 62, 64–66, 69, 75, 78, 188, 284

Commercial use, 5, 132, 154–162, 274

Common heritage of mankind, 2, 15–18, 110, 151, 159, 197, 203–204, 206–208, 217, 219, 226–227, 229, 234–235, 241, 244–247, 287, 290, 305–306, 318–321

Convention on the Law of the Sea, 15–166, 99, 110, 168–169, 203–204, 244–256, 319, 321

Cooperation Declaration, 117, 123–124

Copper, 4, 28–30, 36, 42, 44, 57, 90, 205, 236, 250, 279

COPUOS, 99, 101, 103, 105, 108, 110–111, 123, 172, 256, 258–260, 263, 267, 274

COSPAR, 81, 174

Customary international law, 103, 105, 107, 115, 117–118, 121–122, 124–125, 129, 131, 153–155, 169–172, 197–198, 226, 235, 296, 300

D

Deep seabed, 1, 7, 15–17, 49, 89, 110, 140, 183–184, 194, 203–204, 245–254, 318–320

Deimos, 25, 54, 69, 73, 80

Developing States, 14–17, 30, 119–120, 156, 171, 177, 210, 220, 247–251, 254, 256, 269, 276–282, 288–293, 301, 318–319

E

Earth, 1–2, 6–13, 19–27, 77–82, 112, 123, 140–141, 171–176, 201–202, 283–287, 304, 306–307, 315–317, 320–321

Earth orbit, 6, 22, 24, 49, 51–52, 61, 67, 71, 73–75, 81, 86, 88, 90, 93, 95, 108, 123, 176, 201–202, 285–286

Earth's crust, 1, 4, 6, 21, 27–28, 41, 45, 48, 50, 58, 77, 290, 292, 296, 306, 315–316, 320

ECAS, 55–56

ECSL, 109, 270

Entropy, 27

ESA, 127

Escape velocity, 25, 51, 71, 73

Europe, 5, 219, 227, 243

F

Feldspar, 52, 56–58

Finance, 92, 199, 221, 241, 254–255, 279, 306

Flyby, 55, 62, 74, 78–80, 89

Fuel cell, 45, 47

G

GATT, 222–223, 225, 227, 250

General Assembly, 96, 101, 102–103, 105, 108–124, 136, 138–139, 210, 220–229, 256, 259, 266–267

GEO, 24, 171, 208, 215, 223, 236, 253

H

High seas, 14, 103–104, 133, 142–143, 176, 229–233, 235, 245, 319

Hohmann transfer, 9, 72, 74–75, 316

Hubbert, 31, 34, 42

Hydrocarbons, 8, 63–64, 68

Hydrogen, 21, 45–47, 50–51, 53, 57, 64, 68

Hydrogen economy, 35, 45–47, 50

Hyperbolic velocity, 71, 74

I

IAU, 190

IMF, 221, 225, 228, 292–293, 298–299, 307

Implementation Agreement, 16–17, 280, 295–296, 301, 303, 307, 309, 313, 320

Industrialised States, 14–17, 29–30, 45, 171, 204, 219–220, 222, 224, 242–243, 276–278, 319

INMARSAT, 127, 135–136, 161, 290, 298–299

Innovation, 6, 9, 22, 34–35, 40–41, 42, 48, 90, 295

INTELSAT, 127, 161, 290–291, 298–299

International Court of Justice, 101, 116, 131, 300, 307–311

International Space Development Authority, 18, 277

International Space Station, 23, 97, 127

Iron, 4, 28, 42, 44–45, 52–53, 56–60, 63, 205

ISA, 246–251, 253, 255

ITLOS, 307–311

J

Japan, 5, 23–24, 81, 97, 105, 107, 113, 149, 160, 170, 209–211, 236, 243, 255, 258, 278, 281, 292

Jupiter, 8–9, 53, 62, 64–66, 68, 78, 173, 187–188, 284, 316

Jus cogens, 125–126, 172, 231

K

Kerogens, 63–64

Kuiper Belt, 65, 284

L

Lander, 80–81, 89

Launching States, 12, 106, 108–109, 135–136, 140, 142–144, 146–148, 151

Legal Sub-Committee, 101, 103, 105, 108–111, 121, 138, 173, 177, 258, 262, 267

LEO, 6, 24

Liability, 2, 10, 12, 92–93, 97, 102, 105–107, 112–113, 129–130, 136–142, 143–145, 147, 150–151, 189, 228, 321

Liability Convention, 11–13, 98, 101, 105–108, 113, 118, 123, 129, 132, 134, 136–150, 257, 293

M

Magnesium, 4, 28, 52, 58–59

Main Belt, 7–9, 60, 62, 65, 79, 316

Malthus, 33

Manganese, 28, 36, 42, 44, 236–237, 239–240, 250–251

Mars, 8–9, 25, 52–55, 61, 65, 69, 73, 173, 316

Mercury, 25, 28, 42, 52–53

Meteorite, 50, 56, 59, 62–64, 190, 284

Mineral reserves

    conditional reserves, 29–31, 38, 48

    hypothetical reserves, 31–32

Index 371

proven reserves, 28–31, 38, 41, 44
speculative reserves, 32
Mineral resources, 1–2, 4, 6–8, 52–53, 93, 96, 197–202, 245–246, 249–250, 315–317, 320
Moon, 4–6, 8, 14–17, 22, 24–25, 48–52, 73–74, 82, 88, 96, 98–99, 101, 157–161, 193–208, 216–218, 256–269, 317–321
Moon Agreement, 12, 15–17, 99, 101, 109–111, 157–159, 161, 179–187, 256–269, 296, 298, 300–301, 317–321

N

NASA, 25, 62, 96
NEAR, 55, 62, 80
Near Earth Asteroids, 2, 8–9, 22, 50–52, 56, 60–66, 71–74, 77–82, 90, 94, 201, 315–317
Neptune, 48, 53, 66, 188
NIEO, 14–15, 204, 219–229, 237–238, 318
Nitrogen, 57, 63, 68
Non-appropriation, 2, 13–14, 18, 112, 151, 153, 165–166, 170–172, 178, 198–199, 203, 233, 298, 317–318
Nuclear Power Sources Principles, 105, 117, 121–123, 147–148
Nuclear Test Ban Treaty, 98–99

O

Oil, 21, 29, 31–32, 34–35, 42, 45, 63, 143, 205, 214, 221–222, 238
Olivine, 55–61
Orbiter, 78, 80–81
Outer Space Treaty, 11–13, 98–105, 107–108, 128–129, 133–136, 138, 140, 142, 147, 162–169, 171–181, 216–219, 282–285, 300–301

P

Palladium, 45, 57
Petroleum, 7, 32, 35, 42, 89, 228, 238, 328
Phobos, 25, 54, 69, 73, 80
Phosphorus, 28, 56–57
Planetesimal, 9, 63–64
Platinum, 8, 21, 45–47, 49–50, 57, 84, 87, 90, 205, 237
Platinum group metals, 45–47, 49–50, 57, 87, 90
Polar regions, 1, 89, 163, 212, 214
Population, 1, 4, 6, 33, 40, 49, 51, 60, 65, 291, 322

Principles Declaration, 101, 112, 116, 118–119, 134, 138, 154
Propulsion, 9, 24–26, 48, 51, 75, 81, 83, 85, 93, 189, 191, 201, 276, 280, 285–287, 295, 316
Prospecting, 6–7, 13, 18, 25, 67, 77, 80, 83–85, 89–90, 159–160, 163, 182–184, 194–196, 202, 213, 255, 277, 283, 289, 293
Pyrolysis, 63
Pyroxene, 55–61

R

Registration Convention, 98, 101, 107–109, 113, 123, 135–136, 139
Regolith, 54, 61–62
Remote Sensing Principles, 117, 120, 122, 161
Rendezvous, 55, 61–62, 70–73, 75, 78, 80–81
Rescue Agreement, 98, 101, 103–105, 113, 117, 123, 147–148
Resource base, 27–28, 30, 38, 48–51
Resource depletion, 38, 40, 48, 219
Resource exhaustion, 26, 33, 38, 40–41, 43, 49
Ricardo, 33, 36, 90
Risk, 84, 87–89, 91–92, 122, 147, 199, 202, 268, 278, 284–285, 287–288, 307
Russian Space Agency, 23

S

Satellites, 4–6, 8–9, 13, 52, 54–55, 63, 78, 95, 97, 109, 116, 121, 136–137, 165, 174–176, 179, 187, 217
Saturn, 9, 53, 65, 68, 188
Scarcity, 20, 22, 31–33, 35–38, 40, 43, 51, 93, 290, 315–316
Short period comets, 9, 66, 67–68, 75
Solar System, 5, 7–9, 21, 25, 48–49, 51–54, 58, 60, 63–65, 69, 77, 94, 109–110, 179, 182–183, 186, 191, 201, 256, 258, 284–285, 316, 321
Soviet Union, 4–5, 95–96, 101, 102, 109, 136, 139, 149, 210, 226, 234, 236, 243, 252, 258–259, 263, 277
Space tourism, 23
Spinel, 58–59
SSPS, 22
Sun, 8–9, 48, 51, 53–54, 59, 65–67, 72
Synodic period, 73, 75

T

Taurid Complex, 64
Transfer orbit, 70, 72–73, 81, 85, 88
Trojan, 74, 79

U

UNCLOS, 236–239, 241–243, 246, 248, 249, 253, 256, 259, 269, 277, 300
UNCTAD, 221, 223
UNESCO, 126, 301–303
UNIDROIT, 13
United Nations, 12, 97–99, 101–111, 113–115, 120, 122–124, 126, 128, 140, 153, 236, 252–253, 298–303
United States, 5, 34–35, 95–97, 101, 135–139, 236–246, 296
Uranium, 4, 27
Uranus, 48, 53, 65, 188

V

Venus, 25, 52–53
Volatiles, 63–64, 67, 75

W

Water, 45, 51, 53–54, 56–57, 59, 63–64, 67, 87, 98, 203, 209, 283, 286
Wellington Convention, 14, 16, 183–184, 194, 205–207, 211–215
WIPO, 126
WTO, 298, 302, 307–311